AF323167

Progress in Drug Research
Fortschritte der Arzneimittelforschung
Progrès des recherches pharmaceutiques
Vol. 32

Progress in Drug Research
Fortschritte der Arzneimittelforschung
Progrès des recherches pharmaceutiques
Vol. 32

Edited by / Herausgegeben von / Rédigé par
Ernst Jucker, Basel

Authors / Autoren / Auteurs
David R. Webb · Vijendra K. Singh and H. Hugh Fudenberg ·
Roger W. Russell · L. Öhman, G. Maluszynska, K.-E. Magnusson
and O. Stendahl · Peter A. Lambert · David H. P. Streeten and
Gunnar H. Anderson Jr · Vickram Ramkumar, George Pierson and
Gary L. Stiles · Bernard Testa and Joachim M. Mayer ·
Duncan Stewart-Tull · Michael Williams and Gary L. Neil ·
Peter Hambleton, Stephen D. Prior and Andrew Robinson ·
Jed F. Fisher and Paul A. Aristoff

1988 Birkhäuser Verlag
Basel · Boston · Berlin

Contents · Inhalt · Sommaire

Foreword

In the first years of the existence of this series of monographs, during the so-called "Golden Age" of drug research, the majority of the papers published were mainly concerned with the traditional domains of drug research, namely chemistry, pharmacology, toxicology and pre-clinical investigations. The series' aim was to give coverage to important areas of research, to introduce new active substances with therapeutic potential and to call attention to unsolved problems.

This objective has not changed. The table of contents of the present volume makes evident, however, that the search for new medicines has become increasingly complex, and additional, new disciplines have entered the research arena. The series now includes reviews on biochemical, biological, immunological, physiological and medicinal aspects of drug research. Researchers actively engaged in the various scientific fields forming the entity of drug research can benefit from the wealth of knowledge and experience of the respective authors, and will be assisted in their endeavour to discover new pharmaceutical agents. Those simply wanting to keep abreast of new developments in the complex, multi-discipline science can turn to the "Progress in Drug Research" volumes as an almost encyclopaedic source of information without having to consult the innumerable original publications.

Volume 32 contains 12 reviews, a subject index, an index for the close to 400 articles published in the series so far, and an author and titles index for all 32 volumes.

I should like to thank all the authors for their willingness to prepare the reviews and for sharing their knowledge with the readers. Thanks are also due to L. Koechlin, H. P. Thür and A. Gomm of Birkhäuser Publishers for their most valuable help in the preparation of this volume.

Basel, October 1988 Dr. E. JUCKER

Vorwort

Die Gründung dieser Monographien-Reihe fiel in das «goldene Zeitalter» der Arzneimittelforschung. Eine nicht mehr zu überblickende Zahl von chemischen, pharmakologischen und klinischen Publikationen manifestierte die Suche nach neuen Medikamenten, und der aktive Forscher lief Gefahr, den Überblick zu verlieren. Die «Fortschritte der Arzneimittelforschung» hatten denn auch zum Zweck, in größeren Referaten wichtige Forschungsgebiete zusammenfassend darzustellen, neue Wirkstoffe vorzustellen und auf ungelöste Probleme hinzuweisen.

An dieser Zielsetzung hat sich nichts geändert; wie das Inhaltsverzeichnis des vorliegenden 32. Bandes jedoch illustriert, hat eine Verschiebung der Akzente in Richtung Biochemie, Biologie, Immunologie, Physiologie und Medizin stattgefunden. Damit verbunden ist gezielteres Forschen und vertieftes Verständnis der Wirkungsweise neuer Arzneimittel. Die Vielzahl der an dieser Forschung beteiligten Disziplinen dokumentiert auch die Komplexität dieses Arbeitsgebietes.

Der 32. Band der «Fortschritte» enthält 12 Übersichtsreferate, ein Stichwortverzeichnis des Bandes, einen Artikel- sowie Autoren- und Artikelindex aller bisher erschienenen 32 Bände. Weit über 300 Artikel der Reihe und die ihnen angegliederten unzähligen Hinweise auf die Originalliteratur sowie die erwähnten Verzeichnisse erlauben müheloses Auffinden der relevanten Angaben.

Der aktive Forscher wird direkten Nutzen für seine eigenen Arbeiten ziehen und Anregungen für neue Untersuchungen erhalten. Die Fülle des bisher publizierten Materials erfüllt indessen noch einen anderen Zweck: Wer sich über die Entwicklungen der Arzneimittelforschung rasch, umfassend und zuverlässig informieren will, wird in den «Fortschritten» eine nützliche, fast enzyklopädische Stütze finden, ohne sich in der Flut der Originalpublikationen umsehen zu müssen.

Nach diesen einleitenden Ausführungen verbleibt mir nur noch die angenehme Pflicht, den Autoren dieses und auch der vorangegangenen Bände für ihre große Arbeit zu danken. Dank gebührt auch den Mitarbeitern des Birkhäuser Verlages, vor allem Frau Koechlin und den Herren Thür und Gomm; ohne ihren großen persönlichen Einsatz wäre die Herausgabe der «Fortschritte» nicht möglich gewesen.

Basel, Oktober 1988 Dr. E. JUCKER

Antigen-specific T-cell factors and drug research

By David R. Webb
Synthex Research, 3401 Hillview Avenue, Palo Alto,
California 94303, USA

1   Introduction

The central feature of the immune system is the specificity of antigen-binding proteins for the ligand that induced their synthesis. It is this specificity that has intrigued medical scientists for the last 100 years and has led to repeated attempts to use this feature to combat disease. One of the earliest, successful uses of immune specificity involved the production of specific anti-toxins to combat a variety of ills from bacterial infections to snake bites [1]. In addition, advantage was taken of the fact that by providing appropriate antigens in the proper form (eg. viruses, bacteria), specific immune responses could be elicited [2]. Thus, over the last 80 years or so the primary focus of the therapeutic use of antigen-specific proteins has been the antibody molecule [3]. With the development of monoclonal antibody methodology, there has developed a renewed interest in the use of specific single antibodies as therapeutic agents; "magic bullets" as they are often termed [4]. In addition, the capacity to produce monospecific reagents is providing invaluable tools to probe the structure and function of other proteins; to use in the specific detection of cellular or tissue specific antigens and for purifying and identifying a wide variety of normal body components (eg. peptides, lipids, etc.) and/or ingested drugs [5]. In the early 1970s several groups of investigators reported that, in addition to antibodies, there exist other antigen-binding moieties that appear to function primarily as immune regulators. These were termed antigen-specific helper factors and antigen-specific suppressor factors [6–8]. Since that time an extensive literature has developed as more and more immune response models have been studied. At the present time we have developed a fair degree of understanding of the biology and biochemistry of antigen-specific suppressor factors while progress in studies of antigen-specific helper factors has stalled [9]. The first portion of this article will briefly review what is known about the antigen-specific T-cell factors before discussing the possible use of these molecules in drug research.

2   Antigen-specific helper factors

The antigen-specific helper factors were originally reported in the early 1970s [8]. Since that time, many studies have documented the production of non-specific helper factors (e.g.; interleukin 1 and interleukin 2) [10, 11]. However, a thorough-going elucidation of the nature of antigen-specific helper factors has proved elusive. Many different laboratories have reported such factors [6, 8] but a complete biochemical characterization of even one T-helper factor has not been done. To date we must content ourselves with the accumulated evidence suggesting that biochemically these factors may resemble antigen-specific suppressor factors [9]. The missing key elements so far have been the lack of cloned T-cell lines that consistently produce such factors and/ or the construction of stable T-cell hybridomas that make antigen-specific helper factors. The fact that it is easy to produce such lines that make non-specific helper factors has raised doubts about the existence and/or relevance of antigen-specific helper factor producing cells. It may well be, as is true of antigen-specific T-suppressor cells, that the growth factor requirements for antigen-specific helper factor producing cells are quite different from non-specific helper factor producing cells and, therefore, the latter cells are most often observed because they are the easiest to grow and study. This being the case, I will have little more to say about helper factors except that their potential utility in drug development would be similar to that of antigen-specific suppressor factors, i.e., the specific manipulation of selected immune responses.

3 Antigen-specific suppressor factors (TsF)

By contrast with the antigen-specific helper factors, the literature on suppressor factors is vast; since 1975 there are nearly 3,000 primary publications on suppression that have been located by a literature search. In addition, there are many fine reviews and monographs that have summarized the work done in this area [6–9, 12–21]. Despite the large number of laboratories that have studied the problem of antigen-specific suppression, few groups have extensively characterized the factors themselves in terms of their physico-chemical properties [9]. Nevertheless, our own studies coupled with those of other laboratories have yielded a fairly complete picture of the factors themselves [22–38]. The major stumbling block to a complete understanding of these molecules has been our inability, so far, to establish the nature of the genes that code for the antigen-specific TsF [34, 38]. What follows is a summary of what is known about antigen-specific TsF in terms of their biology and biochemistry.

Upon antigenic challenge, the native antigen is bound by B-cells (antibody-producing cell precursors) and macrophages. The macrophages process the antigen by enzymatic cleavage and display antigenic fragments on their surface in conjunction with class II antigens of the major histocompatibility complex [39]. Although Ts are the only T-cell population capable of binding unprocessed antigen [6], under normal circumstances it appears that they probably respond to processed fragments. The initial cell activated in the suppressor cascade is a suppressor-inducer cell (Tsi). In primary immune responses this cell is usually Lyl^+2^-, $L3T4^-$, $I-J^+$ in terms of its cell surface phenotype [40]. The product of this cell is an antigen-specific factor that induces the activation of a second cell in conjunction with antigen. There are two prominent opinions as to the biochemical characteristics of these suppressor-inducer factors. The one view exemplified by Green and his colleagues [12] holds that the suppressor inducer factor is composed of two polypeptide chains (one antigen binding, the other $I-J^+$) that are synthesized by two separate cells. The evidence they have developed is based primarily on the use of monoclonal antibodies and biological assays to monitor the factor activity. The second view of suppressor-inducer factors is that based on work in my own laboratory in collaboration with Drs J. A. Kapp, C. M. Sorensen and C. W. Pierce at Washington University in St. Louis. We have purified suppressor-inducer factors (TsF_1) from several different hybridomas to chemical homogeneity [22–25]. Using one of these TsF_1, we have recently analyzed directly the antigen-binding capacity and tested directly for the presence of an I–J determinant (I–J is a serologically detected determinant found on Ts and suppressor factors [6–9]). The results show that both the antigen-binding site and the I–J determinant are present on the same polypeptide [22–25]. The TsF_1 polypeptide chain exists primarily as a homodimer of mol.wt 66,000 and is, so far as we can tell, non-glycosylated. These results suggest there may be at least two types of suppressor-inducer factors produced by Tsi depending on the nature of the inducing antigen and whether the immune response is the result of a single antigen challenge (primary) or multiple antigen challenges (secondary). Recently, we have used monoclonal anti-TsF_1 antibodies that have broad cross-reactivity to show that the TsF_1, described by us operates as a suppressor-inducer in cellular responses to defined haptens such as tri-methyl ammonium (tyrosine) or 4-hydroxy-3-nitrophenyl-acetyl (NP) as well as in the response to allogeneic class I major histocompatibility complex antigens [Webb et al. submitted for publication; Webb, unpublished observations; Devens and Webb, submitted for publication].

The TsF_1, suppressor-inducer factor is so named because it induces a second suppressor cell. In our own studies, the second suppressor cell is called the suppressor-effector cell (Tse) [40]. This Tse can interact directly with a T-helper cell and block its ability to make or release lymphokines that help B cells differentiate to become antibody-producing cells. The product of the Tse cell is termed TsF_2. It also has been purified to chemical homogeneity; it has a mol.wt of 66,000 and is composed of two heterologous polypeptide chains joined via a disulfide bridge. In this case, both chains are synthesized by a single cell following activation by TsF_1 + antigen. The acidic α-chain bears the I–J determinant and the basic β-chain can bind directly to antigen coupled to a solid support [27, 28]. This molecule is glycosylated and bears a striking resemblance to the T-cell antigen receptor (see below).

In several other models of antigen-specific suppression, the second cell is not necessarily an effector cell or even antigen-specific [7, 13]. It is possible for TsF_1 to induce an anti-idiotype TsF_2 that in turn may stimulate either an antigen-specific or non-specific tertiary level suppressor cell that may then interact directly to suppress T-helper cell or B-cell function [7, 13]. Very little is known about the biochemistry of such factors except that many of them appear to be similar to the disulfide linked heterodimeric structure described earlier [13]. In fact, we hypothesize that there are, indeed, only two fundamental types of antigen-specific proteins made by Ts-cells; one is the homopolymer (TsF_1) type that always serves as an inducer of antigen-specific suppression, and the second disulfide-bonded heterodimer which may be either a suppressor-inducer or the product of a suppressor-effector cell involved in the direct interaction with a target cell type.

At the molecular genetic level, it is clear that Ts do not use classical β-chain genes since there are no productive rearrangements of such genes in any of the suppressor factor producing cells save one [38]. It is not yet clear whether α-, δ- or γ-chain genes are involved [32, 34, 38]. Further, the accessory proteins of the CD 3 type that have been identified in both mouse and human T-cells as the transducing elements for the T-cell receptor, are yet to be thoroughly investigated. These results strongly imply that antigen-specific suppressor T-cells may represent a separate lineage from those T-cells expressing T-helper cell activity,

cytotoxic T-cells or other lymphokine producing T-cells. In support of
this notion are studies showing that Ts have different growth factor requirements [J. Trial, personal communication] than T-helper cells. For
example, IL-2, a growth factor for T-helper cells, will not stimulate the
growth of most Ts; rather IL-2 stimulates the production of TsF [41].
We have examined this issue directly by measuring the effects of IL-2
on the production of TsF_1 and TsF_2 by T-cell hybridomas that consti-
tutively produce low levels of these factors. In both cases IL-2 stimulates increased synthesis of TsF_1 or TsF_2 without any measurable effect
on overall protein synthesis, RNA synthesis or DNA synthesis [J.
Freire-Moar et al., in preparation].

In sum, antigen-specific suppressor factors represent an apparently
unique group of antigen-binding proteins that in addition to recognizing foreign antigens, perform discrete biological functions namely;
binding to a specific target cell and secondly inducing in that cell an
appropriate response.

Suppressor cells are also induced in response to cellular antigens such
as tumor cells [42, 43] and have been shown to regulate autoimmune T-
cells [31, 41]. In fact it has often been postulated that Ts are responsible
for the maintenance of tolerance to self antigens [6–8]. The data to support such a hypothesis have not been unambiguous. To date all that
can be confidently said is that there are many experimental models
that suggest a role for Ts-cells in preventing anti-self responses [45–53].
To what extent these models reflect a normal physiological role for Ts
in tolerance and autoimmunity remains an open issue.

4   Antigen-specific suppressor factor derived probes as tools in drug research

4.1   Monoclonal anti-suppressor cell (factor) antibodies as probes for drug research

A major development in the study of antigen-specific suppressor factors has been the generation of panels of monoclonal antibodies that
recognize antigen-specific suppressor factors [28, 54, 55]. These antibodies have helped to validate the role that suppressor cells play in the
physiological regulation of immunity. From a drug development perspective they have raised the possibility of routine standardized assays
for the presence of TsF in biological fluids. At least two laboratories
have now developed or have under development, ELISA type assays

that bypass the need for biological testing. This means that it should soon be possible to assess what is happening to TsF levels in a variety of animal models of disease. Moreover it suggests that it will soon be feasible to attempt to modulate TsF levels experimentally while monitoring those levels in tissues or fluids. In addition the monoclonal anti-TsF antibodies themselves may be used as models of drugs that can specifically recognize TsF and neutralize TsF activity; thus one could examine the consequences of specific immunosurgery on a given disease model or a particular immune response as has been done with anti-Ig antibodies or anti-idiotypic antibodies [4].

Another area where monoclonal anti-TsF antibodies can play a role is as model antagonists. As mentioned earlier, monoclonal antibodies have been used to probe the physiological role of TsF by blocking or modulating factor activity. A recent example of this approach to drug development has been the demonstration in NZB/NZW mice that the administration of anti-γ-interferon antibodies will delay the onset of autoimmune disease [57]. As applied to antigen-specific TsF it should be possible to ask whether these factors play any role in allograft maintenance following immunosuppressive therapy for example. In response to tumors or chronic infection it is possible to now establish whether specific or non-specific suppression plays any role in the pathogenesis of these diseases by using specific anti-TsF antibodies as antagonists. Thus not only might we gain a better understanding of the pathophysiology of certain diseases, we also may be able to validate an appropriate approach to therapy.

4.2   The use of specific DNA probes

In concert with studies using monoclonal antibody probes, DNA probes complimentary to some or all of the coding sequences of antigen-specific TsF also provide highly useful diagnostic probes. Since we are discussing antigen-specific factors certain assumptions are possible. The first is that the genes involved in specifying these factors will be organized in a fashion similar to other genes that specify antigen-binding proteins (e.g.; the T-cell receptor and immunoglobulins). That is, they will have variable (V region) joining (J region) and constant (C region) segments or minigenes [58]. A major question, still unresolved, is whether antigen-specific TsF use the same set of genes as is used for the T-cell receptor. As mentioned earlier, evidence now suggests that

TsF do not come from the genes coding for the β-chain of the T-cell receptor. This leaves the α-, γ- and recently described δ-chains as possibilities. Should it turn out that TsF uses one or more of these genes, then the strategy used to detect and differentiate TsF gene expression from T-cell receptor gene expression will have to be modified. For example, should the same C region genes be used by Ts and Th, then simple analysis of tissue samples will not suffice to establish whether TsF genes have been induced; instead one may have to employ V region genes that may be more specific to a given factor.

The second assumption is that the TsF genes are inducible and not in a permanently "switched on" mode. This assumption has recently been called into question by preliminary data suggesting that at least one gene product related to TsF derives from a mRNA expressed in virtually all tissues but which may be translated only in certain lymphocyte subpopulations [Keven Moore and Christine Martins, personal communication]. The same may be said to be true in cells expressing the T-cell antigen-receptor where there is a constant level of synthesis in mature cells [58]. However, we expect that for at least some of the TsFs, as is true for other lymphokines, the genes will be inducible so that analysis by Northern blot would reveal increased mRNA expression.

A third assumption related to the second is that gene expression will result in the production of a product – the TsF. This assumption is testable under some circumstances in that we can analyze for bioactivity and look for protein synthesis using specific immunoaffinity reagents (i.e., antibody-bound to Sepharose). This will work in those cases where a good antibody is available and where an unambiguous bioassay exists. However there are, particularly in human studies, often situations where the bioassay is not available and an antibody may not exist. In these cases a good DNA probe may provide the initial lead that would require a more extensive study to demonstrate the role of TsF in pathogenesis.

The last major consideration is in some ways the most important, and that is we assume that the level of mRNA expression will be within reasonable limits for detection. For most of the lymphokines it has proven true that if their genes are induced the level of mRNA expression rises sufficiently so as to allow detection [11]. It is not known if this is true for TsF mRNA except to point out that it is possible to isolate TsF mRNA from T-cell hybridomas that can be detected by its capacity to synthesize bioactive TsF in cell-free systems [9, 23].

It is also necessary to point out, however, that the TsF studied by us have very high specific activities [9, 26] and thus a very small amount of mRNA will appear to produce a good bioactivity profile. Nevertheless, we expect to be able to detect TsF gene expression under appropriate conditions. Initially, these will include mainly animal models of immune responses or diseases where the cellular immunology is better understood. For example if one follows the pattern of cytotoxic T-cell (CTL) appearance after a virus infection in the mouse, the CTL peak at around days 21–30 and decline to normal levels some time thereafter [59]. It has been shown that Ts can be demonstrated in the later stages of the infection. With specific DNA probes it should be possible to follow the course of TsF expression back to its earliest stages since the appearance of mRNA usually occurs hours earlier than one can detect its specified protein [23]. If one couples the use of TsF DNA with DNA probes specific for other known mediators (IL-1, IL-2, interferons, etc.) then a rather complete pattern of cell activation and interaction may be inferred based on what we know about the cellular immunology.

5   **Antigen-specific suppressor factor derived probes in pathology**

There are numerous reports in the literature [45–53] that implicate antigen-specific TsF in disease. The major difficulty with these reports relates to the unambiguous determination of specific types of factors, their antigenic specificity, and their role in the disease process. It is therefore important to be able to develop both antibody and DNA probes that could begin to remove the above-mentioned ambiguities. Whether the antibody probes currently available will be useful in this regard is problematical in the case of human diseases. However, in the case of the mouse, evidence is beginning to develop that suggests that there are now several monoclonal antibodies as well as polyclonal antisera that have broad class specificity for TsF. Several of the anti-TsF$_1$ and anti-TsF$_2$ monoclonal antibodies produced by Kapp, Pierce and Webb and their colleagues have recently been shown to have very broad reactivity with TsF generated in mixed lymphocyte cultures [Devens, Sorensen, Kapp and Webb, in press]. These cultures would be expected to produce TsF of varying antigen-specificities thus the potential utility of the monoclonals for detection of the factors is clear. An ELIZA assay based on these antibodies is under development [Kapp et al.]. Flood and his associates [54] have also developed antibodies that have broad specificity as has the laboratory of J. A. Kapp et al. [55]. Thus it is possible to construct a panel of monoclonal antibodies that, at least *in vitro* may be used in neutralization or cytotoxicity assays to establish the presence of specific types of TsF. The cross-referencing tests with these antibodies will also aid in establishing a consensus concerning the nature of TsF produced in normal immune responses as well as in models of human diseases. Still to be developed are a similar set of antibodies to putative human TsF; also cross-reactivity between anti-murine TsF and human TsF cannot be ruled out. The most useful probes for human pathology will be those antibody and DNA probes that allow tissue localization. With tissue localization will come the capacity to associate not only particular tissues with TsF production but will also allow the development of strategies to approach treatment. This could take two paths; on the one hand, failure to detect TsF in tissues where their presence might be of benefit would suggest the need to stimulate endogenous TsF synthesis or provide TsF exogenously. On the other hand the existence of TsF under circumstances where increased responsiveness is desirable (anti-tumor responses for example) would require strategies to decrease TsF production. Lastly these probes would also allow clinical investigators to monitor TsF during drug therapy which would establish whether a given regimen might have positive or negative effects on an important segment of the immune system. Used in conjunction with probes that detect stimulatory factors a very complete picture of the effect of a given drug on the immune system could be developed.

6   Antigen-specific factors as drugs

The most speculative area for discussion concerns the use of TsF themselves as drugs. Many researchers both past and present have cast a covetous eye on the antigen-specificity of, for example, immunoglobulins as potential vehicles for the highly specific eradication of tumor cells or other undesirable cellular products [4]. Currently considerable effort is being expended to use antibodies to tumor antigens to deliver covalently-linked cytotoxic drugs to a very specific target [4]. The antigen-specificity of the TsF also lends itself to this type of strategy. For example in transplantation it might be possible to specifically block only the response to transplantation antigens leaving the bulk of the individual's immune system intact. This would be a tremendous advantage over current therapies that suppress non-specifically all immunity thereby rendering treated individuals susceptable to a number of undesirable sequelae such as cytomegalovirus infection, lymphomas and increased susceptability to common infectious organisms [60].

Another area where specific therapy might be helpful are the autoimmune diseases and rheumatic diseases. In these instances much more would need to be known about the nature of the antigens involved in order to develop the appropriate TsF therapy. At present these approaches remain in the distant future. More immediately TsF specific probes will certainly be useful in understanding more completely the overall picture of the immune system in specific diseases and following drug therapy.

References

1   H. J. Parish: Victory with Vaccines. Livingston, Edinburgh 1968.
2   I. Raitt: Essential Immunology. Blackwell, Oxford 1977.
3   F. M. Burnet: Cold Spring Harbor Symposia on Quantitative Biology. *32*, 1 (1967).
4   E. S. Vitetta, R. J. Fulton, R. D. May, M. Till and J. W. Uhr: Science *238*, 1098 (1987).
5   G. Köhler and C. Milstein: Nature *256*, 495 (1975).
6   T. Tada and K. Okumura: Adv. Immun. *26*, 1 (1980).
7   R. N. Germain and B. Benacerraf: Scand. J. Immun. *13*, 1 (1981).
8   M. Taussig: Immunology *41*, 759 (1980).
9   D. R. Webb, J. A. Kapp and C. W. Pierce: A. Rev. Immun. *1*, 423 (1983).
10 K. A. Smith: Immun. Rev. *51*, 337 (1980).
11 C. A. Dinarello and J. W. Mier: N. Engl. J. Med. *317*, 940 (1987).
12 D. R. Green, P. M. Flood and R. Gershon: A. Rev. Immun. *1*, 439 (1983).
13 M. E. Dorf and B. Benacerraf: A. Rev. Immun. *2*, 127 (1984).
14 S. Strober: A. Rev. Immun. *2*, 219 (1984).
15 G. L. Asherson, V. Colizzi and M. Zembala: A. Rev. Immun. *4*, 37 (1986).
16 P. B. Hausmann, D. H. Sherr and M. E. Dorf: Concepts Immunopath. *3*, 38 (1986).
17 N. K. Damle: Year Immun. *2*, 60 (1986).
18 C. E. Hayes: Bioessays *4*, 278 (1986).
19 M. Taniguchi and T. Sumida: Immun. Rev. *83*, 125 (1985).
20 N. A. Mitchison: Nature *316*, 676 (1985).
21 K. Ishizaka: CRC Crit. Rev. Immun. *5*, 229 (1985).
22 K. Krupen, B. A. Araneo, L. Brink, J. A. Kapp, S. Stein, K. J. Wieder and D. R. Webb: Proc. natl Acad. Sci. USA *79*, 1254 (1982).
23 K. Wieder, B. A. Araneo, J. A. Kapp and D. R. Webb: Proc. natl Acad. Sci. USA *79*, 3599 (1982).
24 C. T. Healy, J. A. Kapp and D. R. Webb: J. Immun. *131*, 2843 (1983).
25 C. M. Sorensen, C. W. Pierce and D. R. Webb: J. exp. Med. 158, 1253 (1983).

26   K. Krupen, C. W. Turck, S. Stein, J. A. Kapp and D. R. Webb: Meth. Enzymol. *116*, 325 (1985).
27   C. W. Turck, J. A. Kapp and D. R. Webb: J. Immun. *135*, 3232 (1985).
28   C. W. Turck, J. A. Kapp and D. R. Webb: J. Immun. *137*, 1904 (1986).
29   M. Taniguchi, I. Takei and T. Tada: Nature *283*, 227 (1980).
30   M. Taniguchi, T. Saito, I. Takei and T. Takuhisa: J. exp. Med. *153*, 1672 (1982).
31   J. K. Steele, A. T. Staumers, J. G. Levy and J. D. Waterfield: Cell Immun. *102*, 386 (1986).
32   K. Imai, M. Kanno, H. Kimoto, K. Shigemoto, S. Yamamoto and M. Taneguchi: Proc. natl Acad. Sci. USA *83*, 8708 (1986).
33   P. M. Flood, C. Waltenbaugh, T. Tada, B. Chue and D. B. Murphy: J. Immun. *137*, 2237 (1986).
34   K. Sugimura, N. Yamasaki, M. Matsuura and T. Watanabe: Eur. J. Immun. *15*, 873 (1985).
35   G. S. Jendrisak, J. Trial and C. J. Bellone: Cell Immun. *97*, 419 (1986).
36   M. J. Daley, M. Nakamura and M. L. Gefter: J. exp. Med. *163*, 1415 (1986).
37   M. Taniguchi, M. Kanno and T. Saito: Meth. Enzymol. *116*, 311 (1985).
38   D. Ballinari, C. Castelli, C. Traversari, M. A. Pierotti, G. Parmiani, G. Palmieri, P. Ricciardi-Castagnoli and L. Adorini: Eur. J. Immun. *15*, 855 (1985).
39   G. Ada (Ed.): Immun. Rev. *98* (1987).
40   C. W. Pierce, C. M. Sorensen and J. A. Kapp: J. Immun. *134*, 29 (1985).
41   J. Trial, D. C. Shreffler and J. A. Kapp: Lymphokine Res. *5*, 275 (1986).
42   H. Okamoto and M. L. Kripke: Proc. natl Acad. Sci. USA *84*, 3841 (1987).
43   L. K. Roberts: J. Immun. *136*, 1908 (1986).
44   T. Noma and J. Yata: J. Immun. *138*, 3345 (1987).
45   A. J. Sultan, A. S. Jawad, H. Berry and J. Sharp: Clin. Rheumat. *5*, 450 (1986).
46   H. Murata and J. Yata: Asian Pac. J. Allergy Immun. *4*, 95 (1986).
47   M. Estrin, C. Smith and S. Huber: Am. J. Path. *125*, 578 (1986).
48   T. Sakane, S. Takada, N. Suzuki, T. Tsuchida, Y. Nurakawa and Y. Ueda: J. Immun. *137*, 3809 (1986).
49   G. J. Watt, C. J. Elson, D. G. Healey, A. Oryan and D. C. Hooper: Eur. J. Immun. *16*, 1131 (1986).
50   D. Lohmann, J. Krug, E. F. Lampiter, B. Bierwolf and H. J. Verlohren: Diabetologia *29*, 421 (1986).
51   D. A. Wilson: Cell. Immun. *96*, 312 (1985).
52   G. J. Watt, J. Russell and C. J. Elson: Scand. J. Immun. *24*, 39 (1986).
53   P. E. Jensen and J. A. Kapp: J. Immun. *136*, 1309 (1986).
54   P. M. Flood, C. Waltenbaugh, T. Tada, B. Chue and D. B. Murphy: J. Immunol. *137*, 2237 (1986).
55   J. A. Kapp, C. M. Sorensen and C. W. Pierce: Meth. Enzymol. *116*, 303 (1985).
56   N. Suzuki, T. Saito, T. Tokuhisa and M. Taniguchi: Int. Arch. Allergy appl. Immun. *77*, 300 (1985).
57   C. O. Jacob, P. H. Van der Meide and H. O. McDewitt: J. exp. Med. *166*, 798 (1987).
58   G. Möller (Ed.): T Cell Receptors and Genes, Immun. Rev. *81* (1984).
59   R. M. Zinkernagel and K. L. Rosenthal: Immun. Rev. *58*, 131 (1981).
60   R. Schindler (Ed.): Cyclosporin in Autoimmune Diseases. Springer-Verlag, New York 1985.

Implications of immunomodulant therapy in Alzheimer's disease

By Vijendra K. Singh and H. Hugh Fudenberg
Department of Microbiology and Immunology, Medical University of
South Carolina, 171 Ashley Avenue, Charleston, SC 29425, USA

This work was supported by funds from the Advanced Immunotherapeutics and Immunohematology Research Foundation

1   Introduction

Alzheimer's Disease (AD) is a degenerative disorder of the central nervous system (CNS), generally characterized by the impairment of memory (especially, the memory of recent events), intellect and cognitive functions in elderly individuals. The deterioration of these mental activities is attributed to the degeneration of nerve cells specifically localized in certain brain regions, e.g. basal forebrain, neocortex, hippocampus and amygdala [1–3]. About 6 to 8 % of individuals past the age of 60 years suffer from senile dementia and approximately 56 % of them are known to be afflicted with AD.

The etiology and the pathogenesis of the disease is unknown, and thus far, there is no effective treatment available for the management of patients with AD. There is, however, general concensus that AD is comprised of a heterogeneous group of patients in which the constellation of clinical symptoms may be the result of factors such as an infectious agent (perhaps an unconventional slow virus), genetic predisposition as in "familial" AD, dysfunction of immune system or immunoincompetence, functional disturbance of neurotropic factors or exposure to toxic substances like aluminum [4–7].

We recently hypothesized that AD is not a single disease but a syndrome of different subsets each with a different etiology, and also that at least one subset of AD is immunologic in origin [7, 8]. Several lines of investigations suggest the importance of studying immune system function in AD. These are: (a) both CNS and immune system display an intriguing interrelationship with each other [9–11]; (b) AD is a disease of the aging, and there is a general decline of function of immune cells with aging accompanied by an increased incidence of diseases, especially of autoimmune origin [12, 13]; (c) Down's Syndrome (DS), a well-known high-risk factor in AD, has been considered as an appropriate model of both accelerated aging [14] and primary immunodeficiency [15] in humans, and (d) immune-mediated mechanisms play an important role in the etiology and pathophysiology of other neurologic diseases (e.g. myasthenia gravis and multiple sclerosis), and patients with these disorders respond to immunotherapy [16–18]. We describe herein the evidence that at least one subset of Alzheimer's "Syndrome" is immunologically-derived and responsive to therapy with appropriate immunomodulators. As suggested previously [19, 20], we believe that different therapeutic modalities (e.g., immunotherapy for immunologic subset, therapy based on neurotransmitter deficiency for a neuro-chemical subset, etc.) will be necessary for different subsets of AD, much like the therapeutic strategies developed for other "diseases" such as anemia and diabetes.

2   Neuropathology

The morphological examinations of AD brains show variable degrees of symmetrical cortical atrophy, particularly in the frontal and temporal lobes. The hallmark features of the neuropathology of AD brains are commonly described as the "neurofibrillary tangles" (NFT) and the "neuritic plaques" (NP) [3, 21, 22]. The NFT are twisted pairs of fine nerve fibers (composed of paired helical filaments or PHF) accumulating in the cell bodies of neurons whereas the NP consist of clusters of neurites, enlarged nerve-endings and unmyelinated axons around a core of fibrous material called "amyloid". The Hirano bodies represent another form of abnormal inclusion prominently increased in AD brains and it contains actin and actin-associated proteins [23]. The exact biochemical nature of NFT and NP is not known but certain proteins have been extracted from each of these structures. The antiserum to neurofilaments of intermediate size shows immunoreactivity with NFT [24, 25] which also contains a polypeptide of 9–12 kDa molecular weight, probably a fragment of a larger molecule [26] and another protein known as "tau" protein is increased in phosphorylated form [27]. Additionally, the presence of ubiquitin protein has been detected by sequence analysis in PHF [28]. The major constituent of amyloid core is a protein known as amyloid A4 protein or β-amyloid protein of 4 kDa molecular weight, which is derived from a membrane-associated glycoprotein of 92 kDa molecular weight, namely the amyloid-precursor protein [29]. Recently, the gene encoding for amyloid A4 protein was localized on chromosome 21, the chromosome which also contains the gene linked to the rare autosomal-dominant familial form of AD (FAD). Although the two genes are located on chromosome 21 in close proximity with each other, they are not responsible for either FAD or more common sporadic form of AD [30–31].

The role of glial cells is poorly defined in the neuropathophysiology of AD. The number of GFAP+ astrocytes is considerably increased in the neocortex of AD patients [32]. A small increase (1.3-fold) in the mitotic activity of rat cortical astrocytes was reported in response to stimulation by cortical extracts from AD patients relative to aged subjects [33]. A recent immunohistochemical study [34] demonstrated the presence of HLA-DR antigen on the surface membrane of reactive microglia (GFAP$^-$), which correlated in number with NP but not with choline acetyltransferase (ChAT) activity. Such observations should kindle newer interest in the role of glial cells in AD.

3 Neurochemical dysfunction

It is now well established that certain specific groups of neurons degenerate in the basal forebrain, neocortex, limbic system and brain stem of AD patients [2]. Severe degeneration of neurons was found in the nucleus basalis of Meynert (nbM) which appears to be the primary source of cholinergic innervation to the cerebral cortex and hippocampus [35–37]. In addition to these cholinergic nerve-terminals, there are some cholinergic neurons in the cerebral cortex as shown by ChAT-immuno histochemistry [38]. Some of the prominent neurochemical

Table 1

Summary of various neurochemical changes in Alzheimer's Disease analyzed in the brain, CSF and blood immunocytes (References are cited in the text).

Parameter	Brain (Cerebral Cortex)	CSF	Blood (Immunocytes)
ACh	Decreased		
ChAT	Decreased		
AChE	Decreased		Decreased
ACh-receptor	Decreased		
Serotonin	Decreased		
GABA	Unchanged		
Glutamate/Aspartate	Decreased		
Glutamate-receptor	Decreased or unchanged		
Somatostatin	Decreased	Decreased	
CRF	Decreased	Decreased	
CRF-receptor	Increased		Decreased
Oxytocin	Decreased		
VIP Cholecystokinin	Normal		
Neuropeptide Y	Decreased		
NFT/PHF	Increased	Increased	
Alz-50	Increased		
Calcium Uptake			Decreased
Membrane Fluidity			Increased (in platelets)

and related changes in AD are summarized in Table 1. One of the most consistent biochemical abnormalities in AD brains is the lack of cholinergic activity (acetylcholine or ACh), specifically in brain regions that are associated with memory and cognitive functions. This particular deficit is the direct result of marked decrements (50–90 %) in the activity of ChAT [1–3], the enzyme responsible for the synthesis of neurotransmitter ACh. Moreover, variable degrees of reduction in the activity of acetylcholinesterase (AChE) [36] and ACh-muscarinergic receptors (M2 type) [37] have also been reported in different cortical regions of AD brains. The activity of AChE is also decreased in the mononuclear cells of the blood from patients with senile dementia of AD type [39], supporting our hypothesis that peripheral blood immunocytes can be used to study neuropsychiatric disorders [9].

The levels of other neurotransmitter substances like gamma-aminobutyric acid (GABA), dopamine and norepinephrine are generally within normal range in AD patients compared to age-matched non-AD controls [40]. Serotonin [41] and glutamate [42] systems, at least in terms of receptor binding, are decreased in the cerebral cortex of AD patients relative to aged controls; but detailed analyses are necessary to draw any firm conclusions about their importance in AD pathology. Certain neuropeptides, e.g. somatostatin [43, 44] and corticotropin-releasing factor (CRF) [45, 46] are consistently reduced in the brain as well as in the cerebrospinal fluid (CSF). Moreover, the CRF receptor binding is increased in certain cortical regions of AD brains [46]. It is noteworthy, however, that we recently found a significant decrease in the CRF receptor binding to peripheral blood lymphocytes of AD patients compared to non-AD controls (Fig. 1). This finding further substantiates our hypothesis that blood mononuclear cells can be used to detect abnormal function of the counterpart cells in the CNS [9]. The difference in these results of CRF receptor binding may be related to differences of tissues used: the brain analysis showed [46] increments in the occipital and parietal cortical areas (which are not relevant to memory function) unlike our work with blood lymphocytes. The brain concentrations of neuropeptide Y [48] and oxytocin [49] peptide hormones were also lower in AD patients than matched controls.

In addition to the reduced levels of somatostatin and CRF neuropeptides, the concentration of two polypeptides is significantly elevated in AD. The CSF concentration of PHF derived from NFT is about 2-fold greater in AD patients than the non-AD aged controls [50]. The con-

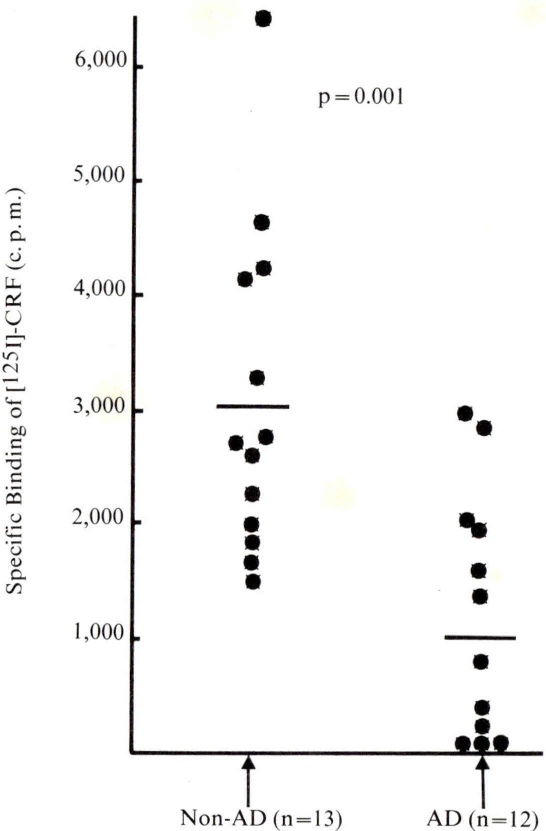

Figure 1

Specific receptor binding of CRF to peripheral blood immunocytes from AD patients and non-AD controls.

The CRF binding assay was performed according to the method of De Souza et al. [46]. The assay mixture, in a final volume of 0.2 ml, contained 0.1 ml of immunocyte membrane fraction (equivalent to 1×10^6 cells), 0.02 ml of human/rat ^{125}I-Tyr-CRF [approximately 35,000 counts per min (CPM); specific activity = 2,200 Ci/mmole purchased from NEN] and 0.08 ml of 0.1 % bovine serum albumin solution in phosphate-buffered saline. The blank binding was obtained by running assays in the presence of a 4-μM final concentration of carrier CRF (purchased from Sigma). After a 2-h incubation in an icebath, 1 ml of cold homogenizing buffer was added to each tube and the tubes were centrifuged for 3 min at 12,000 × g using a Brinkmann Microcentrifuge. The radioactivity of the membrane pellet, after an additional washing with 1.5 ml of the same buffer, was determined using a gamma-counter (Beckman Model Gamma 4000). The specific receptor binding of ^{125}I-CRF was calculated by subtracting the blank binding from the total binding. The results are given as the average CPM of triplicate assays.

centration of "Alz-50", a cytoplasmic protein of 68 kDa molecular weight, is elevated 15 to 30 times in the brain of AD patients compared to aged healthy individuals [51].
Alterations of nucleic acids have also been reported in AD brains. Mann et al. [52] found a significant decrease of cytoplasmic RNA content, nuclear and nucleolar volume in the temporal cortex. Using enzyme micrococcal nuclease, structural alterations of brain chromatin were shown in AD relative to non-AD subjects [53]. Moreover, the changes of RNA were shown to be related with an increase in the activity of alkaline ribonuclease enzyme, presumably due to abnormal ribonuclease-inhibitor complex. Doebler et al. [54], based on RNA metabolism studies of hippocampal pyramidal neurons, suggested that the depletion of RNA is not related to the local loss of neurons or the formation of NFT or NP but rather a generalized disturbance in the hippocampal region of the brain.

4 Studies of peripheral blood immunocytes

Based on the existence of intriguing structural and functional relationships (Table 2) between the CNS and the immune system, we recently hypothesized that peripheral blood immunocytes (a term we use to describe mononuclear cells) can be used to study various neuropsychiatric diseases [9]. Blalock and Smith [55] previously suggested that the immune system represents the body's "mobile brain" while Hall and Goldstein [56] proposed the use of the term "immunotransmitter", synonymously to "neurotransmitter", to refer to the soluble products of the immune system cells. Ader and Cohen [57] showed that the change in behavior or "classical conditioning" can alter immune responses in animals. It is now generally accepted that the cells of the nervous system influence the function of the cells of the immune system, and vice-versa, the cells of the immune system can modify the function of the cells of the nervous system. This type of physiological interrelationship is very important in the understanding of pathophysiology of numerous neuropsychiatric diseases, especially those of unknown etiology.
An anatomical relationship between the nervous system and immune system is shown by the presence of nerve fibers (cholinergic, peptidergic and noradrenergic) in various lymphoid organs [58], e.g. spleen, thymus gland, lymph nodes, gut-associated lymphoid tissue and bone

Table 2

A   Structural Similarities:

1. Thy-1 antigen, a glycoprotein of structural homology with immunoglobulin G molecule, is distributed in both lymphoid (thymus gland) and CNS tissue.
2. MRC OX-2 antigen is a glycoprotein present on thymocytes, dendritic cells, some B lymphocytes, and certain neurons of CNS, and it resembles in structure the immunoglobulin light chain.
3. T suppressor cells and the glial cells (oligodendrocytes) express common antigens as defined by human T cell monoclonal antibody.
4. Purkinje cells share antigens with human T cells.
5. Anti-Leu-7 positive antigen is shared by NK cells and myelin from CNS and peripheral nervous system. A myelin-associated antigen (MAG) is present on a human T cell leukemia cell line.
6. Fc receptors and Ia antigens are found on B lymphocytes and monocytes as well as on astrocytes and microglia; monocyte monoclonal antibody (OKM_1) stains cells in the white matter of the brain.
7. Existence of S-100 brain protein in monocyte-derived Langerhans cells.
8. Various ion-channels (K^+, Na^+, and Ca^{++}) are found on neuronal and lymphoid cells.
9. Occurrence of receptor entities for neurotransmitter substances, e.g., acetylcholine, dopamine, serotonin and neuropeptides (e.g., endorphins, somatostatin, substance P) on various types of immunocytes.
10. Presence of interleukin-1 (IL-1) receptors on brain cells.
11. Expression of CD_4 antigen by brain astrocytes.
12. Presence of receptors for Corticotropin-Releasing Factor (CRF), a hypothalamic neurohormone, on both brain cells and immune cells (especially monocytes and T cells).

B   Functional Relationship:

1. Immune response evokes firing of CNS noradrenergic neurons.
2. Expression of Ia and MHC-restricted H_2 antigens by normal brain cells after exposure to interferon.
3. Electrolytic lesion or simulation of brain modified immune responses directly or indirectly via neuroendocrine system.
4. Modification of single cell recording in different brain regions after the microiontophoretic application of γ-interferon, a soluble product of lymphocytes.
5. Both monocytes and astrocytes present "antigen" to T cells.
6. Production of interleukin-1 (IL-1) lymphokine by monocytes and also by astrocytes; the monocyte-derived IL-1 has a potent mitogenic effect on astroglial proliferation.
7. Synthesis of endorphin- and corticotrophin-like neuropeptides by lymphocytes.
8. Modulation of immunologic function *in vitro* by neuropeptides like encephalin, substance P, somatostatin.
9. Stimulation of proliferation of and differentiation of oligodendroglial cells by interleukin-2 (IL-2) lymphokine.
10. Recombinant IL-1 stimulates secretion of CRF by the hypothalamic cells.
11. Suppression of leucocyte interferon-γ production by β-endorphin neuropeptide.

Revised after V. K. Singh and H. H. Fudenberg, Ref. #9

marrow. As summarized in Table 2, several antigenic proteins of the nervous system have been localized in the cells of the immune system [59, 60]. Certain neuropeptides, like endorphins and corticotropins, are natural products of both the CNS and immune system. The AChE enzyme, a marker of cholinoceptic nerve cells, is present in the peripheral blood mononuclear cells [39]. The high-affinity binding receptor sites for more than one neurotransmitter (ACh, serotonin, dopamine, histamine) [61], several neuropeptides (substance P, somatostatin, endorphins, CRF) ([9] and Fig 1.) and the sigma-opiate receptors for phencyclidine or PCP [62] have been identified in the membranes of peripheral blood lymphocytes. Conversely, the receptor sites for immune cell products, like IL-1 of monocytes [63] and CD4 antigen of T helper cells [64], have been detected on brain cells.

Functionally speaking, several neuropeptides modify the function of lymphocytes *in vitro* [11]. The brain astrocytes present antigens to T cells in a similar fashion as monocytes do [65]. The lymphokine substances (IL-1 or IL-2), the products of lymphoid cells, function as the supplementary growth factors for brain cells [66, 67]. Additionally, thymic hormones (e.g. thymosin-α_1) increase levels of corticosteroids by the stimulation of hypothalamic and pituitary cells [68], and similarly, lymphokine IL-1 stimulates the secretion of CRF [69–71], a hypothalamic hormone in the brain. These observations, undoubtedly, formulate a complete loop between nervous system (via neurotransmitters, neuropeptides and other soluble products) and immune system (via thymic hormones, lymphokines and related factors).

The aforementioned findings suggest the presence in the brain of cells as the possible counterpart cells of the immune system, e.g. astrocytes may correspond to monocytes. Consequently, neuropsychiatric disorders can be studied by the use of peripheral blood mononuclear cells rather than the brain biopsy or autopsy tissue as we hypothesized previously [9]. In specific terms, some recent findings support our hypothesis. These are: (a) we found deficiency of CRF receptors in peripheral blood mononuclear cells from AD patients compared to non-AD controls (Fig. 1); (b) the AChE activity of blood lymphocytes from AD patients is lower than the age-matched controls [39]; (c) the mitogen-induced uptake of calcium by blood lymphocytes is depressed in AD patients [72]; and (d) the membrane fluidity of blood platelets is increased in AD patients and their family members compared to non-AD controls [73]. Clearly then, it is possible to use peripheral blood mononuclear cells, instead of brain biopsied or autopsied material, for the studies of neuropsychiatric disorders.

5   Immunologic factors

We and others have found abnormal distribution of more than one immune parameter in AD. In patients with presenile dementia of AD type, the concentration of serum IgG, IgA and IgD have been shown to be normal, but IgM was found to be decreased [74]. This observation differs from that of Cohen and Eisdorfer [75] who reported significantly elevated levels of IgG and IgA and unchanged IgM concentration; however, their study population was a heterogenous group of cognitively-impaired elderly individuals, many with senile dementia due to reasons other than AD (e.g., cerebral arteriosclerosis). We recently quantified serum concentrations of various subclasses of IgG [76], and found no significant difference in the distribution of IgG_1, IgG_2 and IgG_4 in AD patients compared to age-matched controls. However, 45 % of the AD patients had an IgG_3 level significantly higher than aged normals, and of the 9 patients with elevated IgG_3, 8 had serum antibodies for brain neurons as tested by the indirect immunoflourescent technique; the concentration of IgG_3 isotype in the remaining patients was moderately reduced and they were negative for brain auto-antibodies [20, 76]. Skias et al. [77] reported an increased production of IgG *in vitro* in SDAT patients relative to elderly controls, but the synthesis of IgG subclasses was not measured.

The role of autoimmunity in the pathogenesis of AD has long been suspected. This is based on several circumstantial observations: (a) familial or genetic incidence of AD; (b) late onset of AD in concordance with increase of autoimmune diseases with aging; (c) amyloid formation, presumably of immune origin, in cerebral plaques; and (d) generation of circulating autoantibodies reacting with brain tissue, hereafter also referred to as "anti-brain". Nandy [78] reported a significant difference between senile dementia patients and age-related controls. In another study [79], the anti-brain staining could not be distinguished between patients and parallel controls; however, the pattern of staining was similar to nonspecific autofluorescence. We recently found a high incidence of "specific" brain autoantibodies in 57 % of sera of AD patients [80], but not in the sera from young or aged healthy controls and several other disease controls including young or elderly patients with Down's syndrome. This is the first evidence that Down's syndrome, a high risk factor in AD, and with the same neuropathology as in AD, differs from AD. Since our report, others [81, 82] have provided confirmatory results of circulating brain antibodies in AD. Additionally, it was shown that the immunoreactivity was directed, at least in some cases, against ChAT-containing cholinergic neurons [81, 82] and it was mainly of IgG_3 isotype [82]. The latter finding confirms our result that a subset of AD patients (45 % of the total) had elevated level of IgG_3 and nearly all were positive for brain autoantibody [20, 76]. Since one of the biological properties of IgG_3 isotype of antibody is to effectively fix complement [83], it is conceivable that brain auto-antibodies have an important pathogenetic role via complement-fixing immune complexes in the destruction of certain neurons as seen in AD. We have recently initiated such investigations.

In addition to antibodies reacting with brain tissue, we and others have detected autoantibodies reacting with neuron-axon filament proteins (200 kDa) in AD sera [20, 84]; such autoantibodies, however, may not be restricted to AD since they are also detected with high incidence in other neuropsychiatric diseases like autistic syndrome [20], and especially those with evidence of a slow virus infection [84]. Others have found serum autoantibodies to human pituitary (prolactin) cells [85] which wait for further characterization. Moreover, antibodies reacting with neurofibrillary tangles (NFT) were produced *in vitro* by the EBV-transformed B cells from both AD patients and normal blood donors; the two groups, however, differed in the frequency (6.3 % for AD vs. 1.6 % for the controls) of production of EBV-transformed B cell lines secreting NFT antibodies [86].

Immunologic dysfunction has also been described in AD as measured in terms of cell-mediated immunity (T cell function). We recently reported that the DNA synthesis of peripheral blood lymphocytes *in vitro* in response to stimulation with T cell mitogens (PHA, anti-T_3, Con A) was significantly lower in AD patients compared to age-matched controls [8]. We also found that the T cells expressing orosomucoid, a glycoprotein with immunomodulator function [87, 88], and the T cell stimulation by non-T cells in the autologous mixed leukocyte reaction (AMLR) were abnormally distributed in many AD patients. The deficiency of T cell function in AD has also been demonstrated by reduced activity of $T8^+$ suppressor cells [77] and by significantly lower than normal response of Con A-stimulated lymphocyte DNA synthesis and BCG vaccine-induced delayed-type hypersensitivity [89].

6   Pharmacological approach to therapy

One of the major problems in trying to develop any form of treatment for AD patients is the lack of appropriate biological markers for the clinical diagnosis of the disease; 15 to 25 % of the total patients are currently misdiagnosed. Furthermore, the patient population is a very heterogenous population, at least in terms of the constellation of the symptoms. In addition to the loss of memory, especially of the recent events and functions, there are many other disease-related disturbances. These include retardation of intelligence, depression, agitation, anxiety, language impairment, visual-spatial disturbance, mood changes and behavioral changes. Is AD the same disease as the senile dementia of the AD type? Is a drug for the treatment of early-onset going to be efficacious for the late-onset of AD? All of these and many other considerations are very important in the development of a successful treatment for AD.

The drugs currently prescribed in the United States to treat cognitive disturbances in aged individuals are dihydroergotoxine, vasodilators (e.g., papaverine, isoxsuprine, cyclandelate) and stimulants like methylphenidate and pentylenetetrazol [90]. Hydergine and Vincamine, two substances with cerebrovasodilating properties, have been used in some patients with SDAT. None of these agents have any beneficial effects directly in the improvement of memory, but some seemed to help in alleviating certain secondary symptoms such as depression, irritability, fatigue, etc. However, the therapeutic value of any one of these agents has not been proven by the Food and Drug Administration (FDA) of the U.S.A.

A number of investigational drugs are being clinically tested. The rationale behind the development of these agents is that there exists one or the other type of neurochemical abnormality in the brain of AD patients. Based on the most consistent abnormality of ACh neurotransmitter, there have been several clinical trials [91, 92] with cholinomimetic drugs which are expected to increase the concentration of ACh in the brain, the so-called "cholinergic" model. Initially, ACh precursors like choline and lecithin were tested with failing results either alone or in combination with other agents like piracetam. ACh-muscarinic agonists such as arecoline, RS-86 or bethanechol were also used; there were small positive effects on memory but the large side effects precluded their usefulness. An alternative approach was to design therapy using AChE inhibitors such as physostigmine or tetrahydroaminoacridine (THA). Some patients showed positive short-term response to physostigmine [93], but since this drug has a rapid turnover rate, it must be given in large doses to achieve constant inhibition of AChE. This is impractical and the possibility of severe neurotoxic side effects due to long-term overdose is inevitable. Summers et al. [94], using THA, reported remarkable improvement of clinical symptoms in AD patients. However, not all patients in this study were afflicted with AD, and recently, liver toxicity [95] has been found as one of the major side effects of THA.

Based on the results of animal experiments, certain neuropeptides like ACTH, Org 2766 (ACTH 4-9 analog), naloxone, naltrexone, vasopressin and its derivatives (DDAVP, DGAVP), have been tested in patients with dementing illnesses of AD type [90, 96]. In general, these clinical trials have not resulted into any significant improvement of memory or cognition, but moderate positive effects were seen for mood changes and attention span. Likewise, vasopressin seems to cause non-specific stimulation of neuronal pathways rather than directly enhance cognitive activities. Recently, attempts have been aimed at using opiate antagonist naloxone and/or its analog naltrexone, but the preliminary results of various clinical trials do not offer much success. One of the major problems associated with the use of neuropeptides may be that their concentration at the site of action in the brain is not large enough to produce any beneficial effect. This problem may be the net outcome of blood-brain barrier which could be circumvented, at least in part, by using intracranial infusions.

Other investigational drugs include "nootropic" agents, miscellaneous compounds like zimelidene, 4-aminopyridine, nimodipine, vinpocetine and the aluminum chelator desferrioxamine. The patients with AD do not seem to be benefited by the use of these miscellaneous agents. The nootropics [97], however, are of particular importance since they are virtually free of adverse side effects and appear to display some positive effects in alleviating certain clinical symptoms, but not memory, in AD. Piracetam, the prototype nootropic, is related structurally to GABA neurotransmitter, and it was originally used in Europe for the treatment of children with dyslexia and other learning disabilities [98]. Other piracetam-related compounds of interest are aniracetam, oxiracetam, pramiracetam and CI-911, which show limited success having amnesia-reversal activity. The results of clinical trials with piracetam alone or in combination with lecithin are equivocal (if anything, they are related to short-term memory), and no clear-cut conclusion can be drawn about its efficacy in the treatment of AD patients.

7 New approach to immunotherapy

We recently demonstrated that AD is a syndrome of subsets with different etiologies, one of which is immunologically-derived [7, 8]. As outlined earlier in this article, a unique structural and functional interrelationship exists between the nervous and immune systems [9–11]. We considered this type of knowledge to be very important since it implies that the longitudinal studies of patients with CNS disorders can be performed by using various immune cells in peripheral blood, especially of the effects of therapy, instead of requiring serial brain biopsies. Using this approach in AD, we found not only the abnormal function of cell-mediated immunity and humoral immunity (brain auto-antibodies) but also demonstrated a specific deficiency of immunocyte receptors for CRF – the major hypothalamic neurohormone.

Based on the aforementioned observations and considerations, we have been exploring, for the last 3–4 years, the possibility of developing a new approach to treatment of AD patients using appropriate biologic and synthetic compounds with immunomodulating properties; as reported herein, the responsiveness to immunomodulant therapy is dependent upon the nature of immune deficits. Consequently, we have distinguished at least four subsets of AD by immunologic tests and the patients in two such subsets showed remarkable improvement of certain clinical features, including regain of memory function and sphinctor control. These four subsets are: (a) a subset with defect (e.g., membrane fluidity) of a specific T lymphocyte and response to pyrrolidone therapy; (b) a subset with circulating antibodies to neuron-axon filament proteins (NAFP) of 200 kDa; these patients displayed decrease in severity of clinical symptoms after therapy with Dialyzable Leucocyte Extract containing Transfer Factor (DLE-TF) described in greater detail before [19, 20]; (c) a subset with antibodies to brain antigens (henceforth "autoimmune") for which immunotherapy is not yet developed; and (d) a subset which is probably heterogeneous due to the lymphocyte deficiencies of more than one biochemical factor; once again, the therapy for this subset is not yet investigated.

The basic unit of pyrrolidone compounds is a monomer polyvinyl pyrrolidone (PVP), which is known to stimulate immune responses in animals [99]. The nootropic drugs like piracetam and aniracetam contain basic PVP molecule in their structure and they also act as modulators of immune functions at least *in vitro*. These substances are structurally related to 2-pyrrolidinone which is a naturally occurring beta-lactam of GABA, an inhibitory neurotransmitter in the brain. In the body, 2-pyrrolidinone is synthesized from polyamine putrescine and it is metabolized to succinimide via 5-hydroxy-2-pyrrolidinone as outlined by Bandle et al. [100]. As described above, several AD patients in one subset, referred to as the "pyrrolidone subset", responded to experimental therapy with aniracetam [10, 19, 20]. Eleven of the forty AD patients treated with aniracetam showed improvement in both laboratory immune parameters and clinical symptoms in either the first or second arm. A few patients who were responders in the initial drug trial also participated in the second study, and these patients again responded with both laboratory and clinical improvement in one or another arm. On the other hand, in those patients in the Piracetam Study [101] who failed to show improvement and who also entered the second study, no amelioration of clinical symptoms was seen in either arm of the second study. Of the 40 patients, those who improved clinically in one or another arm had laboratory improvement prior to clear-cut positive effect of the therapy. Those who improved in either arm, after completion of both arms, then received 6 months of pyrrolidone without placebo. Thus, those who received aniracetam in the second arm received continuous therapy for 9 months whereas those who had received aniracetam in the first arm received only 6 months of uninterrupted therapy. The degree of improvement was not related to duration or severity or age of onset of illness but was closely related to duration of therapy.

The precise mechanism by which pyrrolidones work is not known, but we hypothesized that one of the enzymatic steps of the metabolic pathway of 2-pyrrolidinone bypasses a block [19, 20], presumably genetic in origin. This situation resembles the example of adreno-cortical hyperplasia in which deficiency of any one of the 5 enzymes produces identical clinical symptoms [102]. The products of the metabolism of 2-pyrrolidinone are necessary for the energy metabolism via Krebs cycle of both the nerve and immune cells. The AD patients in this subset may have abnormal membrane fluidity (or flexibility) and pyrrolidones may bypass the enzymatic "block" and thereby restore normal membrane function. In this regard, it is worthy of note that a deficiency of membrane receptors, in particular ACh neurotransmitter and CRF neurohormone, exists in AD (see Table 1) and pyrrolidones might normalize these deficient receptor functions. Our unpublished preliminary data showed that the addition of pyrrolidone to the cultures of nerve cells and immune cells produced 25–35 % increase in the uptake of neurotransmitter substances; additional experiments are needed to confirm these results.

Additionally, we treated 6 AD patients in a similar fashion as those having infections [103] with Dialyzable Leucocyte Extract (DLE) which contains Transfer Factor (TF). Although various attempts to characterize its precise biochemical nature have met with failures, however, TF appears to be an RNA-peptide of molecular weight ranging between 2 to 5 kDa [104]. The DLE containing TF is prepared from the immune cells isolated from the blood of highly selected donors, and since this is a crude preparation it may contain more than one molecule which control T cell-mediated immune responses. The mode of administration of TF is the intramuscular injection of a predetermined dose. In our experimental study of treatment of AD patients, each of the 6 patients received 10 units of TF (1 unit is the amount of TF derived from 500 million leucocytes) the first day, 20 units the second day, 30 units the third day, and 40 the fourth day. They were then followed for 30 days without additional injection.

Laboratory assessment before and after TF treatment showed improvement of defective cell-mediated immunity present in all 6 patients, but only those with circulating antibodies to neuron-axon filament proteins (NAFP) had any appreciable degree of clinical response. As summarized in Table 3, the two patients without anti-NAFP did not respond. These 6 patients were in stages IV to VI; three had lost the ability to ambulate and also had lost rectal and bladder sphincter control. By the 4th day of therapy, 3 patients were able to walk unaided, regain of sphincter control took place and their spouses and friends saw noticeable improvement of cognitive functions (e.g., resumption of reading ability which was lost before therapy). These clinical benefits generally lasted for about 30 days (no additional injection of TF was given during this period), but thereafter disabilities gradually returned. At 8 weeks, readministration of TF on the same schedule as given above resulted in the improvement of clinical symp-

Table 3
Results of various immune parameters and response to therapy with DLE.

Patient code	Disease stage	DNA Synthesis* (+PHA) (CPM)	Anti-NAFP**	Response to DLE therapy
D.M.	VI	35,135/38,483	+	+ + + +
V.H.	V	31,309/49,148	+	+ + +
A.R.	VI	102,348/196,334	+	+ +
B.G.	V	23,011/63,212	−	+
S.H.	V	50,210/57,281	+	+
E.M.	IV	44,366/196,334	−	−

  * Incorporation of ^{3}H-thymidine into lymphocyte DNA (average of triplicate assays) reported for patient/control.
** Screened by the indirect immunofluorescent technique using frozen sections of rat spinal cord.

toms to a degree similar to that observed after the first course.
In addition, we have identified an autoimmune subset of AD which is discerned by the detection of serum antibodies against brain tissue [80] and their relationship to increased serum level of IgG_3 isotype of antibody [20, 76]. Although we do not know the nature of the specific antigen involved, we suggest that these brain antibodies may be directed against receptors for neurotransmitters and/or the enzymes of neurotransmitter synthesis; there is preliminary evidence to show that IgG_3 subclass of antibody reacts with ChAT-containing cholinergic neurons in some AD patients [81, 82]. Our hypothesis would explain, at least in part, the lack of efficacy of cholinergic therapy in AD patients since brain autoantibodies may bind to these neurons and impair their normal function. This model, therefore, appears to be similar to autoimmune myasthenia gravis in which circulating antibodies, mostly of IgG_3 isotype [105], bind to nicotinic ACh receptor and block post-synaptic release of the neurotransmitter ACh at the neuromuscular junction. Thus, we suggest that Cyclosporin A, a fungal metabolite with well-known immunosuppressive activity, may be used to treat patients in this autoimmune subset of AD.
By studies of peripheral blood immunocytes, additional subsets of AD might be discerned by as yet unknown biochemical defects that may exist not only in the CNS but also in the cells of the immune system. For example, a genetic defect may occur in enzyme(s) of the oxidative metabolism in one subpopulation of blood monocytes equivalent of the microglial cells in the brain; this situation is analogous to the en-

zyme deficiencies (e.g., NADH-oxidase) found in monocyte-derived cells of peripheral blood, liver and skin in chronic granulomatous disease [106]. Moreover, it is conceivable that other monocyte defects in the subset of astrocytes, which are the brain equivalent of the "antigen-presenting" cells [65], exist that might be responsible for the reduced production of proteases predisposing to amyloid formation (enhanced production of anti-protease or protease inhibitor would have the same effect). Such a possibility may be reflected by the recent observation that α-1-antichymotrypsin, a protease inhibitor which reacts with several proteases in addition to its most preferred substrate chymotrypsin, is localized in the cerebral amyloid [107]. Furthermore, certain other non-immune parameters like the biochemical dysfunction of neurotransmitters, e.g. ACh [39] and/or neuropeptides, e.g., CRF ([108] and Fig. 1), might also be important factors contributing to the pathophysiology of AD.

8   Summary

The recent findings of our own and those of others demonstrated that the immune system function is one of the important paradigms associated with the etiology and pathophysiology of AD. Based on our current immunologic research, we conclude that AD is not a single disease but rather a syndrome of heterogenous subsets involving multiple etiological factors. One of these subsets is immunologically-derived since aberrations of both T cell-mediated and humoral immunity (in particular, the autoimmunity to CNS antigens) exist. Moreover, appropriate immunomodulant therapy in each patient depends upon the nature of immune deficit and predicts the responsiveness, and based upon these criteria we discerned at least two sub-groups of immunologically-defined subset. Our result of treatment clearly implicates the importance of immunotherapy for AD patients. Therapies for each subset will differ, just as is the situation with each subtype of anemia or diabetes. We furthermore conclude that the regain of function with immunotherapy in many AD patients implies that the neurons relevant to memory and cognitive functions are not "dead" but merely "atrophied" or physiologically "inactive", and the agents having immunomodulating properties might act by the stimulation of these functionally suppressed CNS cells *in vivo*.

References

1   R. D. Terry and P. Davis: A. Rev. Neurosci. *3*, 77 (1980).
2   D. L. Price, P. J. Whitehouse, R. G. Strubble, A. W. Clark, J. T. Coyle, M. R. DeLong and J. C. Herdeen: Neurosci. Commentaries *1*, 84 (1982).
3   P. Davis: A. Rev. Neurosci. *9*, 489 (1986).
4   H. M. Wisniewski, G. S. Merzx and R. I. Carp: Interdiscipl. Topics Geront. *19*, 45 (1985).
5   G. G. Glenner: Human Path. *16*, 433 (1985).
6   Z. S. Khachaturian: Neurobiol. Aging *7*, 537 (1986).
7   H. H. Fudenberg, H. D. Whitten, P. Arnaud and N. Khansari: Clin. Immun. Immunopath. *32*, 127 (1984).
8   V. K. Singh, H. H. Fudenberg and F. R. Brown, III: Mech. Age Develop. *37*, 257 (1987).
9   V. K. Singh and H. H. Fudenberg: J. clin. Psychiatr. *47*, 592 (1986).
10 V. K. Singh and H. H. Fudenberg: Prog. Drug Res. *30*, 345 (1986).
11 J. E. Blalock: Immunopath. Immunother. Lett. *2*, 3 (1987).
12 M. E. Weksler: Proc. Soc. expl. Biol. Med. *165*, 200 (1980).
13 C. S. Vissinga, C. J. Dirven, F. A. Dirven, V. A. Steinmeyer, R. Bonner and W. J. A. Boersma: Cell. Immun. *108*, 323 (1987).
14 R. L. Walford, T. C. Gossett, F. Naeim, C. F. Tam, J. L. Van Lancker, E. V. Bennett, D. Chia, R. S. Sparkes, J. L. Fahey, C. Spina, R. A. Gatti, M. A. Medici, H. Grossmann, H. Hibrawi and M. Motola, in: Immunological Aspects of Aging, p. 479. Eds D. Segre and L. Smith. Marcel Dekker, New York 1981.
15 G. R. Burgio, A. Ugazio, L. Nespoli and R. Maccario: Birth Defects *19*, 325 (1983).
16 R. L. Dawkins and M. J. Garlepp, in: The Autoimmune Diseases, p. 591. Eds N. R. Rose and I. R. Mackay. Academic Press, New York 1985.
17 B. G. Arnason, in: The Autoimmune Diseases, p. 399. Eds N. R. Rose and I. R. Mackay. Academic Press, New York 1985.
18 M. P. Dabrowski, B. K. Bernstein, A. Stasiak, K. Gajkowski and S. Korniluk: Ann. N. Y. Acad. Sci. *496*, 697 (1987).
19 H. H. Fudenberg and V. K. Singh, in: Immunological Aspects of Alzheimer's Disease and Brain Amyloidosis. IPSEN Foundation, Angers, France, in press 1988.
20 H. H. Fudenberg and V. K. Singh, in: Autoimmunity: Its Role in Alzheimer's Disease and Other Behavioral Disorders. Alan R. Liss, Inc., in press 1988; V. K. Singh et al.: Ann. N. Y. Acad. Sci., in press 1988.
21 R. D. Terry, H. K. Gonates and M. Weiss: Am. J. Path. *44*, 269 (1964).
22 M. Kidd: Brain *87*, 307 (1964).
23 P. M. Galloway, G. Perry and P. Gambetti: J. Neuropathol. expl. Neurol. *46*, 185 (1987).
24 S. H. Yen, F. Gaskin and S. M. Fu: Am. J. Path. *113*, 373 (1983).
25 P. Gambetti, L. Autilio-Gambetti, G. Perry, G. Shecket and R. C. Crane: Lab. Invest. *49*, 430 (1983).
26 B. H. Anderton: Nature *325*, 658 (1987).
27 I. Grundke-Iqbal, K. Iqbal, Y. C. Tung and H. M. Wisniewski: Proc. natl. Acad. Sci., USA *83*, 4913 (1986).
28 H. Mori, J. Kondo and Y. Ihara: Science *235*, 1641 (1987).
29 J. Kang, H. G. Lemaire, A. Unterbeck, J. M. Salbaum, C. L. Masters, K. H. Grzeschik, G. Multhaup, K. Beyreuther and B. Muller-Hill: Nature *325*, 733 (1987).
30 C. Van Broeckhoven, A. M. Genthe, A. Vandenberghe, B. Horsthemke, H. Backhovens, P. Raemaekers, W. Van Hul, A. Wehnert, J. Gheuens, P. Cras, M. Bruyland. J. J. Martin, J. M. Salbaum, G. Multhaup, C. L. Masters, K.

Beyreuther, H. M. D. Gurling, M. J. Mullan, A. Holland, A. Barton, N. Irving, R. Williamson, S. J. Richards and J. A. Hardy: Nature *329*, 153 (1987).

31   R. E. Tanzi, P. H. St. George-Heslop, J. L. Haines, R. J. Polinsky, L. Nee, J. F. Foncin, R. L. Neve, A. I. McClatchey, P. M. Conneally and J. F. Gusella: Nature *329*, 156 (1987).

32   G. L. Mancardi, B. H. Liwnicz and T. I. Mandybur: Acta Neuropath. (Berl.) *61*, 76 (1983).

33   M. Nieto-Sampedro: Neurobiol. Aging *8*, 249 (1987).

34   P. L. McGeer, S. Itagaki, H. Tago und E. G. McGeer: Neurosci. Lett. *79*, 195 (1987).

35   R. T. Bartus, R. L. Dean III, B. Beer and A. S. Lippa: Science *217*, 408 (1982).

36   J. T. Coyle, D. L. Price and M. R. DeLong: Science *219*, 1184 (1983).

37   D. C. Mash, D. D. Flynn and L. T. Potter: Science *288*, 1115 (1985).

38   P. L. McGeer, E. G. McGeer, V. K. Singh and W. M. Chase: Brain Res. *81*, 373 (1974).

39   E. Bartha, J. Szelenyi, K. Szilagyi, V. Venter, N. T. Thu Ha, P. Paldi-Haris and S. Hollan: Neurosci. Lett. *79*, 190 (1987).

40   P. T. Francis, A. M. Palmer, N. R. Sims, D. M. Bowen, A. N. Davison, M. M. Esiri, D. Neary, J. S. Snowden and G. K. Wilcock: N. Engl. J. Med. *313*, 7 (1985).

41   D. M. Bowen, S. J. Allen, J. S. Benton, M. J. Goodhardt, E. A. Haan, A. M. Parmer, N. R. Sims, C. C. T. Smith, J. A. Spillane, M. M. Esiri, D. Neary, J. S. Snowden, G. K. Wilcock and A. N. Davison: J. Neurochem. *41*, 266 (1983).

42   E. G. McGeer, E. A. Singh and P. L. McGeer: Neurobiol. Aging *8*, 219 (1987).

43   C. A. Tamminga, N. L. Foster, P. Fedio, E. D. Bird and T. N. Chase: Neurology *37*, 161 (1987).

44   K. J. Reinikainen, P. J. Riekkinen, J. Jolkkonen, V. M. Kosma and H. Soininen: Brain Res. *402*, 103 (1987).

45   G. Bissette, G. P. Reynolds, C. D. Kilts, E. Widerlov and C. B. Nemeroff: J. Am. med. Assoc. *254*, 3067 (1985).

46   E. B. De Souza, P. J. Whitehouse, M. J. Kuhar, D. L. Price and W. W. Vale: Nature *319*, 593 (1986).

47   C. May, S. I. Rapoport, T. P. Tomai, G. P. Chrousos and P. W. Gold: Neurology *37*, 535 (1987).

48   M. F. Beal, M. F. Mazurek and M. A. McKee: Neurosci. Lett. *79*, 201 (1987).

49   M. M. Mazurek, M. F. Beal, E. D. Bird and J. B. Martin: Neurology *37*, 1001 (1987).

50   P. D. Mehta, L. Thal, H. M. Wisniewski, I. Grundke-Iqbal and K. Iqbal: Lancet *2*, 35 (1985).

51   B. L. Wolozin, A. Pruchnicki, D. W. Dickson and P. Davies: Science *232*, 648 (1986).

52   D. M. A. Mann, D. Neary, P. O. Yates, J. Lincoln, J. S. Snowden and P. Stanworth: J. Neurol. Neurosurg. Psychiatr. *44*, 97 (1981).

53   E. M. Sajdel-Sulkowska and C. A. Marotta: Science *225*, 947 (1984).

54   J. A. Doebler, W. R. Markesbery, A. Anthony and R. E. Rhoads: J. Neuropath. expl. Neurol. *46*, 28 (1987).

55   J. E. Blalock and E. M. Smith: Immun. Today *6*, 115 (1985).

56   N. R. Hall, J. P. McGillis, B. L. Spangelo and A. L. Goldstein: J. Immun. *135*, 806s (1985).

57   R. Ader and M. Cohen: Psychosom. Med. *37*, 333 (1975).

58   D. Felton, S. Y. Felton, S. L. Carlton, J. A. Olschowska and D. Livnat: J. Immun. *135*, 755s (1985).

59 E. M. Smith and J. E. Blalock: Proc. natl Acad. Sci., USA *78*, 7530 (1981).
60 E. M. Smith, D. Harbour-McMenamin and J. E. Blalock: J. Immun. *135*, 779 (1985).
61 R. G. Coffey and J. W. Hadden: Fedn. Proc. *44*, 112 (1985).
62 V. K. Singh and H. H. Fudenberg, in: Sigma and Phencyclidine-like Compounds as Molecular Probes in Biology, p. 653. Eds E. F. Domino and J. F. Kamenka. NPP Books, Ann Arbor, Michigan 1988.
63 W. L. Ferrar, P. L. Kilian, M. R. Ruff, J. A. Hill and C. B. Pert: J. Immun. *139*, 459 (1987).
64 I. Funke, A. Hahn, E. P. Rieber, E. Weiss and G. Reithmuller: J. expl Med. *165*, 1230 (1987).
65 A. Fontana, W. Fierz and H. Wekerle: Nature *307*, 273 (1984).
66 D. Giulian and L. B. Lachman: Science *228*, 497 (1985).
67 E. N. Benveniste and J. E. Merrill: Nature *321*, 610 (1986).
68 B. L. Spangelo, N. R. Hall and A. L. Goldstein: Ann. N. Y. Acad. Sci. *496*, 196 (1987).
69 E. W. Bernton, J. E. Beach, J. W. Holaday, R. C. Smallridge and H. G. Fein: Science *238*, 519 (1987).
70 R. Sapolsky, C. Rivier, G. Yamamoto, P. Plotsky and W. Vale: Science *238*, 522 (1987).
71 F. Berkenbosch, J. van Oers, A. del Rey, F. Tilders and H. Besedovsky: Science *238*, 524 (1987).
72 G. E. Gibson, P. Nielsen, K. A. Sherman and J. P. Blass: Biol. Psychiatr. *22*, 1079 (1987).
73 G. S. Zubenko, M. Wusylko, B. M. Cohen, F. Boller and I. Teply: Science *238*, 539 (1987).
74 B. Tavolato and V. Argentiero: J. Neurol. Sci. *46*, 325 (1980).
75 D. Cohen and C. Eisdorfer: Br. J. Psychiatr. *136*, 40 (1980).
76 V. K. Singh and H. H. Fudenberg: Fedn Proc. *46*, 1335 (1987).
77 D. Skias, M. Bania, A. T. Luchins and J. P. Antel: Neurology *35*, 1635 (1985).
78 K. Nandy: Aging *7*, 503 (1978).
79 H. Watts, P. G. E. Kennedy and M. Thomas: J. Neuroimmun. *1*, 107 (1981).
80 V. K. Singh and H. H. Fudenberg: Immun. Lett. *12*, 277 (1986).
81 H. Fillit, V. N. Luine, B. Reisberg, R. Amador, B. McEwen and J. B. Zabriskie, in: Senile Dementia of the Alzheimer Type, p. 307. Eds J. T. Hitton and A. D. Kenny. Alan R. Liss., Inc., New York 1985.
82 A. McRae-Degueurce, S. Booj, K. Haglid, L. Rosengren, J. E. Karlsson, I. Karlsson, A. Wallin, L. Svennerholm, C. G. Gottfries and A. Dahlstrom: Proc. natl Acad. Sci., USA *84*, 9214 (1987).
83 L. Hammarstrom and C. I. E. Smith, in: Immunoglobulins in Health and Disease, p. 31. Ed. M. A. H. French. MTP Press, London 1986.
84 B. H. Toh, C. J. Gibbs, Jr., D. C. Gajdusek, J. Goudsmit and D. Dahl: Proc. natl Acad. Sci. USA *82*, 3485 (1985).
85 A. Pouplard and J. Emile: Interdiscipl. Topics Geront. *19*, 62 (1985).
86 F. Gaskin, B. S. Kingsley and S. M. Fu: J. exp. Med. *165*, 245 (1987).
87 V. K. Singh and H. H. Fudenberg: Immun. Lett. *14*, 9 (1986).
88 V. K. Singh and H. H. Fudenberg: Clin. Immun. Newslett. *8*, 164 (1987).
89 R. M. Torack: Neurosci. Lett. *71*, 365 (1986).
90 T. Crook: Ann. N. Y. Acad. Sci. *444*, 428 (1985).
91 E. Hollander, R. C. Mohs and K. L. Davis: Br. med. Bull. *42*, 97 (1986).
92 D. E. Moss and L. A. Rodriguez: Tex. Med. *83*, 32 (1987).
93 L. J. Thal, D. M. Masur, N. S. Sharpless, P. A. Fuld and P. Davies: Prog. Neuropsychopharmac. Biol. Psychiatr. *10*, 627 (1986).
94 W. K. Summers, L. V. Majovski, G. M. Marsh, K. Tachiki and A. Kling: New Engl. J. Med. *315*, 1241 (1986).

95   J. L. Marx: Science *238*, 1041 (1987) Editorial.
96   J. P. Moreau and F. V. DeFeudis: Life Sci. *40*, 419 (1987).
97   F. S. Abuzzahad, G. E. Merwin, R. L. Zimmerman and M. C. Sherman: Pharmakopsychiatr. *10*, 49 (1977).
98   C. Wilsher, G. Atkins and P. Manfield: Psychopharmacology *65*, 107 (1979).
99   A. M. Van Buskirk and H. Braley-Mullen: J. Immun. *139*, 1400 (1987).
100   E. F. Bandle, G. Wendt, U. B. Ranalder and K. H. Trautmann: Life Sci. *35*, 2205 (1984).
101   H. H. Fudenberg, H. D. Whitten, P. Arnaud, N. Khansari, K. Y. Tsang and C. G. Hames: Biomed. Pharmacother. *38*, 290 (1984).
102   P. C. White, M. I. New and B. Dupont: New Engl. J. Med. *316*, 1519 (1987).
103   H. H. Fudenberg: Clin. Immun. Newslett. *5*, 109 (1984).
104   H. H. Fudenberg and H. D. Whitten: A. Rev. Pharmac. Toxic. *24*, 147 (1984).
105   A. Rodgaard, F. C. Nielsen, R. Djurup, F. Somnier and S. Gammeltoft: Clin. expl. Immun. *67*, 82 (1987).
106   A. J. Ammann and H. H. Fudenberg, in: Basic and Clinical Immunology, p. 391. Eds H. H. Fudenberg, D. P. Stites, J. L. Caldwell and J. W. Wells. Lange Med. Publ., Los Altos, California 1976.
107   D. J. Selkoe, in: Immunological Aspects of Alzheimer's Disease and Brain Amyloidosis. IPSEN Foundation, Angers, France, in press 1987.
108   V. K. Singh and H. H. Fudenberg: Immunol. Lett., in press 1988.

Behavioral correlates
of presynaptic events in the cholinergic
neurotransmitter system

by Roger W. Russell, Ph. D., D. Sc.
Department of Pharmacology, School of Medicine and
Brain Research Institute, University of California, Los Angeles,
CA 90024–1735, USA

1   **Introduction**

A quarter century ago in the Epilogue to his critical review of information about "Drug and Chemical Stimulation of the Brain" Myers (215a) stated: "Most impressive is the singular fact that ACh is the only substance that can influence every physiological or behavioral response thus far examined." The present quarter century has seen our knowledge about the cholinergic neurotransmitter system develop by quantal steps arising from imaginative new concepts and from innovative techniques for testing hypotheses generated by them. The new knowledge is upgrading our basic understanding of the roles of the cholinergic system in functions, both centrally and peripherally, within the organism. It is contributing very significantly to our understanding of ways in which the system may malfunction and enter into abnormal states of the organism. Such information is essential to uncovering the etiology of human behavioral disorders and the invention of treatments for them. These are disorders that affect human health and happiness and may, through illness, incapacitation and demand for health services, have an important impact on a society's economy. The paragraphs that follow are viewed within this broad context. They focus on how presynaptic events in the cholinergic system are involved as living organisms cope with their physical and psychosocial environments.

1.1   The "integrated" organism

Coping with the environment ultimately involves the *behavior* of the "integrated" organism. To study such integration effectively requires some overall strategy for viewing interactions among the basic properties of living organisms – biochemical, electrophysiological, morphological and behavioral. The present discussion is concerned primarily with interactions between biochemical (neurochemical) events and behavior and with manipulation of the former using pharmacological "tools" (drugs). For our purposes, drugs are defined as any chemical agents, endogenous or exogenous in origin, that affect biological processes. Events occurring from the release of a transmitter or the administration of a drug to its behavioral consequences are summarized schematically in Figure 1. Effects produced by endogenous or exogenous chemicals are consequences of physicochemical interactions with functionally important molecules, "receptors", within the

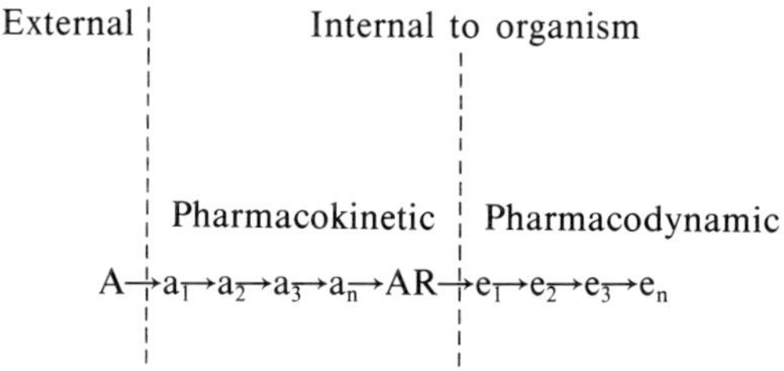

Figure 1

Schematic representation of events involved in modulating cholinergic activity and eventual effects on behavior. An endogenous or exogenous chemical, A, is transported to its site of action, $a_1 --- a_n$, where it binds to its specific receptor, R, initiating an extensive series of events, $e_1 --- e_n$, leading to effects on behavior.

body. Events $(a_1 --- a_n)$ preceding the formation of the chemical-receptor complex, AR, are involved in processes by which the chemical reaches its site of action. Effects $(e_1 --- e_n)$ following formation of the AR complex are independent of those preceding it (306), but are influenced by the state of the organism and by other processes which interact with those stimulated by AR. It is logical that the further an end point is from its receptor activation, the greater is the possibility that other events may influence the nature of the effect. "Some forms of behavior are linked more directly to their biochemical correlates than others. Where the linkage is direct (e.g., sensory-reflexive responses), changes in biochemical events are reflected in specific changes in behavior. But where the linkage is diffuse (e.g., cognitive behaviors), changes in biochemical events may affect a variety of behavior patterns" (264). In terms of this schema our present concern is, basically, with events involved in the synthesis and release of the neurotransmitter, acetylcholine (ACh), events that precede the binding of the transmitter to its postsynaptic receptor, yet are related in very important ways to subsequent effects on behavior.

There will be three exceptions to this general plan. One will be the role played by presynaptic *autoreceptors* in the feedback regulation of ACh release. Secondly, because the events in which we are presently interested occur within specific morphological structures, it will also be important to examine the roles played by *neurotrophic factors* in the generation and regeneration of cholinergic neurons. Thirdly, because neurons containing different transmitters innervate each other in the brain, it would be expected that a property as complex as behavior would generally involve *interactions* among transmitter systems. Impairment of the synthesis or release of one neurotransmitter would be expected to alter the functioning of other transmitters. For example, "It is undeniable that dopamine receptor stimulation changes cholinergic function, and that behaviorally there are clear and substantial dopamine/acetylcholine interactions" (175).

1.2 The cholingergic system

The general features of *acetylcholine* (ACh) synthesis and hydrolysis are diagrammed in Figure 2. ACh synthesis occurs in nerve terminals

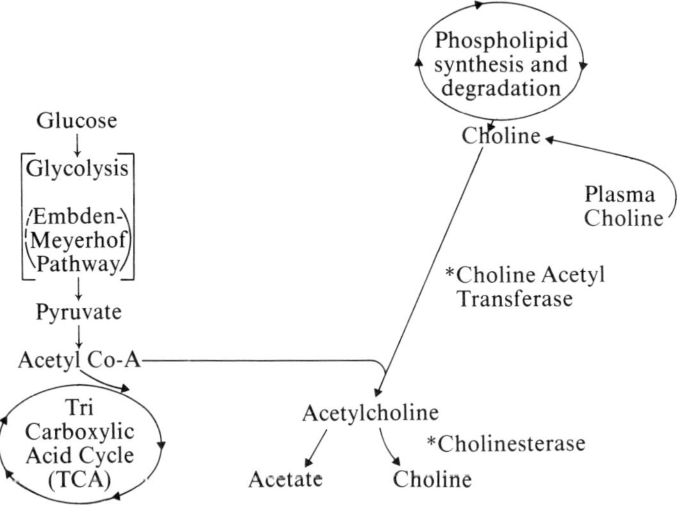

Figure 2
Metabolic pathways involved in the biosynthesis and hydrolysis of acetylcholine.

and depends upon the availability of its two precursors, *choline* (Ch) and *acetyl-coenzyme A* (AcCoA). Ch may be supplied from phospholipid breakdown or from plasma; AcCoA comes from oxidative metabolism in mitochondria. Ch is transported to the proper site by a high affinity, sodium-dependent *uptake system*, which may function both as a mechanism to ensure adequate supplies of Ch and as a "gate" regulating the synthesis of ACh (146; 215). The enzyme catalyzing the synthesis of ACh is *choline acetyltransferase* (ChAT), an enzyme specific for cholinergic neurons but not generally rate-limiting being present in great excess relative to the maximum rate at which the transmitter is normally formed. Upon release, ACh may interact with both pre-synaptic (auto-) and post-synaptic *muscarinic receptors* (mAChR), the former involved in regulating release and the latter in activating the downstream neuron. The enzyme, *acetylcholinesterase* (AChE), located both pre- and postsynaptically, hydrolyzes ACh with Ch and acetate as the products. The detailed information upon which this series of events is based has been reviewed on a number of occasions during the past decade, some very recent (120; 145; 152; 326; 327). Details relevant to our present discussion will be considered as specific topics and developed in the paragraphs that follow. Roles of the cholinergic system as substrates for both normal and abnormal behaviors are considered, evidence being drawn from preclinical and clinical research. Because special attention is being given to primary degenerative dementia and particularly to Alzheimer disease (DAT), results of studies of DAT will be used frequently to illustrate effects of cholinergic dysfunctions in human disorders involving memory loss, cognitive deficits and other behavioral impairments.

2   Precursors

The rate of ACh synthesis appears to be dependent upon the availability of its precursors – choline and acetylcoenzyme A. Neither of these is produced within the cytosol in any significant amounts. Choline enters from extracellular sources, being carried by two transport systems, one coupled to ACh synthesis and the other, to the synthesis of phospholipids. Acetylcoenzyme A is formed in mitochondria by the enzyme complex, pyruvate dehydrogenase (PDH), as illustrated in Figure 2. The paragraphs that follow discuss ways in which variations in these two precursors affect behavior.

2.1   Choline

Cholinergic neurons use Ch both as a precursor in the synthesis of ACh and as a membrane constituent. The maintenance of a supply of Ch adequate to meet the needs of these two roles is essential to normal functioning of the cholinergic system and, hence, of behaviors that are cholinergically coded. Because Ch is not synthesized within the brain, it must be taken up from plasma, where it is present in both free and bound forms. Free Ch is supplied from the diet and is also synthesized in the liver and is rapidly taken up by brain. Lipid-bound Ch also enters the brain and can be used for the synthesis of ACh. These two sources are represented in Figure 2. Further details about each of these processes are available in several summaries (e.g., 152; 154; 327; 355). The Ch used selectively for ACh synthesis in mammalian brain is transported to the site of synthesis in nerve terminals by a sodium-dependent, high-affinity transport (uptake) system (HAChT). The first suggestion of its existence was an outcome of studies involving pharmacological inhibition of ACh synthesis (184). Since then it has been the subject of several reviews (152; 168; 302). Evidence from experiments using inhibitors of the high affinity carrier and others using acute septal lesions suggested a mechanism by which HAChT could vary with changing demands for ACh and thereby serve a regulatory (rate-limiting) role in ACh synthesis. ACh release, itself, has been shown both *in vivo* and *in vitro* to result in subsequent activation of Ch uptake (188). Because of the importance of HAChT as a rate-limiting step in the synthesis of ACh, it will receive more detailed attention later in our discussion.

2.1.1 Choline enrichment and impoverishment

In seeking to understand how Ch may be involved in behavioral outputs, investigators have used the traditional experimental approach of manipulating the level of Ch as their independent variable, while measuring various behaviors as dependent variables. Effects of such manipulations on ACh synthesis appear to be contingent upon the level of neuronal activity (324; 338). Among the variables affecting the state of the cholinergic system are changes associated with aging and with behavioral abnormalities, which, as will become apparent below, have received special attention in much of the research reported.

2.1.1.1 Preclinical studies

Typical of experimental designs using animal models is a report of studies that provided ". . . the first clear evidence that the manipulation of cholinergic precursor availability can significantly alter behavior" (15). Mice served as subjects. Groups of senescent animals were placed on a Ch-enriched or Ch-deficient diet or were maintained on a normal diet. Young mice served as comparison groups. A one-trial passive avoidance procedure provided measures of retention at 24 hours or 5 days after the learning trial. The results showed a "... dramatic decrease in retention of the task in senescent mice ..." compared with control animals. Those on the enriched diet showed significantly better memory; those on the deficient, significantly poorer retention. Among the former, senescent animals on the enriched diet performed as well as young mice. The effects produced by the deficient diet were "... qualitatively and quantitatively similar to those occurring naturally across the life span of the mouse." Later studies have raised the question of whether such conclusions can be generalized to other types of learning and memory. Other investigators have confirmed the finding that chronic Ch enrichment enhances memory of the kind required by the passive avoidance paradigm (e.g., 334). On the other hand, 14 weeks of enriched or impoverished diet resulted in no significant differences in the acquisition of an operant discrimination task nor in its reversal (20). The latter study presented evidence that the release of cortical ACh, as measured by the traditional cortical cup technique, was significantly depressed in animals on the Ch impoverished and slightly elevated in those on supplement Ch diets when compared with controls. That Ch enrichment may have differential effects on different types of memory of the same task is suggested by the finding that long-term memory was improved, while there was no difference in short-term memory (334). The picture becomes more complicated by the report of these investigators that: "The effect of a chronic choline enrichment appears to be strain-dependent; in comparable experiments with rats of two other inbred strains the choline treatment had no effects on memory." Yet augmentation of memory functions following Ch enrichment has been reported across such a broad phyletic distance as that between primates and the gastroped molluscs (279).

There are basically two sources and metabolic pathways through which free Ch can be derived (85): (a) Ch and phosphatidylcholine (lecithin) in the diet and (b) methylation of phosphatidylethanolamine in the liver to form lecithin. Another substance, deanol (2-dimethylaminoethanol), has been shown to provide a source of methyl groups and to prevent some of the effects of Ch deficiency and, therefore, in these respects must also be considered as a Ch precursor. Effects of both lecithin and deanol on behavior have received attention, particularly because of their possible applications in precursor-loading therapies.

Results of experiments on animal models had shown that increases in rat brain ACh could be induced by administration of Ch or of deanol (121). There is evidence that lecithin can enhance learning of at least some forms of behavior and may be effective only in certain neuronal states of the cholinergic system, e.g., in aging. An example of significant behavioral effects is the superior maze learning of old mice when supplemented with lecithin (108). The possibility that supplements of lecithin during pregnancy might have adverse effects on offspring has recently received some attention (19). Pregnant rats were exposed to a 2 % or 5 % soy lecithin preparation or to a normal diet at three time periods: life long beginning at gestation; during the time preceding gestation only; and, following weaning. All lecithin-treated animals and their offspring had elevated ChAT activity, brain and body weight. Behaviorally, those exposed throughout their life span were hypoactive, had poor postural reflexes and showed attenuated morphine algesia. Soy lecithin had been selected for the study because of its wide non-prescriptive uses in the general population. Although the investigators appreciated the need for a much further exploration of the issue, they suggested that: "Nontherapeutic use of SLP (soy lecithin) in the lay population may be predisposing unborn progeny to subtle neurological and biochemical impairments by alteration of a generalized cellular maturation" (11; 19). These and other experiments indicate that there are limits to the beneficial effects of Ch enrichment and that the limits vary for different forms of behavior.

Under normal conditions the synthesis of ACh appears to be controlled by an efficient regulatory system so that a relatively constant releasable pool is maintained despite wide variations in release rate. Despite the abundance of Ch in normal brain it is possible that pathological or functional states exist in which a temporary or extended deficiency of Ch may affect ACh synthesis. A series of experiments in our laboratory set out to study behavioral effects of precursor loading with deanol on a malfunctioning cholinergic system (148; 271). In large doses deanol increases Ch levels in plasma and brain, but does not normally increase ACh levels. In our experiments deanol administered by itself in otherwise untreated animals produced no change in any of the behaviors measured. However, when injected cerebroventricularly in combination with HC-3, the most potent synthetic inhibitor of HAChT, deanol suppressed the abnormal hyper-reactivity characteristic of HC-3. In other words, when ACh levels were depressed, deanol compensated for abnormal behavioral consequences, illustrating yet again the importance of the state of neuronal activity to effects produced by the Ch precursors.

2.1.1.2   Studies of human subjects

The very basic role that Ch plays in the cholinergic system, coupled with the results of experiments on animal models, has motivated research on normal human subjects. Some of the investigations have focused on persons without medical complaints and have been concerned with normal nutritional requirements, a few designed to answer questions about the role(s) of Ch in normal cognitive behaviors (65; 296). During recent years there has developed a flurry of studies directed toward testing the obvious hypothesis of Ch as a potential therapeutic agent in the treatment of patients with symptomatologies that include impairment of cognitive functions in such disorders as progressive degenerative dementia (PDD). Research with both normal subjects and with patients continues to be at a frontier of investigation. A brief view of the former will provide a baseline against which to compare a later discussion of precursor loading as a therapeutic procedure for the latter.

Effects of Ch or lecithin supplements on the behaviors of normal volunteer subjects have been reported by several investigators. As the following typical examples indicate, some have found positive results and some, no significant effects. In two reports, oral administration of Ch produced selective enhancement of serial learning and of memory when compared to the performance of control subjects (295; 296). In another study involving chronic administration of increasing doses of lecithin, an extensive battery of measures testing both short- and long-term memory and sophisticated statistical analyses of the results, subjects showed elevated plasma Ch levels, but no significant changes in performance (72). Of special interest is the report of an experiment designed to study "... effects of lecithin on memory recall in a group of 'coping' normal geriatric volunteers", using a double-blind random design protocol by which each subject served as his or her own control. Analyses of the results showed that plasma choline was significantly elevated in the lecithin-treated subjects, but that there were no significant differences in performance of six memory tasks. At best the results of such research must be viewed as equivocable. They suggest, as did the preclinical studies, that there is an effective range of precursor levels consistent with optimal behavioral performance and that further "enrichment" either has no effect or may, indeed, impair performance.

2.1.2   Precursor loading

A discussion of "The Neurochemical Basis of Acetylcholine Precursor Loading as a Therapeutic Strategy" (145) has analyzed the intricacies of cholinergic system dynamics that complicate research in this area. A basic assumption underlying the precursor loading approach to treatment of hypocholinergic states is that, within the system, administration of a precursor, i.e., Ch, lecithin, deanol, will increase brain ACh content and, consequently, central cholinergic activity. As has already been pointed out, preclinical experiments using animal models have provided evidence that Ch, administered orally or parenterally, can, in fact, increase ACh levels in brain. Pharmacokinetic studies of the oral administration of ChCl to human subjects show that it increases plasma Ch levels in a dose-dependent manner (133). Given in single doses the levels do not begin to decline for some 6 hours after administration, suggesting advantages in therapeutic usage of chronic treatment consisting of multiple daily doses (114). Blood choline levels double during the daytime when naturally occurring lecithin-rich foods are being consumed as part of the normal diet. Purified Ch or lecithin added to the diet can produce an even greater increase; taken directly, the increase is still greater.

A series of experiments carried out by one group of investigators illustrates effects of Ch on the behaviors of patients presenting a variety of disorders. Included was an experiment with normal subjects which provided a baseline against which to compare other experiments in the series (65). Both short- and long-term memory were measured under treatment conditions in which three days of Ch were interposed between pre and post sessions of 4 days of placebo. Having in mind the rate at which Ch levels remain elevated after a single administration, both Ch and the placebo were given as doses of 4 grams four times daily. The results indicated that neither short- nor long-term memory was affected significantly, although there were individual differences in the latter indicating that some subjects may have responded. Such individual differences were also apparent among patients with Huntington's disease, only some of whom showed improvement and then only at high doses (20 g/day) (63). Chronic treatment of patients diagnosed as schizophrenic produced no change in clinical ratings – although there was some evidence for an actual increase in depression.

Center stage in the study of precursor loading has been its application in senile dementia of the Alzheimer type (DAT). In the main, such studies have reported discouraging results. The same general conclusion has been reached when lecithin has replaced Ch as the precursor. Lecithin raises blood Ch levels in humans to a greater extent and for a longer period than an equimolar dose of Ch (and is more socially acceptable because it does not produce a "foul odor") (77). On the market it is not a "clean" pharmacological agent, which may account for some of the differences between the results it has been reported to produce. When care has been taken to assess the purity of the supply and double blind placebo controlled crossover designs have been used, "the results obtained showed lecithin administration had no remarkable effect on the functions of Alzheimer's patients despite a twofold to fourfold rise in their plasma choline levels" (76). Double blind trials have also been carried out using deanol as the precursor (79). A significant number of patients withdrew in the first five weeks of a two-month program because of side effects that included drowsiness, lowered activity, increase in confusion and a slight increase in blood pressure. Those who remained showed no significant benefit from the drug treatment. The story was more promising when ChCl was administered i. p. to a patient with tardive dyskinesia (62). That the cholinergic system was responding became evidenced after eight days of treatment with high doses of ChCl by increases in salivation and sweating, variables associated with the actions of cholinergic agonists. At this level abnormal movements decreased, only to return to pretreatment baseline within 3 days after the treatment was discontinued. Such concomitant variation suggests a close interaction between the abnormal behavior and the apparent increase in cholinergic activity. The findings in this single case were replicated in later studies (63). There also has been some support for the possibility that lecithin may retard the slow, progressive mental deterioration characteristic of DAT.

The research reported above to illustrate the present state of knowledge about precursor loading is but a part of the total presently available and continuing to grow. Although precursor loading for clinical purposes has been less than impressive, there are reasons to believe that it deserves further testing. The general discrepancy between results of clinical trials and the successes of precursor loading in preclinical studies of behavior in animal models encourages this view. More results from well-controlled studies are needed before substantial conclusions can be drawn about the effectiveness of acetylcholine precursors as therapeutic agents. Parameters of treatment should be explored more systematically, i.e., dose, frequency, duration, development of tolerance. Objectives of precursor treatments – remission, amelioration – should be defined in evaluating results. Purity of the precursor could account for some of the disparate results so far reported and, therefore, should be specified. Anticipating that precursor loading may be justified by future results, it would appear wise to give further consideration to the synthesis of preparations of highly purified lecithins or amino acids and their uses alone or as mixtures. Effects of combinations of precursors with other cholinergic agents are worthy of consideration, e.g., combinations with partial agonists which act as presynaptic antagonists and postsynaptic agonists, an approach we will consider more fully below.

2.1.3   Competition for available choline: "autocannibalism"

Before beginning consideration of the second precursor in the synthesis of ACh, a new view about the involvement of Ch in the pathogenesis of such disorders as the PDD deserves attention. The selective vulnerability of cholinergic neurons in the basal forebrain cholinergic system in these disorders has raised questions about the mechanism(s) involved. Recently a hypothesis has been suggested in which competition for available Ch is the central element (147; 352; 353). Cholinergic neurons use Ch for both the synthesis of ACh and, via phospholipid metabolism, as a structural component in the membrane. When there, exists a deficiency in the normal supply of Ch for use in the two biochemical pathways involved, cholinergic neurons may break down membrane phosphatidyl Ch in order to maintain the Ch concentration required for ACh synthesis. This could lead to what has been imaginatively termed "autocannibalism" of the neuron (352). Conditions under which such degeneration could occur include either a relatively short supply of Ch or the overuse of Ch for ACh synthesis (see 353 for fuller details). It has been suggested that: "This competition could be precipitated by a number of inherited or acquired characteristics, is likely to be age-related and could result in failure of synaptic transmission, of mechanisms for cell membrane renewal or both (147)." Although this hypothesis has not yet been put to adequate test, it is consistent with a number of facts already available (353). One approach to an experimental test has involved the replacement of Ch with an analog that uses the same metabolic pathways as Ch, but is utilized less efficiently than Ch and results in the production of a model hypocholinergic syndrome (148 a; 219). (This model will be considered below in a section on "False Transmitters").

2.2   Carbohydrate metabolism: AcCoA and the cholinergic system

The normal source of the second direct precursor of ACh, AcCoA, is from the oxidation of pyruvate derived from glucose via glycolysis and catalyzed by the multi-enzyme complex, pyruvate dehydrogenase. The general sequence of events is outlined in Figure 2. Pyruvate or compounds giving rise to pyruvate are the most efficient precursors of AcCoA and, hence, of the acetyl group in ACh (242). Under normal circumstances less than 1 % of the glucose oxidized is incorporated into the group. This fact suggests that ACh synthesis would not be impaired unless there occurred a severe deficiency of carbohydrate metabolism. However, the system is now known to be much more sensitive: even a relatively small deficiency leads to a proportional fall in the synthesis of ACh. In terms of biological activity at a synaptic level, experiments in which oxygen or glucose was removed from the bathing medium provided evidence of very significant impairment in the response of sympathetic ganglia to transmission across cholinergic synapses (67). Under the same conditions the response of axonal conduction to direct stimulation remained unchanged. Such results suggest that the loss of transmission resulting from deficiency in carbohydrate metabolism is pre- and not postsynaptic. That the cholinergic system is involved at this level is further indicated by the fact that addition of exogenous ACh restores this response. Direct measurement of the rate of synthesis of ACh has shown that it is exceedingly sensitive to conditions that impair cerebral oxidative metabolism (25; 26; 325). Indeed, the brain has one of the highest metabolic rates in the human body, accounting for 20 % of the resting total body oxygen consumption. It follows that such sensitivity could be reflected in behaviors with cholinergic substrates. Indeed, clinical investigators have implicated impairment of oxidative metabolism in disorders with behavioral components such as hypoglycemia, hypoxia, Friedreich's ataxia, Wernicke-Korsakoff syndrome, as well as in behavioral effects produced by a number of neurotoxins (e.g., carbon monoxide, cyanide, heavy metals). It is possible that in some instances, nerve cell loss "... may be the result of a primary defect in neuronal glucose metabolism" (33).

2.2.1   Carbohydrate metabolism and behavior

Evidence that behavior is, indeed, very sensitive to changes in cerebral carbohydrate oxidation and that such changes may underlie behavioral abnormalities comes from several sources. The research involved has included studies of animal models, of normal human subjects and of patients with behavioral disorders. Carbohydrate metabolism has been manipulated by several techniques, including restriction of cerebral blood flow, of oxygen uptake and of glucose availability. A look at some typical reports will provide a more detailed view of the directions in which this area of research is progressing.

2.2.1.1   Preclinical studies

Because ACh is involved in such a wide range of behaviors, it is understandable that its synthesis becomes a candidate for the role of the "pacemaker step". Even a single exposure of rats to early postnatal anoxia can cause long-lasting dysfunctions of the cholinergic system, e.g., gradual decline in ChAT activity throughout development and maturity (300). Experiments with rats as the model and using radiolabelled glucose and Ch have demonstrated that even mild hypoxia impairs ACh synthesis and have suggested that the results were "... consistent with the hypothesis that the behavioral changes in man associated with similar degrees of hypoxia may be caused by impaired acetylcholine synthesis" (97). However, the synthesis of several other neurotransmitters (catecholamines and 5-HT) are also directly dependent on glucose oxidation, as is synthesis of putative amino acid neurotransmitters: "The high correlation coefficients between the decreases in amino acid synthesis and ACh formation suggest that their syntheses decline in parallel in various forms of hypoxia" (101). Changes in a measure of behavior also occurred under the same experimental conditions, but, clearly, no reasonable logic could be invoked to decide which among the various neurochemical variables, if any, was responsible. Appreciating this, the team of investigators involved took a further step toward unravelling the problem by attempting to correct the impairment by elevating ACh levels using the anticholinesterase, physostigmine, and achieved partial success (98). Another approach to this problem has used a developmental design in which rats were exposed to 25 min of anoxia within 24 h of birth, resusitated, and effects of the treatment traced over an extended period of time thereafter (131). Behaviorally the animals showed changes, including hyperactivity, that reached a maximum at 20–25 days and returned to normal at 40 days. At 60–80 days the experimental animals had recovered their performance of relatively simple tasks, but still were impaired in the more complex. Other techniques for manipulating carbohydrate metabolism in animal models have also been applied in the search for relations to behavior. Insulin has been used to induce hypoglycemia, thus reducing normal levels of glucose and depressing metabolism in the CNS. Activity levels of enzymes or coenzymes essential to normal carbohydrate metabolism have been lowered, e.g., thiamine, vitamin B_1, deficiency that leads to decreases as great as 50 % to 60 % in the utilization of glucose by nervous tissue. Advantage has been taken of the fact that the aging process per se is associated with decreases in rates of glucose utilization in specific brain regions (299). In principle such searches for concomitant variation between neurochemical and behavior variables is a worthy endeavor. The selection of variables to be studied and the means of eliminating others of potential significance remain problems, albeit familiar ones in the biomedical sciences. Surprisingly little attention has been given to the study of conditions under which ACh synthesis may be limited by the availability of glucose. Research on behavioral effects of hypoglycemia have used insulin as the agent for reducing the normal levels of glucose, thus depressing metabolism in the CNS. Early experiments (46) looked for effects of such treatment on learning and retention, using such test situations as a maze task that provided information about long-term ("reference") memory. Hypoglycemic animals took significantly more trials to reach a criterion of learning and were significantly slower in making decisions. More recent studies (95) have assayed effects of insulin-induced hypoglycemia on the functioning of the cholinergic system in the brain and have related the effects to behavioral reactivity and to levels of general activity. When glucose concentrations in the cerebral cortex, striatum and hippocampus were reduced to less than 10 % of normal, ACh concentrations and the incorporation of Ch into ACh were very significantly reduced. Such reductions were associated with corresponding impairment of motor activity and of reactivity to stimulus changes in the external environment (severe hypoglycemia inducing a stuporous state). These behavioral signs were counteracted by supplements of glucose. Such experimental evidence has indicated that interference with oxidative metabolism at its source impairs functioning of the cholinergic neurotransmitter system and thereby contributes to the impairment of adaptive behavior.

Among other techniques for studying interactions between carbohydrate metabolism and behavior is the manipulation of its enzymes or coenzymes. For example, effects of thiamine (vitamin B_1) on behavior have received attention for many years. In thiamine deficiency the utilization of glucose by nervous tissue may be decreased as much as 50 % to 60 %, neurons frequently showing changes that are characteristic of poor nutrition. Thiamine deficiency may also cause degeneration of myelin sheaths, giving rise in peripheral nerves and spinal cord to polyneuritis and to severe muscular weakness. Early experimentation showed that, as with hypoxia, only "... a moderate impairment of cerebral carbohydrate catabolism ..." was necessary to impair ACh synthesis (187). Furthermore, a deficiency that was minimal under resting conditions produced profound effects when metabolic or physiological demands increased, a finding confirmed by several later studies. The early experiments that included behavioral variables were equivocal in their findings. However, these were followed by studies which overcame research design problems and reached the conclusion that the behaviors they measured were not affected by thiamine deficiency until the latter was sufficiently severe to produce the symptoms of polyneuritis. Research in our laboratory at the time confirmed earlier reports that less severe deficiency did not affect the efficiency of conditioned responses when thiamine-deficient animals were compared with their paired caloric controls (159). However, the results did indicate that, under conditions of "conflict" (when a conditioned response that had led to the avoidance of punishment was now punished) thiamine-deficiency was associated with greater response rigidity and abnormal displacement activity. More recently evidence has been reported that associates such behavioral effects more specifically with muscarinic deficiency in the CNS (11). Thiamine-deficient rats treated with the centrally-acting antiChE, physostigmine, recovered from behavioral impairments in a dose-dependent manner. The peripherally-acting AChE inhibitor, neostigmine, had no effect and the centrally-acting cholinergic antagonist, atropine, blocked the beneficial effect of physostigmine. Furthermore, nicotine had no effect, while the direct muscarinic agonist, arecholine, was as effective as physostigmine. Such evidence points clearly to the involvement of a central muscarinic "biochemical lesion" (264) in the behavioral changes measured.

The previous discussion has been almost entirely oriented toward effects of deficiencies in carbohydrate metabolism on behavior, although a mention was made of an effort to counteract such effects by the use of physostigmine to elevate ACh levels centrally. Recently more attention has been turned to investigations of potential methods for ameliorating metabolic deficits once they have appeared (e. g., 87). Encouraging results have been reported of success in modulating memory in animal models by the direct procedure of administering glucose s. c. in a wide range of doses, i. e., of 1, 10, 100 or 500 mg kg^{-1} (109). Retention of a passive avoidance response was measured 24 hr later. The results took the form of an inverted-U: animals receiving 10 or 100 mg kg^{-1} showed retention significantly better than controls; those injected with 1 or 500 mg kg^{-1} had no significant effect. Subjects tested 1 hr after the glucose injection were not different from controls, indicating that glucose must have been having its facilitating effect by modulating the memory storage process, which was occurring between the 1 hr and 24 hr assays. Results of other research have shown that decrements in glucose associated with aging and with lesions of the nucleus basalis magnocellularis can be reversed pharmacologically by treatment with ACh agonists, e. g., oxotremorine, or with metabolic enhancers, e. g., co-dergocrine (182). Whether or not such treatments *per se* can improve cognitive performance in the intact CNS remains a question.

2.2.1.2   Studies of human subjects

There is a relatively long history (316) of experimental studies of behavioral effects of hypoxia produced in hypobaric chambers (263) and of field studies of hypoxic effects at sites of high altitude (202). The results of these studies were consistent in concluding that: (a) there were definite deteriorating effects on behavior associated with reduced oxygen pressure; (b) the effects were more marked in mental than in physical work; (c) behaviors well-learned were more resistant; and (d) there occurred a very significant development of tolerance as the duration of exposure increased. Such observations left unanswered questions about the mechanisms by which these several effects were mediated and by what means impairments might be counteracted. The search for answers has involved the testing of several hypotheses.

Impaired carbohydrate metabolism is a pathophysiological mechanism found commonly in many types of human brain disorders of diverse etiologies (27), including cerebrovascular disease, heart or lung failure, tumors and other conditions inducing cerebral edema. Under such circumstances when the supply of oxygen or of glucose is subnormal, the transport of active acetyl moieties across the mitochondrial membrane may replace HAChT as the limiting factor in ACh synthesis. As shown by research on animal models described above, complex cognitive behaviors are especially susceptible. Results of research into the effects of various forms of hypoxia have suggested that: "Biosynthetic activities are more sensitive to graded hypoxia than are energy-producing processes" (26). This is evidenced in the fact that "... decreased synthesis of acetylcholine from [U^{14}C]glucose occurs with hypoxia so mild that there is no change in lactate, or in cyclic AMP..." (cyclic adenosine 3'5'-monophosphate). Comparisons between elderly normal subjects and patients with dementia using positron emission tomography and objective measures of behavior have shown high correlations between the rate of glucose utilization and cognitive impairment (78). High correlations ($r_s = -0.63$, $p < 0.007$) have also been reported between the synthesis of ACh from [U^{14}C]glucose and ratings of cognitive impairment in patients with DAT (83). An association has been found between disturbances of carbohydrate metabolism resulting from poor diabetic control and impaired performance in measures of intellectual functioning and of reaction time (134). Comparisons of regional cerebral blood flow using a 133 xenon inhalation technique showed markedly greater reductions in patients with behavioral impairment than in the age-dependent decreases in normal aging (203). Pyruvate dehydrogenase activity, so vital to the formation of AcCoA, has been shown to decrease significantly in the cerebral cortex of DAT patients, the decrease being related to loss in ChAT activity (236; 291).

It is clear from the extensive fund of information now available that the behavior of the integrated organism is vitally affected by presynaptic events involving the precursors of ACh. The relationship is selective, some behaviors being significantly impaired when malfunctions occur while others carry on normally. The capability to cope behaviorally with situations requiring complex adjustments appears to be most sensitive. It is reasonable to assume that this is one of the stages in cholinergic function where treatments for behavioral abnormalities should be sought, despite the very limited successes so far achieved.

2.3   False precursor – false transmitter

Some half-century ago the possibility that a false precursor leading to the synthesis of a false transmitter might serve as a means for examining neurotransmitter systems at a molecular level began to receive attention. *In vivo* studies of Ch analogs began to appear in scientific journals (42; 199). Criteria that must be satisfied if a compound is to be accepted as a false transmitter were established (163). The first direct demonstration that a Ch analog, triethylcholine, could be acetylated and released in a cholinergic synapse was reported (136). Later it was appreciated that the synthesis of false transmitters might provide valid animal models for studying clinical states involving dysfunctions in the CNS. A few experiments were designed to use such models for measuring effects on both normal (104) and abnormal (198) behaviors. The outcomes of this research approach have been reviewed by Collier et al. (47), by Whittaker and Lugmani (345) and, most recently, by Newton and Jenden (219).
A major difficulty faced by the earlier studies was to demonstrate that chronic dietary administration of a Ch analog did in fact result in functionally significant replacement of Ch in the synthesis of an endogenous analog of ACh. Ways in which the availability of quantitative analytical techniques eventually solved this problem are illustrated in a series of experiments in our laboratories reported during the past five years, involving the false precursor, N-amino-N, N-dimethylaminoethanol (N-aminodeanol, NADe) (218). NADe is taken up by the Ch transport system in competition with Ch. It is acetylated by ChAT, stored as 0-acetyl-N-aminodeanol (ANADe) in vesicles and released on stimulation. ACh stores are depleted as they are replaced with ANADe. Upon release ANADe interacts with both muscarinic and ni-

cotinic receptors and is hydrolyzed by AChE. Because ANADe's potency at these receptors is only 4 % and 17 % of ACh, respectively, there occurs a profound interference with cholinergic transmission, particularly at muscarinic sites. Replacement of Ch with NADe in the diet of weanling rats for periods of 60 to 120 days results in the replacement of 85 % to 95 % of free Ch by free NADe in brain, plasma and peripheral tissues. ChAT is reduced in the cortex, hippocampus, striatum and ileum, suggesting the possible loss of cholinergic neurons. This evidence showed that NADe satisfied the neurochemical requirements of a false precursor, leading to the synthesis of a false cholinergic transmitter and enabled the study of its behavioral and physiological consequences.

General observations during the experiments described above had failed to show any gross neurobehavioral toxicity, although when handled the animals showed a discernable hypertonia. This did not, of course, mean that the replacement of Ch by NADe had no concomitant behavioral or physiological effects. For this reason an extensive series of experiments has been carried out using quantitative assays of variables known to be cholinergically "coded". Initially these variables were measured only between 25 and 32 days on the NADe diet, yet, even after this relatively short time, they showed some significant effects (217). A more extensive series of experiments (as yet unpublished) has now been completed, measuring many more variables periodically as the level of replacement of Ch by NADe progressively increased over a period of 120 days (see 148a for a brief summary of the results). The results provide the data necessary to relate magnitudes of behavioral and physiological changes to levels of the false transmitter. There were no significant effects on total caloric intake, on the maintenance of body fluid balance, or on core body temperature, results which strongly suggest that the behavioral effects described below are unlikely to be attributable to imbalances in basic homeostatic mechanisms. The behavioral changes concomitant with increasing replacement of Ch by NADe may be summarized in three major categories. The first includes measures that are primarily *sensory-reflexive* in nature, i. e., innate, appearing without the necessity for learning (e. g., reflexive responses to electric shock and acoustic stimuli). Such behaviors were affected significantly, being evidenced in sensory hypersensitivity and motoric hyper-reactivity. The latter was so pronounced as to produce convulsions in some animals with very low Ch levels. The effects resembled symptoms often referred to as "psychomotor agitation" in DAT and in some affective disorders. There have been clinical reports of seizures in such patients. Clearly apparent in DAT and in other degenerative disorders of aging are symptoms of a *sensory-perceptual* nature: e. g., failure to "understand" events in the physical and psychosocial environments; spatial disorientation; and, eventually a general loss of response to most stimuli. These behaviors go well beyond reflexive responding, involving coordination at higher levels in the CNS. Our present experiments include several measures of such behaviors, measures of behavior in novel stimulus conditions or of the eliciting of a previously learned response upon presentation of stimulus conditions associated with the initial learning. Analyses of such measures showed decrements in the performance of NADe animals when compared with their controls. The effects were reminiscent of those observed in earlier experiments in which HAChT was impaired by HC-3 administered intracerebroventricularly (272) and of those found in septally-lesioned animals with reduced cholinergic activity. The third category of behavorial effects included learning, memory and other *cognitive processes*. Because these are among the early signs of dysfunction in degenerative disorders involving the cholinergic system, they received particular attention and were studied in several different test situations. Measures that are intrinsically dependent upon the functioning of the cholinergic system showed a consistent pattern of effects: NADe animals took more trials to learn, made more errors, were slower in their response times and had poorer memories. Furthermore, these effects increased progressively as the available supplies of Ch decreased.

At an earlier point in the present discussion the hypothesis of "auto-cannibalism" was introduced as a possible mechanism underlying behavioral disorders associated with hypofunctioning of the cholinergic system. The hypofunctional condition might be a consequence of competition for available Ch supplies between the needs for Ch in maintaining the membrane integrity of cholinergic neurons and in synthesizing ACh. In the present experiments the supply of Ch was progressively replaced by the much less efficient NADe. In the continuing competition both free and lipid-bound Ch were very significantly reduced. Paralleling these reductions were highly significant impairments of behavioral variables, the changes being analogous to those characteristic of such primary degenerative disorders as DAT. Research is now underway to study still further the neurochemical, behavioral and histopathological properties of the model, to determine whether it can be influenced by drugs intended to enhance cholinergic function and to discover the extent to which the various effects may be reversed when the diet is returned to normal. At the present time we believe that it is a potentially useful model for research on DAT and similar human disorders. We also believe that it may have significant implications for their pathogenesis.

3   Choline transport (uptake)

The rate of synthesis of ACh is limited by the availability of its substrates. Because they are not produced within the cytosol in significant amounts they must be transported to the site of synthesis. Choline uptake is accomplished by two carrier systems. A high-affinity sodium-dependent carrier (HAChT) is located specifically at cholinergic nerve terminals and is tightly coupled to ACh synthesis. A low-affinity system associated with phospholipid synthesis appears to be present in all cells, including cholinergic cell bodies. Because HAChT normally is a rate-limiting step in ACh synthesis (12), means have been sought to manipulate it experimentally in animal models in order to study the behavioral, as well as neurochemical consequences.

3.1   Preclinical studies

A number of compunds have been shown to have *in vivo* effects on the rate of choline transport (210; 307). Among these is HC-3 the most potent of a series of hemiacetals containing a Ch-like moiety. It was early noted that the primary action of HC-3 involved "... a presynaptic depressant effect resulting from decreased acetylcholine formation ... not upon the acetylation phase ..." (284). Injected intracerebroventricularly (i.c.v.), HC-3 accumulates in the caudate nucleus and the hippocampus, where it becomes associated with nerve endings and suppresses or inhibits HAChT. Choline is a specific and dramatic antagonist for the effect. HC-3 has been used as [^{3}H]-HC-3 in identifying HAChT sites in brain (186; 280).
HC-3 has also served as a pharmacological tool to manipulate HAChT in several studies of relations between the carrier and its behavioral effects. Dominant features of the results in the early studies were: (a) hy-

per-reactivity to changes in a variety of environmental stimuli (84; 86), and (b) occasional occurrence of seizures, both overt (clonic-tonic) and evidenced by changes in the EEG (288; 297). A series of experiments in our laboratory helped to define such relations in fuller detail (272; 273). Doses of HC-3 used in our studies reduced ACh content to a low of 19 % of normal 4 hr after injection and recovering to 76 % by 24 hr (84). Analyses of the several behaviors measured showed that neither levels of motivation nor motoric capabilities were affected. During peak effects of the treatment, subjects at high dose levels, i.e., low ACh, evidenced a very marked hyper-reactivity analogous to manic states. When learning and memory were tested, inhibiting HAChT and, hence, ACh synthesis affected the central processing of information. Incoming sensory information arrived centrally and was stored in brain for later use. This was evidenced in animals that did not learn when HAChT was suppressed, yet whose performance following recovery of the HAChT system showed a very considerable memory for their earlier experiences. Put in other terms, suppression of HAChT in brain was associated with deficiences in the use (retrieval) of information and not with sensory input and storage nor with motor output. The effects were clearly selective, rather than nonspecific.

The associations between HAChT and behavior observed in studies involving experimental manipulation of the carrier have been reinforced by research employing a type of correlational approach: behavior has been measured and, subsequently, the state of the carrier determined. This approach is exemplified in a report on choline uptake and permanent memory storage (250). Animals were selected for their retention of a brightness discrimination task. HAChT was then studied *in vitro* on hippocampal slices from brains of those with differences in long-term memory. [^{3}H]-Ch uptake was higher in slices from animals with better retention. The findings suggest that individual differences in the activity level of the hippocampal cholinergic system do exist and are capable of influencing that very important feature of behavior, memory. Other investigators using the same general research design have studied aged animals. It has been reported that, among the several presynaptic cholinergic parameters, ". . . only Ch uptake seems to be affected by aging" (191). In aged animals with documented memory impairments HAChT in the hippocampus was only 22 % of that in young animals without memory deficits (290). This reciprocal interaction between HAChT and behavior is dramatically illustrated by reports of experiments in which HAChT levels were measured following a variety of different behavioral training procedures (243; 340): "The pattern of results . . . indicates that experience in many different kinds of discrimination tasks is sufficient to elevate (HAChT) in the hippocampus, while simple sensory-motor experience (as in the treadmill) is not" (340).

The possibility that preloading with a precursor could compensate for decreases in HAChT has received limited testing. One study reported that the hyper-reactive state induced by the intracerebroventricular injection of HC-3 can be antagonized in a dose-dependent way by simultaneous i.c.v. or systemic administration of deanol (148). This finding led to a second series of experiments of similar design, but including a wider variety of behavioral measures (271). Administered by itself deanol produced no changes in any of the behaviors. However, where injected i.c.v. in combination with HC-3, deanol suppressed changes in behavior that would otherwise have been produced by HC-3. The most likely explanation for this antagonism appears to be an elevation of Ch levels in brain following deanol (153).

3.2 Studies of human subjects

The behavioral effects of manipulating HAChT discussed above have analogies to the behavioral symptoms in patients with dementia in whom Ch uptake has been shown to be significantly reduced. Comparisons of HAChT in synaptosomes prepared from fresh necropsy-brain of patients with DAT and of "mentally" normal controls have reported a decrease of about 50 % in the frontal cortex and 80 % in the hippocampus of the former (276). In patients for whom DAT was histologically confirmed, HAChT in neocortical biopsy samples was found to be reduced by at least 40 % below mean control values (293). In the latter report: "Samples from other demented patients, in whom the histological features of Alzheimer's disease were not detected, produced values for all three biochemical parameters (ACh synthesis, ChAT activity, HAChT) which were similar to controls" (293). Such evidence supports the view that changes in these cholinergic markers differentiate clinical states of individuals presenting with different behavioral symptomatologies and suggest that the latter are consequences, at least in part, of malfunction(s) in the presynaptic cholinergic system.

4   ChAT and the synthesis of acetylcholine

In our earlier discussion of presynaptic events in the cholinergic system, changes in activity of the synthesizing enzyme, ChAT, have received frequent attention. The synthesis of ACh from its two direct precursors is dependent upon the level of this activity. For those interested in the full details of the regulation of ACh synthesis in the brain two excellent reviews are available (326; 327). A brief outline of the major features will serve as background for our present consideration. ACh is synthesized by the reaction shown in Figure 3, catalyzed by

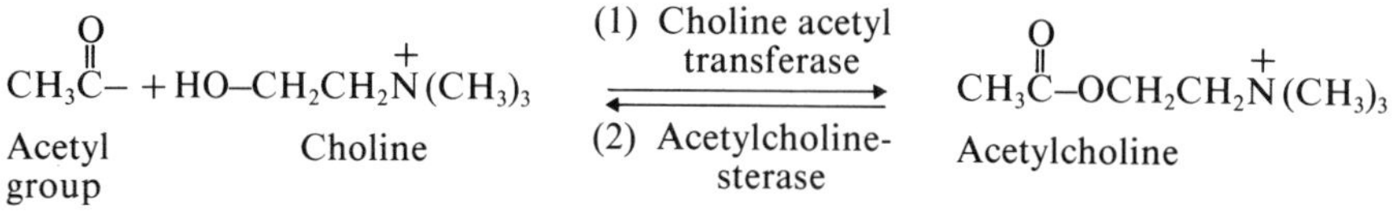

$$CH_3\overset{O}{\overset{\|}{C}}- \ + HO{-}CH_2CH_2\overset{+}{N}(CH_3)_3 \quad \underset{\text{(2) Acetylcholine-}}{\overset{\text{(1) Choline acetyl transferase}}{\rightleftharpoons}} \quad CH_3\overset{O}{\overset{\|}{C}}{-}OCH_2CH_2\overset{+}{N}(CH_3)_3$$

Acetyl group   Choline   (2) Acetylcholinesterase   Acetylcholine

Figure 3

Immediate reactions involved in the biosynthesis and hydrolysis of acetylcholine. Synthesis is catalyzed by the enzyme, choline acetyltransferase (1), which mediates the transfer of an acetyl group from acetyl coenzyme A to choline (shown from left to right). After its release from a presynaptic nerve terminal, acetylcholine is rapidly inactivated by the hydrolyzing enzyme, acetylcholinesterase (2) (shown from right to left).

ChAT. In the mammalian brain the presence of ChAT appears to be specific for neurons in which ACh is the synaptic transmitter, a fact that makes the enzyme a specific marker for the cholinergic system (205). ChAT is synthesized in neuronal perikarya and transported to presynaptic nerve terminals where ACh is mainly synthesized. Within nerve terminals ChAT is concentrated in the cytoplasm, the primary location of newly synthesized ACh. ACh is transported into vesicles by a mechanism which serves both for its storage and for its release by exocytosis. A "panel" of monoclonal antibodies specific to ChAT has been produced as tools for studies of cholinergic neurons (176). On *a priori* grounds, the key role of ChAT in the normal functioning of the cholinergic system and the widespread involvement of that system in the normal behavior of the integrated organism would suggest that abnormalities in behavior could be manifestations of abnormalities in the synthesis of ACh. More than a quarter century of research using animal models had shown that experimental manipulation of the synthesis can, indeed, impair normal behavior. A decade ago evidence began to appear documenting the presence of extensive losses of ChAT activity in cases of primary degenerative dementia and other human disorders with major behavioral characteristics.

4.1 Preclinical studies

Several studies using animal models have manipulated ACh synthesis genetically to observe effects on behavior. Some have used the technique of comparing relations between behavior and synthesis in different inbred strains of animals (139; 140). The results have shown, for example, that strains with higher levels of genetically-induced ChAT activity had much greater capabilities for long-term memory. It was suggested that the effect could be a result of the release of greater amounts of ACh in the hippocampus immediately after a learning session, i.e., during the memory consolidation period when interference with cholinergic functioning can impair memory. A second genetic approach to studying relations between ACh synthesis and behavior has involved selective breeding for differences in cholinergic function. In one extensive series of experiments selective breeding for sensitivity to an anti-ChE resulted in the development of two lines of rats, one of which was significantly more sensitive than the other in both behavioral and physiological variables and more sensitive to cholinergic agonists (227). When compared with a resistent line and to randomly-bred animals, the sensitive line: (a) had elevated rates of ACh synthesis in the cortex; (b) had increased densities of muscarinic receptors in the striatum and hippocampus; (c) were hyperactive; (d) had better memories in passive avoidance situations; and, (e) became more immobile when exposed to mild stressors (227; 229; 266). Clearly, in experiments using these lines, the differences in cholinergic function provided a genetically-determined base for studying effects of experimental manipulations of the cholinergic system. Because of their behavioral, neurochemical and physiological characteristics, the sensitive line is serving as an animal model of cholinergic supersensitivity and affective disorders (228).
Relatively few behavioral studies using animal models have manipulated ChAT activity by pharmacological means. Illustrative of these are experiments with the ChAT inhibitor, trans-4-(1-naphthyl-vinyl) pyridinium methiodate (NVP) and its hydroxymethyl bromide analog, by means of which ACh synthesis was suppressed. Interpretation of the results depends upon the specificity with which NVP acts, evi-

dence for which is equivocal. However, in one series of experiments comparisons were made between effects of bilateral lesions in the dorsal hippocampus and of i.c.v. injection of NVP (105): decreases in ChAT activity in both treatments were associated with significant impairments of passive avoidance learning and memory. Other investigators have reported significant effects of NVP on behaviors involving interactions between animal subjects: NVP produced reversible blocks of muricidal behavior in rats and of fighting behavior in mice (332). A limitation of NVP is its impermeability through the blood-brain barrier (BBB), necessitating i.c.v. administration. Recently a series of water soluble ChAT inhibitors have been synthesized that are capable of crossing the BBB (204). One of these, BW813u (Burroughs-Wellcome) which has no significant effects on AChE or on monoamine oxidase, has been used in experiments designed to study interactions between ChAT activity and memory (342). Single injections of the compound produced rapid decreases in ChAT activity throughout the brain ranging from 50 % to 70 % following a 25 mg kg^{-1} i.p. dose. With a dose of 50 mg kg^{-1} these decreases persisted for at least 5 days and had still not attained control level 19 days later. Under these conditions spatial memory (choice accuracy in radial maze performance) "... was not impaired at any time following the injection" (342). The investigators suggested that the decrease in ACh production was not sufficient to affect behavior. Other manipulations of the cholinergic system have reached similar conclusions about the "margin of safety" built into the system and the differences in threshold levels required for normal performance of different behaviors (275). That this concept may be applicable to the results just described seems likely because of the large excess of ChAT present in all cholinergic tissues, excesses well beyond their maximum turnover rates (146).

Chronic treatment with lithium (LiCl) on cholinergic function in rat brain has provided a means of experimentally stimulating cholinergic activity in certain brain regions: HAChT of [^{2}H$_4$] Ch and its conversion to [^{2}H$_4$] ACh was reported to be 131 % of control values in synaptosomes isolated from brains of 10-day treated rats (151). This information was used to stimulate cholinergic activity in the brains of animals, while consequent effects on a variety of behaviors were measured (274). The results clearly indicated that increased cholinergic activity was associated with differential effects on behavior, behaviors affected being those characterized by changes in reactivity to environmental stimulation. Reactivity was suppressed when compared with control levels, without complete inhibition. The extent of the suppression was equivalent to the level of reactivity reached by control subjects after habituation to the environmental stimulation. As would be predicted, the behavioral effects were opposite to those reported for conditions under which ACh activity was reduced.

A third experimental procedure for manipulating ChAT activity and ACh synthesis in animal models has involved the use of morphological lesions in cholinergic regions of the brain. In one study, unilateral injection of the excitotoxin, ibotenic acid, destroyed cell bodies at the site of injection in the nucleus basalis of Meynert, but not damaging axons of passage. The immediate neurochemical effect was to produce significant decreases in ChAT activity and ACh production (341). After lesioning, ChAT levels gradually increased, being about 60 % at one week and reaching control levels by 3 months. The cognitive behavior of lesioned animals in radial maze performance was significantly impaired relative to controls only during the first ten days of post lesion measurement, suggesting the existence of a functional threshold for this behavior at somewhat above 60 % of normal ChAT activity. The similarity between the extent of reduction of ChAT activity following structural lesions and that after injection of the specific ChAT antagonist, BW813u, discussed above is striking. Yet the former produced significant behavioral effects and the latter did not. Other studies involving lesioned animals have provided analogous results (129; 214). To test effects of interaction between the two procedures for reducing ChAT activity, the investigators (342) treated lesioned and non-lesioned animals with the BW813u after behavioral recovery from lesioning: "... in spite of the dramatic decrease in regional levels of ChAT activity ... choice accuracy in this spatial memory task (radial maze) was not impaired" (342). The basis of this discrepancy is a matter of speculation at present. An obvious hypothesis is that structural lesions created by IBO damaged other neurotransmitter systems interacting with the cholinergic system in normal spatial memory, while BW813u was specific to ChAT activity. One direction from which answers may come is through the use of neurotrophic factors for regenerating neural structures following lesions: nerve growth factor mediated recovery of AChE-positive fibers following lesioning has been shown to be directly related to recovery of ChAT activity (107; 167).

4.2   Studies of human subjects

The general finding in studies of ChAT activity in brain tissues from human subjects is consistent with the results from experiments with animal models. Significant decreases of ChAT activity accompanied by reductions in ACh synthesis have been found in both demented and non-demented elderly patients (61; 235; 293). Comparisons between patients with and without histological verification of DAT have reported significant loss of ChAT activity in the former, but not in the latter (32). These changes in enzyme activity and in ACh synthesis are also related to reduction in Ch uptake, the three markers being consistent with the hypothesis that DAT and its behavioral correlates are results of functional loss at presynaptic cholinergic nerve endings. Cognitive impairment has been shown to increase as the functional loss increases (237; 347): "In the most severely affected patients (behaviorally) synthesis (of ACh) was reduced to only 33 % of control" (34). Several mechanisms have been considered by which the functional loss could be produced. An obvious possibility is the neuronal degeneration found in advanced cases of DAT: ACh synthesis could be affected by the loss of sites in which it normally takes place. This hypothesis assumes that neuronal loss precedes the observed decreases in ChAT, Ch uptake and ACh synthesis, an assumption challenged by the results of studies of DAT patients that have led to the proposal of the alternative hypothesis, i. e., reduction of ChAT levels precedes the loss of neurons (234). Results of the latter studies suggest ". . . that in the majority of elderly cases of Alzheimer's disease the almost total loss of ChAT in the nucleus of Meynert is not paralleled by a similar loss of the putative cholinergic neurons". From such observations it has been hypothesized that neuron loss is a secondary feature subsequent to the downregulation in production of the transmitter-specific enzyme. Clearly here is another focal point for further research.

Results of preclinical studies comparing genetic differences in cholinergic system-behavior interactions were discussed earlier. Analogous research in human behavioral genetics is ". . . but for a modest beginning, virtually unexplored" (144). Judging by publications in *Nature* and in *Science* during 1987, considerable excitement has been generated by the report that, in sporadic DAT, there occurs duplication of the chromosomal region around the locus of the amyloid protein gene on chromosome 21 (66 a). This finding suggested the hypothesis that predisposition to DAT is due to genetically determined over-expression of the β-amyloid protein or to the expression of an abnormal variant of it. The flurry of tests of this hypothesis has failed to provide support for it (e. g., 238 a). Similar conclusions have also been drawn about the possible involvement of the amyloid β-protein gene in familial DAT (e. g., 318 a).

5   Acetylcholine: storage and release

Following its synthesis ACh is immediately absorbed into vesicles within the presynaptic cell and stored, awaiting changes in the cell that will release it into the synaptic cleft. The subsequent association of ACh with specific binding sites on the postsynaptic cell membrane initiates dynamic events that may eventuate in significant effects on behavior of the integrated organism (see Figure 1). Regulation of ACh release is one of the key processes in this chain of events. Essential to the regulation is the existence of muscarinic autoreceptors on central presynaptic nerve endings. mAChRs mediate the inhibition of ACh release by negative feedback modulation. Because impairment at any point in these processes could result in malfunction(s) of the cholinergic system that could be reflected in behavior, regulation of the storage and release of ACh has received much attention, using both animal models and clinical material for research purposes (117; 168; 171; 190; 303; 337).

5.1   ACh storage (vesicles)

Although relations between behavior and events associated with the roles presynaptic vesicles play in the dynamics of the cholinergic system have yet to be studied directly, the strategic importance they have in ACh storage and release should be recognized in the present discussion. Cholinergic synaptic vesicles have been extensively characterized as to their structure, molecular composition and functional properties (162). However, in a 1978 review of neurosecretion and the "vesicle hypothesis" appeared the statement: "We presently seem close to validating the common, textbook assertion that neurosecretion occurs by exocytosis of synaptic vesicles" (35). Yet to be answered appear to be such questions as: Are the ACh quanta discharged during exocytosis derived from vesicles (208); and ". . . how the acetylcholine

(ACh) balance is affected when synaptic vesicles pump Ca^{++}" (36). These unanswered questions have given rise to several alternative hypotheses to account for what is known about their roles in ACh storage and release (160). Despite such ambiguities, recent concepts of ways in which presynaptic cholinergic events may be associated with behavior assume that the hypothesis has, indeed, been validated. The sequence of events may be outlined briefly. Newly synthesized ACh enters vesicles and is stored. Release into the synaptic cleft occurs by exocytosis, the vesicles fusing with the synaptic membrane and release being induced by depolarization on the entry of Ca^{++} into the terminal (41). Following this the vesicular portions of the membrane invaginate and pinch off to form new vesicles. In order to replace those lost in the process, additional new vesicles are continually transported to the presynaptic terminal. Neurochemical studies suggest that functional but empty vesicles may be formed at the membrane, that newly synthesized ACh may be preferentially loaded into vesicles and that newly synthesized ACh is preferentially released at exocytosis. "Fusion and fision must then occur in less than 200 µs so that the transmitter is liberated" (35).

Experiments with *Torpedo* electromotor nerve terminals first suggested that vesicles exist in two subpopulations, VP_1 and VP_2, that differ in size and in density. VP_2 appears to be the "recycling" subpopulation, being smaller in size and accumulating in the terminal cytoplasm as stimulation continues. It is VP_2 vesicles within which ACh is preferentially absorbed. Recently evidence has been presented that a similar dichotomy in the structure and function of vesicles is to be found in mammalian cortical cholinergic terminals (1). Studies addressed to the question of the mechanism by which ACh is taken up by the vesicle have led to the hypothesis that ". . . an ATPase pumps protons into the cholinergic synaptic vesicle to produce an internally acidic and positively charged proton gradient that is linked to acetylcholine uptake" (3). New pharmacological tools are becoming available which are potent inhibitors of ACh transport in ways suggesting that they block ACh storage by synaptic vesicles *in vivo* (230). Research with human subjects suggests ". . . that in the aging human brain two populations of neurons coexist side by side. One undergoing regressive changes leading to cell death and a second surviving and undergoing continuous and dynamic growth of processes" (96). Changes in the former would be associated with behavioral changes attributable to cholinergic functions. Another series of experiments involving measurements of *in vitro* release of [³H]-ACh from slices of human postmortem frontal cortex has confirmed a decrease of ACh release in DAT patients compared with non-demented subjects (221). Of special interest was the finding that physostigmine had no significant effect on release from control subjects but increased evoked release from DAT patients by more than a factor of three, returning it almost to normal level. Such neuropharmacological effects could underlie the beneficial effects of physostigmine on the cognitive behavior of DAT subjects discussed earlier.

5.2   Autoreceptors

In reviewing our knowledge of autoreceptors, one author (303) commented: "Presynaptic receptors are an amazingly diversified mechanism for the physiological regulation of neurons. They are an even more diversified mechanism for pharmacological manipulation – a challenge to pharmacological analysis" (303). Our main present interest is in the roles that presynaptic muscarinic receptors (autoreceptors) play in regulating the release of ACh (170; 315), although we will wish to see how they may be modulated by other endogenous compounds in the process. The general hypothesis that feedback regulation of ACh release could be exerted by muscarinic autoreceptors had been clearly expressed in early reports of *in vitro* studies of atropine's stimulating effect on ACh release by cerebral cortical slices from rat brain (239). It then became the task to establish that the natural transmitter, ACh, in fact inhibits its own release and to clarify the characteristics of the autoreceptors. Direct demonstration of the former came from research using isolated hippocampal synaptosomes prelabeled with [³H]Ch (189). Depolarization with KCl resulted in a very significant increase in release of radioactivity, which was inhibited when Ca^{++} was absent from the medium. Addition of exogenous ACh to the medium produced a dose-dependent inhibition of this release, an effect that was counteracted by the cholinergic antagonist, atropine. Differential effects of various muscarinic agonists suggested that pre- and post-synaptic mAChRs differ from each other (241). Detailed exploration of each of these steps in the overall process has now provided experimental data in support of the negative feedback model. Further evidence that presynaptic regulation is exerted by mACh autoreceptors localized in cholinergic nerve endings came, for example, from the *in vitro* demonstration that feedback regulation of [³H]ACh release could occur in nerve endings isolated in a cell-free preparation, i.e., feedback inhibition does not require intact neurons or neuronal loops (223). Muscarinic autoreceptors in cortical nerve endings were shown to be pharmacologically similar to those in the hippocampus (247).

5.2.1   Subpopulations

Evidence from more detailed analyses now indicate the existence of at least two discernable mAChR types, M_1 receptors being distinguishable from M_2 on the basis of: (a) higher affinity of M_1 for the antagonist, pirenzepine, or (b) by higher affinity of M_2 for cholinergic agonists, e.g., carbachol. The presynaptic modulation of [³H]ACh release appears to be mediated by M_2, but not by M_1 receptors (207). These and other results (195) are consistent with the conclusion that M_1 receptors are located on postsynaptic cells and facilitate cellular excitation, while M_2 receptors function in cholinergic nerve terminals to regulate the release of ACh. The M_2 receptor activation by agonists has been found to be sensitive to changes in membrane potential (207). Cerebral cortex tissue samples from individuals with and without DAT show the former to have decreases of 20%–25% in total mAChR, with downregulation of M_2 but no changes in M_1 receptors. The same DAT subjects had decreases of 55%–80% in ChAT activity levels (195). Analagous results have been reported for tissues from rat brain after ibotenic acid lesions in the ventral and medial globus pallidus. Functionally, the activation of these autoreceptors appears to be characterized by the "threshold" phenomenon already considered at several points in our discussion (141), i.e., a threshold level of stimulation is necessary to activate the autoreceptors. Studies of cortical presynaptic cholinergic processes have shown no age-related decrement in HAChT affecting subsequent Ch acetylation. However the results ". . . suggest that acetylcholine release is decreased in synaptosomes prepared from old rats, although the presynaptic muscarinic regulation of release is functional . . . the principal synaptosomal cholinergic dysfunction associated with aging . . . is at the level of depolarization release coupling" (206).

5.2.2   Modulation of ACh autoreceptors

Several *in vitro* studies have manipulated this overall storage and release process at its various stages in a search for conditions under which it may be modulated. For example, the effects of an HAChT blocker at the subcellular level have suggested that reduction in the amount of ACh released from central cholinergic nerve terminals in response to depolarization may occur ". . . through a combination of effects: (1) it may facilitate the breakdown or loss of ACh stored in the vesicular fraction; (2) it may also block the transport of newly synthesized ACh into the vesicular fractions" (39). Results of experiments with adenosine have been interpreted as suggesting that it depresses ACh release at the level of the nerve terminal (138) and that it ". . . inhibits neurotransmitter release by suppressing the presynaptic influx of calcium ion during depolarization of the cholinergic nerve terminals" (292). Low oxygen tension has been shown to decrease very significantly the release of ACh *in vitro* (99). Several polypeptide neurotoxins act at presynaptic nerve terminals to alter the storage and release of neurotransmitters (135). The evidence indicates that the neurotoxins bind rapidly and irreversibly to specific sites on the plasma membrane of nerve terminals to produce a wide variety of effects that have still to be fully elaborated but include the blocking of ACh release with behavioral consequences and, eventually, destruction of synaptic vesicles (321). For example, studies (unpublished) in our laboratory contrasting behavioral effects of botulinum neurotoxin and β-bungarotoxin have produced striking differences related, presumably, to their different modes of action: at high dose levels, administered intraventricularly, with peripheral protection by antiserum i.p., botulinum neurotoxin produced relatively subtle behavioral effects apparent as impaired learning and memory, while very low doses ivt of β-bungarotoxin induced marked behavioral changes culminating in tremor and convulsions. Such information indicates that the storage and release of ACh can be affected by a wide variety of conditions, effects being evidenced in behavior as well as in other biological properties.

5.3   Interactions with behavior

The fact that interactions occur between various behaviors and the level of activation of the cholinergic system makes the negative feedback control of ACh release of particular interest to neuropsychopharmacology. The study of relations between ACh release and its behavioral effects has been complicated by problems in differentiating the relative contributions made by pre- and post-synaptic mAChRs. In general two major research designs have been used: (a) behavior has been modified and concomitant changes in release measured or (b) release has been manipulated and consequent changes in behavior observed. As with studies of roles for other presynaptic events, both animal models and human subjects have been involved. Examples of each of the two major approaches will illustrate their general features and the kinds of information they have provided.

5.3.1   Manipulation of behavior – changes in ACh release

Capitalizing on advantages in using relatively simple animal models have been experiments involving behavioral sensitization (conditioning) of the gill-withdrawal reflex of *aplysia* (161). The behavior observed was the acquisition of a prolonged enhancement of the response to a specific stimulus that initially did not produce the response. Enhancement (learning and memory) was accompanied by an increased output of transmitter from the presynaptic terminals of the sensory neurons and could be ". . . accounted for by an increase in the voltage-sensitive calcium current that underlies the action potential and is necessary for transmitter release" (161). In another example, experiments were designed to observe changes in ACh release during acquisition of new responses with different reinforcement schedules. ACh was collected in cups implanted epidurally over the visual and sensory motor cortex of rabbits (249). ACh release increased over baseline levels under all behavioral conditions, but was significantly greater during periods of positive than during negative reinforcement. An extensive series of experiments in our own laboratory was designed to study changes in the dynamics of the cholinergic system during the development of behavioral tolerance, one of the processes by which living organisms adjust to changes in their internal and external environments (267; 268; 269). In our experiments tolerance was observed as the return of a variety of behaviors to pretreatment baseline performance during chronic treatment with the anti-AChE, DFP (diisopropylfluorophosphate), when brain AChE activity remained at a constant, low level. Our early experiments failed to find changes in ACh synthesis or in Ch uptake. One of our colleagues showed that down-regulation of mAChR paralleled the behavioral changes (281), an observation confirmed by several other investigators at about the same time. Our interest in defining the mechanism more completely led us, first, to search for subcellular changes that might occur concomitantly with the development of tolerance (269). The concentrations of ACh and Ch, the high-affinity transport of Ch, and the rate of synthesis of ACh were measured in synaptosomes prepared from the brains of rats, tissue samples being taken at critical times during the development of behavioral tolerance to the anti-AChE. No statistically significant effects were found in any of the variables measured. Our next effort was to look for possible changes in total and regional brain ACh levels and dynamics (268). Again the results were negative: no effects were seen in ACh synthesis or Ch uptake, indicating that the behavioral tolerance observed was not due to end-product inhibition of ACh synthesis. It appeared obvious at this point that we should look for possible changes in presynaptic ACh release (267). Techniques were not then available to differentiate the involvement of pre- and post-synaptic receptors in brain, but the myenteric plexus preparation had been established as a peripheral model (160). Assays carried out at critical periods in the development of behavioral tolerance showed that a significant disinhibition of ACh-evoked release from the presynaptic neurons occurred with chronic depression in AChE activity. This was consistent with the model in which muscarinic autoreceptors control ACh release. The results were also consistent with time parameters for maximum decreases in muscarinic binding sites in brain and for the behavioral changes observed. To summarize the apparent chain of events, the behavioral changes developed following decreases in AChE activity which led to greater release of ACh, further modulating the downregulation of postsynaptic mAChRs.

5.3.2 Manipulation of ACh release – changes in behavior

In the second general approach to studying behavioral phenomena of these kinds the investigator has manipulated the release of ACh and measured the behavioral consequences. Our interest in the potential value of compounds with selective affinities for presynaptic receptors has led to experiments that illustrate this design. The studies have involved several new compounds that are potent and specific muscarinic agonists, binding irreversibly to mAChRs (254). Experiments on the effects of these compounds using the myenteric plexus model have shown that they produce a highly significant inhibition of evoked ACh release from presynaptic neurons (74). Although the relative contribution of this decrease in ACh release could not be distinguished from other changes occurring simultaneously in the intact animal, it clearly was a major component in behavioral and physiological effects produced by the compounds (270; 275). These variables were affected differentially. Physiological ("vegetative") variables (tremor, chromodacryorrhea, salivation, core body temperature) showed effects in less than 5 minutes after injection and returned to their pretreatment baselines within minutes thereafter. Nociceptive thresholds, dependent on sensory-perceptual processes, showed that the animals were hypoalgesic, peak changes and return to normal occurring within hours. Motoric responses (drinking to maintain body fluid balance and general activity), very significantly impaired initially, recovered in 3–4 days. Cognitive behaviors (learned responses and those requiring temporal discrimination) took 8–11 days to recover and were the only responses apparently requiring full or nearly full return of the cholinergic system to normal functioning (as judged by [³H](–)QNB binding). Such hierarchial effects suggest that normal functioning of different behavioral and physiological processes have different requirements for levels of activity in the cholinergic system.

From the point of view of maximally enhancing cholinergic transmission an ideal compound would be one that, simultaneously, acts as a presynaptic antagonist and as a postsynaptic agonist. The possibility that such a compound might be found among partial agonists (6) has motivated several experiments with a muscarinic ligand, compound BM-5   (N-methyl-N-(1-methyl-4-pyrrolidino-2-butynyl)-acetamide) (59). Initial experiments established that, *in vitro*, BM-5 met the requirements (222). Later, evidence was produced that BM-5 also serves as a partial agonist *in vivo* (75; 224). During the course of these experiments several measures of behavior were recorded. Among these, activity in a standardized "open field" situation was significantly suppressed after all doses of BM-5, a well-established effect of cholinergic agonists. Measures of both learning and memory showed effects that appeared to be dose-dependent: low doses of BM-5 ". . . provoked a *tendency toward improved performance*, reminiscent of muscarinic receptor agonists . . .", although the effects were not statistically significant (224). Within the general range of doses used "... BM-5 enhances in a dose-dependent manner the ACh release from the cerebral cortex *in vivo* ..." (40). In our own studies of this partial agonist we expanded the range of BM-5 doses administered and included a wider range of physiological and behavioral measures. Behaviorally, chronic administration of BM-5 has produced rather disappointing results, i.e., significant hypothermia, hypoalgesia and hypoactivity, but only trends without statistical significance in measures of cognitive behavior. In our opinion, despite such results, it would be premature to abandon research efforts toward further investigation of this approach to enhancing cholinergic transmission and, hence, to ameliorating impairments in cholinergically-coded behaviors.

5.3.3   Studies of human subjects

Studies of relations between ACh release and behavior in human subjects have focused on post-mortem assays of brain tissue from individuals with and without DAT. Among the latter have been studies of changes during aging that have been interpreted as indicating that "... the most consistent and pronounced presynaptic change in cholinergic function during senescence appears to be a decrease in ACh release" (55). Comparisons between subjects with and without DAT have indicated that, when the presynaptic marker ChAT was significantly reduced, total muscarinic receptor numbers were downregulated by 20 % to 25 % and that this was due almost entirely to M_2 receptors (195). It will be recalled that M_2 receptors are associated with presynaptic inhibitory events regulating the release of ACh. Such results have encouraged speculation about their implications for therapeutic purposes: "... if presynaptic high-affinity agonist muscarinic subtypes of receptors are inhibitory, selective M_1 postsynaptic agonists or M_2 antagonists may be more effective in improving memory function" (343). This is reminiscent of the interest in the partial muscarinic agonist, BM-5, discussed earlier. Furthermore, it implies that to increase synthesis, e.g., by precursor loading, will not be sufficient in treating progressive degenerative dementia – that increase in ACh release may be necessary (55). It may also be that treatments with such compounds as physostigmine, which appear to have some beneficial effects, may do so by increasing ACh efflux (221).

5.4   Interactions between neurotransmitter systems

In the "integrated organism" neurotransmitter systems interact and are mutually regulating. This characteristic is clearly apparent when regulation of ACh release is considered. There is evidence that cholinergic neurons may acquire adrenergic neurotransmitters even after their cholinergic coding has been established (53). Experimental results have demonstrated ". . . the regulation of a receptor population by a nerve supply that apparently does not release the corresponding transmitter" (262). Functional interactions exist presynaptically between the cholinergic and noradrenergic systems that have inhibitory effects on ACh release (18; 336); ". . . 5-HT inhibits ACh release via a receptor probably located on the terminals (presynaptic receptor) . . ." (102); and, ". . . dopaminergic modulation of acetylcholine release appears to be mediated by a presynaptic dopamine receptor situated on the terminals of cholinergic interneurons . . ." (377).
Several theoretical models have been proposed of ways in which interactions of these kinds may be involved in normal and abnormal behaviors. An example that has received considerable attention is a cholinergic-adrenergic hypothesis of affective disorders in which balance between the behavioral-suppressing effects of the cholinergic system and the behavioral-enhancing effects of the adrenergic are viewed as determining the behavioral state of the individual at any particular time (142). Overbalancing in either direction would lead to shifts in affective state, sufficient to become incapacitating if excessive. "It is possible that additional examples of local neurotransmitter interactions through presynaptic receptor systems will be found in the near future because these mechanisms appear to be involved in the physiological modulation of neurotransmission in the central nervous system" (171). Other examples have been reported, e. g., changes in the central noradrenergic system in addition to central cholinergic effects in the pathophysiology of DAT (31). Although such interactions add significantly to the complexity of the roles presynaptic cholinergic events play in behavior, the possibility that malfunctions in their normal relations may be important contributions to behavioral aberrations makes them very worthy areas for further study.

6   Cholinesterases

The role of AChE is normally associated with the inactivation (hydrolysis) of ACh once the transmitter has been released into the synaptic cleft and, therefore, beyond the presynaptic side of the cholinergic synapse. However, evidence has been accumulating that AChE also appears to be involved presynaptically (180; 213). Results of recent experiments have demonstrated that presynaptic AChE and HAChT are localized very close to each other on the cholinergic terminal membrane, suggesting some functional relationship between the two mechanisms (248). Furthermore, the existence of molecular heterogeneity and the different cellular distribution of AChE support the view that the enzyme may have more functions than hydrolyzing ACh postsynaptically (112). Although the role of presynaptic AChE has not yet been fully elaborated, it has been proposed that "... a coupling between hydrolysis of ACh and uptake of Ch may represent a synaptic homeostatic mechanism assuring efficient and rapid recovery of Ch for ACh synthesis. In this case, the Ch produced by postsynaptic AChE would be available for both high- and low-affinity uptake, whereas the Ch produced by presynaptic AChE would be preferentially 'sent' to the HACU system" (248).

6.1   Preclinical studies of AChE and behavior

The possibility that manipulation of presynaptic AChE independently from the postsynaptic enzyme might produce differential effects on behavior awaits the invention of techniques for varying the one without the other in the intact organism. Meanwhile, it is relevant to the development of our present review to summarize briefly the nature of behavioral effects when available antiAChEs are employed as pharmacological tools.

Early experiments (264) using animal models concluded that exposure to antiAChEs produced differential effects on behavior, some behavior patterns being affected and others not. Behaviors affected involved the extinction of old responses which were no longer appropriate in coping with new environmental demands. Flexibility in the use of new responses is essential to learning and memory. More recent research has made it quite clear that such cognitive behaviors are particularly sensitive to manipulation of AChE activity. Dose-effect studies have revealed a critical brain AChE activity level at about 45 % of normal, below which behavior is impaired exponentially. Furthermore, the typical cumulative normal population curves that represent dose-effect relations indicate that, behaviorally, the cholinergic system is capable of adaptive changes within limits defined by the rectilinear components of the curves, i. e., when response output increases systematically as dose increases (Figure 4). Behavioral subsensitivity characterizes levels below this "normal" range and supersensitivity, levels above it. AChE activity levels at both extremes may be associated initially with non-adaptive responses. However, prolonged changes in AChE activity at these extremes may initiate compensatory mechanisms within the cholinergic system, e. g., downregulation of mAChRs, that are paralleled by the return of behavior to normal limits. Such "tolerance development" is an important form of behavioral homeostasis and also has significant implications for the use of antiAChEs, e. g., physostigmine, as therapeutic agents.

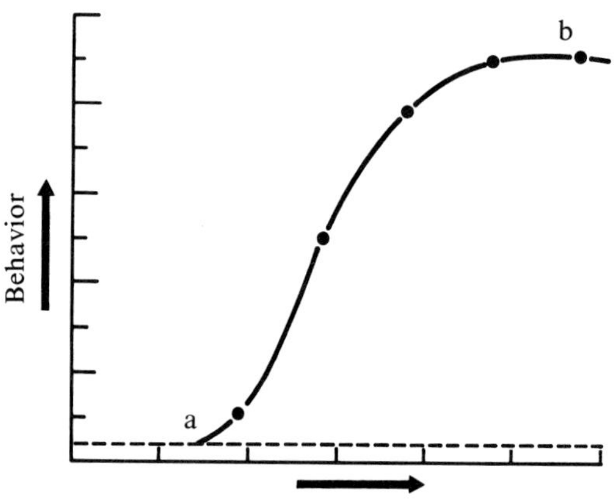

Figure 4

Relations between cholinergic activity (abscissa) and behavioral output (ordinate). Taken from experimental results, the curve in this figure shows three general characteristics of such relations: (1) A "basal" threshold (a) below which the behavior is not initiated; (2) monotonic increase in the magnitude of behavioral output as cholinergic activity increases; and, (3) a "terminal" threshold (b). "Normal" plasticity of the system appears in "(2)" within the limits of the basal and terminal thresholds. Cholinergic activity below the basal threshold or above the terminal threshold is associated with behavioral abnormalities.

6.2   AChE in behavioral disorders

Several references have already been made to the evidence that AChE
activity is decreased in DAT. ". . . the first report of an altered distribution of acetylcholinesterase molecular forms in a disease of the central
nervous system" distinguished three forms in post-mortem tissues
from both the normal and DAT neocortex (8). Losses in activity levels
appeared selectively in the intermediate form assayed in DAT samples. Involvement of presynaptic AChE in human behavioral disorders is further suggested by the fact that the immature and mature
plaques found in DAT have been shown to be rich in AChE (238; 308;
346). As the plaques matured, the activity of AChE decreased. The results of these investigations were interpreted as indicating ". . . that
changes in cortical cholinergic innervation are an important feature in
the pathogenesis and evolution of the neuritic plaque" (308). The discovery of senile plaques and neurofibrillary tangles in the entorhinal
cortex and/or the hippocampus of patients who died between 55 and
64 years of age without the behavioral signs of progressive degenerative dementia suggests the possibility that they may represent early
stages in the development of the disorder (328). In any event, their
etiology appears to be coupled at least in part, to events in the cholinergic system.

6.3   AntiAChE in therapy

The possibility that antiChEs, alone or in conjunction with other
means for manipulating the cholinergic system when it is hypofunctional might serve therapeutic purpose has been under consideration
for several years.

6.3.1   Physostigmine

Indeed, physostigmine continues to be a therapeutic strategy by which
the half-life of ACh in the synaptic cleft is prolonged by decreasing its
hydrolysis (inactivation). The varied success obtained, as well as the
difficulties (i.e., short half-life, peripheral side effects, very narrow
therapeutic window) involved in the clinical application of this particular compound are reflected in a number of reports during the past
decade (e.g., 44; 64; 320). A possible mechanism by which it may produce its effects has been discussed above.

6.3.2   Tetrahydroaminoacridine (THA)

At present considerable attention is being given to another antiChE, THA, with extensive clinical trials being planned. Early preclinical studies of THA established that it is a reversible inhibitor of ChE both *in vivo* and *in vitro* (287). One of its actions which has not been given much consideration in the recent literature is its role as a "More potent inhibitor of BuChE (butyrylcholinesterase) than AChE" (128). Later comparisons of the enzymatic actions of THA reported that it ". . . is approximately 100 times more potent an inhibitor of pseudocholinesterase than it is of acetylcholinesterase" (132). BuChE, ("nonspecific", "pseudo" cholinesterase) is dissociated from AChE activity in cortical tissue. In the autopsied brains of subjects with no history of neuropsychiatric or neurological disorders the distribution of the BuChE molecular forms, in contrast to the distribution of AChE, has been found not to vary markedly from brain region to brain region (7). There appears to be a lack of major changes in BuChE activity during development and during the degeneration of brain tissue. These findings have led investigators to the conclusion ". . . that in the CNS BuChE is not related to cholinergic neurotransmission" (7). The possibility has been suggested that it may be associated with a more general metabolic function, e. g., proteolysis, control of the metabolism of the higher esters of Ch. It would seem, therefore, that any therapeutic efficacy of THA would be related to its less potent effect in inhibiting AChE. Such an interpretation is consistent with the observations that ". . . inhibition of BuChE at most sites produces no apparent functional derangements" (200), that practically all pharmacological effects of antiChE compounds are due to inhibition of AChE.

Recent reports of the *in vitro* release of ACh have confirmed the antiChE activity of THA and have shown electrophysiologically that THA inhibits the slow outward potassium current from neurons, thus increasing the duration of action potentials (69). Based upon its antiChE mechanism of action, THA has been used in the treatment of intoxications produced by psychotropic agents known to have strong anticholinergic effects (309). In a preliminary clinical trial to test the hypothesis that combined precursor loading (lecithin) and antiChE (THA) treatment would improve memory functions, memory was assessed using several well-standardized psychometric procedures (158). Normal functioning was not restored in the ten patients, 51 to 71 years of age, with progressive degenerative dementia, although the less impaired did respond positively on some of the measures. More recently preliminary trials of the effects of oral THA treatment of "moderate to severe" DAT patients have been reported to produce significant improvement in global assessment of their condition and in psychometric tests of orientation and memory (310; 311). The investigators concluded that: "These encouraging initial results suggest that THA may be at least temporarily useful in the long-term palliative treatment of patients with Alzheimer's disease" (310). It was clear from these reports that THA needs much fuller examination, particularly for its dose- and time-effect characteristics. Clinical trials have been underway and more planned to provide such information. At the time of this writing those being conducted in the United States were on "temporary hold" because of some indication that elevation in liver enzymes, transaminases, may be an undesireable side effect at high dose levels.

7   Aging

Although it is understandable that attention has been directed toward behavioral changes and their underlying mechanisms in such clinically prominant disorders as DAT, it has become recognized that hypofunctioning of the cholinergic system is also involved in normal aging. During the present century changes in life expectancy have increased the numbers of individuals in the "aged" category and the number of persons suffering from primary degenerative dementia. Research into the nature of both neurochemical and behavioral changes occurring during aging has greatly increased. It has followed the traditional biomedical approach of using both animal models and human material. The results have been analyzed in a number of general summaries (29; 56; 110; 117; 304; 314; 331). Involvement of the cholinergic system in cognitive changes has been recognized by the proposal of "the cholinergic hypothesis of geriatric memory dysfunction" (14); by special attention to the role of the cholinergic system in Alzheimer disease (38); and, by a review of the present status of our knowledge about neurotransmitter alterations in the aging brain (255).

7.1   The neural basis of aging

Longitudinal studies using animal models have shown that during aging, significant changes occur in the dynamics of cholinergic function. Synaptosomes from older animals show significantly less K^+-stimulated release of ACh that cannot be accounted for in terms of changes in presynaptic muscarinic receptors (206). Pharmacological manipulations have produced similar effects: "... the most consistent and pronounced presynaptic change in cholinergic function during senescence appears to be a decrease in ACh release" (55). In general terms, such research with animal models helps to localize hypofunctioning of the cholinergic system during aging to persistent presynaptic dysfunction. The search for *loci* of changes within presynaptic events has established that incorporation of both the acetyl and choline moieties into ACh is decreased in aged animals, indicating reduced ACh synthesis (29; 100). Aging *per se* is related to decreases in the rates of glucose utilization in specific brain regions, including the hippocampus and parietal cortex (299). The effect of such changes in precursor incorporation is a decrease in the synthesis of ACh. The emergent picture is that the cholinergic terminal is selectively vulnerable, with presynaptic mechanisms being more affected than postsynaptic. Such findings, consistent with the hypothesis that cholinergic neurotransmission is age-dependent, are analogous to changes occurring in the normal aging of human subjects and exaggerated in DAT patients (68; 88).

Success in the uses of brain transplants of cholinergically-rich tissues to ameliorate, at least in part, behavioral deficits following brain lesions encouraged attempts to discover their effectiveness in treating impairments in normal aging, using animal models. An initial step was to characterize presynaptic cholinergic mechanisms in brain of aged rats with memory impairments. The neurochemical findings were "... consistent with an age-related decrease in hippocampal cholinergic neuronal activity without an actual loss in cholinergic neuron number" (290). Homologous tissues from fetal animals have been implanted as dissociated cell suspensions intrahippocampally and intrastriatally into rat brain with subsequent (2.5–3.0 month later) improvement in learning and memory (91; 92). The results have led to the conclusion that such implants "... can reinnervate their target areas in the absence of any denervating lesions ..." and to the suggestion that they "... may be able to restore age-related impairments in neurotransmission in specific neuronal systems" (92).

7.2   Behavioral changes during aging

Given the roles the cholinergic system is known to play as substrates of behavior, it would be expected that these age-associated changes in that system would be reflected in changes in behavior (173). The general approach to the study of such interactions has been to compare age-dependent changes in one with an analogous series in the other. Having outlined changes in presynaptic cholinergic events during aging, we should now do the same for changes in behavior.

7.2.1   Preclinical studies

Results of a research program using the rat as an animal model and designed to study normal changes during aging that occur in behavioral (sensory, motor and cognitive) functions provide useful benchmarks with which to compare effects of variations in presynaptic cholinergic events (37). It is clear that the pattern of changes during aging are not uniform, varying from behavior to behavior as do patterns of change from one neurochemical system to another. For example, decline in sensory functions varies significantly from one modality to another, visual sensitivity being most affected and nociceptive sensitivity least in the rat model. Sensory-reflexive responses to electric shock are less impaired with age than is the acoustic startle response. Such results strongly suggest that peripheral sensory and motor systems decline more rapidly than behaviors with major CNS involvement. At the cognitive level ". . . it is evident that the effects of aging, at least in the rat model, are selectively related to the level of information processing required . . . sensory memory is the least affected . . . and long-term memory the most" (37).
The existence of a concordance of events does not assure that the variables involved are necessarily related as antecedent and consequence ("cause" and "effect"). In order to put such relations to fuller test some researchers have manipulated the hypothesized antecedent, usually a specific presynaptic event, and measured consequent changes in behavior. Other investigators have carried out cross-sectional studies, measuring cholinergic and behavioral variables at different times in the aging process. Research of these kinds has been systemically analyzed in a recent review of presynaptic approaches to enhancing central cholinergic transmission (55). Taken as a whole, the results of pre-

clinical studies lend strong support to the involvement of the presynaptic cholinergic system in age-associated behavioral deficits.

7.2.2   Studies of human subjects

Concomitant with changes in the cholinergic system are age-related deficits in several important aspects of human behavior. These are analogous to deficits following experimental and traumatic lesions in cholinergically-coded pathways in the brain. Although there are significant individual differences in onset, the prominant symptoms in aging humans include impairments in such cognitive functions as: short-term memory, in the use of language (aphasia), in coordinated motor activity (apraxia) and in perception. More detailed descriptions of the behavioral deficits have led to the conclusion that performance of virtually all the cognitive functions that can be measured decreases with age (150). The decreases may affect job performance and have unfortunate effects on personal interaction within a family or other social groups, often leading to such secondary psychiatric symptoms as prolonged depression. The obvious importance of these effects has encouraged the search for pharmacological, as well as other forms of therapy (192). As with pharmacotherapies for other purposes, specific procedures may arise by serendipity, but the systematic approach to their invention depends upon knowledge of the etiology and mechanism(s) involved in a disorder. Included is the need to know much more about genetic factors in aging: "... such factors have been demonstrated whenever an adequate search for etiologic components has been undertaken" (144). Treatment development strategies are presently influenced by the state of our knowledge about the involvement of presynaptic events in the aging process and in such special conditions as DAT (38; 55; 143). Strategies tried so far have led to criticism that " ... 'therapeutic' drugs are at present more often the cause of pseudodementia than they are the cure ..." (313). The failure to achieve therapeutic successes emphasizes the need for more basic research on the aging of the brain and its relations to behavior.

8   Nootropic drugs

"The treatment of senile cognitive decline is one of the greatest challenges in the health sciences today" (130). Gaining recent attention in taking up the challenge has been research on the effectiveness of a heterogeneous group of compounds of diverse chemical composition, which allegedly facilitate learning und memory – the "nootropic drugs". The term, "nootropic", was proposed by Giurgea and colleagues in 1972 ". . . as a class of psychoactive drugs that selectively improve efficiency of higher telencephalic integrative activities" (103). These compounds appear to have no sedative or stimulatory effects and have low toxicities. Physiologically they increase blood flow and brain metabolism, both of which could influence oxidative metabolism. It has been reported that piracetam, the prototypic compound, has a protective effect against cerebral hypoxia (220). Its adenylate kinase stimulating property increases the conversion of ADP to ATP, elevating the ATP/ADP ratio in brain and thus increasing energy reserves (231).

There is experimental evidence that the nootropics improve performance of normal animals or protect against a variety of impairments in the learning and memory (acquisition, consolidation, or retrieval) of behaviors under conditions of reward or punishment (93; 282). Piracetam improves learning in young mice, ". . . but the drug is remarkably more active in improving the performance of old mice" (333). Other investigators using aged rats as their models have reported "profound effects" on memory enhancement of combining piracetam with Ch, the combination said to be several times better than either administered alone (16). In these latter experiments, repeated injections were more effective than acute. Neurochemical assays showed Ch to be elevated in the hippocampus, striatum and cortex; however, the pattern of changes in ACh levels was not consistent. Enhancement of performance has also been reported following administration of structural analogs of piracetam, aniracetam (58; 240) and etiracetam (351). Studies have demonstrated favorable effects in other animal models in addition to rodents, including primates. Administered for one month to healthy middle-aged human volunteers, using a double-blind, intra-individual crossover design, piracetam has been reported to improve significantly performance in a broad range of behaviors, ranging from visual acuity to cognitive problem solving (209). It has also been found that a nootropic agent can improve cognitive functions which have been impaired by several different experimental procedures and in different phases of the learning and memory process. In general, results of research using animal models ". . . indicate that experimentally induced cognitive dysfunction in animals can be attenuated with drug treatment" (282).

Unfortunately, the promising story still lacks substantial validity when extended to the human clinic. Nootropic compounds are still listed as "investigational drugs" for treatment of age-related cognitive symptoms (56). Results of early studies of the effectiveness of piracetam given in treatment of a variety of human disorders have been reviewed by Schneck (283). Some of the studies suffered from inadequate research design. Some of those satisfactory in this regard provided positive results and others, negative. Positive effects, although sometimes statistically significant, were small. For example, one well-designed study using 43 psychometric measures to evaluate effects of piracetam in elderly outpatients with mild to moderate memory impairment has reported significant changes only in three of the measures (252). Therapeutic windows of effective doses appear to be narrow. Investigators undertaking clinical trials have experienced the frustrating technical problems familiar to research of this kind (181). Some effort has been made to test hypotheses about possible effectiveness of combinations of treatments, e. g., piracetam in combination with Ch loading (89), again only with small improvements in behavioral deficits. It is clear that, taken together, the present evidence does not indicate a therapeutic breakthrough. On the other hand, it is not sufficient to warrant discarding the basic concept, particularly in light of the results of research on animal models. Several investigators have stated or implied a need for great care in research design, for improved compounds, and for much more complete data about dose- and time-effects. "Too few controlled studies are as yet available and the validity of study parameters is always called into question when a clinical, or therapeutic, response is demonstrated for vasoactive or nootropic agents" (127).

9   Neurotrophic factors

The dynamic biochemical events we have been discussing take place in defined morphological sites within the body – the majority of them in presynaptic neurons. There is a rapidly growing body of knowledge describing mechanisms by which these sites are themselves influenced by biochemical processes. It appears that neurotrophic factors determine which neurons will survive during ontogeny and thereby regulate the development of neural pathways. Present results also indicate: (a) that there may be a large class of neurotrophic factors, each specific to particular neuronal populations; (b) that they regulate neuronal survival during adulthood as well as during earlier development; and, (c) that inadequate neurotrophic activity may lead to neurological malfunctions arising from nerve cell death (60). It has been proposed that, phylogenetically, "... specific heritable, trophic interactions during development, which determine cell survival and pathway size, form a substrate for neural evolution" (24). Clearly the growth of neuronal cell structures to synapse on other, "target" tissues is a process of great importance as a presynaptic event.

Award of the 1986 Nobel Prize for Physiology or Medicine to Rita Levi-Montalcini and Stanley Cohen for their discoveries of neuronotrophic factors called special attention to the potential significance of these agents not only neurologically, but also behaviorally. Results of Levi-Montalcini's early research (177) had suggested the possibility that interactions might occur between behavior and biochemical events involved in the synthesis and/or release of nerve growth factors. "It seems a reasonable hypothesis that there may exist a mechanism by which, during the coding of new behavioral patterns, the effects of information input on protein synthesis increase the production of protein molecules capable of modifying the structure of nerve or glial cells. The structural modifications would then serve as engrams for long-term memory storage" (265). When we examine more recent research results, it will become clear that some special relationship exists between the "nerve growth factor (NGF)" discovered by Levi-Montalcini and the cholinergic neurotransmitter system in which we are presently particularly interested. It will be very tempting to consider a theoretical model that incorporates behavior as a third among three partners in this relationship.

9.1   "Trophic"

The term "trophic" has been defined as "... any relatively long-term influence that passes from one cell or tissue to another either during development or in the mature state" (335). Interest in the remarkable ways in which neurons form synapses and become affiliated with their eventual target cells and tissues during embryonic, fetal and neonatal growth raised questions about the processes involved. Early theoretical models hypothesized that growth of axons and of dendrites occurred along the flow of bioelectric currents, neurons being "electrically polarized": "stimulogenous fibrilation" (28); "neurobiotaxis" (155). As research continued, doubt was cast that electrical fields could play any major role in neuronal patterning within the brain (339). Further evidence led to models more closely akin to those presently favored. For example, another Nobel Laureate, R. W. Sperry, proposed that during maturation of the nervous system, genetically coded differentiation . . . continues until in many nuclei it approaches the level of the individual nerve cell . . . thus refined specification of the neurons makes possible the formation of selective synaptic linkages on the basis of a chemoaffinity" (301). Although this concept was on a fruitful path, it was not until somewhat later that the current model took a definite shape, with the hypothesis that factors with neurotrophic effects might be produced by other "target" tissues (178). Initially the physiological role and the distribution of NGF was well characterized peripherally as synthesized by target tissues innervated by sympathetic and certain sensory neurons and taken up by those neurons to be transported retrogradely to their cell bodies.

More recently, new evidence supports the involvement of NGF in several CNS systems (164). NGF receptors indistinguishable from those in the peripheral nervous system (PNS) have been detected in mammalian whole brain and in its cholinergically-coded septal, hippocampal, thalamic and cerebral cortical subregions (318). NGF receptors have been found in human brain, exclusively located in the medial septal nucleus, the diagonal band of Broca, and the nucleus basalis, leading to the suggestion ". . . that NGF acts as a trophic factor for cholinergic neurons in the human brain in a similar way as has been established in recent years for the rat brain" (126).

The search for knowledge about the biosynthesis and mechanism of action of NGF is currently at one of the frontiers in neurobiology (111). In summarizing a 1986 International Symposium on Nerve Growth Factor (NGF) and Related Substances, Rush (261) has commented: "While so much has been achieved so little is understood". Among the achievements have been the characterization of NGF, the isolating of the NGF gene, the detection of mRNA for NGF, the immunohistochemical localization of endogenous NGF, and the generation of antibodies. It is beyond our frame of reference to attempt a full review of what is now known about NGF and its receptor and of the methods by which the information has been discovered. We must focus upon NGF-cholinergic neurotransmitter system interactions and their relevance to the central theme of the present manuscript. Those who wish to explore the nature of NGF in greater detail could begin by reading such references as: 124; 149; 164; 166; 244; 253; 318; 354.

9.2 NGF: Interaction with the cholinergic neurotransmitter system

The fact that the subregions of the brain mentioned above contain innervation from the cholinergic neurotransmitter system suggests some form of interaction between NGF and the functioning of that system. Considerable evidence has accumulated that is consistent with such a hypothesis. The evidence comes from several sources. Endogenous levels of NGF in the brain have been found to be correlated with regions innervated by NGF-responding cholinergic neurons, being relatively high in the hippocampus, cortex, medial septum and nucleus basalis of Meynert (166; 289). Magnocellular cholinergic neurons of the basal forebrain transport NGF retrogradely (286). NGF mRNA is found only in regions to which magnocellular cholinergic neurons project (166). The specific activity of the synthesizing enzyme, ChAT, is increased in a dose-dependent manner when cultures of fetal rat striatum are treated with NGF, an effect that is blocked by anti-NGF antiserum (194). Dose-dependent, selective increases in septal ChAT have been reported to result from intracerebroventricular administration of NGF (125; 211). Lesions in cholinergic pathways (e.g., fornix–fimbria) lead to a significant elevation of NGF content in brain regions affected (e.g., septum, hippocampus). Of particular interest is the recent report that NGF receptors have been found in the medial septal nucleus, the diagonal band of Broca and the nucleus basalis of the human basal forebrain, locations coincident with those of cholinergic cell bodies (124). They have not been seen in non-cholinergic neurons in the same areas. The NGF, receptor-positive cells were also co-stained for AChE. "The findings . . . strongly suggest the concept that NGF is a specific neurotrophic factor for cholinergic neurons of the human basal forebrain" (126).

9.3   Roles in early neuronal development and beyond

Early development of the nervous system is characterized by a very significant overproduction of cells. During a circumscribed phase of embryonic life neuronal degeneration occurs by which a high percent of these cells die, despite the fact that they develop properties of mature neurons (51; 52; 115; 172; 330). The degeneration generally coincides with the arrival of surviving cells in their target area, indicating that developmental survival depends upon NGF derived from target cells. The development of neural pathways and connections are similarly dependent, neurons whose axons have extended to an inappropriate target area being eliminated. It has been suggested that during early development of the mammalian brain the "transient cells" "... function as neurons in a synaptic circuitry that disappears by adulthood" (45). In addition to their roles during early development, NGF may be involved in at least two series of events vital to the behavior of living organisms: (a) the maintenance and viability of a normally-functioning mature nervous system and (b) the modification of neural structures during behavioral adjustments to changing physical and psychosocial environments, e. g., memory. Indeed, it has been suggested "... that NGF may function not only as a trophic agent, but also as a modulator of neurotransmission in the CNS" (253).

9.4   NGF antibodies

Among early studies of NGF were some involving the preparation of an antiserum with selective effects on sympathetic neurons of newborn animals (177). Since then antisera have been raised, in several animal species, to NGF purified from the mouse submaxillary gland. These antisera are approximately equally potent against all mammalian NGFs (118; 119). They destroy specific neuronal populations through NGF deprivation (179).
The possibility that antibodies to cholinergic neurons may be involved in at least some human neurodegenerative disorders was suggested by a report of the presence of anti-neuronal antibodies in Alzheimer disease (216). More recent findings using a sensitive method specific to either cholinergic nerve terminals (synaptosomes) or cell bodies (PK) have lent further credence to this hypothesis: "Our results show that sera from patients with DAT contain specific antibodies directed

against cholinergic PK" (43). That cholinergic cell bodies were selectively affected is consistent with the hypothesis that cholinergic dysfunction in DAT originates in selective degeneration of cholinergic cell bodies in the basal forebrain (e. g., 54). There is a need for experiments designed to test this hypothesis in much greater detail.

9.5   Gangliosides

In 1984 a special issue of *The Journal of Neuroscience Research* summarized evidence that gangliosides may serve an important role in the regulation of morphological and biochemical events in neurons in both the developing and the adult organism (116). The growing interest in gangliosides has since been reflected in the publication of two texts bringing up to date our knowledge about their roles in neuronal plasticity (246; 319). Gangliosides comprise a complex group of glycosphingolipids containing one or more residues of a sialic acid (174). Although they are only minor constituents of membranes, current research is pointing to the significance of the fact that they are highly concentrated in synaptic membranes and are abundant in brain, especially in gray matter. Their amount and complexity vary both ontogenetically and phylogenetically. They show receptor-associated binding capabilities for neurotrophic substances, e. g., NGF, and have a very significant capacity to complex with calcium ions. They also have the capability to cross the blood brain barrier. It has been hypothesized ". . . that variations in the composition of synaptic membrane-bound gangliosides reflect a mechanism of modulating the sensitivity of the membrane-mediated process of transmission" (245). For this reason and because, as we shall see, at least one of the gangliosides appears to affect events in the cholinergic system, it is relevant for us to consider them in the context of the present review.

Research of present interest to us has concentrated on effects of administering GM_1 monosialoganglioside (312) or its antibodies. The basic concept is that beneficial effects of GM_1 after brain injury are dependent upon increased NGF activity at the site(s) of the lesion (350). GM_1 is not, in itself, a growth-promoting agent, rather information now available suggests that it modulates neurotrophic activity induced by brain lesions. Dal Toso et al. (60), in summarizing effects of the modulation given different loci of experimentally-induced lesions, have pointed out that both the position of the damage and the technique used to induce it may affect the neurochemical consequences. Among the latter have been reported effects on nucleus basalis (232) and on hippocampal ChAT and AChE activity levels (225), all involved in the functioning of the cholinergic system. It should be pointed out that GM_1 treatment has also been shown to ameliorate the loss of dopaminergic cell bodies in the substantia nigra after hemi-transection and to facilitate dopaminergic reinnervation of the striatum (319). GM_1 has been shown to have behavioral, as well as biochemical, effects. GM_1 treatment accelerates neuronal maturation in rats and facilitates behavioral recovery after CNS lesions in areas with cholinergic innervation, including such complex responses as alternation behavior (156) and 2-choice footshock learning (278). It facilitates the recovery of behavioral impairments following both unilateral (156) and bilateral (277; 278) brain lesions. In a typical experiment (278), bilateral lesions were induced in the caudate nucleus of adult rats. They were then treated i. p. with GM_1 or saline for 14 days. After a 9-day recovery period, practice began on a 2-choice footshock discrimination problem. When learning to avoid the shock by going to the initally nonpreferred side had reached a criterion of 100 % correct for 2 days, learning to reverse the response began. Saline-lesioned animals were severely impaired in their attempts to learn. GM_1-lesioned subjects learned the initial task as well as non-lesioned control animals and also learned the reversal problem.

To examine the role of GM_1 during development and its significance to behavior, antibodies to the ganglioside have been administered neonatally (157). Five-day-old rats were given a single injection of the antibody into the cisterna magna, while a control group were similarly injected with the antiserum absorbed with pure GM_1 to remove specific antibodies. Both GM_1 and control animals showed normal rates of body growth, normal food and water consumption, normal activity levels, normal sleep/waking cycles and similar proficiency in a test of memory. However, by 60 days of age they were significantly impaired in the performance of a more complicated cognitive task. These results confirmed a general theme running through research with neurotrophic factors: effects were differential in the sense that some variables were affected and others were not, i. e., the effects cannot be attributed to some global debilitation.

The conclusion that "... GM_1 can affect the efficacy of different neurotrophic agents acting on specific neuronal subpopulations ..." has led to the suggestion that gangliosides may serve as pharmacological tools in the treatment of chronic neurodegenerative diseases (60). Indeed, two reports of clinical trials have appeared which have demonstrated that clinical parameters in peripheral and alcoholic neuropathies, are "very responsive to ganglioside therapy" (196; 197). At present this promising possibility must be viewed with caution for not all research results are consistent with the picture painted by the preceding few sentences. Some investigators have reported results that seem to imply that ganglioside treatments do not accelerate onset of proliferation by surviving inputs to the hippocampus after fimbria–fornix transections. However, taken as a whole, current information encourages further time-response studies seeking to discover temporal relations among the various effects of exogenous gangliosides. "Although there may be limits to the capabilities of ganglioside treatments to promote neuroplasticity and recovery of function, the treatments evidently do affect both early and late sequellae of brain lesions" (277).

It is relevant to this discussion that there exist "ganglioside storage diseases" that create "devastating neurologic problems" and are reflected in behavior (294). These diseases are inherited disorders characterized by the accumulation of gangliosides and metabolites within neurons, leading to a progressive neurologic deterioration. There are two major groups, GM_1 and GM_2 gangliosides. Both have a variety of clinical subtypes and are found in humans and in a number of animal species. This "dual personality" of GM_1 is an extension of the pharmacological truism that biologically-active compunds have limited conditions for effective functioning.

9.6   NGF: Recovery from morphological lesions

One of the traditional experimental approaches to the study of relations between the brain and behavior using animal models has been to lesion sites in the former and to measure concomitant changes in the latter. Lesions resulting from trauma or from therapeutic procedures have provided clinical opportunities to observe such interactions in humans. Both approaches have demonstrated the differential sensitivity of behavior depending upon the site of the lesion. They have also shown that recovery of function, including behavior, can occur. Mechanisms involved in recovery have been of particular interest (193). Evidence that central neurons exhibit a capacity for axonal sprouting and generation of new connections in response to injury, not only in the developing but also in the mature CNS, has been accumulating for some time (49; 50; 354). In peripheral models, axotomy is followed by a dramatic increase in the density of NGF receptors in Schwann cells through which axonal regeneration must occur (260). The density diminishes with a time course parallel to that of the axonal regeneration. "The Schwann cells thereby provide trophic support and trophic guidance to growing sensory and sympathetic axons" (317). In the CNS, lesions of the septohippocampal pathway are followed by an increase of NGF content in the hippocampus and septum (165). Injury to the cholinergic system appears to be the mechanism involved in this effect (94). Results of other research suggest that the NGF mRNA in such CNS effects is produced by neurons, in contrast to non-neural cell origin in the peripheral system (253). Central to our present interests is the fact that, both centrally and peripherally, NGF-supported regeneration occurs. It does not logically follow that the promotion of regeneration results in the "correct wiring", i.e., in "functional connections" (2; 322). One line of evidence that regenerated cells are indeed functional is to be found in terms of electrophysiological activity recorded in the hippocampal area of lesioned animals with choline-rich septal implants but not in those lesioned with no transplant (183). Further evidence comes from the recovery of behaviors disrupted by the original lesion. It has been ". . . argued that synaptic growth in the mature nervous system of the type eluted by lesions is related to the same phenomenon caused by changes in the environment or other stimuli" (48). This suggests a means by which information input during an organism's life span may be stored and available, as "memory", to influence future behavior.

9.7   NGF and behavior

A century ago Hermann Ebbinghaus (73) published results of years of highly original experimentation on human memory in a volume, *Über das Gedächtnis*. In it he specified the conditions that must be met for measurement of memory to be possible and described methods for fulfilling the requirements. Among the hypotheses that later developed, as questions arose about the somatic substrate(s) of memory, were those that envisaged memory as represented in morphological (anatomical, structural) changes induced during or following the learning experience (122). As knowledge grew it became clear that the CNS in the several species of animal models studied was capable of morphological and neurochemical plasticity as a result of learning, a plasticity that continued throughout puberty and into adulthood (256). Soon thereafter experimental evidence began to accumulate indicating that inhibition of protein synthesis could interfere with the formation of long-term memory (22) and, indeed, that the formation of memory, regardless of the length of training, could be prevented by the inhibition of cerebral protein synthesis of sufficient duration (13; 80). During the same general period histological studies revealed that, in normal animals, experience in an enriched environment resulted in a consistent pattern of morphological changes: increase in cortical weight paralleled by an increase in protein (21); increase in dendritic branching (113); increase in number of dendritic spines (106); and, the size of synaptic contact areas. All these results are consistent with the hypothesis that an organism's interactions with its environments leaves traces ("memories") in the form of morphological changes in the nervous system, the hypothesis proposed by Ebbinghaus almost a century earlier. Now, however, techniques had been invented by which the hypothesis could be put to test. Furthermore, concepts had developed about the nature of dynamic mechanisms underlying the changes and techniques to examine the concepts had become available. Use of the techniques by many investigators led to the "protein synthesis hypothesis" of memory formation (257). Earlier studies using pharmacological tools to manipulate the cholinergic neurtransmitter system had provided evidence of the system's involvement in the modulation of memory (81; 82; 264). The new techniques now advanced knowledge of this relationship to a much more sophisticated level.

9.7.1   Preclinical studies

Throughout these facts runs a central theme that suggests a role for NGF and its biologic activity in behavior. NGF is a well-characterized protein (111). Endogenous NGF levels in the brain have been found to correlate with the presence of cell bodies and nerve terminals of magnocellular cholinergic neurons in the basal forebrain, high levels being found in the hippocampus, cortex and medial septum. Neuronal degeneration in these areas is followed by impairment in behavior, including such cognitive processes as learning and memory. It is also followed by an increase of NGF content in the sites of injury, in NGF-supported neuronal regeneration and in an increase in ChAT activity. Concomitant with this series of events is full or partial recovery of behaviors impaired by the injury, the final endpoints in which we are presently interested.

9.7.1.1   Recovery from brain lesions

During the past decade reports of research into relations between NGF and behavior have fallen into three major categories. Most frequent have been studies designed to measure the recovery of behavioral impairment following lesions of cholinergic pathways in the brain. Lesions of the cholinergic septohippocampal pathway have been shown to produce severe impairment of behaviors involving sensory-motor processes, learning and memory. Such lesions are followed by a transient increase in hippocampal NGF content that reaches its maximum after two weeks (165). Similar orders of magnitude for time-dependent increases in neurotrophic activity, neurite-promoting activity and sprouting of commissural/associational fibers have been demonstrated following lesions in the entorhinal cortex, lesions that functionally isolate the hippocampus (50). Continuous infusion of exogenous NGF i.c.v. has been found to reduce total neuronal and cholinergic neuronal degeneration measured at a similar time (2 weeks) after fimbria–fornix transection in adult rats, sparing being 50 % in the medium septum and 100 % in the diagonal band of Broca (349). Relations between these events and changes in behavior have been summarized as follows: "... all studies which reported 'positive' effects of NGF treatment on behavioral recovery have found that these effects are transient and seen only during the second and third weeks after NGF injection ..." (348). Although "transient", such observations suggest a concomitance of events between the increase in NGF and the recovery of behavior. They also suggest that behavioral recovery should be monitored for longer periods of time if increased NGF activity were also monitored. Two experimental approaches have been used in testing this hypothesis: (a) repeated (chronic) injections (348) or infusion of NGF (91) and (b) transplantation of neuronal or target tissue (2; 70). Both of these have been reported to have produced at least partial compensation for effects of brain damage, restoring the pattern of cholinergic innervation and producing concomitant improvements in some but not all behavioral patterns. A brief review of some of the experiments providing support for these conclusions will illustrate how they have been designed to study such interactions.

In 1982 Dunnett and colleagues presented results of a prototype program showing ". . . in the first clear demonstration that intracerebral neural grafts can ameliorate lesion-induced impairments in conditioned behaviours" (70). The design permitted comparisons to be made among effects on several behavior patterns of the bilateral implantation of embryonic cells from the septum or from the locus coeruleus, as solid grafts or as dissociated cell suspensions, into the septum or into the locus coeruleus after bilateral fimbria–fornix lesions. Typical of the differential effects observed was behavior in a rewarded alternation task which control animals learned rapidly, lesioned animals performed only at a chance level, and both septal grafts (but not the locus coeruleus graft) reversed the impairment induced by the lesion. By contrast, the septal grafts did not compensate for effects of the lesion on spontaneous activity or spontaneous alternation. When behavior recovered, the improvement correlated significantly with reinnervation in the hippocampal formation, as measured by the presence of AChE-positive fibers.

9.7.1.2 Are the results of neural regeneration functional?

A broad range of behaviors have served as dependent variables in the search for evidence that cellular changes following neural regeneration stimulated by implants or by cell suspensions are functional. Interest in motoric abnormalities in Huntington disease has led to the development of animal models involving lesions produced by such neurotoxins as ibotenic and kainic acids. Bilateral lesions in the striatum followed by bilateral fetal striatal implants reversed the spontaneous motor abnormalities produced by the lesions (66). Severe striatal neuronal cell loss and shrinkage following ibotenic acid lesions of the caudate-putamen produced hyperactivity that was completely compensated by "neural grafting", i.e., implantation of dissociated cell suspension from fetal rat striatum in the lesioned sites (137). Cognitive deficits have been reduced in animals with frontal cortex implants after bilateral damage to the medial frontal cortex (169). Clinical reports of sensorimotor deficits in Alzheimer disease have led researchers using animal models to include assays for sensorimotor conduction in their behavioral batteries (71; 305). Unilateral ibotenic lesions of the nucleus basalic magnocellularis produced "... a significant contralateral sensorimotor neglect and ipsilateral turning bias ...", which was partially compensated by cortical grafts of cholinergically rich ventral forebrain tissue. At a more complex behavioral level particular attention has been given to learning and memory, using a variety of different assays. Retention has been found to be very significantly improved in lesioned animals following implantation of cholinergically rich cells when measured by such well-established techniques as spatial alternation, inhibited (passive) avoidance, and learning and memory in water and radial maze situations.

By contrast, it has been reported that at least under some experimental conditions, impaired acquisition (learning) of new responses may not be influenced by implants, i. e., that only the retention component may be ameliorated by the grafts (71). Because improvement of performance (learning) with practice requires memory as it proceeds, this important finding deserves further study. The recent report that behavioral recovery following bilateral lesions of the nucleus basalis is very significantly affected by postoperative experience also needs further attention (17). It is clear from these examples that exogenous NGF and tissue implants may produce neural regeneration and concomitant full or partial recovery of behaviors impaired by brain lesions. The behavioral effects appear to be differential in the sense that some behaviors may be affected and others, not. Recovery may be transient or long range depending upon the procedure used to induce increases in NGF activity.

9.7.1.3   Effects of aging

Behavioral changes during normal aging have similarities to those observed after lesions in the septo-hippocampal system. Predominant among them are deficits in learning and memory (68). One of the most consistent features of aging are histological changes in cholinergic nerve cells, aggregations of granula or filamentous fragments within cells and concomitant decreases in activity of the two cholinergic neurotransmitter enzymes, ChAT and AChE. With these and related facts in mind, it is not unreasonable to hypothesize that NGF and target tissue transplants would have beneficial effects in ameliorating senile cognitive decline. Typical of the experiments designed to test the hypothesis are those reported by a group of investigators with whose work we are already familiar (91; 92). Fetal septal tissue, rich in cholinergic neurons, was implanted as a dissociated cell suspension into the "depth of the hippocampal formation" in aged rats with severe impairments in spatial learning. The effect was to produce a partial reversal of the learning deficit.

It has been proposed ". . . that denervation of certain CNS or PNS target tissue results in the reactivation of mechanisms for the release of trophic factors that, although active during development, may be dormant during adulthood" (90). We have been considering the roles of NGF in regenerating neural tissue. The question we should also address is that of the aging processes that may be involved in the initial degeneration. A decade ago observations were reported of *brain reactive antibodies* in aging (216). Brain-specific antibodies, gamma globulins in nature, were found to be present in the sera of old mice, but not in the young: "The antibodies began to appear in the blood at 6 to 15 months of age and thereafter increased as a function of age" (216). Circulating antibodies appeared to be separated from brain antigens by the blood brain barrier. Changes in the permeability of the BBB in old mice resulted in antigen-antibody reactions, leading to damage of brain tissue. Antibodies against brain tissues have also been observed in the sera of normal individuals as well as in patients with senile dementia, the sera of the latter having significantly higher levels than control subjects matched for age. More recently research has provided results which show that specific antibodies in the sera of DAT partients are directed against cholinergic cell bodies and not against cholinergic nerve terminals (43).

9.8 NGF: Implications for therapy

"The rush is on to treat neural disorders with brain implants (323)." "Parkinson's Disease – A New Therapy?" (212). "Former world heavy-weight boxing champion . . . is considering whether to undergo this surgery" reported an Australian newspaper. Such headline news burst forth following publication of a preliminary report describing the treatment of two patients with intractable and incapacitating Parkinson's disease (185). Other investigators using animal models had found that chromaffin cells from the adrenal medulla cultured in the presence of NGF (329) or placed as tissue implants (226) are capable of developing nerve processes. Experiments using animal models had discovered that, following destruction of the magnostriatal system, implants of fetal dopamine-containing neurons adjacent to the caudate nucleus showed good survival and axonal outgrowth and significantly reduced motor abnormalities (233). Because chromaffin cells are a source of dopamine, ". . . such cells could theoretically be transplanted into the brains of patients with Parkinson's disease" (185). Indeed, in 1985 clinical trials had been carried out, using a cannula inserted into the caudate nucleus, with reported success (9). The later trials used a technique for ". . . grafting chromaffin cells in direct contact with both the cerebrospinal fluid and the caudate nucleus . . .", which ". . . produced excellent amelioration of most of the clinical signs of Parkinson's disease in our two patients" (185). It is understandable that such reports have gained so much attention. Might the proper tissue implants also be successful in the treatment of human disorders involving the cholinergic system?

In 1981 a very stimulating paper had presented "A Unifying Hypothesis for the Cause of Amyotrophic Lateral Sclerosis, Parkinsonism, and Alzheimer Disease", the thesis of which was ". . . that each of these disorders is due to lack of a disorder-specific neurotrophic hormone" (4). The hypothesis postulated that the failure of target tissues to supply necessary neurotrophic factors is the primary manifestation of the disorders: ". . . in each system, the lack of an appropriate hormone released from postsynaptic cells would impair the viability of the presynaptic cells" (4). More specifically, in Alzheimer disease the failure would be in the production of NGF by hippocampal and cortical cells, resulting in a gradual deterioration of septal and basal nuclei and associated decreases in ChAT activity and ACh synthesis (5; 123). Because our knowledge of such processes was derived primarily from research on animal models, a basic step toward testing this hypothesis was to establish the presence of NGF in the human brain. The evidence now indicates ". . . that NGF acts as a trophic factor for cholinergic neurons in the human brain in a similar way as has been established in recent years for the rat brain" (124). Furthermore, "The similarity between the response in rodents to entorhinal cell loss and that in AD patients indicates that studies using the rodent model may be directly applicable to AD" (49). To summarize, the validity of such generalizations gains additional credence from the facts that: (a) NGF receptors have been localized in the human basal forebrain, but not in adjacent neurons (124); (b) neurons of the nucleus basalis of Meynert undergo profound and selective degeneration in DAT, providing a pathological substrate for cholinergic malfunction (344) and (c) sera from patients with DAT contain specific antibodies directed against cholinergic cell bodies (43).

Accepting the relevance of our knowledge of NGF from all sources, what hypotheses about effective treatment of DAT and related disorders have been generated? Attention appears to have been directed toward three possibilities. The overwhelming evidence that NGF is capable of inducing neural regeneration in brain leads logically to the suggestion that it be considered as a therapeutic agent (123). Because NGF does not cross the blood brain barrier, it, itself, is not a useful candidate, unless techniques for administration directly into brain are considered. Efforts to synthesize new compounds that can cross the barrier and that are not affected by biotransformation on the way might lead to treatments applicable through peripheral routes. A second possibility, already put to preliminary clinical tests for another transmitter system, could involve the implantation of cholinergically-rich tissue in affected areas. A third approach has been suggested by the mechanism of action of GM_1 (60): "If a partial deficit or inhibition of the neuronotrophic activity occurs in these (neurodegenerative) diseases, GM_1 by acting synergistically with the neuronotrophic factor, may decrease progressive presynaptic neuronal cell death and favor beneficial adaptive responses." Clearly much more knowledge about CNS target-derived neurotrophic factors and about their role(s) in neuronal cell survival and in neurodegenerative diseases, particularly in adult organisms, is necessary before the validities of such treatments are adequately understood. Despite the understandable desire to put our present results to practical applications, we are cautioned that "...they do not support using this technique (grafting) for treating cognitive disorders of neurological origin, at short survival times" (285). The need for more basic and clinical research has priority for other treatments as well.

10   In perspective

The context for our present discussion was set initially in terms of the concept of the "integrated organism", in which behavior is one of the basic properties. It will be well to return to that concept in making a few concluding observations about interactions between behavior and presynaptic events in the cholinergic system.

10.1   Cholinergic substrates of normal and abnormal behavior

Hulings Jackson, an early giant among pioneer neurologists, once commented that ". . . the highest nervous processes are potentially the whole organism." Developments since his day have impressed those who have studied relations between neurochemical events and behavior with the predictability of interactions between the two. The pages above have described a number of examples of concomitant variations between measures of various events in the presynaptic cholinergic system and measures of "normal" behavior. The general observation is that, within limits, this covariation is essential to an organism's success in adjusting to its constantly changing environments. Beyond these limits, changes in presynaptic events can result in behavioral abnormalities. Within the framework of our present discussion, hypotheses about how the two are related must consider two major elements. One, parametric in nature, involves the concept of threshold limits, i.e., the extent of change in the neurochemical event(s) before the threshold limit(s) of normal behavioral adjustment are exceded and impairments of behavior occur. The second defines the particular presynaptic event(s) affected. Although our attention is now focused on the cholinergic system, recognition has been given to the fact that interactions with other neurotransmitter and neuroendocrine systems modulate its overall effects on behavior. This point is emphasized in the statement: ". . . we believe that defective cholinergic transmission always results in dementia, but dementia is not always due to defective cholinergic transmission" (60a).
One threshold model from which hypotheses can be generated views the focal or "pacemaker" step as the binding of ACh to postsynaptic mAChRs, thus initiating the sequence of events leading ultimately to behavioral effects. The binding is dependent upon the sequence of presynaptic processes by which ACh is synthesized and released. As we have seen, malfunctions, sometimes referred to as "biochemical lesions" (264), of any one event in the sequence may affect the pacemaker step, one of the consequences being the development of behavioral abnormalities. Because each presynaptic event operates through the common pacemaker step, it follows that the production of the same or

similar behavioral effects may result from any particular presynaptic malfunction. One corollary of such a hypothesis is that what appears to be a single population of behavioral disorders may, in fact, be the end result of primary impairments in any one of the presynaptic events. Under these circumstances, adequate treatment of the behavior disorders (s) would require identification of the particular malfunction involved and the selection of an appropriate therapy specific to it. In our discussion above, we have noted examples of therapies, e. g., Ch precursor loading, which have been successful with some but not all patients with overtly similar behavioral symptomatologies. One interpretation of this fact could be that the malfunction in those treated successfully involved a lack of adequate Ch, while the others suffered from malfunctions of other kinds. Significant correlations between cortical and hippocampal ChAT activity and measures of cognitive functioning suggest that impairment of ACh synthesis as a result of decreased enzyme activity is another example of such malfunctions in cases of dementia. In both these examples the neurochemical changes are viewed as exceeding thresholds of normal variation before behavioral impairment occurs.

10.2   Homogeneity or heterogeneity?

Special attention is being given in the current literature to the question of whether such behavioral disorders as primary degenerative dementia are unitary ("homogeneous") or multiple ("heterogeneous") in nature. Perceived from the position presented in the immediately preceding paragraphs, answers to this question would favor "heterogeneous": any one of several presynaptic cholinergic events could be the source of malfunction. Again using DAT as an example, the consensus of views expressed in the literature would clearly support the heterogeneous model (258). Several suggestions have been made about criteria upon which multiple categories of DAT may be based: clinical evaluations of differences in behavior, histopathological variants, neurochemical impairment, genetic predisposition, infectious disease, and age of onset. A significant difficulty in any classification system is to establish mutually exclusive categories. Evaluation of the literature indicates that the suggested number of "subsets" or "subpopulations" in DAT varies with the criterion used and, even within the same criterion, from investigator to investigator (e. g., 23; 30; 201; 251; 259; 298).

Such differences may be significant handicaps for diagnosis, for predicting outcomes and for selecting therapies: "A pharmacological treatment . . . may prove effective in one but not in another subgroup of a disorder" (259). They may also influence directions taken in the search for the etiologies of behavioral disorders and for new or improved therapies (10; 57).

10.3   An exciting frontier for research

Among the publications referred to in the preceding pages, the very high percent appearing in the last three-year period indicates that the search for cholinergic substrates of behavior is now at a particularly active frontier of the biomedical sciences. That there remains much still to be learned is reflected in a recent comment concerning the status of our knowledge about primary degenerative dementia: "The scientific study of PDD has been hampered by (1) the lack of an early, reliable diagnostic method, (2) an unknown etiology, (3) little knowledge about the homogeneity or heterogeneity of the disease, and (4) the absence of effective therapeutic agents and appropriate animal models" (130). There remains ample scope for "Progress in Drug Research" directed both toward basic research leading to a much fuller understanding of mechanisms by which malfunctions in the cholinergic neurotransmitter system are related to abnormalities of behavior and clinical research designed to study means by which malfunctions may be prevented or corrected once they have appeared.

Acknowledgement

In the few pages of this review it is not possible to do full justice to the contributions of the colleagues with whom I have worked in different laboratories. Nor is it possible to discuss all the innovative research of the many scientists and clinicians who have found the topic as fascinating as I have. I wish to acknowledge them all. I am especially indebted to Karen L. Roberts, who made my poor penmanship intelligible, and to Dr. Robert Pechnick, Ruth A. Booth, Sharlene D. Lauretz and Kay Russell who were constructively critical of my efforts. I also wish to acknowledge the support provided for my own research by USPMS grant MH-17691 to Donald J. Jenden.

References

1   D. V. Agoston, J. W. Kosh, J. Lisziewicz, P. E. Giompres and V. P. Whittaker: Separation of recycling and reserve synaptic vesicles from cholinergic nerve terminals of the myenteric plexus of guinea-pig ileum, p. 509–517. In: Dynamics of Cholinergic Function. Ed. I. Hanin. Plenum Press, New York (1986).

2   A. J. Aguayo, P. M. Richardson, S. David and M. Benfey: Transplantation of neurons and sheath cells – a tool for the study of regeneration, p. 91–105. In: Repair and Regeneration of the Nervous System. Ed. J. G. Nicholls. Springer-Verlag, Heidelberg (1982).

3   D. C. Anderson, S. C. King and S. M. Parsons: Proton gradient linkage to active uptake of [^{3}H]acetylcholine by *Torpedo* electric organ synaptic vesicles. Biochemistry 21, 3037–3043 (1982).

4   S. H. Appel: A unifying hypothesis for the cause of amyotrophic lateral sclerosis, Parkinsonism, and Alzheimer disease. Ann. Neurol. 10. 499–505 (1981).

5   S. H. Appel, Y. Tomozawa and R. Bostwick: Trophic factors and neurologic disease, p. 75–85. In: Alzheimer's and Parkinson's Diseases. Eds. A. Fisher, I. Hanin and C. Lachman. Plenum Press, New York (1986).

6   E. J. Ariens: Intrinsic activity: partial agonists and partial antagonists. J. Cardiovasc. Pharmacol. 5 (Suppl. 1), S8–S15 (1983).

7   J. R. Atack, E. K. Perry, J. R. Bonham, J. M. Candy and R. H. Perry: Molecular forms of acetylcholinesterase and butyrylcholinesterase in the aged human central nervous system. J. Neurochem. 47, 263–277 (1986).

8   J. R. Atack, E. K. Perry, J. R. Bonham, R. H. Perry, B. E. Tomlinson, G. Blessed and A. Fairbairn: Molecular forms of acetylcholinesterase in senile dementia of the Alzheimer type: selective loss of the intermediate (10s) form. Neurosci. Lett. 40, 199 (1983).

9   E. O. Backlund, P. O. Grandberg, B. Hamberger, E. Knutsson, A. Matensson, G. Seduall, A. Seiger and L. Olson: Transplantation of adrenal medullary tissue to striatum in Parkinsonism: first clinical trials. J. Neurosurg. 62, 169–173 (1985).

10 C. C. Bagne, N. Pomara, T. Crook and S. Gershon: Alzheimer's disease: strategies for treatment and research, p. 585–638. In: Treatment Development Strategies for Alzheimer's Disease. Eds. T. Crook, R. T. Bartus, S. Ferris and S. Gershon. Mark Pawley Associates, Madison, Connecticut (1986).

11 L. L. Barclay, G. E. Gibson and J. P. Blass: Cholinergic therapy of abnormal open-field behavior in thiamine-deficient rats. J. Nutr. 112, 1906–1913 (1982).

12 L. A. Barker: Choline availability – choline high affinity transport and the regulation of acetylcholine synthesis, p. 515–531. In: Brain Acetylcholine and Neuropsychiatric Disease. eds. K. C. Davis and P. A. Berger. Plenum Press, New York (1979).

13 S. H. Barondes and M. E. Jarvik: The influence of actinomycin-D on brain RNA synthesis and on memory. J. Neurochem. 11, 187–195 (1964).

14 R. T. Bartus, R. L. Dean, B. Beer and A. S. Lippa: The cholinergic hypothesis of geriatric memory dysfunction. Science 217, 408–417 (1982).

15 R. T. Bartus, R. L. Dean, J. A. Goas and A. S. Lippa: Age-related changes in passive avoidance retention: modulation with dietary choline. Science 209, 301–303 (1980).

16 R. T. Bartus. R. L. Dean, K. A. Sherman, E. Friedman and B. Beer: Profound effects of combining choline and piracetam on memory enhancement and cholinergic function in aged rats. Neurobiol. Aging 2, 105–111 (1981).

17   R. T. Bartus, M. J. Pontecorvo, C. Flicker, R. L. Dean and J. C. Figueiredo: Behavioral recovery following bilateral lesions of the nucleus basalis does not occur spontaneously. Pharmacol. Biochem. Behav. 24, 1287–1292 (1986).

18   L. Beani, S. Tanganelli, T. Antonelli and C. Bianchi: Noradrenergic modulation of cortical acetylcholine release is both direct and gamma-aminobutyric acid mediated. J. Pharmacol. Exp. Ther. 236, 230–236 (1986).

19   J. M. Bell and P. K. Lundberg: Effects of a commercial soy lecithin preparation on development of sensorimotor behavior and brain biochemistry in the rat. Dev. Psychobiol. 18, 59–66 (1985).

20   R. J. Beninger, S. A. Tighe and K. Jhamandas: Effects of chronic administrations of dietary choline on locomotor activity, discrimination learning and cortical acetylcholine release in aging adult Fisher 344 rats. Neurobiol. Aging 5, 29–34 (1984).

21   E. L. Bennett, M. C. Diamond, D. Krech and M. R. Rosenzweig: Chemical and anatomical plasticity of brain. Science 146, 610–619 (1964).

22   E. L. Bennett, A. Orme and M. Hebert: Cerebral protein synthesis inhibition and amnesia produced by scopolamine, cycloheximide, streptovitacin A, anisomycin and emetine in rats. Fed. Proc. 31, 838 (1972).

23   D. F. Benson: Alzheimer's disease: the pedigree, p. 1–7. In: the Biological Substrates of Alzheimer's Disease. Eds. A. B. Scheibel, A. F. Wechsler and M. A. B. Brazier. Academic Press, New York (1986).

24   I. B. Black: Trophic molecules and evolution of the nervous system. Proc. Natl. Acad. Sci. USA 83, 8249–8252 (1986).

25   J. P. Blass and G. E. Gibson: Cholinergic systems and disorders of carbohydrate catabolism, p. 791–803. In: Cholinergic Mechanisms and Psychopharmacology. Ed. D. J. Jenden. Plenum Press, New York (1978).

26   J. P. Blass and G. E. Gibson: Consequences of mild, graded hypoxia. Adv. Neurol. 26, 229–253 (1979).

27   J. P. Blass, G. E. Gibson, M. Shimada, T. Kihara, M. Watanabe and K. Kurimoto: Brain carbohydrate metabolism and dementias, p. 121–134. In: Biochemistry of Dementia. Ed. P. J. Roberts. John Wiley and Sons, London (1980).

28   S. T. Bok: Die Entwicklung der Hirnnerven und ihrer zentralen Bahnen. Die stimulogene Fibrillation. Folia Neuro-biol. 9, 475 (1915).

29   W. Bondareff: The neural basis of aging, p. 95–112. In: Handbook of the Psychology of Aging. Eds. J. E. Birren and K. W. Schaie. Van Nostrand Reinhold Co., New York (1985).

30   W. Bondareff, C. Q. Mountjoy, M. Roth, M. N. Rossor, L. L. Iversen and G. P. Reynolds: Age and histopathologic heterogeneity in Alzheimer's disease: evidence for subtypes. Arch. Gen. Psychiatry 44, 412–417 (1987).

31   W. Bondareff, C. Q. Mountjoy, M. Roth, M. N. Rossor, L. L. Iversen, G. P. Reynolds and D. L. Hauser: Neuronal degeneration in locus coeruleus and cortical correlates of Alzheimer Disease. Alzheimer Disease Assoc. Disorders (in press).

32   D. M. Bowen, J. S. Benton, J. A. Spillane, C. C. Smith and S. J. Allen: Choline acetyltransferase activity and histopathology of frontal neocortex from biopsies of demented patients. J. Neurol. Sci. 57, 191–202 (1982).

33   D. M. Bowen and A. N. Davison: Neurotransmitter and neurophysiological changes in relation to pathology in senile dementia or Alzheimer's disease, p. 331–346. In: The Psychobiology of Aging: Problems and Perspectives. Ed. D. G. Stein. Elsevier/North Holland, New York (1980).

34   D. M. Bowen, P. T. Francis and A. M. Palmer. Cholinergic and non-cholinergic neurotransmitter hypotheses for Alzheimer's disease, p. 53–61. In: Alzheimer's and Parkinson's Diseases. Eds. A. Fisher, I. Hanin and C. Lachman. Plenum Press, New York (1986).

35 A. F. Boyne: Neurosecretion: integration of recent findings with the vesicle hypothesis. Life Sci. 22, 2057–2066 (1978).

36 A. F. Boyne and T. E. Phillips: Biochemical implications of the synaptic vesicle organization in quick frozen electric organ nerve terminals, p. 519–527. In: Dynamics of Cholinergic Function. Ed. I. Hanin. Plenum Press, New York (1986).

37 B. A. Campbell, E. E. Krauter and J. E. Wallace: Animal models of aging: sensory-motor and cognitive function in the aged rat, p. 201–226. In: The psychobiology of aging: problems and perspectives. Ed. D. G. Stein. Elsevier/North Holland, New York (1980).

38 J. M. Candy, E. K. Perry, R. H. Perry, J. R. Court, A. E. Oakley and J. A. Edwardson: The current status of the cortical cholinergic system in Alzheimer's disease and Parkinson's disease. Eds. D. F. Swaab, E. Fliers, M. Mirmiran, W. A. Van Gool and F. Van Haaren. Elsevier, New York (1986).

39 P. T. Carroll: The effect of the acetylcholine transport blocker 2-(4-phenyl-piperidine)cyclohexanol (AH5183) on the subcellular storage and release of acetylcholine in mouse brain. Brain Res. 358, 200–209 (1985).

40 F. Casamenti, C. Cosi and G. Pepeu: Effect of BM-5, a presynaptic antagonist–postsynaptic agonist, on cortical acetylcholine release. Eur. J. Pharmacol. 122, 288–290 (1986).

41 B. Ceccarelli and W. P. Hurlbut: Vesicle hypothesis and the release of quanta of acetylcholine. Physiol. Rev. 60, 396–441 (1980).

42 H. J. Channon and J. A. B. Smith: The dietary prevention of fatty livers, triethyl-β-hydroxyethylammonium hydroxide. Biochem. J. 30, 115–120 (1936).

43 J. Chapman, A. D. Korczyn, M. Hareuveni and D. M. Michaelson: Antibodies to cholinergic cell bodies in Alzheimer's disease, p. 329–336. In: Alzheimer's and Parkinson's Diseases. Eds. A. Fisher, I. Hanin and C. Lachman. Plenum Press, New York (1986).

44 J. E. Christie, A. Shering, J. Ferguson and A. I. M. Glen: Physostigmine and arecoline: effects of intravenous infusions in Alzheimer's presenile dementia. Br. J. Psychiat. 138, 46–50 (1981).

45 J. J. M. Chun, M. J. Nakamura and C. J. Shatz: Transient cells of the developing mammalian telencephalon are peptide-immunoreactive neurons. Nature 325, 617–620 (1980).

46 S. J. Clayson: Effect of hypoglycemia on T-maze learning in rats. Physical Therapy 51, 991–999 (1971).

47 B. Collier, P. Boska and S. Loviat: False cholinergic transmitters. Prog. Brain Res. 49, 107–125 (1979).

48 C. W. Cotman and M. Nieto-Sampedro: Brain function, synapse renewal, and plasticity. Annu. Rev. Psychol. 33, 371–401 (1982).

49 C. W. Cotman, M. Nieto-Sampedro and J. W. Geddes: Synaptic plasticity in the hippocampus: implications for Alzheimer's disease, p. 99–117. In: Treatment Development Strategies for Alzheimer's Disease. Eds. T. Crook, R. T. Bartus, S. Ferris and S. Gershon. Mark Powley Associates, Madison, Connecticut (1986).

50 C. W. Cotman, M. Nieto-Sampedro and E. Harris: Synapse replacement in the nervous system of adult vertebrates. Physiol. Rev. 61, 684–784 (1981).

51 W. M. Cowan: Neuronal death as a regulative mechanism in the control of cell number in the nervous system, p. 19–41. In: Development and Aging in the Nervous System. Eds. M. Rockstein and M. L. Sussman. Academic Press, New York (1973).

52 W. M. Cowan, J. W. Fawcett, D. D. M. O'Leary and B. B. Standfield: Regressive events in neurogenesis. Science 225, 1258–1265 (1984).

53 J. N. Coulombe and M. Bronner-Fraser: Cholinergic neurones acquire adrenergic neurotransmitters when transplanted into an embryo. Nature (Lond.) 324, 569–572 (1986).

54 J. T. Coyle, D. L. Price and M. R. De Long: Alzheimer's disease: a disease of cortical cholinergic innervation. Science 219, 1184–1190 (1983).

55 F. T. Crews, E. M. Meyer, R. A. Gonzales, C. Theiss. D. H. Otero, K. Larsen, R. Raulli and G. Calderini: Presynaptic and postsynaptic approaches to enhancing central cholinergic neurotransmission, p. 385–419. In: Treatment Development Strategies for Alzheimer's Disease. Eds. T. Crook, R. T. Bartus, S. Ferris and S. Gershon. Mark Powley Associates, Madison, Connecticut (1986).

56 T. Crook: Clinical drug trials in Alzheimer's disease. Ann. N. Y. Acad. Sci. 444, 428–436 (1985).

57 T. Crook, R. T. Bartus, S. Ferris and S. Gershon: Treatment Development Strategies for Alzheimer's Disease. Mark Powley Associates, Madison, Connecticut (1986).

58 R. Cumin, E. F. Bandie, E. Gamzu and W. E. Haefely: Effects of the novel compound aniracetam (Ro 13–5057) upon impaired learning and memory in rodents. Psychopharmacology 78, 104–111 (1982).

59 R. Dahlbom: Structure and steric aspects of compounds related to oxotremorine, p. 621–638. In: Cholinergic Mechanisms: Phylogenetic Aspects, Central and Peripheral Synapses, and Clinical Significance. Eds. G. Pepeu and H. Ladinsky. Plenum Press, New York (1981).

60 R. D. Dal Toso, L. Cavicchioli, S. Calzolari, A. Leon and G. Toffano. Are gangliosides a rational pharmacological tool for the treatment of chronic neurodegenerative diseases?, p. 293–309. In: Treatment Development Stragies for Alzheimer's Disease. Eds. T. Crook, R. Bartus, S. Ferris and S. Gershon. Mark Poweley Associates, Madison, Connecticut (1986).

60a P. Davies: A critical review of the role of the cholinergic system in human memory and cognition. Ann. N. Y. Acad. Sci. 444, 212–217 (1985).

61 P. Davies and A. J.P. Maloney: Selective loss of central cholinergic neurons in Alzheimer's disease. Lancet 2, 1403 (1976).

62 K. L. Davis, P. A. Berger and L. E. Hollister: Choline for tardive dyskinesia. New Engl. J. Med. 293, 152 (1975).

63 K. L. Davis, L. E. Hollister, P. A. Berger and A. L. Vento: Studies on choline chloride in neuropsychiatric disease: human and animal data. Psychopharmacol. Bull. 14(4), 56–58 (1978).

64 K. L. Davis and R. C. Mohs: Enhancement of memory processes in Alzheimer's disease with multiple-dose intravenous physostigmine. Am. J. Psychiat. 139, 1421–1424 (1982).

65 K. L. Davis, R. C. Mohs, J. R. Tinklenberg, L. E. Hollister, A. Pfefferbaum and B. S. Kopell: Cholinomimetics and memory: the effect of choline chloride. Arch. Neurol. 37, 49–52 (1980).

66 A. W. Deckel, R. G. Robinson, J. R. Coyle and P. R. Sandberg: Reversal of long-term locomotor abnormalities in the kainic acid model of Huntington's disease by day 18 fetal striatal implants. Eur. J. Pharm. 93, 287–288 (1983).

66a J. M. Delabar, D. Goldgaber, Y. Lamour, A. Nicole, J. L. Huret, J. de Grouchy, P. Brown, D. C. Gajdusek and P. M. Sinet. Beta amyloid gene duplication in Alzheimer's disease and karyotypically normal Down syndrome. Science 235, 1390–1392 (1987).

67 M. Dolivo: Metabolism of mammalian sympathetic ganglia. Fed. Proc. 33, 1043–1048 (1974).

68 D. A. Drachman and B. J. Sahakian: Memory, aging, and pharmacosystems, p. 347–368. In: The Psychobiology of Aging: Problems and Perspectives. Ed.: D. G. Stein. Elsevier/North Holland, New York (1980).

69   B. Drukarch, K. S. Kits, E. G. Van Der Meer, J. C. Lodder and J. C. Stoof:
     9-Amino-1,2,3,4-tetrahydroacridine (THA), an alleged drug for the treat-
     ment of Alzheimer's disease, inhibits acetylcholinesterase activity and slow
     outward K+ current. Eur. J. Pharmacol. 141, 153–157 (1987).

70   S. B. Dunnett, W. C. Low, S. D. Iversen, U. Stenevi and A, Björklund: Sep-
     tal transplants restore maze learning in rats with fornix fimbria lesions.
     Brain Res. 251, 335–348 (1982).

71   S. B. Dunnett, G. Toniolo, A. Fine, C. N. Ryan, A. Bjorklund and S. D.
     Iversen: Transplantation of embryonic ventral forebrain neurons to the
     neocortex of rats with lesions of nucleus basalis magnocellularis. II. Sen-
     sorimotor and learning impairments. Neuroscience, 16, 787–797 (1985).

72   M. W. Dysken, P. Foual, C. M. Harris, A. Noronha, D. Bergen, T. Hoepp-
     ner and J. M. Davis: Lecithin administration in patients with primary de-
     generative dementia and in normal volunteers, p. 385–392. In: Alzheimer
     Disease: A. Report of Progress in Research. Eds. S. Corkin, K. L. Davis,
     J. H. Growdon, E. Usdin and R. J. Wurtman. Raven Press, New York
     (1982).

73   H. Ebbinghaus: Über das Gedächtnis. Memory: a contribution to experi-
     mental psychology. Trans. H. A. Ruger and C. E. Bussenius. Columbia
     University Press, New York (1913).

74   F. J. Ehlert and D. J. Jenden: Alkylating derivatives of exotremorine have
     irreversible actions on muscarinic receptors, p. 1177–1185. In: Dynamics of
     Cholinergic function. Ed. I. Hanin. Plenum Press, New York (1986).

75   C. Engström, A. Unden, H. Ladinsky, S. Consolo and T. Bartfai: BM-5, a
     centrally active partial muscarinic agonist with low tremorogenic activity:
     in vivo and in vitro studies. Psychopharmacology 91, 161–167 (1987).

76   P. Etienne: Treatment of Alzheimer's disease with lecithin, p. 353–354. In:
     Alzheimer's Disease. Ed. B. Reisberg. Free Press, New York (1983).

77   P. Etienne, D. Dastoor, S. Gauthier, R. Ludwick and B. Collier: Alzheimer
     Disease: lack of effect of lecithin treatment for 3 months. Neurology 31,
     1552–1554 (1981).

78   S. H. Ferris, M. J. de Leon, A. P. Wolf, T. Farkas, D. R. Christmann, B.
     Reisberg, J. S. Fowler, R. MacGregor, A. Goldman, A. E. George and S.
     Rampal: Positron emission tomography in the study of aging and senile de-
     mentia. Neurobiol. Aging 1, 127–131 (1980).

79   M. Fisman, H. Mersky and E. Helmes: Double-blind trial of 2-dimethyl-
     aminoethanol in Alzheimer's disease. Amer. J. Psychiat. 138, 970–972
     (1981).

80   J. F. Flood, E. L. Bennett, M. R. Rosenzweig and A. E. Orme: The influ-
     ence of duration of protein synthesis inhibition on memory. Physiol. Behav.
     10, 555–562 (1973).

81   J. F. Flood, M. E. Jarvik, E. L. Bennett, A. E. Orme and M. R. Rosenzweig:
     The effect of stimulants, depressants and protein synthesis inhibition on
     retention. Beh. Biol. 20, 168–183 (1977).

82   J. F. Flood, D. W. Landry and M. E. Jarvik: Cholinergic receptor interac-
     tions and their effects on long-term memory processing. Brain Res. 215,
     177–185 (1981).

83   P. T. Francis, A. M. Palmer, N. R. Sims, D. M. Bowen, A. N. Davison. M.
     Esiri, D. Neary, J. S. Snowden and G. K. Wilcock: Neurochemical studies
     suggest treatment strategies for early-onset Alzheimer's disease. New Eng.
     J. Med. 313, 7–11 (1985).

84   J. J. Freeman, R. L. Choi and D. J. Jenden: The effect of hemicholinium on
     behavior and on brain acetylcholine and choline in the rat. Psychopharma-
     col. Commun. 1, 15–27 (1975).

85   J. J. Freeman and D. J. Jenden: Minireview: the source of choline for ace-
     tylcholine synthesis in brain. Life Sci. 19, 949–962 (1976).

86   J. J. Freeman, J. R. Marci, R. L. Choi and D. J. Jenden: Studies on the behavioral and biochemical effects of hemicholinium in vivo. J. Pharmacol. Exp. Therap. 210, 91–98 (1979).

87   B. Frieder and C. Allweis: Prevention of hypoxia-induced transient amnesia by post-hypoxic hyperocia. Physiol. Behav. 29, 1065–1069 (1982).

88   E. Friedman, M. J. Brennan, B. E. Lerer, K. A. Sherman, J. W. Schweitzer and J. Kuster: Neurochemical and behavioral alterations in aging and in animal models of Alzheimer's disease, p. 393–405. In: Alzheimer's and Parkinson's diseases: Strategies for Research and Development. Eds. A. Fisher, I. Hanin and C. Lachman. Plenum Press, New York (1986).

89   E. Friedman, K. A. Sherman, S. H. Ferris, B. Reisberg, R. T. Bartus and M. K. Schneck: Clinical response to choline plus piracetam in senile dementia: relation to red-cell choline levels. N. Eng. J. Med. 304, 1490–1491 (1981).

90   F. H. Gage and A. Björklund: Intracerebral grafting of neuronal cell suspensions. Factors affecting survival and growth. Neuroscience 17, 89–98 (1986).

91   F. H. Gage, A. Björklund, U. Stenevi, S. B. Dunnett and P. A. T. Kelly: Intrahippocampal septal grafts ameliorate learning impairments in aged rats. Science 225, 533–536 (1984).

92   F. H. Gage, S. B. Dunnett, U. Stenevi and A. Björklund: Aged rats: recovery of motor impairments by intrastriatal nigral grafts. Science 221, 966–969 (1983).

93   E. Gamzu: Animal behavioral models in the discovery of compounds to treat memory dysfunction. Ann. N. Y. Acad. Sci. 444, 370–393 (1985).

94   U. E. Gasser, G. Weskamp. U. Otten and A. R. Dravid: Time course of the elevation of nerve growth factor (NGF) content in the hippocampus and septum following lesions of the septohippocampal pathway in rats. Brain Res. 376, 351–356 (1986).

95   J. B. Ghajar, G. E. Gibson and T. E. Duffy: Regional acetylcholine metabolism in brain during acute hypoglyccmia and recovery. J. Neurochem. 44, 94–98 (1985).

96   E. Giacobini, T. Mussini and T. Mattio: Aging of cholinergic synapses: fiction or reality?, p. 177–190. In: Dynamics of Choline Function. Ed. I. Hanin. Plenum Press, New York (1986).

97   G. E. Gibson and T. E. Duffy: Impaired synthesis of acetylcholine by mild hypoxic hypoxia or nitrous oxide. J. Neurochem. 36, 28–33 (1981a).

98   G. E. Gibson, C. J. Pelmas and C. Peterson: Cholinergic drugs and 4-aminopyridine alter hypoxic-induced behavioral deficits. Pharmacol. Biochem. Behav. 18, 909–916 (1983).

99   G. E. Gibson and E. Peterson: Decreases in the release of acetylcholine in vitro with low oxygen. Biochem. Pharmacol. 31, 111–115 (1982).

100 G. E. Gibson, C. Peterson and D. J. Jenden: Brain acetylcholine synthesis declines with senescence. Science 213, 674–676 (1981 b).

101 G. E. Gibson, C. Peterson and J. Sansone: Decreases in amino acid and acetylcholine metabolism during hypoxia. J. Neurochem. 37, 192–201 (1981 c).

102 G. Gillet, S. Ammor and G. Fillion: Serotonin inhibits acetylcholine release from rat striatum slices: evidence for a presynaptic receptor-mediated effect. J. Neurochem. 45, 1687–1691 (1985).

103 C. Giurgea and M. Salama: Nootropic drugs. Prog. Neuro-Psychopharmacol. 1, 235–247 (1977).

104 S. D. Glick. A. M. Crane, L. A. Barker and T. W. Mittag: Effects of N-hydroxyethylpyrollidinium methiodide, a choline analogue, on passive avoidance behavior in mice. Neuropharmacology 14, 561–564 (1975).

105 S. F. Glick, T. W. Mittag and J. P. Green: Central cholinergic correlates of impaired learning. Neuropharmacology 12, 291–296 (1973).

106 A. Globus, M. R. Rosenzweig, E. L. Bennett and M. C. Diamond: Effects of differential experience on dendritic spine counts. J. Physiol. Comp. Psychol. 82, 175–181 (1973).

107 H. Gnahn, F. Hefti, R. Heumann and H. Thoenen: NGF-mediated increase in choline acetyltransferase (ChAT) in the neonatal rat forebrain: evidence for a physiological role of NGF in the brain? Brain Res. 285, 45–52 (1983).

108 J. A. Golczewski, R. N. Hiramoto and V. K. Ghanta: Enhancement of maze learning in old 657 BL 16 mice by dietary lecithin. Neurobiol. Aging 3, 223–226 (1982).

109 P. E. Gold: Glucose modulation of memory storage processing. Beh. Neural Biol. 45, 342–349 (1986).

110 C. G. Gottfries (Ed.): Normal Aging, Alzheimer's Disease and Senile Dementia, p. 308. University of Brussels Press, Brussels (1985).

111 L. A. Greene and E. Shooter: The nerve growth factor: biochemistry, synthesis and mechanisms of action. Annu. Rev. Neurosci. 3, 353–402 (1980).

112 S. Greenfield: Acetylcholinesterase may have novel functions in the brain. TINS 7, 364–368 (1984).

113 W. T. Greenough and F. R. Volkmar: Pattern of dendritic branching in occipital cortex of rats reared in complex environments. Exper. Neurol. 40, 491–504 (1973).

114 J. H. Growdon, E. L. Cohen and R. J. Wurtman: Effects of oral choline administration on serum and CSF choline levels in patients with Huntington's disease. J. Neurochem. 28, 229–231 (1977).

115 R. W. Guillery and H. P. Killackey: Disappearing developing cells. Nature 325, 578–579 (1978).

116 B. Haber and A. Gorio: Special issue on neurobiology of gangliosides. J. Neurosci. Res. 12, 2–3 (1984).

117 I. Hanin (Ed.): Dynamics of Cholinergic Function, p. 1273. Plenum Press, New York (1986).

118 G. P. Harper and H. Thoenen: Nerve growth factor: biological significance, measurement, and distribution. J. Neurochem. 34, 5–16 (1980).

119 G. P. Harper and H. Thoenen: Target cells, biological effects, and mechanism of action of nerve growth factor and its antibodies. Ann. Rev. Pharmacol. Toxicol. 21, 205–229 (1981).

120 D. R. Haubrich and T. J. Chippendale: Regulation of acetylcholine synthesis in nervous tissue. Life Sci. 20, 1465–1478 (1977).

121 D. R. Haubrich, D. F. L. Wang, D. E. Clody and D. W. Wedeking: Increase in rat brain acetylcholine induced by choline and deanol. Life Sci. 17, 975–980 (1975).

122 D. O. Hebb: The Organization of Behavior, p. 335. Wiley, New York (1949).

123 F. Hefti: Is Alzheimer's disease caused by a lack of nerve growth factor? Ann. Neurol. 13, 109–110 (1983).

124 F. Hefti: Nerve growth factor promotes survival of septal cholinergic neurons. Eds. A. Fisher, I. Hanin and C. Lachman. Plenum Press, New York (1986).

125 D. Hefti. A. Dravid and J. Hartikka: Chronic intraventricular injections of nerve growth factor elevate hippocampal choline acetyltransferase activity in adult rats with partial septohippocampal lesions. Brain Res. 293, 305–311 (1984).

126 F. Hefti, J. Hartikka, A. Salvatierra, W. J. Weiner and D. C. Mash: Localization of nerve growth factor receptors in cholinergic neurons of the human basal forebrain. Neurosci. Lett. 69, 37–41 (1986).

127 H. Heidrich (Ed.): Proof of Therapeutical Effectiveness of Nootropic and Vasoactive Drugs. Springer-Verlag, Berlin (1985).

128 E. Heilbronn: Inhibition of cholinesterases by tetrahydroaminoacridine. Acta Chem. Scand. 15, 1386–1390 (1961).

129 D. J. Hepler, G. L. Wenk, B. L. Cribbs, D. S. Olton and J. T. Coyle: Memory impairments following basal forebrain lesions. Brain Res. 346, 8–14 (1985).

130 F. M. Hershenson and W. H. Moos: Drug development for senile cognitive decline. J. Med. Chem. 29, 1125–1130 (1986).

131 M. Hershkowitz, V. E. Grimm and Z. Speiser: The effects of postnatal anoxia on behavior and on the muscarinic and beta adrenergic receptors in the hippocampus of the developing rat. Develop. Brain Res. 7, 147–155 (1983).

132 A. K. Ho and S. E. Freeman: Anticholinesterase activity by tetrahydroaminoacrine and succinylcholine hydrolysis. Nature 205, 1118–1119 (1965).

133 L. E. Hollister, D. J. Jenden, J. R. Doamaral, J. D. Barchas, K. L. Davis and P. A. Berger: Pharmacokinetic studies of choline chloride: a preliminary report, p. 533–539. In: Brain Acetylcholine and Neuropsychiatric Disease. Eds. K. L. Davis and P. A. Berger. Plenum Press, New York (1979).

134 C. S. Holmes: Neuropsychological profiles in men with insulin dependent diabetes. J. Consult. Clin. Psychol. 54, 386–389 (1986).

135 B. D. Howard and C. B. Gundersen: Effects and mechanisms of polypeptide neurotoxins that act presynaptically. Annu. Rev. Pharmacol. Toxicol. 20, 307–336 (1980).

136 D. Ilson, B. Collier and P. Boksa: Acetyltriethylcholine: a cholinergic false transmitter in cat superior cervical ganglion and rat cerebral cortex. J. Neurochem. 28, 371–381 (1977).

137 O. Isacson, P. Brundin, F. H. Gage and A. Björklund: Compensation of lesion-induced changes in cerebral metabolism and behavior by striatal neural implants in a rat model of Huntington's disease, p. 519–535. In: Brain Plasticity, Learning and Memory. Eds. B. E. Will, P. Schmitt and J. G. Dalrymple-Alford. Plenum Press, New York (1985).

138 R. Jackisch, H. Strittmatter, L. Kasakov and G. Hertting: Endogenous adenosine as a modulator of hippocampal acetylcholine release. Naunyn Schmiedebergs Arch. Pharmacol. 327, 319–325 (1984).

139 R. Jaffard, C. Destrade, T. Durkin and A. Ebel: Memory formation as related to genotypic or experimental variations of hippocampal cholinergic activity in mice. Physiol. Behav. 22, 1093–1096 (1979).

140 R. Jaffard, A. Ebel, C. Desterade, T. Durkin, P. Mandel and B. Cardo: Effects of electrical stimulation on the long-term memory and on cholinergic mechanisms in three inbred strains of mice. Brain Res. 133, 277–289 (1977).

141 M. K. James and L. X. Cubeddu: Pharmacologic characterization and functional role of muscarinic autoreceptors in the rabbit striatum. J. Pharmacol. Exptl. Therap. 240, 203–215 (1987).

142 D. S. Janowsky, M. K. El-Yousef and J. M. Davis: A cholinergic adrenergic hypothesis of mania and depression. Lancet 2, 632–635 (1972).

143 L. F. Jarvik and S. Gerson: Outcome of drug treatment in depressed patients over the age of fifty, p. 29–36. In: Treatment of Affective Disorders in the Elderly. Ed. C. A. Shamoian. American Psychiatric Press, Washington, D. C. (1985).

144 L. F. Jarvik and S. S. Matsuyama: Genetic aspects of aging, p. 68–91. In: Clinical Geriatrics. Ed. I. Rossman. J. B. Lippincott, Philadelphia (1986).

145 D. J. Jenden: The neurochemical basis of acetylcholine precursor loading as a therapeutic strategy, p. 483–513. In: Brain Acetylcholine and Neuropsychiatric Disease. Eds. K. L. Davis and P. A. Berger. Plenum Press, New York (1979).

146 D. J. Jenden: Regulation and acetylcholine synthesis and release, p. 3–15. In: Psychopharmacology and Biochemistry of Neurotransmitter Receptors. Eds. H. I. Yamamura, R. W. Olsen and E. Usdin. Elsevier, New York (1980).

147 D. J. Jenden: The pharmacology of cholinergic mechanisms and senile brain disease, p. 205–215. In: The Biological Substrates of Alzheimer's Disease. Eds. A. B. Scheibel and A. P. Wechsler. Academic Press, New York (1986).

148 D. J. Jenden, J. Macri, M. Roch and R. W. Russell: Antagonism by deanol of some behavioral effects of hemicholinium. Commun. Psychopharm. 1, 575–580 (1977).

148a D. J. Jenden, R. W. Russell, R. A. Booth, S. D. Lauretz, B. J. Knusel, M. Roch, K. M. Rice, R. George and J. J. Waite: A model hypocholinergic syndrome produced by a false choline analog N-aminodeanol. J. Neural Transm. 24 (Suppl.), 325–329 (1987).

149 D. Johnson, A. Lanahan, C. R. Buck, A. Sehgal, C. Morgan, E. Mercer, M. Bothwell and M. Chao: Expression and structure of the NGF receptor. Cell 47, 545–554 (1986).

150 J. Jolles: Cognitive, emotional and behavioral dysfunctions in aging and dementia, p. 15–39. In: Aging of the Brain and Alzheimer's Disease. Eds. D. F. Swaab, E. Fliers, M. Mirmiran, W. A. Van Gool and F. Van Haaren. Elsevier, New York (1986).

151 R. S. Jope: Effects of lithium treatment *in vitro* and *in vivo* on acetylcholine metabolism in rat brain. J. Neurochem. 33, 487–495 (1979a).

152 R. S. Jope: High affinity choline transport and acetylCoA production in brain and their roles in the regulation of acetylcholine synthesis. Brain Res. Revs. 1, 313–344 (1979b).

153 R. S. Jope and D. J. Jenden: Dimethylaminoethanol (deanol) metabolism in rat brain and its effect on acetylcholine synthesis. J. Pharmac. Exp. Ther. 211, 472–479 (1979).

154 R. S. Jope and D. J. Jenden: The utilization of choline and acetyl coenzyme A for the synthesis of acetylcholine. J. Neurochem. 35, 318–325 (1980).

155 C. U. A. Kappers, G. C. Huber and E. C. Crosby: The Comparative Anatomy of the Nervous System of Vertebrates Including Man. Vol. I. Macmillan, New York (1936).

156 S. E. Karpiak: Ganglioside treatment improves recovery of alternation behavior after unilateral cortex lesion. Exptl. Neurol. 81, 330–339 (1983).

157 E. J. Kasarskis, S. E. Karpiak, M. M. Rapport, P. K. Yu and N. H. Bass: Abnormal maturation of cerebral cortex and behavioral deficit in adult rats after neonatal administration of antibodies to ganglioside. Develop. Brain Res. 1, 25–35 (1981).

158 W. H. Kaye, H. Sitram, H. Weingartner, M. H. Ebert, S. Smallberg and J. C. Gillan: Modest facilitation of memory in dementia with combined lecithin and anticholinesterase treatment. Biol. Psychiat. 17, 275–280 (1982).

159 M. Khairy, R. W. Russell and J. Yudkin: Some effects of thiamine deficiency and reduced caloric intake on avoidance training and on reactions to conflict. Quart. J. Exp. Psychol. 9, 190–205 (1957).

160 H. Kilbinger: Muscarinic modulation of acetylcholine release from the myenteric plexus of the guinea pig small intestine, p. 401–410. In: Cholinergic Mechanisms and Psychopharmacology. Ed. D. J. Jenden. Plenum Press, New York (1977).

161 M. Klein and E. R. Kandel: Presynaptic modulation of voltage-dependent Ca^{++} current: mechanism for behavioral sensitization in *Aplysia californica*. Proc. Natl. Acad. Sci. USA 75, 3512–3516 (1978).

162 R. L. Klein, H. Lagercrantz and H. Zimmerman (Eds.): Neurotransmitter Vesicles, p. 241–359. Academic Press, New York (1982).

163   I. J. Kopin: False adrenergic transmitters. Annu. Rev. Pharmacol. 8, 377–394 (1968).

164   S. Korsching: The role of nerve growth factor in the CNS. TINS, 570–573 (1986).

165   S. Korsching, R. Heumann, H. Thoenen and F. Hefti: Cholinergic denervation of the rat hippocampus by fimbrial transection leads to a transient accumulation of nerve growth factor (NGF) without change in mRNA NGF content. Neurosci. Lett. 66, 175–180 (1986).

166   S. Korsching, G. Auburger, R. Heumann, J. Scott and H. Thoenen: Levels of nerve growth factor and its mRNA in the central nervous system of the rat correlate with cholinergic innervation. EMBO J. 4, 1389–1393 (1985).

167   L. F. Kromer, A. Björklund and U. Stenevi: Regeneration of the septohippocampal pathway is promoted by utilizing embryonic hippocampal implants as bridges. Brain Res. 210, 173–200 (1980).

168   M. J. Kuhar and L. C. Murrin: Sodium-dependent high affinity choline uptake. J. Neurochem. 30, 15–21 (1978).

169   R. Labbe, A. Firl, E. J. Mufson and D. G. Stein: Fetal brain transplants: reduction of cognitive deficits in rats with frontal cortex lesions. Science 221, 470–472 (1983).

170   S. Z. Langer: Presynaptic receptors and their role in the regulation of transmitter release. Br. J. Pharmacol. 60, 481–497 (1977).

171   S. Z. Langer: Presynaptic receptors and neurotransmission. Med. Biol. 56, 288–291 (1978).

172   T. H. Large, S. C. Bodary, D. O. Clegg, G. Weskamp, U. Otten and L. F. Reichardt: Nerve growth factor gene expression in the developing rat brain. Science 234, 352–355 (1986).

173   A. Larue, C. Dessonville and L. F. Jarvik. Aging and mental disorders, p. 664–702. In: Handbook of the Psychology of Aging. Eds. J. E. Birren and K. W. Schaie. Van Nostrand Reinhold, New York (1985).

174   R. W. Ledeen, R. K. Yu, M. M. Rapport and K. Suzuki: Ganglioside Structure, Function and Biomedical potential. Plenum Press, New York (1984).

175   D. Lehman and S. Z. Langer: The striatal cholinergic interneuron: synaptic target of dopamine terminals? Neuroscience 10, 1105–1120 (1983).

176   A. I. Levey, D. M. Armstrong, S. F. Atweh and B. H. Wainer: Monoclonal antibodies to choline acetyltransferase: production, specificity and immunochemistry. J. Neurosci. 3, 1–9 (1983).

177   R. Levi-Montalcini: Growth control of nerve cells by a protein factor and its anti-serum. Science 143, 105–110 (1964).

178   R. Levi-Montalcini (Ed.): Nerve Cells, Transmitters and Behavior, p. 679. Elsevier, Amsterdam (1980).

179   R. Levi-Montalcini and P. U. Angeletti: Nerve growth factor. Physiol. Rev. 48, 534–569 (1968).

180   Z. Y. Li and C. Bon: Presence of a membrane-bound acetylcholinesterase form in a preparation of nerve endings from *Torpedo marmorata* electric organ. J. Neurochem. 40, 338–349 (1983).

181   S. Lloyd-Evans, J. C. Brocklehurst and M. C. Palmer: Piracetam in chronic brain failure. Curr. Med. Res. Opin. 5, 351–357 (1979).

182   E. D. London: Reversal of cerebral metabolic deficits in rodent models for aging and Alzheimer's disease, p. 407–417. In: Alzheimer's and Parkinson's Diseases: Strategies for Research and Development. Eds. A. Fisher, I. Hanin and C. Lachman. Plenum Press, New York (1986).

183   W. C. Low, P. R. Lewis, S. T. Bunch, S. B. Dunnett, S. R. Thomas, S. D. Iversen, A. Björklund and U. Stenevi: Functional recovery following neural transplantation of embryonic septal nuclei in adult rats with septohippocampal lesions. Nature (Lond.) 300, 260–262 (1982).

184 F. C. MacIntosh, R. I. Birks and P. B. Sastry: Pharmacological inhibition of acetylcholine synthesis. Nature (Lond.) 178, 1181 (1956).

185 I. Madrazo, R. Drucker-Colin, V. Diaz, J. Martinez-Mata, C. Torres and J. J. Becerril: Open microsurgical autograph of adrenal medulla to the right caudate nucleus in two patients with intractable Parkinson's disease. N. Engl. J. Med. 316, 831–834 (1987).

186 S. Manaker: Identification of sodium-dependent, high affinity choline uptake sites in rat brain with [^{3}H]hemicholinium-3. J. Neurochem. 46, 483–488 (1986).

187 P. J. G. Mann and J. H. Quastel: Vitamin B$_1$ and acetylcholine formation in isolated brain. Nature 145, 856–857 (1940).

188 R. M. Marchbanks: The activation of presynaptic choline uptake by acetylcholine release. J. de Physiologie (Paris) 78, 373–378 (1982).

189 M. Marchi, P. Paudice and M. Raiteri: Presynaptic autoreceptors controlling acetylcholine release in synaptosomes from rat hippocampus. Brit. J. Pharmacol. 74, 220 P (1981a).

190 M. Marchi, P. Paudice and M. Raiteri: Autoregulation of acetylcholine release in isolated hippocampal nerve endings. Eur. J. Pharmacol. 73, 75–79 (1981 b).

191 M. Marchi, P. Paudice and M. Raiteri: Choline uptake, acetylcholine synthesis, and regulation of acetylcholine release through autoreceptors in the aging rat cortex. Aging 22, 191–195 (1983).

192 G. R. Marsh: Psychopharmacological issues, p. 155–220. In: Aging in the 1980s: Psychological Issues. Ed. L. W. Poon. American Psychological Association, Washington, D. C. (1980).

193 J. F. Marshall: Brain function: neural adaptations and recovery from injury. Ann. Rev. Psychol. 35, 277–308 (1984).

194 H. J. Martinez, D. P. Dreyfus, G. M. Jonakait and I. B. Black: Nerve growth factor promotes cholinergic development in brain striatal cultures. Proc. Natl. Acad. Sci. USA 82, 7777–7781 (1985).

195 D. C. Mash, D. D. Flynn and L. T. Potter. Loss of M$_2$ muscarinic receptors in the cerebral cortex in Alzheimer's disease and with experimental cholinergic denervation. Science 228, 1115–117 (1985).

196 M. Massarotti: Ganglioside treatment of alcoholic neurophaties: experimental and clinical aspects. Pharmac. Biochem. Behav. 18, Suppl. 51–54 (1983).

197 M. Massarotti: Ganglioside therapy of peripheral neurophaties: a review of clinical literature, p. 465–479. In: Gangliosides and Neuronal Plasticity. Eds. G. Tettamanti, R. W. Ledeen, K. Sandihoff, Y. Nagai and G. Toffano. Springer-Verlag, Berlin (1986).

198 R. T. Matthews and C. Y. Chiou: Effects of diethylcholine in two animal models of parkinsonism tremors. Eur. J. Pharmacol. 56, 159–162 (1979).

199 G. A. Maw and V. Du Vigneaud: An investigation of the biological behavior of the sulfur analogue of choline. J. Biol. Chem. 176, 1029–1036 (1948).

200 S. E. Mayer: Neurochemical transmission and the autonomic nervous system, p. 56–90. In: The Pharmacological Basis of Therapeutics. Eds. A. G. Gilman, L.S. Goodman and A. Gilman. Macmillan, New York (1980).

201 R. Mayeux, Y. Stern and S. Spanton: Heterogeneity in dementia of the Alzheimer's type: evidence of subgroups. Neurology 35, 453–461 (1985).

202 R. A. McFarland: Psychophysiological studies at high altitudes in the Andes. Part III. Mental and psychosomatic responses during acclimatization. J. Comp. Psychol. 24, 147–188 (1937).

203 E. Melamed, M. Globus and B. Mildworf: Regional cerebral blood flow in dementing disorders, p. 225–232. In: Alzheimer's and Parkinson's. Diseases: Strategies for Research and Development. Eds. A. Fisher, I. Hanin and C. Lachmann. Plenum Press, New York (1986).

204   N. B. Menta, D. L. Musso and H. L. White: Water soluble choline acetyltransferase inhibitors: SAR studies. Eur. J. Med. Chem. 20, 443–446 (1985).

205   M. M. Mesulam, L. Volicer, J. Marquis, E. J. Mufson and R. C. Green: Systematic regional differences in the cholinergic innervation of the primate cerebral cortex: distribution of enzyme activities and some behavioral implications. Ann. Neurol. 19, 144–151 (1986).

206   E. M. Meyer, E. St. Onge and F. T. Crews: Effects of aging on rat cortical presynaptic cholinergic processes. Neurobiol. Aging 5, 315–317 (1984).

207   E. M. Meyer and D. H. Otero: Pharmacological and ionic characterization of the presynaptic and muscarinic modulation of acetylcholine release. J. Neurosci. 5, 1202–1207 (1985).

208   R. Miledi, P. C. Molenaar and R. L. Polak: Evoked release of acetylcholine at the motor endplate, p. 489–497. In: Dynamics of Cholinergic Function. Ed. I. Hanin, Plenum Press, New York (1986).

209   D. Mindus, B. Cronholm, S. E. Levander and D. Schalling: Piracetam-induced improvement of mental performance. Acta Psychiatr. Scand. 54, 150–160 (1976).

210   J. S. Mistry, D. J. Abraham and I. Hanin: Structural analogs of AP64A: synthesis and their effects on high affinity choline transport and QNB binding, p. 639–644. In: Alzheimer's and Parkinson's Diseases. Eds. A. Fisher, I. Hanin and C. Lachman. Plenum Press, New York (1986).

211   W. C. Mobley, J. L. Rutkowski, G. I. Tennekoon, K. Buchanan and M. V. Johnston: Choline acetyltransferase activity in striatum of neonatal rats increased by nerve growth factor. Science 229, 284–287 (1985).

212   R. Y. Moore: Parkinson's disease – a new therapy? N. Engl. J. Med. 316, 872–873 (1987).

213   N. Morel and P. Dreyfus: Association of acetylcholinesterase with the external surface of the presynaptic plasma membrane in *Torpedo* electric organ. Neurochem. Int. 4, 283–288 (1982).

214   C. L. Murray and H. C. Fibiger: Learning and memory deficits after lesions of the nucleus basalis magnocellularis: reversal by physostigmine. Neuroscience 14, 1025–1032 (1985).

215   L. C. Murrin, R. N. De Haven and M. J. Kuhar: On the relationship between (^{3}H)choline uptake activation and (^{3}H)acetylcholine release. J. Neurochem. 29, 681–687 (1977).

215a R. D. Myers: Handbook of Drug and Chemical Stimulation of the Brain: Behavioral, Pharmacological and Physiological Aspects. Van Nostrand Reinhold, New York (1974).

216   K. Nandy: Brain reactive antibodies in aging and senile dementia, p. 503–514. In: Alzheimer's Disease: Senile Dementia and Related Disorders. Eds. R. Katzman, R. D. Terry and K. L. Bick. Raven Press, New York (1978).

217   M. W. Newton, R. D. Crosland and D. J. Jenden: Effects of chronic dietary administration of the cholinergic false precursor n-amino-N, N-dimethylaminoethanol on behavior and cholinergic paramters in rats. Brain Res. 373, 197–204 (1986).

218   M. W. Newton and D. J. Jenden: Mechanism and subcellular distribution of N-amino-N, N-dimethylaminoethanol (N-aminodeanol) in rat striatal synaptosomes. J. Pharmacol. Exp. Ther. 235, 135–146 (1985).

219   M. W. Newton and D. J. Jenden: False transmitters as presynaptic probes for cholinergic mechanisms and function. Trends Pharmacol. Sci. 7, 316–320 (1986).

220   V. J. Nicholson and O. L. Wolthuis: Effect of the acquisition-enhancing drug piracetam on rat cerebral energy metabolism. Comparison with naftidrofuryl and methamphetamine. Biochem. Pharmacol. 25, 2241–2244 (1976).

221 L. Nilsson, A. Nordberg, J. Hardy, P. Wester and B. Winbland: Physostigmine restores ^{3}H-acetylcholine efflux from Alzheimer brain slices to normal level. J. Neural Transm. 67, 275–285 (1986).

222 A. Nordström, P. Alberts, A. Westlind, A. Unden and T. Bartfai: Presynaptic antagonist–postsynaptic agonist at muscarinic cholinergic synapses: N-methyl-N-(1-methyl-4-pyrolidino-2-butynyl)-acetamide. Mol. Pharmacol. 24, 1–5 (1983).

223 O. Nordström and T. Bartfai: Muscarinic autoreceptor regulates acetylcholine release in rat hippocampus: *in vitro* evidence. Acta Physiol. Scand. 108, 347–353 (1980).

224 O. Nordström, A. Unden, U. Grimm, B. Frieder, H. Ladinsky and T. Bartfai: *in vivo* and *in vitro* studies on a muscarinic presynaptic antagonist and postsynaptic agonist: BM-5, p. 405–414. In: Dynamics of Cholinergic Function. Ed. I. Hanin. Plenum Press, New York (1986).

225 B. Oderfeld-Nowak, M. Skup, J. Ulas, M. Jezierska, M. Gradkowska and M. Zaremba: Effect of GM_1 ganglioside treatment on postlesion responses of cholinergic enzymes in rat hippocampus after various partial deafferentation. J. Neurosci. Res. 122, 409–420 (1984).

226 L. Olson, A. Seiger, R. Freedman and B. Hoffer: Chromaffin cells can innervate brain tissue: evidence from intraocular double grafts. Exp. Neurol. 70, 414–426 (1980).

227 D. H. Overstreet and R. W. Russell: Selective breeding for sensitivity to DFP: behavioural effects of cholinergic agonists and antagonists. Psychopharmacology 78, 150–155 (1982).

228 D. H. Overstreet, R. W. Russell, A. D. Crocker and J. C. Gillin: Genetic and pharmacological models of cholinergic supersensitivity and affective disorders. Adv. Biol. Psychiat. (in press).

229 D. H. Overstreet, R. W. Russell, A. D. Crocker and G. D. Schiller: Selective breeding for difference in cholinergic function: pre- and postsynaptic mechanisms involved in sensitivity. Brain Res. 294, 327–332 (1984).

230 S. M. Parsons, D. C. Anderson, B. A. Bahr and G. A. Rogers: A. new pharmacological tool to study acetylcholine storage in nerve terminals, p. 1169–1176. In: Dynamics of Cholinergic Function. Ed. I. Hanin. Plenum Press, New York (1986).

231 J. P. Pede, L. Shimpfessel and R. Crokaert: The action of piracetam on oxidative phosphorylation. Arch. Int. Physiol. Biochem. 79, 1036–1037 (1971).

232 F. Pedeta, L. Giovanelli and G. Pepeu: GM_1 ganglioside facilitates the recovery of high-affinity choline uptake in the cerebral cortex of rats with a lesion of the nucleus basalis magnocellularis. J. Neurosci. Res. 12: 421–427 (1984).

233 M. J. Perlow, W. J. Freed, B. J. Hoffer, A. Seiger, L. Olson and R. J. Wyatt: Brain grafts reduce motor abnormalities produced by destruction of nigrostriatal system. Science 204, 643–647 (1979).

234 A. Perry, J. M. Candy, E. K. Perry, D. Irving, G. Blessed, A. F. Fairbairn and B. E. Tomlinson: Extensive loss of choline acetyltransferase activity is not reflected by neuronal loss in the nucleus of Meynert in Alzheimer's disease. Neurocsi. Lett. 33, 311–315 (1982).

235 E. K. Perry: The cholinergic system in old age and Alzheimer's disease. Age Aging 9, 1–8 (1980a).

236 E. K. Perry, R. H. Perry, B. E. Tomlinson, G. Blessed and P. H. Gibson: Coenzyme A-acetylating enzymes in Alzheimer's disease: possible cholinergic compartment of pyruvate dehydrogenase. Neurocsci. Lett. 18, 105–110 (1980b).

237 E. K. Perry, B. E. Tomlinson, G. Blessed, K. Bergman, P. H. Gibson and R. H. Perry: Correlation of cholinergic abnormalities with senile plaques and mental test scores in senile dementia. Br. Med. J. 2, 1457–1459 (1978).

238 R. H. Perry, G. Blessed, E. K. Perry and B. E. Tomlinson: Histochemical observations on cholinesterase activities in the brains of elderly normal and demented (Alzheimer-type) patients. Age aging 9, 9–16 (1980).

238a M. B. Podlisny, G. Lee and D. J. Selkoe: Gene dosage of the amyloid β precursor protein in Alzheimer's disease. Science 238, 669–671 (1987).

239 R. L. Polak: The stimulating action of atropine on the release of acetylcholine by rat cerebral cortex in vitro. Brit. J. Pharmacol., 14, 600–606 (1971).

240 M. J. Pontecorvo and H. L. Evans: Effects of amiracetam on delayed matching-to-sample performance of monkeys and pigeons. Pharmacol. Biochem. Behav. 22, 745–752 (1985).

241 L. T. Potter, D. O. Flynn, H. E. Hanchett, D. L. Kalinoski, J. Luber-Harod and D. C. Mash: Independent M_1 and M_2 receptors: ligands, autoradiography and functions. TIPS 5 (Suppl.), 22–31 (1983).

242 J. H. Quastel: Source of the acetyl group in acetylcholine, p. 411–430. In: Cholinergic Mechanisms and Psychopharmacology. Ed. D. J. Jenden. Plenum Press, New York (1977).

243 W. G. M. Raaijmakers: High-affinity choline uptake in hippocampal synaptosomes and learning in the rat, p. 373–385. In: Neuronal Plasticity and Memory Formation. Eds. C. A. Marsan and H. Matthies. Raven Press, New York (1982).

244 M. J. Radeke, T. P. Misko, C. Hsu, L. A. Herzenberg and E. M. Shooter: Gene transfer and molecular cloning of the rat nerve growth factor receptor. Nature 325, 593–597 (1987).

245 H. Rahmann: Possible functional role of brain gangliosides in adaptive neuronal processes, p.159–177. In: Neural Transmission, Learning and Memory. Eds. R. Caputto and C. Ajmone Marsan. Raven Press, New York (1983).

246 H. Rahmann (Ed.): Gangliosides and Modulation of Neuronal Functions. Springer, New York (1987).

247 M. Raiteri, R. Leardi and M. Marchi: Heterogeneity of presynaptic muscarinic receptors regulating neurotransmitter release in rat brain. J. Pharmacol. Exp. Ther. 228, 209–214 (1984).

248 M. Raiteri, M. Marchi and A. M. Caviglia: Studies on a possible functional coupling between presynaptic acetylcholinesterase and high-affinity choline uptake in the rat brain. J. Neurochem. 47, 1696–1699 (1986).

249 D. Rasmusson and J. C. Szerb: Cortical acetylcholine release during operant behaviour in rabbits. Life Sci. 16, 683–690 (1975).

250 C. H. Rauca, E. Kammerer and H. Matthies: Choline uptake and permanent memory storage. Pharmac. Biochem. Behav. 13, 21–25 (1980).

251 S. L. Read, G. W. Small and L. F. Jarvik: Dementia syndrome, p. 211–226. In: Psychiatry Update: American Psychiatric Association Annual Review, Vol. 4. Eds. R. E. Hales and A. J. Frances. American Psychiatric Press, Washington, D. C. (1985).

252 B. Reisberg, S. H. Ferris, M. K. Schneck, J. Gorwin, P. Mir, E. Friedman, K. A. Sherman, M. McCarthy and R. T. Bartus: Piracetam in the treatment of cognitive impairment in the elderly. Drug Devel. Res. 2, 475–480 (1982).

253 P. D. Rennert and G. Heinrich: Nerve growth factor mRNA in brain: localization by *in situ* hybridization. Biochem. Biophys. Res. Commun. 138, 813–818 (1986).

254 B. Ringdahl, B. Resul, F. J. Ehlert, D. J. Jenden and R. Dahlbom: The conversion of 2-chloroalkylamine analogues of oxotremorine to aziridinium ions and their interactions with muscarinic receptors in the guinea pig ileum. Mol. Pharmacol. 26, 170–179 (1984).

255 J. Rogers, W. J. Shoemaker and F. E. Bloom: Neurotransmitter alterations in the aging brain, p. 61–101: In: Proof of Therapeutical Effectiveness of Nootropic and Vasoactive Drugs. Springer-Verlag, Berlin (1985).

256 M. R. Rosenzweig: Environmental complexity, cerebral change, and behavior. Amer. Psychologist 21, 321–332 (1966).

257 M. R. Rosenzweig: Experience, memory and the brain. Amer. Psychologist 39, 365–376 (1984).

258 M. N. Rossor, L. L. Iverson, G. P. Reynolds, C. Q. Mountjoy and M. Roth: Neurochemical characteristics of early- and late-onset types of Alzheimer's disease. Br. Med. J. 288, 961–964 (1984).

259 M. Roth: Evidence on the possible heterogeneity of Alzheimer's disease and its bearing on future inquiries into etiology and treatment, p. 251–271. In: The Aging Process: Therapeutic Implications. Eds. R. N. Butler and A. G. Bearn. Raven Press, New York (1985).

260 R. A. Rush: Immunohistochemical localization of endogenous nerve growth factor. Nature 312, 364–167 (1984).

261 R. A. Rush: Nerve growth factor, forty years on. TIBS 11, 493 (1986).

262 R. A. Rush, M. F. Crouch, C. P. Morris and B. J. Gannon: Neural regulation of muscarinic receptors in chick expansor secundariorum muscle. Nature 296, 569–570 (1982).

263 R. W. Russell: The effects of mild anoxia on simple psychomotor and mental skills. J. Exp. Psychol. 38, 178–187 (1948).

264 R. W. Russell: Effects of "biochemical lesions" on behavior. Acta Psychol. 14, 281–294 (1958).

265 R. W. Russell: Biochemical substrates of behavior, p. 185–246. In: Frontiers in Physiological Psychology. Ed. R. W. Russell. Academic Press, New York (1966).

266 R. W. Russell: Genetic and early environmental factors in response variability to psychotropic drugs, p. 97–124. In: Response Variability to Psychotropic Drugs. Ed. W. Janke. Pergamon Press, Oxford (1983).

267 R. W. Russell, R. A. Booth, D. J. Jenden, M. Roch and K. M. Rice: Changes in presynaptic release of acetylcholine during development of tolerance to the anticholinesterase, DFP. J. Neurochem. 45, 293–299 (1985).

268 R. W. Russell, V. G. Carson, R. A. Booth and D. J. Jenden: Mechanisms of tolerance to the anticholinestererase, DFP: acetylcholine levels and dynamics in the rat brain. Neuropharmacology 20, 1197–1201 (1981).

269 R. W. Russell, V. G. Carson, R. S. Jope, R. Booth and J. Macri: Development of behavioral tolerance: a search for subcellular mechanisms. Psychopharmacology 66, 155–158 (1979).

270 R. W. Russell, A. D. Crocker, R. A. Booth and D. J. Jenden: Behavioral and physiological effects of an aziridinium analog of oxotremorine (BM 130). Psychopharmacology 88, 24–32 (1986a).

271 R. W. Russell and D. J. Jenden: Behavioral effects of deanol, of hemicholinium and of their interaction. Pharmac. Biochem. Behav. 15, 285–288 (1981).

272 R. W. Russell and J. Macri: Some behavioral effects of suppressing choline transport by cerebroventricular injection of hemicholinium-3. Pharmac. Biochem. Behav. 8, 399–403 (1978).

273 R. W. Russell and J. Macri: Central cholinergic involvement in behavioral hyperreactivity. Pharmac. Biochem. Behav. 10, 43–48 (1979).

274 R. W. Russell, R. Pechnick and R. S. Jope: Effects of lithium on behavioral reactivity: relation to increases in brain cholinergic activity. Psychopharmacology 73, 120–125 (1981).

275 R. W. Russell, C. A. Smith, R. A. Booth, D. J. Jenden and J. J. Waite: Behavioral and physiological effects associated with changes in muscarinic receptors following administration of an irreversible cholinergic agonist (BM 123). Psychopharmacology 90, 308–315 (1986b).

276   R. J. Rylett, M. J. Ball and E. H. Colhoun: Evidence for high affinity choline transport in synaptosomes prepared from hippocampus and neocortex of patients with Alzheimer's Disease. Brain Res. 289, 169–175 (1983).

277   B. A. Sabel, G. L. Dunbar, B. Fass and D. G. Stein: Gangliosides, neuroplasticity and behavioral recovery after brain damage, p. 481–493. In: Brain Plasticity, Learning and Memory. Eds. B. E. Will, P. Schmidt and J. C. Dalrymple-Alford. Plenum Press, New York (1985).

278   B. A. Sabel, M. D. Slavin and D. G. Stein: GM_1 ganglioside treatment facilitates behavioral recovery from bilateral brain damage. Science 225, 340–342 (1984).

279   C. L. Sahley, S. R. Barry and A. Gelperin: Dietary choline augments associative memory function in Limax maximus. J. Neurobiol. 17, 113–120 (1986).

280   K. Sandberg and J. T. Coyle: Characterization of [^{3}H]hemicholinium-3 binding associated with neuronal choline uptake sites in rat brain membranes. Brain Res. 348, 321–330 (1985).

281   G. D. Schiller: Reduced binding of [^{3}H]-quinuclidinyl benzilate associated with chronically low acetylcholinesterase activity. Life Sci. 24, 1159–1164 (1979).

282   W. Schindler, D. K. Rush and S. Fielding: Nootropic drugs: animal models for studying effects on cognition. Drug Dev. Res. 4, 567–576 (1984).

283   M. K. Schneck: Nootropics, p. 362–368. In: Alzheimer's Disease, Ed. B. Reisberg. Free Press, New York (1983).

284   F. W. Schueler: The mechanism of action of the hemicholiniums. Int. Rev. Neurobiol. 2, 77–97 (1960).

285   M. Segal and N. W. Milgram: Can septal grafting facilitate recovery from physiological and behavioral deficits produced by fornix transections?, p. 627–637. In: Alzheimer's and Parkinson's Diseases. Eds. A. Fisher, I. Hanin and C. Lachman. Plenum Press, New York (1986).

286   M. Seiler, and M. E. Schwab: Specific retrograde transport of nerve growth factor (NGF) from neocortex to nucleus basalis in the rat. Brain Res. 300, 33–39 (1984).

287   F. H. Shaw and G. A. Bentley: Pharmacology of some new anticholinesterases. Aust. J. Exp. Biol. Med. Sci. 31, 573–576 (1953).

288   M. K. Shellenberger and E. F. Domino: Observations in i. v. t. hemicholinium-3 induced EEG seizures. Int. J. Neuropharmacol. 6, 283–291 (1967).

289   D. L. Shelton and L. F. Reicherdt: Studies on the expression of the β-nerve growth factor (NGF) gene in the central nervous system: level and regional distribution of NGF mRNA suggest that NGF functions as a trophic factor for several distinct populations of neurons. Proc. Natl. Acad. Sci. USA 83, 2714–2718 (1986).

290   K. A. Sherman, J. E. Kuster, R. L. Dean, R. T. Bartus and E. Friedman: Presynaptic cholinergic mechanisms in brain of aged rats with memory impairments. Neurobiol. Aging 2, 99–14 (1981).

291   K. F. R. Sheu, Y. Y. Kim, J. P. Blass and M. E. Weksler: An immunochemical study of the pyruvate dehydrogenase deficit in Alzheimer's desease brain. Ann. Neurol 17, 444–449 (1985).

292   K. Shinozuka, T. Maeda and E. Hayashi: Effects of adenosine on ^{45}Ca uptake and [^{3}H]acetylcholine release in synaptosomal preparation from guinea-pig ileum myenteric plexus. Eur. J. Pharmacol. 113, 417–424 (1985).

293   N. R. Sims, D. M. Bowen, S. J. Allen, C. C. T. Smith, D. Neary, D. J. Thomas and A. N. Davison: Presynaptic cholinergic dysfunction in patients with dementia. J. Neurochem. 40, 503–509 (1983).

294   H. S. Singer: Animal models of ganglioside storage diseases, p. 199–222. In: Animal Models of Dementia: A. Synaptic Neurochemical Perspective. Ed. J. T. Coyle. Alan R. Liss, New York (1987).

295 N. Sitram, H. Weingartner, E. D. Caine and J. C. Gillin: Choline: selective enhancement of serial learning and encoding of low imagery words in man. Life Sci. 22, 1555–1560 (1978a).

296 N. Sitram, W. Weingartner and J. C. Gillin: Human serial learning: enhancement with arecoline and choline and impairment with scopolamine. Science 201, 274–276 (1978b).

297 D. Slater: The effects of triethylcholine and hemicholinium-3 on the acetylcholine content of rat brain. Int. J. Neuropharmacol. 7, 421–427 (1968).

298 G. W. Small and S. S. Matsuyama: HLA-A_2 as a possible marker for early onset of Alzheimer disease in men. Neurobiol. Aging 7, 211–214 (1986).

299 C. B. Smith, C. Goochee, S. I. Rapoport and L. Sokoloff: Effects of aging on local rates of cerebral glucose utilization in the rat. Brain 103, 351–365 (1980).

300 Z. Speiser, G. Sharafan, S. Gitter, S. Cohen, B. Gonen and M. Rehavi: Dysfunction of central cholinergic system in hyperkinetic rats, following postnatal anoxia, p. 487–494. In: Alzheimer's and Parkinson's Diseases: Strategies for Research and Development. Eds. A. Fisher, I. Hanin and C. Lachman. Plenum Press, New York (1986).

301 R. W. Sperry: Mechanisms of neural maturation, p. 236–280. In: Handbook of Experimental Psychology. Ed. S. S. Stevens. John Wiley and Sons, New York (1951).

302 R. C. Speth and H. I. Yamamura: Sodium dependent high affinity neuronal choline uptake, p. 129–139. In: Nutrition and the Brain. Eds. A. Barbeau, J. H. Growden and R. J. Wurtman. Raven Press, New York (1979).

303 K. Starke: Presynaptic receptors. Ann. Rev. Pharmacol. Toxicol. 21, 7–30 (1981).

304 D. G. Stein (Ed.): The psychology of aging: problems and perspectives, p. 446. Elsevier/North-Holland, New York (1980).

305 D. G. Stein, R. Labbe, M. J. Attelia and H. A. Rakowsky: Fetal brain tissue transplants reduce visual defects in adult rats with bilateral lesions in the occipital cortex. Behav. Neural Biol. 44, 266–277 (1985).

306 R. P. Stephenson: A modification of receptor theory. Br. J. Pharmacol. 11, 379–393 (1956).

307 G. H. Sterling: Inhibition of high-affinity choline uptake and acetylcholine synthesis by quinuclidinyl and hemicholinium derivatives. J. Neurochem. 46, 1170–1175 (1986).

308 R. G. Struble, L. C. Cork, P. J. Whitehouse and D. L. Price: Cholinergic innervation in neuritic plaques. Science 216, 413–415 (1982).

309 W. K. Summers, K. R. Kaufman, F. L. Altman and J. M. Fischer: THA – a review of the literature and its use in treatment of five overdose patients. Clin. Toxicol. 16, 269–281 (1980).

310 W. K. Summers, L. U. Majovski, G. M. Marsh, K. Tachiki and A. Kling: Oral tetrahydroaminoacridine in treatment of senile dementia, Alzheimer type. N. Engl. J. Med. 315, 1241–1245 (1986).

311 W. K. Summers, J. O. Viesselman, G. M. Marsh and K. Candelora: Use of THA in treatment of Alzheimer-like dementia: pilot study in twelve patients. Biol Psychiat. 16, 145–153 (1981).

312 L. Svennerholm: Chromatographic separation of human brain gangliosides. J. Neurochem. 10, 613–623 (1963).

313 D. F. Swaab and E. Fliers: Clinical strategies in the treatment of Alzheimer's disease, p. 413–427. In: Aging of the Brain and Alzheimer's Disease. Eds. D. F. Swaab, E. Fliers, M. Mirmiran, W. A. Van Gool and F. Van Haaren. Elsevier, New York (1986).

314 D. F. Swaab, E. Fliers, M. Mirmiran, W. A. Van Gool and F. Van Haaren. Aging of the Brain and Alzheimer's Disease, p. 525. Elsevier, New York (1986).

315   J. C. Szerb: Autoregulation of acetylcholine release, p. 293–298. In: Presynaptic Receptors. Eds. S. Z. Langer, K. Starke and M. L. Dubocovich. Pergamon Press, Oxford (1979).

316   K. Tanaka: Experimental study on the effects of low barometric pressures and oxygen deprivation upon the efficiency of mental and physical work. Rep. Aero. Res. Inst. Tokyo Imperial Univ. 3, 128–230 (1928).

317   M. Taniuchi, H. B. Clark and E. M. Johnson, Jr.: Induction of nerve growth factor receptor in Schwann cells after axotomy. Proc. Natl. Acad. Sci. USA 83, 4094–4098 (1986a).

318   M. Taniuchi, J. B. Schweitzer and E. M. Johnson, Jr.: Nerve growth factor receptor molecules in rat brain. Proc. Natl. Acad. Sci. USA 83, 1950–1954 (1986b).

318a R. E. Tanzi, P. H. St. George-Hyslop, J. L. Haines, R. J. Polinsky, L. Nee, J.-F. Foncin, R. L. Neve, A. I. McClatchey, D. M. Conneally and J. F. Gusella. The genetic defect in familial Alzheimer's disease is not tightly linked to the amyloid β-protein gene. Science 329, 156–157 (1987).

319   G. Tettamanti, K. Sandhoff, R. W. Ledeen, Y. Nagai and G. Tofanno (Eds.): Gangliosides and Neuronal Plasticity. Springer, New York (1986).

320   L. J. Thal, P. A. Fuld, D. M. Masur and N. S. Sharpless: Oral physostigmine and lecithin improve memory in Alzheimer disease. Ann. Neurol. 13, 491–496 (1983).

321   S. Thesleff: Reptile toxins and neurotransmitter release, p. 19–25. In: Neurotoxins: Fundamental and Clinical Advances. Eds. I. W. Chubb and L. B. Geffen. Adelaide University Press, Adelaide, South Australia (1979).

322   H. Thoenen and D. Edgar: Neurotrophic factors. Science 229, 238–242 (1985).

323   Time Magazine. Genetic clues: a DNA test for Alzheimer's. 2 March, 1987.

324   B. A. Trommer, D. E. Schmidt and L. Wecker: Exogenous choline enhances the synthesis of acetylcholine only under conditions of increased neuronal activity. J. Neurochem. 39, 1704–1709 (1982).

325   S. Tucek: Acetylcoenzyme A and the synthesis of acetylcholine in neurons: review of recent progress. Gen. Physiol. Biophys. 2, 313–324 (1983).

326   S. Tucek: Problems in the organization and control of acetylcholine synthesis in brain neurons. Prog. Biophys. Mol. Biol. 44, 1–46 (1984).

327   S. Tucek: Regulation of acetylcholine synthesis in the brain. J. Neurochem. 44, 11–24 (1985).

328   J. Ulrich: Alzheimer changes in nondemented patients younger than sixty-five: possible early stages of Alzheimer's disease and senile dementia of Alzheimer type. Ann. Neurol. 17, 273–277 (1985).

329   K. Unsicker, B. Krisch, U. Otten and H. Thoenen: Nerve growth factor-indiced fiber outgrowth from isolated rat adrenal chromaffin cells: impairment by glucocorticoids. Proc. Natl. Acad. Sci. USA 75, 3498–3502 (1978).

330   K. Unsicker, S. D. Skaper and S. Varon: Developmental changes in the responses of rat chromaffin cells to neuronotrophic and neurite-promoting factors. Develop. Biol. 111, 425–433 (1985).

331   E. Valenstein: Age-related changes in the human central nervous system, p. 87–106. In: Aging: Communication Processes and Disorders. Eds. D. S. Beasley and G. A. Davis. Grune and Stratton, New York (1981).

332   L. Valzelli: The "isolation syndrome" in mice. Psychopharmacologia 31, 305–320 (1973).

333   L. Valzelli, S. Bernasconi and A. Sala: Piracetam activity may differ according to the age of the recipient mouse. Int. Pharmacopsychiatry 15, 150–156 (1980).

334   F. J. Van Der Staay, W. G. M. Raaijmakers and T. H. Collijn: Spatial discrimination and passive avoidance behavior in the rat: age-related changes and modulation by chronic dietary choline enrichment, p. 603–608. In: Alz-

heimer's and Parkinson's Diseases: Strategies for Research and Development. Eds. A. Fisher, I. Hanin and C. Lachman. Plenum Press, New York (1986).

335   S. S. Varon and R. P. Bunge: Trophic mechanisms in the peripheral nervous system. Ann. Rev. Neurocsci. 1, 327–361 (1978).

336   E. S. Vizi: Interaction between adrenergic and cholinergic systems: presynaptic inhibitory effect of noradrenaline on acetylcholine release. J. Neural. Transm. Suppl. 11, 61–78 (1974).

337   E. S. Vizi: Presynaptic modulation of neurochemical transmission. Prog. Neurobiol. 12, 181–290 (1979).

338   L. Wecker: Neurochemical effects of choline supplementation. Can. J. Physiol. Pharmacol. 64, 329–333 (1986).

339   P. A. Weiss: Nerve patterns: the mechanics of nerve growth. Third Growth Symposium 5, 163–203 (1941).

340   G. L. Wenk, D. Helpler and D. Olton: Behavior alters the uptake of [^{3}H]choline into acetylcholinergic neurons of the nucleus basalis magnocellularis and medial septal area. Behav. Brain Res. 13, 129–138 (1984).

341   G. L. Wenk and D. S. Olton: Recovery of neocortical choline acetyltransferase activity following ibotenic acid injection into the nucleus basalis of Meynert in rats. Brain Res. 293, 184–186 (1984).

342   G. Wenk, J. Sweeney, D. Hughey, J. Carson and D. Olton: Cholinergic function and memory: extensive inhibition of choline acetyltransferase fails to impair radial maze performance in rats. Pharmacol. Biochem. Behav. 25, 521–526 (1986).

343   P. J. Whitehouse: Neuronal loss and neurotransmitter receptor alterations in Alzheimer's disease, p. 85–94. In: Alzheimer's and Parkinson's Diseases. Eds. A. Fisher, I. Hanin and C. Lachman. Plenum Press, New York (1986).

344   P. J. Whitehouse, D. L. Price, R. G. Struble, A. W. Clark, J. T. Coyle and M. R. DeLong: Alzheimer's disease and senile dementia: loss of neurons in the basal forebrain. Science 215, 1237–1239 (1982).

345   V. P. Whittaker and Y. A. Luqmani: False transmitters in the cholinergic system: implications for the vesicle theory of transmitter storage and release. Gen. Pharmacol. 11, 7–14 (1980).

346   G. K. Wilcock and M. M. Esiri: Plaques, tangles and dementia: a quantitative study. J. Neurol. Sci. 56, 343–356 (1982a).

347   G. K. Wilcock, M. M. Esiri, D. M. Bowen and C. C. T. Smith: Alzheimer's disase: correlation of cortical choline acetyltransferase activity with the severity of dementia and histological abnormalities. J. Neurol. Sci. 57, 407–417 (1982b).

348   B. Will and F. Hefti: Behavioural and neurochemical effects of chronic intraventricular injections of nerve growth factor in adult rats with fimbria lesions. Behav. Brain Res. 17, 17–24 (1985).

349   L. R. Williams, S. Varon, G. M. Peterson, K. Wictorin, W. Fischer, A. Björklund and F. H. Gage: Continuous infusion of nerve growth factor prevents basal forebrain neuronal death after fimbria fornix transection. Proc. Natl. Acad. Sci. USA 83, 9231–9235 (186).

350   M. Wojcik, J. Ulas and B. Oderfeld-Nowak: The stimulatory effect of ganglioside injections on the recovery of acetyltransferase and acetylcholinesterase activities in the hippocampus of the rat after septal lesions. Neuroscience 7, 495–499 (1982).

351   O. L. Wolthuis: Behavioural effects of etiracetam in rats. Pharmacol. Biochem. Behav. 15, 247–255 (1981).

352   R. J. Wurtman, J. K. Blusztajn and J.-C. Maire: "Autocannibalism" of choline-containing membrane phospholipids in the pathogenesis of Alzheimer's disease – a hypothesis. Neurochem. Int. 7, 369–372 (1985).

353   R. J. Wurtman, J. K. Blausztajn and J.-C. Maire: The "autocannibalism" of choline-containing membrane phospholipids in the pathogenesis of Alzheimer's disease, p. 69–73. In: Alzheimer's and Parkinson's Diseases: Strategies for Research and Development. Eds. A. Fisher, I. Hanin and C. Lachman. Plenum Press, New York (1986).

354   B. A. Yankner and E. M. Shooter: The biology and mechanism of action of nerve growth factor. Annu. Rev. Biochem. 51, 845–868 (1982).

355   S. H. Zeisel: Dietary choline: biochemistry, physiology and pharmacology. Annu. Rev. Nutr. 1, 95–121 (1981).

Surface interaction between bacteria and phagocytic cells

By L. Öhman, G. Maluszynska, K.-E. Magnusson and
O. Stendahl
Department of Medical Microbiology, Faculty of Health Sciences,
Linköping University, S-581 85 Linköping, Sweden

1   Introduction

The interactions between bacteria and mammalian cells are important
events both for harmless and necessary colonization of mucosal membranes, and in the pathogenesis of infections. Colonization, invasion
and tissue damage depend both on bacterial virulence factors and de-
fense mechanisms of the challenged host. This encounter involves not
only specific ligand-receptor interactions determining the tropism of
the infection, but also non-specific, physicochemical factors.
The phagocytic process plays an important role both during colonization and elimination of bacteria. Although executed by many types of
cells, only professional phagocytes, like granulocytes and macrophages, recognize and ingest a diversity of prey with great efficiency. It
is evident that these phagocytic cells have a protective role in eliminating harmful invaders, debris and other foreign material.
A prominent feature of the surface interaction between the phagocyte
and the bacterium is its discriminatory character. Whether a bacterium
is accepted or not by the phagocyte, permitting attachment and subsequent ingestion, is determined by the surface properties of both the
prey and the phagocytic cell [1]. Over the last years much knowledge
has accumulated about the mechanisms of bacteria-phagocyte interaction. In several systems, identification and characterization at the molecular level of bacterial ligands recognizing specific "receptor" molecules and the subsequent ligand-receptor interaction [2, 3, 4], have
greatly increased our understanding of host–parasite interactions.
This review will discuss the interplay between phagocytic cells and
bacteria, focusing on the specific and non-specific recognition mechanisms governing the uptake of certain well-characterized bacteria,
e.g. enterobacteriae, and the subsequent activation of the phagocyte.
Less attention will be paid to the intracellular mechanisms leading to
microbial killing.

2   The phagocytic process

The phagocytic process is initially a surface phenomenon that can be
described in physicochemical terms [5, 6].
However, when a phagocytic cell encounters a foreign particle like a
bacterium, a multi-step process is initiated. First, the particle attaches
to the cell membrane via an energy-independent process, where non-

specific and specific ligand-receptor interactions are operative. Then, if the bacterium is properly recognized, several effector mechanisms, including increased adhesiveness of the phagocytes, stimulation of the respiratory burst, phagocytosis, degranulation and generation of inflammatory mediators, will be elicited via series of biochemical transduction signals.

When a particle or a soluble ligand binds to the cell surface, a signal for action on the part of the cell is elicited. How this signal is propagated has recently been clarified in several cell systems [7], such as neurotransmission, hormone action and phagocytic activation. The signaling system involves the phospholipid metabolism in the cell membrane and Ca^{++} mobilization. Via a GTP-binding protein, a phosphodiesterase (phospholipase C) is activated that catalyzes the breakdown of phosphorylated phosphatidylinositol (PIP_2) to inositoltrisphosphates (IP_3) and diacylglycerol (DAG). IP_3 isomers raise the level of intracellular calcium by direct action on intracellular calcium stores, and DAG interacts with protein kinase C, causing its activation and phosphorylation of certain regulatory proteins in the cell [8]. Protein kinase C is translocated from the cytosol to the membrane [9] and activates NADPH-oxidase, the key enzyme in the respiratory burst [10]. IP_3 in turn can be metabolized to IP_4 by a specific kinase, which can open plasma membrane channels for influx of Ca^{++} [11]. This system of signal transduction and amplification in the phagocytic cell is involved in such various functions of the cell as chemotaxis, phagocytosis, degranulation and production of free oxygen radicals and other inflammatory mediators. However, the great diversity and complexity of the system offers the possibility for different cellular responses, depending on the nature of the specific ligand-receptor interaction and the involvement of different physicochemical surface properties. For phagocytic ingestion to occur, sequential ligand-receptor interactions allow the pseudopods to enclose the attached particle [12]. Inability to ingest a particle may thus depend on insufficient number of interacting entities. Furthermore, several bacterial toxins may interfere or modulate the signal-transduction process of the cell [13]. Cholera and pertussis toxin interfere with the GTP-binding proteins, whereas botulinum toxin interferes with Ca^{++} mobilization and actin polymerization [14]. Both processes are necessary for phagocytic uptake and degranulation. Bacterial proteases may also directly inactivate certain surface receptors. Direct or indirect interference with the receptor expression and signal-transduction may in fact be important virulence mechanisms for certain bacteria [13], but much more has to be learned about how virulent bacteria actively interfere with the cellular response of phagocytic cells.

3   Recognition

The interaction between a bacterium and the phagocyte is facilitated by specific opsonins, e.g. IgG and the complement factors C3b and C3bi, lectins and hydrophobic properties of the cell surfaces, all of which recognize surface molecules and specific receptors on the phagocytic cell (Fig. 1). Phagocytic recognition and subsequent inges-

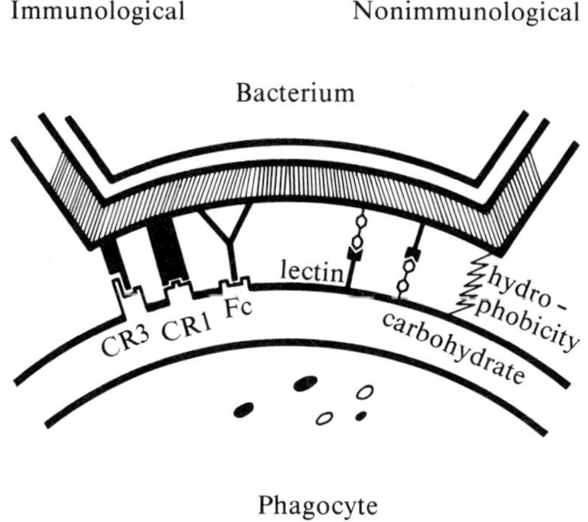

Figure 1
Immunological and non-immunological surface interactions between phagocytic cells and bacteria.

tion can furthermore be promoted by increasing both the number of receptors and their efficiency. Several environmental factors influence this process. Fibronectin, although not operating as an opsonin, will enhance phagocytosis by "activating" the C3b/bi receptors (= CR1/CR3) [15]. Recently it has also been shown that tumor necrosis factor (TNF), produced by macrophages after endotoxin stimulation, enhances phagocytosis [16]. Chemotactic factors, C5a, leukotriene B4 and the tripeptide f-methionyl-leucyl-phenylalanine (fMLP), increase the number of CR1 and CR3, thus enhancing phagocytosis [17]. As is discussed in this review, it is becoming increasingly clear, that the outcome of the phagocytic recognition, leading to eradication of the invasive microorganism or inflammation and tissue damage, depends on (I) physicochemical surface properties of the interacting entities, (II) the specific ligand-receptor interaction, and (III) the expression and specific activities of the involved receptors.

4   Anti-recognition properties of the bacteria

Virulent bacteria have the capacity to avoid phagocytic recognition or intracellular killing. Capsular antigens (K-antigens, slime, other polysaccharides with or without proteins) [18, 19 and Maluszynska, unpubl. obs.] and lipopolysaccharides (smooth LPS) confer negative charge and hydrophilicity to the enterobacteriae, e.g. *Salmonella* and *Escherichia coli* [20, 21]. Both these properties will impair phagocytic recognition, but will also influence the binding and expression of opsonins on the bacterial surfaces [22]. Also the growth conditions will drastically influence the surface properties of the bacteria. When grown under anaerobic conditions *Salmonella* and *E.coli* bacteria express more hydrophobic surface properties, thus conferring increased phagocytic recognition [23]. In general, hydrophilic properties are advantageous for bacteria when avoiding phagocytosis. The bacteria will favor the water phase and avoid contact with the more hydrophobic mammalian cell surfaces. However, when interaction with its target cell is beneficial for colonization, e.g. of mucosal cells in the urinary tract, hydrophilicity may be a disadvantage.

Other bacterial surface structures that impair the contact with phagocytic cells are the M-protein and hyaluronic acid of *Streptococcus pyogenes*, pili of *Neisseria gonorrhoeae*, and the capsule of *Streptococcus pneumoniae*. Furthermore, *Neisseria meningitidis, Klebsiella pneumoniae, Haemophilus influenzae, Yersinia pestis, Bacillus anthracis, Campylobacter fetus, Cryptococcus neoformans, Pseudomonas aeruginosa, Pasteurella multocida* and *Bacteroides fragilis* resist attachment by impairing opsonic activity. The Fc-binding properties of protein A on *Staphylococcus aureus* may also exhibit antiopsonic properties, although no good correlation is apparent between amount of protein A and virulence. Other microorganisms are able to mask their own antigens by acquisition of host antigens, e.g. Schistosoma worms assume the erythrocyte antigens of the host, thereby avoiding recognition of the immune system [24–27].

The Tamm-Horsfall glycoprotein in urine contains N-linked oligomannose units that binds to type 1-fimbriated *Escherichia coli* and thereby markedly reduces the lectin-sugar interaction with the polymorphonuclear leukocytes [28, 29]. Group A streptococci can bind fibrinogen, thus forming an anti-phagocytic coat [30].

Most of these investigations have been performed with bacteria in suspension. It is, however, becoming increasingly clear that the properties of the supporting matrix surfaces or foreign body surfaces to which the bacteria or the phagocytic cells adhere are of great significance for the outcome of the bacteria–phagocyte interaction [15, 31]. When PMNL and macrophages adhere to fibronectin or laminin-coated surfaces they increase the expression of receptors for complement (CR1, CR3) [15]. How other ligand–receptor interactions mediated by lectin-sugar binding are affected is less clear, but it is evident that metabolic activation [31], chemotactic stimulation and degranulation are modulated after adherence to biological surfaces.

Recently, much interest has been focused on the binding properties of certain bacteria to extracellular matrix molecules such as fibronectin and laminin. Specific receptors have been identified on *Staphylococcus aureus* for both these molecules [32], but also *Streptococcus pyogenes* and certain other strains of streptococci bind to fibronectin [33], putatively in a liptoteichoic acid-inhibitable way. When colonizing different surfaces, the resistance to phagocytosis may be enhanced. Vaudaux et al. [31] showed that attachment of *Staphylococcus aureus* to polymethylmethacrylate increased its resitance to phagocytosis. By contrast attachment to fibronectin-coated surfaces enhanced bacterial elimination [Jaconi & Herrmann, pers. comm.]. It has furthermore been suggested that the opsonic requirements are reduced when adherent bacteria are being phagocytosed [34].

Little attention has so far been paid to the role of the bacterial glycocalyx in the resistance to phagocytosis [35]. It is, however, evident that phagocyte–bacteria interaction has to be studied in more physiological settings, and different models have recently been developed for this purpose [36].

5   Specific "non-immunological" recognition

Non-immunological recognition implies that the bacteria–phagocyte interaction takes place in the absence of specific opsonins, such as IgG and the complement factors C3b and C3bi. This is a possible mechanism for the host to eliminate invading microorganisms during the pre-immune state and at sites where serum proteins are less abundant, e.g. the urinary and respiratory tracts and in the cerebrospinal fluid. The best studied system for this specific non-immunological phagocytosis is the interaction between the mannose-speficic type 1 fimbriae on *Escherichia coli* and phagocytic cells. Type 1 fimbriae is a lectin (= sugar-binding protein) with high affinity for a complementary sugar residue, D-mannose [37]. Almost all *Escherichia coli* as well as many other species of enterobacteria possess the genetic ability to express these and other types of fimbriae on their surface. Whether the bacterial cells are fimbriated or not is under transcriptional control by at least four genes and occurs spontaneously with varying frequency, so-called phase variation. The fimbrial expression is also affected by environmental factors, such as culture time and growth conditions [38–41]. The presence of type 1 fimbriae is associated with increased bacterial adherence to several mammalian cells, including neutrophils and macrophages [42–45]. A good correlation exists between the extent of fimbrial expression and carbohydrate specific adherence, indicating that physicochemical surface properties, e.g. charge and hydrophobicity may have little influence on the fimbriae-mediated attachment [46, 47]. The subsequent interplay between attached bacteria and phagocytic cells is, however, strongly affected by general physicochemical surface properties [45].

Furthermore, the fimbriae have hydrophobic properties and as a protein appendage with a small radius, it can supercede the waterfilm surrounding the host cell or over a mucous membrane, through a hydration mechanism [48]. Recently, Rodriguez-Ortega et al. (1987) isolated a receptor on PMNL for this mannose-specific type 1 fimbriae. The receptor is a mannose-containing glycoprotein and may in fact be identical with the α and β subunits of the complement receptors. It is at present not clear whether type 1 fimbriae bind to the same receptor as C3b. But it is becoming evident that certain adhesion-promoting molecules on leukocytes (LFA-1, Mac-1 and p150,95) belong to a supergene family of cellular adhesion protein, involving C3b, fibronectin and laminin [50]. Significant homology between the molecules does not necessarily mean that the ligands bind to the same receptor, but is rather suggestive of a common origin.

Specific ligand-receptor interactions do not ultimately lead to ingestion and intracellular killing. Almost without exception however will ligand-receptor interactions trigger metabolic activation with release of oxygen metabolites (O_2^-, H_2O_2, $OH\cdot$), granule enzymes and inflammatory mediators from the phagocytes. In fact, a sustained response is achieved when ingestion does not occur [45].

The mannose-sensitive type 1 fimbriae-mediated attachment to granulocytes may lead to ingestion, but this depends on the underlying surface properties of the bacteria. Thus, a rough, hydrophobic surface facilitates phagocytosis, whereas a smooth hydrophilic surface will counteract ingestion [45, 51]. Indeed, uropathogenic *Escherichia coli* with a smooth lipopolysaccharide, acidic capsular antigen (K antigen) and type 1 fimbriae adhere avidly to granulocytes but resist ingestion, thereby causing pronounced and sustained metabolic activation and extracellular release of granule enzymes and oxygen metabolites. Avirulent *E. coli* strains with rough LPS and type 1 fimbriae were, however,

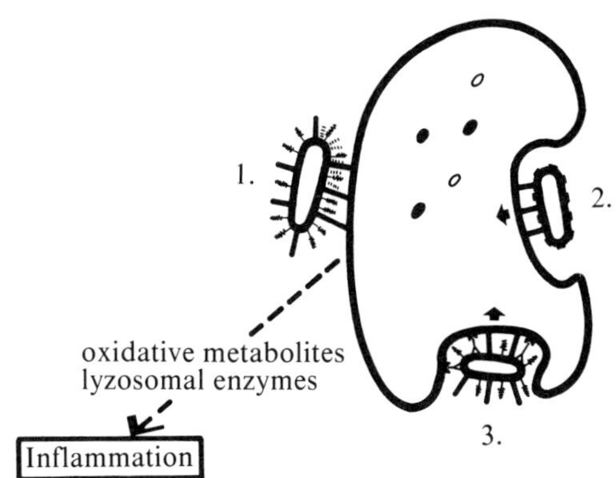

Figure 2

A proposed model of interaction between different type 1 fimbriae-bearing *Escherichia coli* strains and human polymorphonuclear leukocytes:

1. A strain with hydrophilic properties (S-LPS and K-antigen) resisting phagocytosis, but will stimulate the PMNL metabolically with release of potentially toxic oxidative metabolites.

2. A strain with hydrophobic properties (R-LPS) is avidly ingested and killed, with less release of oxidative metabolites.

3. An opsonized, hydrophilic strain is avidly ingested and killed, with less release of oxidative metabolites.

ingested effectively, causing little extracellular release of potentially tissue-damaging metabolites (Lock et al., unpubl. obs.) (Fig. 2).

Both intra- and extracellular microbicidal mechanisms are operative in granulocytes and macrophages [52]. Intracellular inactivation appears more effective since it involves several oxygen-dependent and oxygen-independent microbicidal systems whereas the extracellular killing is oxygen-dependent, and therefore less effective under anaerobic conditions [52]. Moreover, extracellularly located type 1 fimbriae-bearing *E. coli* may activate the granulocyte to release several inflammatory mediators and tissue-damaging products, without being effectively killed. Preliminary results from our own laboratory support these ideas. Attached but not ingested *E. coli* bacteria resist granulocyte-mediated killing under anaerobic conditions more than do ingested *E. coli* strains with type 1 fimbriae. Furthermore, Goetz et al. [53, 54] recently showed that unopsonized *E. coli* with type 1 fimbriae resisted granulocyte-mediated killing, and that this was associated with decreased intraphagosomal myeloperoxidase activity. It can thus be concluded that the abundant mannose-sensitive type 1 fimbriae-mediated interaction serves as a relevant model for specific non-immunological recognition in granulocytes and macrophages. In *E. coli* it is noticeable, however, that the αGal 1,4βGal-specific P fimbriae present on uropathogenic strains do not affect interaction with phagocytic cells, since these cells lack or do not expose the proper receptor molecule (αGal 1,4βGal) [55].

Also reversed ligand-receptor interactions have been proposed, viz. lectin-like, trypsin-sensitive structures on macrophages and granulocytes recognizing different carbohydrate residues (glucose, mannose, N-acetylglucosamine, galactose) on LPS or other surface structures [56, 57].

Outer membrane protein II (OMP II) and leukocyte association factor (LAF) of *Neisserie gonorrhoeae* enhance the association with phagocytic cells [58, 59], and one of the OMP II proteins may be identical with LAF. Recently, it has also been shown that lectin-sugar interaction between certain *Actinomyces* species with type 2 fimbriae specific for β-galactosides, and phagocytic cells can occur [60]. Speert et al. [61] have recently shown that in an unopsonized system, *Pseudomonas aeruginosa* is associated to and phagocytosed by granulocytes, and that this correlated with piliation and hydrophobicity. A non-immunological interaction may in certain settings at the pre-immune state be beneficial for the host. However, Vandenbrouche-Grauls [62] recently showed that opsonized *Staphylococcus aureus* interacting with granulocytes showed less endothelial damage than unopsonized ones. Thus effective phagocytosis may minimize extracellular degranulation and metabolic activation, which minimizes tissue damage.

6   General physicochemical recognition

From studying phagocytic recognition of well-defined particles in the absence of antibodies and complement, it is clear that also more general physicochemical properties of the interacting surfaces influence recognition and the outcome of the interaction; the main properties being surface charge and hydrophobic interaction [63, 64]. Physicochemical analysis of different bacteria showed that increased hydrophobic interaction and decreased negative charge correlated with enhanced phagocytic recognition. This has been shown for smooth and rough *Salmonella typhimurium* bacteria, where the smooth lipolysaccharide increases hydrophilicity and rough LPS hydrophobic surface properties of the bacteria [65, 66] but also for different Salmonella serotypes [21]. A similar pattern was found for LPS mutations in *E. coli,* but in this case the overall hydrophobic interaction was reduced, probably due to the presence of acidic carbohydrate material (K antigen) as well [64]. Thus virulent strains of *E. coli* exposing smooth LPS and acidic K antigen are hydrophilic and negatively charged, thereby resisting phagocytosis [67]. On the other hand, *Neisseria gonorrhoeae* bacteria are more liable to hydrophobic interaction and therefore more easily phagocytosed [68].
Binding of IgG, complement or other serum proteins to the bacterial surface do not only evoke specific ligand-recptor interactions. When Salmonella bacteria were sensitized with anti-Salmonella IgG and complement, hydrophobic interaction was enhanced in parallel with increased phagocytosis [69]. In contrast, non-opsonizing secretory IgA antibodies binding to gram-negative bacteria induced increased hydrophilicity and resistance to phagocytosis. In fact, secretory IgA may counteract the effect of opsonins such as IgG and complement [70, 71]. Thus by affecting the general physicochemical surface properties of the bacterium and the phagocyte, the outcome of their interaction may be determined. It is, however, evident that not one but several recognition systems work in concert during phagocytosis.

7   **Immunological recognition**

To enhance phagocytic uptake it has long been appreciated that different serum proteins, primarily IgG and complement factors C3b and C3bi, can bind to the bacterial surface. These molecules interact with specific receptors on the neutrophils and macrophages. The CR1 receptor molecule extends out up to 100 nm and binds C3b [72]. The CR3 receptor contains two chains, α and β, where the latter is common to other cell surface molecules, LFA-1 and p 150,95. The CR3 expresses two epitopes; one binding to C3bi, the other to the glucosamines of lipid A and β-glucan of zymosan [73]. The Fc receptors (FcR), on the other hand, show less heterogeneity and bind primarily antigen-bound or aggregated IgG1 and IgG3 [74].

Although both sets of receptors, the CR1/CR3 and the FcR, promote attachment, the subsequent cellular response, metabolic activation, ingestion and degranulation, may vary. Binding to the FcR causes metabolic activation and ingestion, whereas binding to the CR1 und CR3 receptors do not ultimately lead to ingestion [75, 76]. It has recently been understood that these differences between FcR and CR1/CR3 may relate to different transduction signals [77], but also to the fact that the CR1 and CR3, but not the FcR, can be upregulated both with respect to number and efficiency [17].

8   **Cooperation between non-immunological and immunological recognition**

In most host-parasite interactions immunological as well as non-immunological recognition mechanisms are operative simultaneously. However, in most systems little attention have been paid to the non-immunological interaction when opsonins are present. When type 1 fimbriated *E. coli* bacteria are exposed to serum, the ingestion of the attached bacteria is greatly enhanced, probably via C3b/bi opsonization. D-mannose still partly inhibited the interaction, indicating that complement factors C3b/bi and lectin cooperate, perhaps by binding to the same receptor [49]. It is, however, difficult to evaluate the specific role of complement in relation to mannose-specific lectins, since there appears to be a direct lectin-complement interaction on the bacterial surface and since complement-coated type 1 fimbriae-bearing *E. coli* autoaggregate (Öhman et al., unpubl. obs.).

With the use of monoclonal antibodies to the fimbriae and the capsular antigen of *E. coli,* Söderström & Öhman [78] were able to probe the cooperative effect of lectin and IgG. Addition of type 1 fimbriae antibodies increased both the association and the subsequent ingestion of the bacteria and metabolic activation of the granulocytes. This antibody-mediated stimulation was still, however, inhibited by D-mannose. One possible explanation for the enhancement of the bacterial interaction by type 1 fimbriae could be aggregation of the fimbriae. It would then be expected that D-mannose should inhibit the interaction also at high antibody concentration. The fact that the inhibitory effect of D-mannose decreased with increasing antibody concentration rather implied that the antibody interacted with the Fc receptor which cooperated with the mannose-specific lectin. In this interplay, the lectin promoted attachment, and the IgG stimulated ingestion. This interpretation was further supported by the fact that the enhanced ingestion induced by anti-capsular IgG was also completely inhibited by D-mannose.

9   Conclusions

Immunoglobulins-FcR, complement C3b/bi-CR1/CR3, lectin-sugar residues and hydrophobicity mediate the recognition and initial contact between microorganisms and phagocytic cells. This offers a broad arsenal for the host to combat microbial challenges. It is, however, evident that interactions, particularly in non-opsonic systems, do not always lead to ingestion and killing, although the phagocytic cell is activated. The cellular response is not an all or none phenomenon, but depends on the stimuli, receptors, and functional state of the cell. The main task for the phagocytic cells is to eradicate the invading microbes, either by extracellular or intracellular killing. If, however, the bacteria interacting with the phagocytes resist ingestion and killing, a sustained activation with release and leakage of toxic oxygen metabolites (O_2^-, H_2O_2), lysosomal enzymes [79] and inflammatory mediators such as leukotrienes and histamine [80] will occur. The ability of bacteria to attach to phagocytic cells may not ultimately be beneficial to the host, but could if the bacteria resist ingestion, in analogy with immune complexes, potentiate the inflammatory response and cause increased tissue destruction. Identification and appreciation of different adhesion molecules on bacteria and host cells, and how these interact with different matrix molecules, will in several ways increase the understanding of host-parasite interaction and homeostasis. Increasing the immunological recognition with specific opsonins and cell activation is one way to combat infection, another is to modulate bacterial adhesion either by specific analogues or by general physicochemical means.

Acknowledgements

This work was supported by grants from The Swedish Medical Research Council (2183, 5968 and 6251), Östergötlands Läns Landsting and King Gustav V's 80 year Foundation. We thank Ingegärd Wranne for expert secretarial and editing assistance.

References

1 O. Stendahl, C. Dahlgren, M. Edebo and L. Öhman: Recognition mechanisms in mammalian phagocytosis. Monographs in Allergy, vol. 17, pp. 12–27. Karger, Basel 1981.
2 F. Lindberg, B. Lund, L. Johansson and S. Normark: Localization of the receptor-binding protein adhesin at the tip of the bacterial pilus. Nature *328*: 84 (1987).
3 N. Sharon: Bacterial lectins, cell-cell recognition and infectious disease. FEB *217*: (2) 145 (1987).
4 C. Svanborg-Edén, L. Hagberg, R. Hull, S. Hull, K.-E. Magnusson and L. Öhman: Bacterial virulence versus host resistance in the urinary tracts of mice. Infect. Immun. *55*:5, 1224 (1987).
5 Elsbach, P.: Cell surfaces changes in phagocytosis. In: The synthesis, assembly, and turnover of cell surface components. Eds G. Poste and G. L. Nicolson. Elsevier/North Holland Biomedical Press, Amsterdam 1977.
6 C. J. Van Oss: Phagocytosis as a surface phenomenon. A. Rev. Microbiol. *32*:19–39 (1978).
7 B. D. Gumperts: Calcium shares the limelight in stimulus-secretion coupling. Trends Biochem. Sci. II: 290 (1986).
8 R. Snyderman, C. D. Smith and M. W. Verghese: Model for leukocyte regulation by chemoattractant receptors: Role of a guanine nucleotide regulatory protein and P I metabolism. J. Leukocyte Biol. *40*:785 (1986).
9 N. O. Christiansen, N. A. Peterslund, C. S. Larsen and V. Esmann: Evidence that bacteria induce translocation of protein kinase C activity in human polymorphonuclear leukocytes. Biochem. biophys. Res. Commun. *147*:2, 787 (1987).
10 A. I. Tauber: Protein kinase C and the activation of the human neutrophil NADPH-oxidase. Blood *69*:3, 711 (1987).
11 R. F. Irvine and R. M. Moor: Microinjection of inositol 1,3,4,5-tetraphosphate activates sear urchin eggs by a mechanism dependent on external Ca^{2+}. Biochem. J. *240*:917 (1986).
12 F. M. Griffin, J. A. Griffin, J. E. Leider and S. C. Silverstein: Studies on the mechanism of phagocytosis. I. Requirements for circumferential attachment of particle-bound ligands to specific receptors on the macrophage plasma membrane. J. exp. Med. *142*:1263 (1975).

13   K.-H. Krause and D. P. Lew: Bacterial toxins and neutrophil activation. Sem. Hematol., in press 1987.

14   F. A. Al-Mohanna, I. Ohishi and M. B. Hallett: Botulinum C2 toxin potentiates activation of the neutrophil oxidase. FEBS *219*:1, 40 (1987).

15   E. J. Brown: The role of extracellular matrix proteins in the control of phgocytosis. J. Leukocyte Biol. *39*:579 (1986).

16   J. W. Larrick, D. Graham, K. Toy, S. L. Lin, G. Senyk and B. M. Fendly: Recombinant tumor necrosis factor causes activation of human granulocytes. Blood *69*:640 (1987).

17   J. J. O'Shee, E. J. Brown, B. E. Seligmann, J. A. Metcalf, M. M. Frank and J. I. Gallin: Evidence for distinct intracellular pools of receptors for C3b and C3bi in human neutrophils. J. Immun. *134*:2580 (1985).

18   S. Schwarzmann and J. R. Boring III (1971). I Antiphagocytic effect of slime from mucoid strain of Pseudomonas aeruginosa. Infect. Immun. *3*(6), 762 (1971).

19   G. M. Johnson, D. A. Lee, W. E. Regelmann, E. D. Gray, G. Peters and P. Quie. Interference with granulocyte function by Stafylococcus epidermidis slime. Infect. Immun. *54*(1), 13 (1986).

20   K.-E. Magnusson and O. Stendahl: Partitioning of bacteria, virus and phage. In: Partitioning in aqueous two-phase systems, p. 415. Ed. H. Walter, Academic Press, New York 1985.

21   H. Jiang, K.-E. Magnusson, O. Stendahl and L. Edebo: Physicochemical surface properties and phagocytosis by PMN of different serotypes of Salmonella. J. Gen. Microbiol. *129*:3075 (1983).

22   L. Edebo, M. Edebo, E. Kihlström, K.-E. Magnusson, L. Öhman, T. Skogh and O. Stendahl: Role of antibacterial antibodies in host protection by antiseptic and aseptic mechanisms. In: Antibodies: Protective, destructive and regulatory role, p. 82. Eds Milgrom, Abeyounis and Albini. 9th Int. Convoc. Immunol. Amherst, N. Y., 1984. Karger, Basel 1985.

23   G. Maluszynska, K.-E. Magnusson and O. Stendahl: Interaction between human polymorphonuclear leukocytes (PMNL) and bacteria cultivated in aerobic and anaerobic conditions. Acta Pathol. Microbiol. Immun. Scand. Sect. B *93*:2, 139 (1985).

24   P. Densen and G. L. Mandell: Phagocytic cells versus microbes. In: Infectious disease reviews, vol. 2, p. 43. Ed. W. J. Holloway, Futura, Mt. Kisco, N. Y. 1978.

25   P. Densen and G. L. Mandell: Phagocyte strategy vs. microbial tactics. Rev. Infect. Dis. *2*:5, 817 (1980).

26   J. K. Spitznagel: Microbial interactions with neutrophils. Rev. Infect. Dis. *5*:4, 806 (1983).

27   E. C. Gotschlich: Thoughts on the evolution of strategies used by bacteria for evasion of host defenses. Rev. Infect. Dis. *5*:4, 778 (1983).

28   I. Orskov, A. Ferencz and F. Orskov: Tamm-Horsfall protein or uromucoid is the normal slime that traps type 1 fimbriated *Escherichia coli.* Lancet *19*(1), 887 (1980).

29   S. M. Kuriyama and F. J. Silverblatt: Effect of Tamm-Horsfall urinary glycoprotein on phagocytosis and killing of type 1-fimbriated *Escherichia coli.* Infect. Immun. *51*:193 (1986).

30   E. Whitnack and E H. Beachey: Anti-opsonic activity of fibrinogen bound to M-protein on the surface of group A *streptococci.* J. Clin.Invest. *69*:1042 (1982).

31   P. E. Vaudaux, G. Zulian, E. Huggler and F. A. Waldvogel: Attachment of *Staphylococcus aureus* to polymethylmethacrylate increases its resistance to phagocytosis in foreign body infection. Infect. Immun. *50*(2), 472 (1985).

32   R. A. Proctor: The staphylococcal fibronectin receptor: Evidence for its importance in invasive infections. Rev. Infect. Dis. *9*(4), 335 (1987).

33 W. A. Simpson, H. S. Courtney and I. Ofek: Interactions of fibronectin with Streptococci: The role of fibronectin as a receptor for *Streptococcus pyogenes*. Rev. Infect. Dis. *9*(4), (1987).

34 K. Hayashi, D. A. Lee and P. G. Quie: Chemiluminescent response of polymorphonuclear leukocytes to *Streptococcus pneumoniae* and *Haemophilus influenzae* in suspension and adhered to glass. Infect. Immun. *32*(2), 392 (1986).

35 J. W. Costerton and T. J. Marrie: The role of the bacterial glycocalyx in resistance to antimicrobial agents. In: Medical Microbiology No. 3 "Role of the envelope in the survival of bacteria in infection", p. 63. Academic Press, New York 1983.

36 E. Falcieri, P. Vaudaux, E. Huggler, D. Lew and F. Waldvogel: Role of bacterial exopolymers and host factors on adherence and phagocytosis of *Staphylococcus aureus* in foreign body infection. J. Infect. Dis. *155*(3), 524 (1987).

37 N. Sharon: Bacterial lectins, cell-cell-recognition and infectious disease. FEBS Lett. *217*(2), 145 (1987).

38 C. C. Brinton, Jr: The structure, funtion, synthesis and genetic control of bacterial pili and a molecular model for DNA and RNA transport in gramnegative bacteria. Trans. N. Y. Acad. Sci. *27*:1003 (1965).

39 J. P. Duguid and D. C. Old: In: Bacterial adherence (Ed. E. H. Beachey, p. 185. Receptors and Recognition, Series B, vol. 6), Chapman and Hall, London 1980.

40 B. J. Eisenstein: Genetic control of type 1 fimbriae in *Escherichia coli*. In: Microbiology 1982. Am. Soc. Microbiol. Washington, D. C., p. 308. Ed. D. Schlesinger (1982).

41 L. Öhman, K.-E. Magnusson and O. Stendahl: Mannose-specific and hydrophobic interaction between *Escherichia coli* and polymorphonuclear leukocytes-influence of bacterial culture periods. Acta path. microbiol. scand. Sect. B. *93*:125 (1985).

42 N. Sharon: In: The lectins: Properties, functions and applications in biology and medicine, p. 493. Eds J. E. Liener et al. Academic Press, New York 1986.

43 Z. Bar-Shavit, R. Goldman, I. Ofek, N. Sharon and D. Mirelman: Mannose-binding activity of *Escherichia coli*: A determinant of attachment and ingestion of the bacteria by macrophages. Infect. Immun. *29*:417 (1980).

44 F. J. Silverblatt, J. S. Dreyer and S. Shavel: Effect of pili on susceptibility of *Escherichia coli* to phagocytosis. Infect. Immun. *21*:218 (1979).

45 L. Öhman, J. Hed and O. Stendahl: Interaction between human polymorphonuclear leukocytes and two different strains of type 1 fimbriae-bearing *E. coli*. J. Infect. Dis. 146:751 (1982).

46 L. Öhman, K.-E. Magnusson and O. Stendahl: The mannose-specific lectin activity of *Escherichia coli* type 1 fimbriae assayed by agglutination of glycolipid-containing liposomes and hydrophobic interaction chromatography. FEMS Microbiol. Lett. *14*:149 (1982).

47 T. Wadström, A. Farris, J. Freer, D. Habte, P. Hallberg and A. Ljung: Hydrophobic surface properties of enterotoxigenic *E. coli* (ETEC) with different colonization factors (CFA/1, CFA/11, K 88 and Kpp) and attachment to interstinal epithelial cells. Scand. J. Infect. Dis., Suppl. *24*:148 (1980).

48 H. J. Buscher and A. H. Weerkamp: Specific and non-specific interactions in bacterial adhesion to solid substracts. FEMS Microbiol. Rev. *46*:165 (1987).

49 M. Rodriquez-Ortega, I. Ofek and N. Sharon: Membrane glycoproteins of human polymorphonuclear leukocytes that act as receptors for mannose-specific *Escherichia coli*. Infect. Immun. *55*:968 (1987).

50 T. K. Kishimoto, K. O'Connor, A. Lee, T. M. Roberts and T. A. Springer: Cloning of the B sumunit of the leukocyte adhesion proteins: Homology to an extracellular matrix receptor defines a novel supergene family. Cell *48*: 681 (1987).

51 R. Lock, L. Öhman and C. Dahlgren: Phagocytic recognition mechanisms in human granulocytes and *Acanthamoeba castellanii* using type 1 fimbriated *E.coli* as phagocytic prey. FEMS Microbiol Lett. *44*:135 (1987).

52 J. Weiss, K. Kau, M. Victor and P. Elsbach: Oxygen-independent intracellular and oxygen-dependent extracellular killing of *Escherichia coli* S 15 by human polymorphonuclear leukocytes. J. Clin. Invest. *76*:206 (1985).

53 M. B. Goetz and F. J. Silverblatt: Stimulation of polymorphonuclear neutrophilic leukocyte oxidative metabolism by type 1 pili from *Escherichia coli*. Infect. Immun. *55*:534 (1987).

54 M. B. Goetz, S. M. Kuriyama and F. M. Silverblatt: Phagolysosome formation by polymorphonuclear neutrophilic leukocytes after ingestion of *Escherichia coli* that express type 1 pili. J. Inf. Dis. *156*(1), 229 (1987).

55 C. Svanborg-Edén, L.-M. Bjursten, R. Hull, S. Hull, K.-E. Magnusson, Z. Moldovano and H. Leffler: Influence of adhesins in the interaction of *Escherichia coli* with human phagocytes. Infect. Immun. *44*:672 (1984).

56 N. B. Freimer, H. M. Ögmundsdottir, C. C. Blackwell, I. W. Sutherland, L. Graham and D. M. Weir: The role of cell wall carbohydrates in binding of microorganisms to mouse peritoneal exudate macrophages. Acta path. microbiol. scand. Sect. B 86, 53:58 (1978).

57 R. L. Doolittle, C. H. Packman and M. A. Lichtman: Amino-sugars enhance recognition and phagocytosis of particles by human neutrophils. Blood *62*(3), 697 (1983).

58 P. R. Lambden, J. E. Heckels, L. T. James and P. J. Watt: Variations in surface protein composition associated with virulence properties in opacity types of Neisseria gonorrhoeae. J. Gen. Microbiol. *114*:305 (1979).

59 M. Virji, J. E. Heckels: The effect of protein II and pili on the interaction of *Neisseria gonorrhoeae* with human polymorphonuclear leukocytes. J. Gen. Microbiol. FEB *132* (p. 72), 503 (1986).

60 A. L. Sandberg, L. L. Mudrick, J. O. Cisar, M. J. Brennan, S. E. Mergenhagen and A. E. Vatter: Type 2 fimbrial lectin-mediated phagocytosis of oral *Actinomyces* spp. by polymorphonuclear leukocytes. Infect. Immun. *54*: 472 (1986).

61 D. Speert, B. A. Loh, D. A. Cabral and S. E. Salit: Nonopsonic phagocytosis of non-mucoid *Pseudomonas aeroginosa* by human neutrophils and monocyte-derived macrophages is correlated with bacterial piliation and hydrophobicity. Infect. Immun. *53*(1) 207, (1986).

62 C. M. J. E. Vandenbrouche-Grauls, H. M. W. M. Thijssen and J. Verhoef: Opsonization of *Staphylococcus aureus* protects endothelial cells from damage by phagocytosing polymorphonuclear leukocytes. Infect. Immun. *55*(6), 1455 (1987).

63 L. Edebo, E. Kihlström, K.-E. Magnusson and O. Stendahl: The hydrophobic effect and charge effects in the adhesion of enterobacteria to animal cell surface and the influences of antibodies of different immunoglobulin classes. In: Cell adesion and motility, p. 65. Eds Curtis and Pitts. 3rd Symp. Br. Soc. Cell Biology (Cambridge University Press) Cambridge 1980.

64 K.-E. Magnusson, C. Dahlgren, G. Maluszynska, E. Kihlström, T. Skogh, O. Stendahl, G. Söderlund, L. Öhman and Å. Walan: Non-specific and specific recognition mechanisms of bacterial and mammalian cell membranes. J. Disp. Sci. Techn. *6*(1), 69 (1985).

65 O. Stendahl, K.-E. Magnusson, C. Tagesson, R. Cunningham and L. Edebo: Characterization of mutants of S. typhimurium by countercurrent distribution in a two-polymer aqueous phase system. Infect. Immun. *7*:573 (1973).

66 O. Stendahl: The physicochemical basis of surface interaction between bacteria and phagocytic cells. In: Medical Microbiology, vol. 3, p. 137. Academic Press, London and New York 1983.

67 O. Stendahl, B. Normann and L. Edebo: Influence of O and K antigens on the surface properties of *Escherichia coli* in relation to phagocytosis. Acta pathol. microbiol. scand. Sect. B, *87*:85 (1979).

68 K.-E. Magnusson, J. Davies, T. Grundström, E. Kihlström and S. Normark: Surface charge and hydrophobicity of *Salmonella, E. coli* and *Gonococci*, and association with animal cells. Scand. J. Infect. Dis. Suppl. *24*:44 (1980).

69 O. Stendahl, C. Tagesson, K.-E. Magnusson and L. Edebo: Physicochemical consequences of opsonization of S. *typhimurium* with hyperimmune IgG and complement. Immunology *32*(11) (1977).

70 K.-E. Magnusson, O. Stendahl, I. Stjernström and L. Edebo: Reduction of phagocytosis, surface hydrophobicity and charge of *Salmonella typhimurium* 395MR10 by reaction with secretory IgA (SIgA). Immunology *36*:439 (1979).

71 K.-E. Magnusson and I. Stjernström: Mucosal barrier mechanisms. Interplay between secretory IgA (SIgA), IgA and mucins on the surface properties and association of Salmonellae with intenstine and granulocytes. Immunology *45*:23 (1982).

72 L. B. Klichstein, W. W. Wong, J. A. Smith, J. H. Weis, J. G. Wilson and D. T. Fearon: Human C3b/C4b receptors (CRI). Demonstration of long homologous repeating domains that are composed of the short consensus repeats characteristics of C3/Cu binding proteins. J. Exp. Med. *16G*:1095 (1987).

73 G. D. Ross, J. A. Cain, B. L. Myones, S. L. Newman and P. J. Lachmann: Specificity of membrane complement receptor type three (CR3) for β-Glucans. Complement *4*:61 (1987).

74 P. M. Henson: Membrane receptor on neutrophils. Immun. Comm. *5*:757 (1976).

75 J. Hed and O. Stendahl: Differences in the ingestion mechanisms of IgG and C3b particles in phagocytosis by neutrophils. Immunology *45*:727 (1982).

76 S. L. Newman and R. B. Johnston Jr: Role of binding through C3b and IgG in polymorphonuclear neutrophil function: studies with trypsin-generated C3b. J. Immunol. *123*:1839 (1979).

77 D. P. Lew, T. Andersson, J. Hed, F. Di Virgilio, T. Pozzan and O. Stendahl: Calcium-dependent and calcium-independent phagocytosis in human neutrophils. Nature *315*:6019 (1985).

78 T. Söderström and L. Öhman: The effect of monoclonal antibodies against *E. coli* type 1 pili and capsular polysaccharides on the interaction between bacteria and human granulocytes. Scand. J. Immun. *20*:299 (1984).

79 O. Stendahl and L. Öhman: Interaction of bacterial adhesins with inflammatory cells. Molec. Biol. Microbial Pathogenicity. Role of protein-carbohydrate interactions, p. 275. Academic Press, New York 1986.

80 M. Scheffer, W. König, J. Hacker and W. Goebel: Bacterial adherence and hemolysin production from *E. coli,* induces histamine and leukotriene release from various cells. Infect. Immun. *50*:271 (1985).

The bacterial cell surface and antimicrobial resistance

By Peter A. Lambert
Pharmaceutical Sciences Institute, Aston University, Aston Triangle,
Birmingham B4 7ET, U. K.

1   Introduction

The cell envelope is an essential bacterial component, vital for the survival and proliferation of the cells in hostile or poorly-supportive growth environments. Among its vital functions are:

(I)   support of the fragile underlying cytoplasmic membrane against the high intracellular osmotic pressure

(II)   control of cell shape

(III)   mediation of adhesion to surfaces and interactions with other cells

(IV)   involvement in nutrient uptake and export of cellular products

Figure 1 shows the basic arrangement of the envelopes of Gram-positive and Gram-negative bacteria. Mechanical strength and cell shape are determined by the peptidoglycan layer, which is common to both

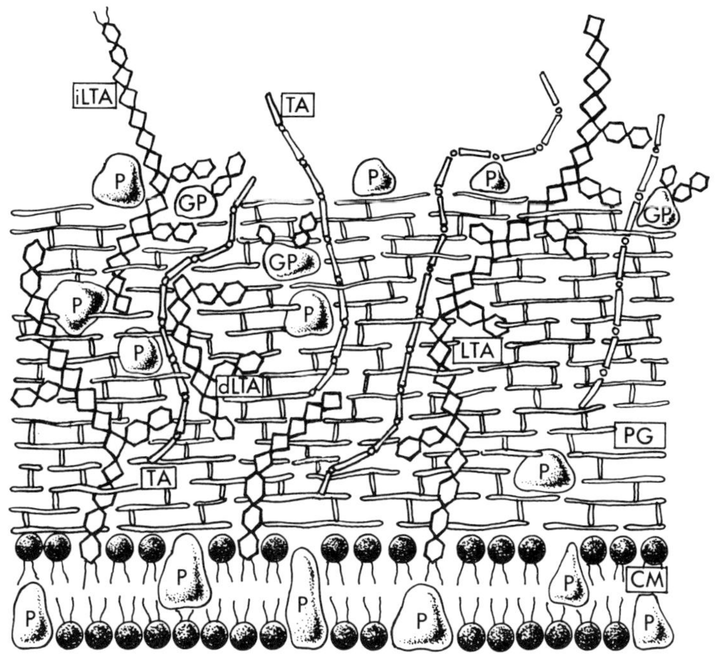

Figure 1a
Schematic representations of cross sections of the envelopes of Gram-positive (a) and Gram-negative bacteria (b).
(a) CM, cytoplasmic membrane; PG, peptidoglycan; P, protein; GP, glycoprotein; TA, teichoic acid or teichuronic acid; LTA, lipoteichoic acid; dLTA, deacylated lipoteichoic acid; iLTA, inverted lipoteichoic acid (glycolipid protruding away from the cell surface).

types of envelope. Apart from this component the two structures are quite different. The main distinguishing feature is the outer membrane (OM), found exclusively in Gram-negative bacteria. Gram-positive envelopes have an open structure which does not impede the diffusion of molecules the size of antimicrobial agents. By contrast, the OM does pose a significant permeability barrier to antimicrobials. This factor is responsible for the generally greater resistance of Gram-negatives to a wide range of agents compared with Gram-positive organisms.

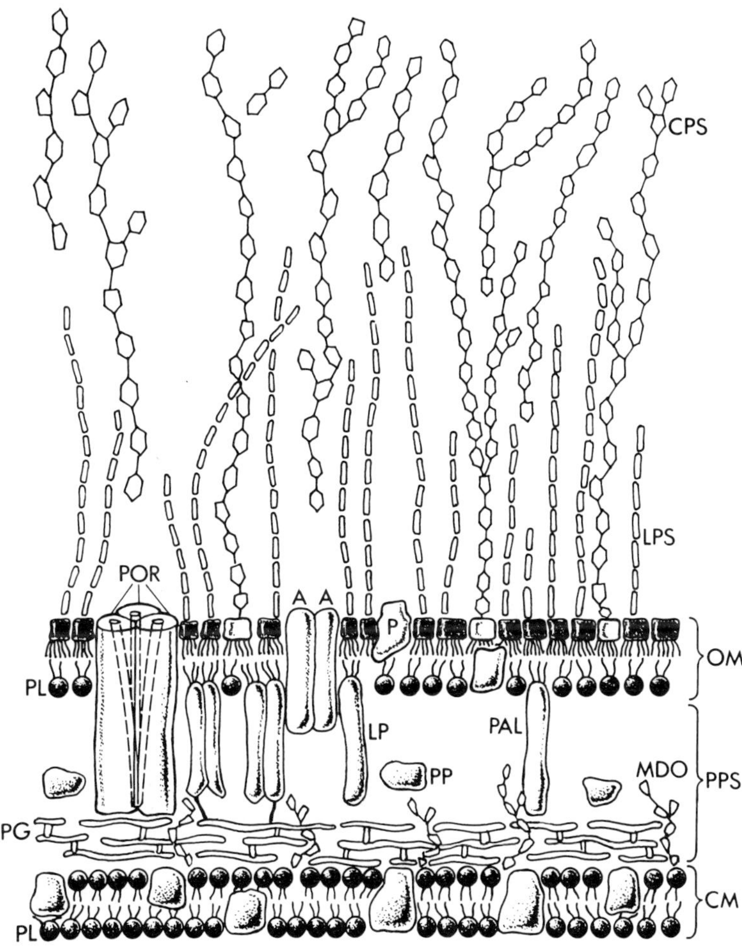

Figure 1 b
(b) CM, cytoplasmic membrane; PPS, periplasmic space; OM, outer membrane; PG, peptidoglycan; P, protein; PP, periplasmic protein; LP, lipoprotein; PAL, peptidoglycan-associated lipoprotein; POR, porin; CPS, capsular polysaccharide; LPS, lipopolysaccharide; MDO, membrane-derived oligosaccharide; A, other outer membrane protein; PL, phospholipid.

Three distinct mechanisms of antimicrobial passage across the OM can be distinguished: the hydrophilic pathway, the hydrophobic pathway, and the self-promoted uptake pathway. The first involves diffusion through the aqueous channels provided by the porins. Small hydrophilic antibiotics cross the OM by this route. Larger hydrophobic agents cannot pass through the porin channels and are only active against rough strains of Gram-negative bacteria, i. e. strains which have no O-antigen on their lipopolysaccharide (LPS). In these strains the hydrophobic molecules can reach the surface of the OM unimpeded by polysaccharide and diffuse across hydrophobic domains where phospholipids are exposed on the surface. The self-promoted uptake mechanism is thought to operate with agents which bind initially to LPS on the cell surface, destabilising the OM and thus allowing access to the underlying regions of the envelope.

The basic principles of bacterial permeability towards antimicrobials were established by Leive [1] who demonstrated that partial removal of the OM rendered Gram-negative cells highly sensitive to agents which could not penetrate untreated cells. Following the discovery of porins by Nakae [2, 3] and Nikaido [4], more detailed information has emerged on the way that small, hydrophilic molecules cross the OM. Properties of charge and shape of antimicrobials, in addition to size and hydrophobicity are now recognised as determinants of ability to penetrate Gram-negatives [for reviews see 5–8]. Although molecular features which confer high permeability are not necessarily the same as those conferring target site activity or resistance to inactivating or modifying enzymes, good penetrability is sought in the design of improved antimicrobials.

Some of the established groups of antibiotics use endogenous transport systems in the cytoplasmic membrane (CM) to gain entry to microbial cells. The aminoglycosides for example, promote their own passage across the OM, then are transported across the CM by an energy-dependent process requiring aerobic respiratory activity [9]. The tetracyclines are also actively-transported across the CM to achieve high intracellular concentrations [10]. In fact, the basis of their selective toxicity is this selective uptake by microbes and exclusion from mammalian cells.

Screening programmes to select new agents with activity against selected antimicrobial targets are usually based upon assays for the isolated target enzymes. The underlying philosophy is that once an active

„lead" compound has been found, substituents can be added or modifications made to ensure that the agent penetrates the intact organism. The concept of „illicit" transport has been demonstrated with the phosphonopeptides which exploit microbial peptide permeases for selective transport [11].

The major cause of resistance to antimicrobials is production of inactivating or modifying enzymes (e. g. β-lactamases and aminoglycoside modifying enzymes). With the introduction of antibiotics which are resistant to this kind of inactivation, the pattern of resistance is changing. Consequently we are now seeing increasing reports of resistance due to changes in the targets themselves, and to exclusion of the antimicrobials [12]. This review will concentrate upon the exclusion mechanism of resistance and will seek to explain the changes in the cell envelope which are responsible.

2 The structure and properties of bacterial cell envelopes
2.1 Gram-positive cell envelopes

These have an open structure composed of a cross-linked peptidoglycan network and covalently associated teichoic and/or teichuronic acids [13]. The latter are linear, anionic polymers which give the wall a strong negative charge [14, 15]. There is as yet no generally accepted model for the organisation of these major components in the Gram-positive envelope. Electron microscopy shows no evidence of discrete layering of peptidoglycan and teichoic acid [16], although after stabilisation with lectins (which bind to glycosyl substituents on teichoic acid) a „fluffy" outer surface has been described on *Bacillus subtilis* [17], suggesting a surface location of the teichoic acid.

The exclusion limit of the peptidoglycan/teichuronic acid network which makes up the wall of *Bacillus megaterium* is estimated in the range 30 000 to 57 000 daltons [18]. This is much larger than any antimicrobial and it is generally assumed that Gram-positive envelopes do not present a physical size barrier to penetration by antimicrobials. The negative charge of the envelope, generated by the acidic groups of the teichoic and teichuronic acids, and to some extent by carboxyl groups in the peptidoglycan does not seem to influence penetration of antimicrobials. Gram-positives are generally as sensitive to cationic agents (e. g. aminoglycosides and quaternary ammonium compounds) as they are to anionic agents (e. g. benzyl penicillin), at least where no other resistance mechanisms are involved. It has been shown that the positively-charged aminoglycosides do bind to the negatively-charged phosphate groups on teichoic acids in Gram-positive envelopes, displacing magnesium ions in the process [19]. Mutants lacking teichoic acid do not bind the aminoglycosides but do not show a significantly different sensitivity to the agents, suggesting that this interaction is of minor importance in the uptake process.

Proteins are also found in Gram-positive envelopes. Examples of proteins with important functions in colonisation of surfaces and survival of organisms in infections are: the M protein of group A streptococci [20]; protein A [21], clumping factor [22], and fibronectin binding protein [23] of *Staphylococcus aureus;* and the glycosyl transferase of *Streptococcus mutans* [24]. There is no evidence that any of these proteins directly influences the uptake of antimicrobials.

Capsular polysaccharides feature as prominent surface-associated material in some Gram-positive organisms. For example, in *Streptococcus pneumoniae* the wide variety of capsular polysaccharide antigens form the basis of serotyping, 84 different types are currently distinguished [25]. Capsules may be more widespread among Gram-positive bacteria than is generally realised, especially since their expression can be phenotypically controlled. Although few strains of *Staph. aureus* are encapsulated, many obtained as fresh isolates from infections do contain a capsule which is rapidly lost on subculture in laboratory medium [26]. The existence of a capsule around an individual cell probably does not influence antimicrobial uptake since the diffusion rate through the highly-hydrated polysaccharide is rapid [27] and ionic binding is negligible under physiological conditions [28]. However, Costerton and coworkers have produced a great deal of evidence that demonstrates the presence of a „glycocalyx" or polysaccharide coating which surrounds bacteria in their natural habitats [29]. This material, which is effectively a capsule or less-ordered form of exopolysaccharide slime, is thought to be produced by the organisms when they adhere to substrates. It binds adjacent cells together forming microcolonies and eventually a biofilm covering the surface being colonised [30]. The biofilm may become extremely thick, perhaps up to 100 cells in depth [31]. Penetration of antimicrobials to the cells deep within the biofilm is impeded and, for this reason, bacteria adhering to surfaces are particularly difficult to eradicate. Microbial biofilms cause major problems whether they are on inanimate objects (e. g.

fouling of pipelines in the oil industry) [32] or on medical prostheses (e. g. indwelling catheters or any other device inserted into the body) [33–36].

Lipoteichoic acid (LTA), also known as membrane teichoic acid, is another form of teichoic acid which is located in the Gram-positive envelope [37, 38]. It is anchored by its glycolipid terminus to the outer face of the CM. The polyglycerolphosphate chain (approximately 30 units long) protrudes through the peptidoglycan/wall teichoic acid matrix of the envelope and is exposed on the outer surface where it can be detected with antibodies. LTA is continuously excreted from the cells during growth [38]. Some of this material is thought to become reorientated in the envelope during release so that the glycolipid end protrudes from the cell surface, making the cells hydrophobic [39]. This „inverted LTA" might play a role in adhesion to surfaces [40]. In some organisms the „inverted LTA" together with deacylated LTA is associated with proteins, forming fibrils or tufts on the cell surface [41]. These surface tufts might fulfill a function comparable to that of fimbriae in Gram-negative organisms [42]. Although no direct involvement for these components in antimicrobial uptake has been demonstrated, their role in colonisation of surfaces and biofilm formation may allow the cells to establish themselves in sites which are inaccessible to antimicrobials.

2.2 Gram-negative cell envelopes

Many accounts have been given of the structure of the Gram-negative envelope [43–46]. In the context of antimicrobial uptake, the most important feature is the OM and its function as a permeability barrier [10, 47–50]. Without the aqueous channels provided by the porins, no small hydrophilic molecules would be able to cross the OM [51]. Since many nutrients fall into this category the porin channels are essential for cell growth. Although mutants apparently deficient in porins have been described [52, 53] it is likely that they do contain very low levels of residual porin molecules, enough to allow the cells to grow under conditions where nutrients are freely available [51].

Porins have been studied most thoroughly in enteric bacteria. They are non-covalently associated with the peptidoglycan, although it is not known if this association occurs *in vivo* or whether it occurs during isolation and purification of envelopes [45]. Individual porin molecules have molecular weights in the range 32 000–42 000 daltons. They are associated in groups of three [54], each porin molecule providing an aqueous channel on the outer face of the OM. The three channels coalesce in the centre of the trimers to give a single channel on the inner face of the OM. Evidence for this unexpected configuration has come from Fourier transform electron diffraction studies of lipid bilayers containing *E. coli* porin [55]. More detailed information on the tertiary structure of the polypeptide chains is eagerly awaited from X-ray crystallography of the porins.

More than one species of porin occurs in the OM, in *Escherichia coli* two porins are normally found, designated OmpF and OmpC. They constitute major OM components, around 100 000 molecules per cell, occupying up to 60 % of the cell surface [43]. Their expression is phenotypically controlled: OmpF increases and OmpC decreases when the cells are grown in peptone containing no fermentable source of carbon [56]; the reverse effect occurs when glucose or lactate are added to the medium or when sucrose or sodium chloride are added in sufficient quantities to increase the osmotic strength [57]. *E. coli* is capable of producing several other distinct porins. PhoE is produced under phosphate starvation [58]; LamB is induced by maltose [59]; protein K appears to be involved with capsule production, although its role is unknown [60]. Properties of the channels formed by the various species of porin are not identical. OmpF produces channels with a radius of 0,58 nm, whilst the OmpC channels are slightly smaller (0,54 nm radius). These pore sizes are just big enough to allow molecules with molecular weights up to 600–700 daltons to pass through. Shape, charge and relative hydrophobicity are important properties which determine how easily molecules can infiltrate the channels. Most small, hydrophilic antibiotics preferentially use the OmpF channels in *E. coli* and OmpF mutants isolated in the laboratory are somewhat resistant to antibiotics, particularly the β-lactams [61].

Porins have been studied in a wide range of pathogenic Gram-negative bacteria: *Salmonella typhimurium* [4]; species of *Proteus, Enterobacter, Morganella* and *Providencia* [62, 63]; protein I in *Neisseria gonorrhoeae* [64]; proteins F [65], P [66] and D1 [67] in *Pseudomonas aeruginosa;* *Haemophilus influenzae* [68]; *Legionella pneumophila* [69]; *Pseudomonas cepacia* [70]; *Bordetella pertussis* [71]; *Yersinia pestis* [72] and *Chlamydia trachomatis* [73]. For many other organisms the major peptidoglycan-associated OM proteins in the 32 000–42 000 dalton range are assumed to be porins, although their function has yet to be demonstrated directly.

Channel-forming properties of the F porin in *P. aeruginosa* have received special attention since they were originally described as having a much higher exclusion limit than their counterparts in enteric organisms. The F porin channels were shown to allow molecules with molecular weights of 3000–9000 daltons to pass [74, 75]. This was incompatible with the recognised impermeable nature of the envelope of this organism and was further complicated by studies which showed no major difference in the porin or OM composition between a mutant with a highly permeable envelope and its parent strain [76]. The initial explanation was that only 1% of the F porin channels are functional in normal strains and a higher proportion are functional in the permeability mutant [77, 78]. Other studies clearly showed the channels to be small [79–81], consistant with the low OM permeability of the organism [82]. The current view is that the porin can adopt two functional conformations; one produces the large channels (1 nm radius) and the other much smaller channels [83]. It has been suggested the smaller channels exist most of the time, and that the conformational change giving the two different pore sizes is caused by altering the arrangement of sulphide bridges between the four cysteine residues in the protein [84].

Bacteroides fragilis is reported to possess porin channels with an exclusion limit in the order of 250 daltons [85, 86]. The porins of *P. cepacia* also produce small channels, giving an OM permeability around tenfold lower than that of *E. coli* [70]. This restricted permeability would significantly reduce the permeability towards a range of hydrophilic antimicrobials.

Other components of the Gram-negative envelope which influence permeability are the O-antigen chains of the LPS. The presence of the O-antigen polysaccharide chains on the surface of smooth strains presents a barrier to the penetration of hydrophobic antimicrobials such as rifampicin, novobiocin and actinomycin D but not to hydrophilic antimicrobials. Rough strains lacking O-antigen are more sensitive than smooth strains to hydrophobic agents because there is no barrier preventing these agents from reaching the outer face of the OM and diffusing across by the hydrophobic pathway. The correlation between LPS composition and sensitivity to hydrophobic agents was first made by Nikaido using a set of *S. typhimurium* mutants with varying LPS core and O-antigen side chains [87, 88]. Organisms such as *N. gonor-rhoeae* which do not have an O-antigen on their LPS are naturally sensitive to hydrophobic agents such as fatty acids [89].

Capsular polysaccharides do not present a permeability barrier to hydrophobic agents, possibly because they are highly hydrated gels with an open structure allowing molecules to diffuse. The negative charges present on many capsules might be expected to influence the penetration of charged molecules, either by presenting an electrostatic barrier to anionic molecules or by binding cationic molecules like an ion exchange resin. Although mucoid (alginate-producing) strains of *P. aeruginosa* can be selected on media containing carbenicillin (an anionic β-lactam) [90] some mucoid strains are hypersensitive to carbenicillin [91]. It appears that in this case the anionic alginate does not inhibit uptake of the anionic β-lactam. As discussed for the Gram-positive capsules earlier, the case for ionic trapping of cationic antimicrobials by anionic capsular polysaccharides is not supported by measurements which show that, under physiological conditions of ionic strength, no significant binding takes place [28]. Mathematical models of the diffusion of charged molecules through capsules indicate that the diffusion barrier is not sufficient to affect antimicrobial sensitivity [27].

Like Gram-positives, Gram-negative bacteria exhibit extensive glycocalyx production and biofilm formation *in vivo* [27–29]. Restricted penetration of antimicrobials through a thick biofilm poses a problem in eradication of organisms growing on surfaces, both in the medical and industrial contexts.

3   Uptake of antimicrobials
3.1   Access to targets

The sites of action of the majority of antimicrobials are located in the cytoplasm. Therefore inhibitors of protein synthesis (aminoglycosides, tetracyclines, chloramphenicol, erythromycin, lincomycin/clindamycin, fusidic acid), chromosome function (quinolones and rifampicin) and folate metabolism (sulphonamides and trimethoprim) all have to cross the CM as well as the outer envelope to reach their targets. The β-lactams, vancomycin and teichoplanin act upon the final stages of peptidoglycan assembly which take place on the outer face of the CM. They only need to cross the envelope to reach their targets.

Penetration of the envelope to the CM is essentially a diffusion process. In Gram-negatives, small hydrophilic molecules pass through the porin channels and hydrophobic agents diffuse across the hydrophobic lipid bilayer of the OM, provided there is no O-antigen polysaccharide barrier blocking their approach. Agents which promote their own uptake presumably damage the OM sufficiently to penetrate to the CM. In Gram-positives, all agents are assumed to diffuse virtually unhindered through the envelope structure, regardless of size, charge or hydrophobicity.

3.2   Properties of antimicrobials affecting penetration through porin channels

3.2.1   Hydrophobicity

The broad distinction between „hydrophilic" and „hydrophobic" agents used to explain the exclusion properties of the OM is based upon the partition coefficients (P) between octanol and phosphate buffer, pH 7 at 25 °C. Table 1 lists P values for some agents; those with va-

Table 1
Molecular weight and hydrophobicity of some antimicrobials.

Compound	Molecular weight	Partition coefficient
Actinomycin D	1255	> 20
Novobiocin	613	> 20
Crystal violet	408	14
Chloramphenicol	323	12
Rifampicin	823	10
Minocycline	457	1.5
Doxycycline	444	0.9
Nafcillin	414	0.3
Oxytetracycline	460	0.09
Oxacillin	418	0.07
Tetracycline	444	0.05
Benzyl penicillin	334	0.02
Carbenicillin	378	< 0.01
Ampicillin	349	< 0.01

Partition coefficient, octanol/0.05 M phosphate buffer, pH 7, 25° C.

lues above 0,1 are considered as hydrophobic and cannot readily penetrate the OM of smooth Gram-negative strains. There are some important exceptions, particularly chloramphenicol, which although notably hydrophobic, is active against smooth Gram-negatives. Minocycline, likewise is active against Gram-negatives, despite its hydrophobicity which is significantly greater than that of the other tetracyclines [92, 93]. Possibly its reduced rate of penetration of the OM is offset by its greater target site activity compared with the other tetracyclines [10, minocycline [95]. Among the penicillins, P values vary from < 0,01 for ampicillin and carbenicillin to 0,3 for nafcillin. As expected, nafcillin is only active against Gram-positives.

Much of the published work relating penetration of the OM by β-lactams to hydrophobicity is complicated by the effects of charge and availability of alternative porins [48]. However, a clear correlation between hydrophobicity and rate of penetration is shown in a group of monoanionic cephalosporins where a 10-fold increase in P produces a 5- to 6-fold decrease in rate of penetration through the OmpF channel of *E. coli* K12 [96, 97]. In a study of cephalosporins and their 1-oxa congeners, an increase in the penetration rate of each of the 1-oxacephalosporins compared with their respective 1-sulphur cephalosporins was matched by a decrease in hydrophobicity [98].

3.2.2   Size and shape

Antimicrobials of a size larger than 600 daltons are too big to penetrate the porin channels in enteric bacteria and consequently are inactive against them. Molecules smaller than the exclusion limit of the pores pass through them at rates dependent upon their hydrodynamic radii [99, 100]. Molecules which are close to the exclusion limit are impeded by interaction with groups lining the pore walls. Within a group of similarly charged β-lactams the measured rates of diffusion across the OM [100] agree quite closely with those predicted by the Renkin equation, which relates diffusion rate to the diameter of the pore and the molecules passing through it [101]. The size of the porin channels is critical in determining the rate of diffusion. Glucose, with a molecular weight of 320 daltons and a Stokes radius of 0,42 nm, penetrates the OmpC channel of *E. coli* at only 62 % of the rate at which it penetrates the OmpF channel [100], even though the channels differ in size by only 8 %. Because of the size restriction, most β-lactams preferentially use the OmpF channel to cross the OM of *E. coli* [100]. The largest cephalosporins, cefoperzone (644 daltons), ceftriaxone (552 daltons) and ceftazidime (552 daltons) do not pass through the OmpC channel at a measurable rate [48, 97].

The influence of molecular shape upon penetrability through porin channels is difficult to evaluate. The compactness of the fused ring system of cephalosporins allows them to penetrate faster than oligosaccharides of similar molecular weight. Cephaloridine (415 daltons) penetrates the OmpF channel of *E. coli* at nearly twice the rate of lactose (342 daltons) [48, 100]. A study with sterically-restricted (2,3)-methylene penicillin G derivatives shows that the „open" (2,3)-α-methylene penam penetrates faster than the „closed" (2,3)-β-methylene penam and that both penetrate faster than penicillin G itself [102]. Cephalexin penetrates the OmpF channel of *E. coli* three times faster than the analogous penicillin, ampicillin, despite its greater size [48, 97].

3.2.3   Charge

Charge is the most important property controlling penetration of porin channels by β-lactams. In general, those with one positive and one negative charge (zwitterionic compounds) penetrate the *E. coli* OmpF channel fastest and those with two negative charges penetrate slowest [97]. The zwitterionic character seems to be more important for penetration of OmpF than hydrophobicity. Cephaloridine penetrates faster than cephalexin, cephaclor and cephaloglycin, despite being the most hydrophobic (and the largest) of this group of zwitterionic cephalosporins [48, 97]. Not all porins are influenced by charge in this way; penetration through the PhoE channel is unaffected by negative charge on the solute [58, 100]. This is perhaps because PhoE is involved with uptake of phosphorylated nutrients [58].

3.3   The self-promoted uptake pathway

Aminoglycosides have been shown to pass through porin channels in an artificial liposome swelling assay [103]. If they enter intact cells via the porin channels, the process must be highly efficient because mutants producing very low levels of porins are as sensitive as the parent strains [103]. It seems more likely that they promote their own uptake across the OM. Several lines of evidence support this view. Aminoglycosides bind strongly to the phosphate groups on the core region of LPS, displacing magnesium ions and destabilizing the OM [9]. The effect is particularly pronounced in *P. aeruginosa,* the LPS of which is highly phosphorylated [9]. Treatment of whole cells with aminoglycosides results in enhanced permeability of the OM to nitrocefin, hydrophobic fluorescent probes and lysozyme [84, 104, 105]. Interestingly, they do not increase the permeability of the OMs of *Pseudomonas cepacia* or *Serratia marcescens* to the fluorescent probes [105]. Possibly in these organisms the aminoglycoside-binding sites on the LPS are not accessible in the OM.

P. aeruginosa produces elevated levels of OM protein H1 when grown in media containing low levels of magnesium [106]. H1 replaces magnesium ions in the OM and stabilises the structure by binding to the LPS core phosphate groups [9]. Cells grown under these conditions are resistant to aminoglycosides [107]. Mutants which naturally overproduce H1 are also resistant, suggesting that aminoglycosides cannot bind to LPS when H1 is present in the OM in place of magnesium ions [107]. This form of resistance has only been reported in *P. aeruginosa*, OM proteins analogous to H1 have not been found in other Gram-negatives.

Another aspect of aminoglycoside action which might contribute to their self-promoted uptake concerns the properties of misread proteins incorporated into the CM following aminoglycoside binding to the ribosomes. These have been reported to form channels across the CM [108].

The molecular basis of the binding of agents to cell surface components is not yet clear. Displacement of magnesium ions from phosphate groups on LPS can be achieved with a magnesium-chelating agent such as EDTA, originally shown by Leive to remove part of the OM and to permeabilize the cells [1]. Polycations such as polylysine, polymyxin and polymyxin nonapeptide [109–111] have similar effects and increase sensitivity to other antimicrobials by allowing increased uptake. A wide range of *P. aeruginosa* OM „permeabilizers" have been identified, including tris, gentamicin, neomycin, poly-L-ornithine, gramicidin S, cetrimide, nitrilotriacetate, L-ascorbate and acetylsalicylate [112, 113]. Exploitation of these „permeabilizers" in synergistic combination with antibiotics has been considered [114] but the applications are likely to be limited; either they are unsuitable for use *in vivo* or they are not effective against key organisms, e. g. *P. cepacia* and *S. marcescens* [84].

3.4   Use of microbial nutrient uptake systems

Interaction of aminoglycosides with the OM is only the initial stage of a complex uptake process [9, 10, 115]. High cytoplasmic concentrations are achieved by transport across the CM which occurs in two distinct energy-dependent phases, EDP I and EDP II [116]. Uptake depends both on electron transport through the membrane-bound respiratory chain and a threshold transmembrane electrochemical potential, $\Delta\Psi$ [115]. Anaerobic organisms do not transport aminoglycosides and are not sensitive to them [117], even though the aminoglycosides presumably cross the OM of anaerobic Gram-negatives by the self promoted pathway.

All hydrophilic antimicrobials which act at cytoplasmic sites presumably cross the CM via a nutrient permease system. Most studied is the transport of tetracyclines [10, 118]. Uptake as a magnesium chelate, a dicarboxylic acid or via the glutamate transport system have each been suggested as mechanisms for tetracycline transport, but the evidence is not convincing [10]. Studies with a variety of inhibitors of membrane energetics [10] show that uptake is energy-dependent but does not require ATP hydrolysis. Little information is available on the mechanisms of transport of other antimicrobials apart from D-cycloserine, which uses the D-alanine transport system [119] and phosphonomycin, which uses the glycerol phosphate system [120]. The quinolones are thought to be transported across the CM by an energy-dependent process [121], although the precise mechanism is not yet clear. Uptake of enoxacin might be different to other quinolones. One report indicates that enoxacin enters *E. coli* and *Bacillus subtilis* by simple diffusion across the CM, in contrast to ciprofloxacin, perfloxacin, amifloxacin and norfloxacin [122].

Hydrophobic agents are assumed to cross the CM by passive diffusion, driven by the concentration gradient generated by binding to their intracellular targets [10, 123]. However, chloramphenicol, generally regarded as a hydrophobic agent and to enter cells by diffusion [10] has been reported to be taken up by *H. influenzae* by an energy-dependent process [124].

4   Resistance due to impaired uptake or exclusion of antimicrobials

4.1 β-lactams

The rate of penetration of β-lactams through porin channels across the OM has only recently become of importance as a determinant of resistance [12, 125]. Periplasmic β-lactamases inactivate sensitive β-lactams before they can bind to their targets, the penicillin-binding proteins (PBPs) which are located on the outer face of the CM. Even the most rapidly-penetrating β-lactamase-sensitive compounds are inactivated. Following the introduction of β-lactamase-resistant agents, particularly the third generation cephalosporins, new groups of resistant organisms have emerged. Resistance due to altered targets [126–128], restricted OM permeability [129–131] and high levels of chromosomal β-lactamase expression, possibly in combination with restricted permeability [132–135] is now being encountered [136]. Restricted OM permeability is often assumed to be the mechanism of resistance in cases where no change is detected in the PBPs or β-lactamase production without direct measurements of permeability being made [137].
A non-hydrolytic „trapping" mechanism, whereby periplasmic β-lactamase molecules bind and immobilise resistant β-lactams, thereby protecting the vulnerable PBP targets, has been proposed as a mechanism of resistance to some third generation cephalosporins [138]. Carefull examination of rates of hydrolysis by β-lactamases and of penetration of β-lactams into the periplasm shows that a combination of slow penetration and slow hydrolysis could account for the observed resistance, rather than trapping alone [139, 140]. The problem is observed particularly with the larger, anionic cephalosporins and organisms in which constitutively-derepressed chromosomal β-lactamase mutations occur at high frequencies. Currently this mechanism of resistance is seen particularly in strains of *P. aeruginosa* and species of *Enterobacter, Acinetobacter, Citrobacter* and *Serratia* [132]. The use of relatively stable third generation cephalosporins selects for the derepressed strains, unless the concentrations are sufficient to kill the mutants [141, 142]. The *ampR* gene product is the key regulatory factor involved in induction and overproduction of the *Citrobacter freundii* chromosomal β-lactamase [143]. Similar control of inducible chromosomal β-lactamase probably occurs in all enterobacteria as well as in *P. aeruginosa* [144]. Some β-lactams are capable of increasing the level

of β-lactamase produced in cells by an induction mechanism [145], presumably involving interaction with the *ampR* gene product.

An „ideal" β-lactam should have high target site activity, allied to good penetration characteristics, β-lactamase resistance and weak inducing properties. It is clearly unlikely that chemical structures can be chosen to optimise all of these properties together with retention of desirable pharmacokinetic performance. In terms of penetration of the OM, small β-lactams with a net positive charge are most desirable [48, 97]. For low inducing activity, a net negative charge appears to be necessary, as in carbenicillin [145].

There are reports of resistance of *P. aeruginosa* to β-lactams in which the organisation of the LPS O-antigen chains appears to influence permeability [146–148]. It is difficult to explain how alterations in the O-antigen presentation on the cell surface could affect uptake of hydrophilic β-lactams such as carbenicillin, moxalactam and cefsulodin. Either some additional barrier is present, or the structure and arrangement of the LPS influences the porin function as proposed for the supersusceptible strain, Z61 [149]. In some cases resistance might be due to subtle alterations in the structure of the F porin, as indicated by studies with monoclonal antibodies and tryptic digestion [150] or to the distribution of the LPS and porin in the envelope [151].

Resistance of *P. aeruginosa* to imipenem has been reported during therapy for severe infections [152–154]. The resistance, which is independent of β-lactamase production or changes in the PBPs, has been attributed to impaired uptake of imipenem. The resistant strains showed no alteration in LPS but lacked two OM proteins, D1 and D2 [155–157]. These studies imply that imipenem uses the glucose-inducible porin, D1 to cross the OM of *P. aeruginosa* [67].

In *E. cloacae* the major cause of resistance towards β-lactams is probably lack of penetration across the OM. Although „trapping" by high periplasmic concentrations of β-lactamase [138] or the presence of a β-lactamase-mediated permeability barrier [158, 159] have been suggested as resistance mechanisms, no direct correlation between β-lactamase levels and resistance to aztreonam, ceftazidime, moxalactam or imipenem could be found [129]. In the case of ceftriaxone, a three- to five-fold increase in resistance of *E. cloacae* was attributed to a combination of increased β-lactamase and decreased permeability [160].

A low level resistance of the fish pathogen *Aeromonas salmonicida* to β-lactams has been related to decreased permeability and altered OM protein patterns, presumably involving porins [161]. Susceptibility to large hydrophobic agents was not affected. Changes in OM protein profiles have also been related to resistance to β-lactams in strains of *S. marcescens* [162, 163] and in species of *Proteus, Morganella* and *Providencia* [164].

Loss of an individual porin has been correlated with a low level of resistance to β-lactams (and other hydrophilic antimicrobials) in a wide range of organisms. Much of the information comes from studies on porin-deficient strains isolated in the laboratory. For example, strains of *E. coli* K12 lacking the OmpF porin are resistant to negatively-charged β-lactams, the minimum inhibitory concentrations being around eight-fold higher than the parent strains [61, 165]. In *P. aeruginosa*, mutants lacking porin F show a six-fold lower permeability to β-lactams [166] and a small increase in resistance [114]. Similarly in *H. influenzae* strains lacking the 40 kdalton porin are more resistant than their porin-containing isogenic partners to a range of β-lactams [68]. There are now many reports of clinical isolates which owe their β-lactam resistance to loss of porins. For example, resistant strains of *P. aeruginosa* [157, 167], *E. coli* [168] and *S. typhimurium* [169] have been isolated during β-lactam therapy. With increasing use of β-lactamase-resistant agents we can expect to see an increased incidence of this form of clinical resistance, alternative strategies or better-penetrating agents will need to be developed.

4.2   Aminoglycosides

Strains of *P. aeruginosa* with altered LPS show some resistance to aminoglycosides [170, 171] due, presumably, to their reduced susceptibility to the self-promoted uptake pathway (see Section 3.3). Clinical failure of aminoglycoside therapy usually results from production of modifying enzymes [172]. However, an example of amikacin-resistant *P. aeruginosa* emerging during therapy of endocarditis has been reported in which resistance was due to restricted uptake [173]. These cells had a small colony phenotype which was unstable on subculture. The resistance could be due to altered respiratory activity and quinone content.

4.3   Tetracyclines

Passage of tetracyclines across the OM by diffusion through the porin channels is influenced by the same properties of hydrophobicity, size and charge as the β-lactams (Section 3.2) [10]. Tetracycline itself is hydrophilic and small enough to diffuse through the porin channels in *E. coli* (Table 1). Strains of *E. coli* K12 lacking the OmpF porin [174] and of *Neisseria gonorrhoeae* with altered principal outer membrane protein (porin) [175] show low-level resistance compared with their respective parent strains. Tetracycline resistance emerging during therapy is plasmid-mediated and is usually associated with defective uptake across the CM [176]. Resistance plasmids in *E. coli* and *Staph. aureus* encode for a number of envelope proteins presumed to be involved in exclusion or to be responsible for defective uptake [10, 125, 177, 178]. Rapid expulsion of tetracycline from the cells is also a mechanism of resistance involving production of plasmid-mediated CM proteins [179, 180]. These proteins promote efflux of tetracycline from the cells and reduce the intracellular concentration to below inhibitory levels.

4.4   Chloramphenicol

There are several reports of resistance to chloramphenicol in which enzymatic inactivation of the drug by acetylation does not appear to be involved. A permeability barrier has been suggested to explain resistance in certain of the Enterobacteriaceae [181], *P. aeruginosa* [182, 183] and *H. influenzae* [184]. Cross resistance to chloramphenicol, nalidixic acid and trimethoprim in strains of *Klebsiella, Enterobacter* and *Serratia* was associated with changes in OM proteins, presumed to involve porins [185].

4.5   Quinolones

Some *E. coli* mutants lacking the OmpF porin show a low level resistance to norfloxacin and reduced accumulation of the fluoroquinolone [186–188]. Norfloxacin resistance in *P. aeruginosa* has been related to altered OM protein composition and permeability [189] and a strain of *P. aeruginosa* lacking porin F has been isolated from a patient undergoing enoxacin therapy [190]. Increased usage of the quinolones will undoubtedly encourage emergence of resistance. It remains to be seen whether this will take the form of altered OM permeability, defective transport across the CM, altered DNA gyrase or mutations affecting other physiological responses [191].

5   Strategies to overcome resistance by exclusion

A number of strategies might be considered to avoid the problem of antibacterial resistance by exclusion. One is the use of OM permeabilizers to act in synergy with antibiotics [112, 113]. Whilst agents such as EDTA-tris drastically increase the uptake of a wide range of antibiotics by bacteria [192], their therapeutic application is obviously limited. Other metal-complexing agents such as nitrilotriacetate and L-ascorbate [112, 113] likewise do not seem suitable in a medical context. The nontoxic polycationic agent, polymyxin nonapeptide does show promising activity *in vitro* but not *in vivo* [109–111]. Recent studies with *P. aeruginosa* [193] indicate that the classical synergy between β-lactams and aminoglycosides [194] is due to enhanced uptake of the aminoglycoside (tobramycin) following action of the β-lactam (ticarcillin or cefsulodin) upon the cells. These results should encourage a more critical examination of synergistic action between antibiotics.
A different approach is to exploit vital microbial uptake systems that operate when the cells grow *in vivo*. Iron uptake is the best example, and several groups of agents act via the iron uptake systems. The sideromycins, e. g. albomycins and ferrimycins [195] are naturally occurring antibiotics which are taken up by bacteria via iron uptake systems then hydrolysed inside the cells to release toxic fragments [196]. Attempts have been made to mimic the sideromycins by coupling sulphonamides to natural and artificial iron chelating agents (siderophores) [197]. Another approach has been to use the natural enterobacterial phenolate siderophore, enterochelin, to transport the toxic metals, scandium and indium into bacteria [198]. Finally, a cephalosporin with a phenolate side chain, designated E-0702, is reported to act only upon cells of *E. coli* which are starved of iron and not upon iron-supplemented cells [199]. The conclusion made from these studies is that this, and closely related cephalosporins cross the OM via an iron transport system. A study of resistant mutants showed the *tonB* gene to be involved, this gene is known to be essential for iron transport in bacteria [200–202].

These examples highlight the advantage of studying properties of microbes grown under conditions which accurately simulate those of the natural growth environment, rather than optimum laboratory culture conditions. The need to recognise and reproduce growth conditions under which antimicrobial agents are to function cannot be overstated [203].

References

1 L. Leive: Ann. N. Y. Acad. Sci. *235*, 109 (1974).
2 T. Nakae: Biochem. Biophys. Res. Commun. *64*, 1224 (1975).
3 T. Nakae and H. Nikaido: J. Biol. Chem. *250*, 7359 (1975).
4 T. Nakae: J. Biol. Chem. *251*, 2576 (1976).
5 H. Nikaido and M. Vaara: Microbiol. Rev. *49*, 1 (1985).
6 H. Nikaido: Pharmac. Ther. *27*, 197 (1985).
7 T. Nakae: Crit. Rev. Microbiol. *13*, 1 (1986).
8 R. E. W. Hancock: J. Bacteriol. *169*, 929 (1987).
9 R. E. W. Hancock: J. Antimicrob. Chemother. *8*, 249 (1981).
10 I. Chopra and P. Ball: Adv. Microb. Physiol. *23*, 183 (1982).
11 P. S. Ringrose. In: Role of the Envelope in the Survival of Bacteria in Infection, Medical Microbiology, vol. *3*, p. 179. Eds C. S. F. Easmon, J. Jeljaszewicz, M. R. W. Brown and P. A. Lambert. Academic Press, London (1983).
12 T. F. O'Brien and the members of Task Force 2: Rev. Infect. Dis. *9*, S244 (1987).
13 S. M. Hammond, P. A. Lambert and A. N. Rycroft. In: The Bacterial Cell Surface, p. 29. Croom Helm, Beckenham (1984).
14 J. B. Ward: Microbiol. Rev. *45*, 211 (1981).
15 G. D. Shockman and J. F. Barrett: Ann. Rev. Microbiol. *37*, 501 (1983).
16 G. R. Millward and D. A. Reaveley: J. Ultrastruct. Res. *46*, 309 (1974).
17 D. C. Birdsell, R. J. Doyle and M. Morgenstern: J. Bacteriol. *121*, 726 (1975).
18 R. Scherrer and P. Gerhardt: J. Bacteriol. *137*, 718 (1971).
19 W. Kusser, K. Zimmer and F. Fiedler: Eur. J. Biochem. *151*, 601 (1985).
20 G. N. Phillips Jr, P. F. Flicker, C. Cohen, B. N. Manjula and V. A. Fischetti: Proc. Natl Acad. Sci. USA *78*, 4689 (1981).
21 J. J. Langone: Adv. Immunol. *32*, 157 (1982).
22 F. Espersen, I. Clemmensen and V. Barkholt: Infect. Immun. *49*, 700 (1985).
23 J.-I. Flock, G. Froman, K. Jonsson, B. Guss, C. Signas, B. Nilsson, G. Raucci, M. Hook, T. Wadstrom and M. Lindberg: EMBO J. *6*, 2351 (1987).
24 H. K. Kuramitsu and L. Ingersoll: Infect. Immun. *20*, 652 (1978).
25 R. B. Roberts. In: Principles and Practice of Infectious Diseases, 2nd Edn, p. 1142. Eds G. L. Mandell, R. G. Douglas and J. E. Bennett. John Wiley and Sons, New York (1985).
26 K. Yoshida, M. Takahashi, T. Ohtomo, Y. Minegishi, Y. Ichiman, K. Haga, E. Kono, and C. L. San Clemente: J. Appl. Bacteriol. *46*, 147 (1979).
27 W. W. Nichols, S. M. Dorrington, M. P. E. Slack and H. L. Walmsley: Antimicrob. Ag. Chemother. *32*, 518 (1988).
28 C. S. Tannenbaum, A. T. Hastie, M. L. Higgins, F. Kueppers and G. Weinbaum: Antimicrob. Ag. Chemother. *25*, 673 (1984).
29 J. W. Costerton, R. T. Irvin and K.-J. Cheng: Ann. Rev. Microbiol. *35*, 299 (1981).

30 J. W. Costerton, K.-J. Cheng, G. G. Geesey, T. I. Ladd, J. C. Nickel, M. Dasgupta and T. J. Marrie: Ann. Rev. Microbiol. *41*, 435 (1987).
31 T. J. Marrie, M. A. Noble and J. W. Costerton: J. Clin. Microbiol. *18*, 1388 (1983).
32 J. W. Costerton: Dev. Ind. Microbiol. *25*, 383 (1984).
33 J. W. Costerton: Rev. Inf. Dis. *6*, S608 (1984).
34 J. C. Nickel, I. Ruseska, J. B. Wright and J. W. Costerton: Antimicrob. Ag. Chemother. *27*, 619 (1985).
35 J. C. Nickel, J. B. Wright, I. Ruseska, T. J. Marrie, C. Whitfield and J. W. Costerton: Eur. J. Clin. Microbiol. *4*, 213 (1985).
36 R. C. Evans and C. J. Holmes: Antimicrob. Ag. Chemother. *31*, 889 (1987).
37 P. A. Lambert, I. C. Hancock and J. Baddiley: Biochim. biophys. Acta *472*, 1 (1977).
38 A. J. Wicken and K. W. Knox: Biochim. biophys. Acta *604*, 1 (1980).
39 E. H. Beachey, W. A. Simpson, I. Ofek, D. L. Hasty, J. B. Dale and E. Whitnack: Rev. Infect. Dis. *5*, S670 (1983).
40 E. H. Beachey and H. S. Courtney: Rev. Infect. Dis. *9*, S475 (1987).
41 I. Ofek, W. A. Simpson and E. H. Beachey: J. Bacteriol. *149*, 426 (1982).
42 P. S. Handley and A. E. Jacob: J. Gen. Microbiol. *127*, 289 (1981).
43 H. Nikaido and T. Nakae: Adv. Microb. Physiol. *20*, 163 (1979).
44 M. J. Osborn and H. C. P. Wu: Ann. Rev. Microbiol. *34*, 369 (1980).
45 B. Lugtenberg and L. van Alphen: Biochim. biophys. Acta *737*, 51 (1983).
46 S. M. Hammond, P. A. Lambert and A. N. Rycroft. In: The Bacterial Cell Surface, p. 57. Croom Helm, London (1984).
47 P. A. Lambert. In: Continuous Culture 8, Biotechnology, Medicine and the Environment, p. 38. Eds A. C. R. Dean, D. C. Ellwood and C. G. T. Evans. Ellis Horwood, Chichester (1984).
48 H. Nikaido: Pharmac. Ther. *27*, 197 (1985).
49 H. Nikaido and M. Vaara: Microbiol. Rev. *49*, 1 (1985).
50 T. Nakae: CRC Crit. Rev. Microbiol. *13*, 1 (1986).
51 R. E. W. Hancock: J. Bacteriol. *169*, 929 (1987).
52 J. Foulds and T.-J. Chai: J. Bacteriol. *133*, 1478 (1978).
53 T. I. Nicas and R. E. W. Hancock: J. Bacteriol. *153*, 281 (1983).
54 T. Nakae, J. Ishii and M. Tokunagu: J. Biol. Chem. *254*, 1457 (1979).
55 A. Engel, A. Massalski, M. Schindler, D. L. Dorset and J. P. Rosenbusch: Nature *317*, 643 (1985).
56 B. Lugtenberg, R. Peters, H. Bernheimer and W. Berendsen: Molec. Gen. Genet. *147*, 251 (1976).
57 L. van Alphen and B. Lugtenberg: J. Bacteriol. *131*, 623 (1977).
58 H. Nikaido, E. Y. Rosenberg and J. Foulds: J. Bacteriol. *153*, 241 (1983).
59 B. A. Boehler-Kohler, W. Boos, R. Dieterle and R. Benz: J. Bacteriol. *138*, 33 (1979).
60 J. Sutcliffe, R. Blumenthal, A. Walter and J. Foulds: J. Bacteriol. *156*, 867 (1983).
61 K. J. Harder, H. Nikaido and M. Matsuhashi: Antimicrob. Ag. Chemother. *20*, 549 (1981).
62 T. Sawai, R. Hiruma, N. Kawana, M. Kaneko, F. Taniyasu and A. Inami: Antimicrob. Ag. Chemother. *22*, 585 (1982).
63 J. Mitsuyama, R. Hiruma, A. Yamaguchi and T. Sawai: Antimicrob. Ag. Chemother. *31*, 379 (1987).
64 J. T. Douglas, M. D. Lee and H. Nikaido: FEMS Microbiol. Lett. *12*, 305 (1981).
65 F. Yoshimura, L. S. Zalman and H. Nikaido: J. Biol. Chem. *258*, 2308 (1983).
66 R. Benz, M. Gimple, K. Poole and R. E. W. Hancock: Biochim. biophys. Acta *730*, 387 (1983).

67   R. E. W. Hancock and A. M. Carey: FEMS Microbiol. Lett. *8*, 105 (1980).
68   J. L. Burns and A. L. Smith: J. Gen. Microbiol. *133*, 1273 (1987).
69   J. E. Gabbay, M. Blake, W. D. Niles and M. A. Horwitz: J. Bacteriol. *162*, 85 (1985).
70   T. R. Parr, R. A. Moore, L. V. Moore and R. E. W. Hancock: Antimicrob. Ag. Chemother. *31*, 121 (1987).
71   S. K. Armstrong, T. R. Parr Jr, C. D. Parker and R. E. W. Hancock: J. Bacteriol. *166*, 212 (1986).
72   R. P. Darveau, W. T. Charnetzsky, R. E. Harlbert and R. E. W. Hancock:
73   P. Bavoil, A. Ohlin and J. Schachter: Infect. Immun. *44*, 479 (1984).
74   R. E. W. Hancock and H. Nikaido: J. Bacteriol. *136*, 381 (1978).
75   F. Yoshimura, L. S. Zalman and H. Nikaido: J. Biol. Chem. *258*, 2308 (1983).
76   B. L. Angus, A. M. Carey, D. A. Caron, A. M. B. Kropinski and R. E. W. Hancock: Antimicrob. Ag. Chemother. *21*, 299 (1982).
77   R. Benz and R. E. W. Hancock: Biochim. biophys. Acta *646*, 298 (1981).
78   T. I. Nicas and R. E. W. Hancock: J. Bacteriol. *153*, 281 (1983).
79   C. A. Caulcott, M. R. W. Brown and I. Gonda: FEMS Microbiol. Lett. *21*, 119 (1984).
80   H. Yoneyama and T. Nakae: Eur. J. Biochem. *157*, 33 (1986).
81   H. Yoneyama, A. Akatsuka and T. Nakae: Biochem. Biophys. Res. Commun. *134*, 106 (1986).
82   F. Yoshimura and H. Nikaido: J. Bacteriol. *152*, 636 (1982).
83   W. A. Woodruff, T. R. Parr Jr, R. E. W. Hancock, L. F. Hanne, T. I. Nicas and B. H. Iglewski: J. Bacteriol. *167*, 473 (1986).
84   R. A. Moore, W. A. Woodruff and R. E. W. Hancock: Antibiot. Chemother. *39*, 172 (1987).
85   Y. Kobayashi and T. Nakae: Biochem. Biophys. Res. Commun. *141*, 292 (1986).
86   Y. Kobayashi, A. Akatsuka and T. Nakae: FEMS Microbiol. Lett. *48*, 325 (1987).
87   H. Nikaido: Biochim. biophys. Acta *433*, 118 (1976).
88   H. Nikaido. In: Bacterial Membranes Biogenesis and Functions, p. 361. Ed. M. Inouye. John Wiley and Sons, New York (1979).
89   P. G. Lysko and S. A. Morse: J. Bacteriol. *145*, 946 (1981).
90   J. R. W. Govan and J. A. M. Fyfe: J. Antimicrob. Chemother. *4*, 233 (1978).
91   R. T. Irvin, J. R. W. Govan, J. A. M. Fyfe and J. W. Costerton: Antimicrob. Ag. Chemother. *19*, 1056 (1981).
92   J. L. Colaizzi and P. R. Klink: J. Pharmaceut. Sci. *58*, 1184 (1969).
93   M. Barza, R. B. Brown, C. Shanks, C. Gamble and L. Weinstein: Antimicrob. Ag. Chemother. *8*, 713 (1979).
94   T. R. Tritton: Biochemistry *16*, 4133 (1977).
95   P. R. Ball, I. Chopra and S. J. Eccles: Biochem. Biophys. Res. Commun. *77*, 1500 (1977).
96   H. Nikaido, E. Y. Rosenberg and J. Foulds: J. Bacteriol. *153*, 232 (1983).
97   F. Yoshimura and H. Nikaido: Antimicrob. Ag. Chemother. *27*, 84 (1985).
98   K. Murakami and T. Yoshida: Antimicrob. Ag. Chemother. *21*, 254 (1982).
99   H. Nikaido and E. Y. Rosenberg: J. Gen. Physiol. *77*, 121 (1981).
100 H. Nikaido and E. Y. Rosenberg: J. Bacteriol. *153*, 241 (1983).
101 E. M. Renkin: J. Gen. Physiol. *38*, 225 (1954).
102 J. S. Chapman and N. H. Georgopapodakou: Antimicrob. Ag. Chemother. *31*, 1994 (1987).
103 R. Nakae and T. Nakae: Antimicrob. Ag. Chemother. *22*, 554 (1982).
104 R. E. W. Hancock, V. J. Raffle and T. I. Nicas: Antimicrob. Ag. Chemother. *19*, 777 (1981).
105 R. A. Moore and R. E. W. Hancock: Antimicrob. Ag. Chemother. *30*, 923 (1986).

106   T. I. Nicas and R. E. W. Hancock: J. Bacteriol. *143*, 872 (1980).
107   T. I. Nicas and R. E. W. Hancock: J. Gen. Microbiol. *129*, 509 (1983).
108   B. D. Davis, L. Chen and P. C. Tai: Proc. Natl Acad. Sci. USA *83*, 6164 (1986).
109   M. Vaara and T. Vaara: Nature *303*, 526 (1983).
110   M. Vaara and T. Vaara: Antimicrob. Ag. Chemother. *24*, 107 (1983).
111   M. Vaara and T. Vaara: Antimicrob. Ag. Chemother. *24*, 114 (1983).
112   R. E. W. Hancock and P. G. W. Wong: Antimicrob. Ag. Chemother. *26*, 48 (1984).
113   R. E. W. Hancock: Antibiot. Chemother. *36*, 95 (1985).
114   R. E. W. Hancock: Ann. Rev. Microbiol. *38*, 237 (1984).
115   H. W. Taber, J. P. Mueller, P. F. Miller and A. S. Arrow: Microb. Rev. *51*, 439 (1987).
116   L. E. Bryan and H. M. Van den Elzen: Antimicrob. Ag. Chemother. *12*, 163 (1977).
117   L. E. Bryan and S. Kwan: J. Antimicrob. Chemother. *8*, Suppl. D, 1 (1981).
118   I. Chopra and T. G. B. Howe: Microbiol. Rev. *42*, 707 (1978).
119   R. J. Wargel, C. A. Shadur and F. C. Neuhaus: J. Bacteriol. *105*, 1028 (1971).
120   F. M. Kahan, J. S. Kahan, P. J. Cassidy and H. Kropp: Ann. N. Y. Acad. Sci. *235*, 304 (1974).
121   J. M. Diver, J. M. Andrew, L. J. V. Piddock and R. Wise: J. Infect. Dis. (in press, 1988).
122   J. Bedard, S. Wong and L. E. Bryan: Antimicrob. Ag. Chemother. *31*, 1348 (1987).
123   R. J. Harvey and A. L. Koch: Antimicrob. Ag. Chemother. *18*, 323 (1980).
124   J. L. Burns and A. L. Smith: Antimicrob. Ag. Chemother. *31*, 686 (1987).
125   I. Chopra: Br. Med. Bull. *40*, 11 (1984).
126   S. Zighelboim and A. Tomasz: Antimicrob. Ag. Chemother. *17*, 434 (1984).
127   T. J. Dougherty, A. E. Koller and A. Tomasz: Antimicrob. Ag. Chemother. *18*, 730 (1980).
128   M. V. Hayes, N. A. C. Curtis, A. W. Wyke and J. B. Ward: FEMS Microbiol. Lett. *10*, 119 (1981).
129   K. Bush, S. K. Tanaka, D. P. Bonner and R. B. Sykes: Antimicrob. Ag. Chemother. *27*, 555 (1985).
130   J. S. Bakken, C. C. Sanders and K. S. Thomson: J. Infect. Dis. *155*, 1220 (1987).
131   A. A. Madeiros, T. O'Brien, E. Y. Rosenberg and H. Nikaido: J. Infect. Dis *156*, 751 (1987).
132   C. C. Sanders and W. E. Sanders Jr: J. Infect. Dis. *151*, 399 (1985).
133   F. Malouin and F. Lamothe: Can. J. Microbiol. *33*, 262 (1987).
134   A. H. Seeberg, R. M. Tolxdorff-Neutzling and B. Wiedemann: Antimicrob. Ag. Chemother. *23*, 918 (1983).
135   A. S. Bayer, J. Peters, T. R. Parr Jr, L. Chan and R. E. W. Hancock: Antimicrob. Ag. Chemother. *31*, 253 (1987).
136   L. J. V. Piddock and R. Wise: J. Antimicrob. Chemother. *16*, 279 (1985).
137   C. C. Sanders and C. Watanakunakorn: J. Infect. Dis. *153*, 617 (1986).
138   R. L. Then and P. Anghern: Antimicrob. Ag. Chemother. *21*, 711 (1982).
139   H. Vu and H. Nikaido: Antimicrob. Ag. Chemother. *27*, 393 (1985).
140   D. M. Livermore: J. Antimicrob. Chemother. *15*, 511 (1985).
141   T. D. Gootz, D. B. Jackson and J. C. Sherris: Antimicrob. Ag. Chemother. *25*, 591 (1984).
142   G. Korfmann, C. Kliebe and B. Wiedemann: J. Antimicrob. Chemother. *18*, Suppl. C, 113 (1986).
143   F. Lindberg, L. Westman and S. Normark: Proc. Natl Acad. Sci. USA *82*, 4620 (1985).

144 F. Lindberg, S. Lindquist and S. Normark: J. Antimicrob. Chemother. *18*, Suppl. C, 43 (1986).
145 D. M. Livermore and Y. J. Yang: J. Infect. Dis. *155*, 775 (1987).
146 A. J. Godfrey, L. Hatlelid and L. E. Bryan: Antimicrob. Ag. Chemother. *26*, 181 (1984).
147 A. J. Godfrey and L. E. Bryan: Antimicrob. Ag. Chemother. *26*, 485 (1984).
148 J. L. Hoekstra, A. J. de Neeling, B. van Klingeren, E. E. Stobberingh and C. P. A. van Boven: Eur. J. Clin. Microbiol. *6*, 22 (1987).
149 A. M. B. Kropinski, J. Kuzio, B. L. Angus and R. E. W. Hancock: Antimicrob. Ag. Chemother. *21*, 310 (1982).
150 A. J. Godfrey and L. E. Bryan: Antimicrob. Ag. Chemother. *31*, 1216 (1987).
151 A. J. Godfrey, M. S. Shahrabadi and L. E. Bryan: Antimicrob. Ag. Chemother. *30*, 802 (1986).
152 S. S. Pedersen, T. Pressler, N. Hoiby, M. W. Bentzon and C. Koch: J. Antimicrob. Chemother. *16*, 629 (1985).
153 J. P. Quinn, E. J. Dudek, C. A. DiVicenzo, D. A. Lucks and S. A. Lerner: J. Infect. Dis. *154*, 289 (1986).
154 M. J. Lynch, G. L. Drusano and H. L. T. Mobley: Antimicrob. Ag. Chemother. *31*, 1892 (1987).
155 W. Cullmann, K. H. Buscher and W. Opferkuch: Infect. Immunol. *14*, 227 (1986).
156 K. H. Buscher, W. Cullmann, W. Dick and W. Opferkuch: Antimicrob. Ag. Chemother. *31*, 703 (1987).
157 K. H. Buscher, W. Cullmann, W. Dick, S. Wendt and W. Opferkuch: J. Infect. Dis. *156*, 681 (1987).
158 C. C. Sanders, W. E. Sanders Jr, and R. V. Goering: Antimicrob. Ag. Chemother. *21*, 968 (1982).
159 A. H. Seeberg, R. M. Toxdorff-Neutzling and B. Wiedemann: Antimicrob. Ag. Chemother. *23*, 918 (1983).
160 B. Marchou, F. Bellido, R. Charnas, C. Lucain and J.-C. Pechere: Antimicrob. Ag. Chemother. *31*, 1589 (1987).
161 S. C. Wood, R. N. McCashion and W. H. Lynch: Antimicrob. Ag. Chemother. *29*, 992 (1986).
162 C. C. Sanders and C. Watanakunakorn: J. Infect. Dis. *153*, 617 (1986).
163 W. H. Traub and D. Bauer: Chemotherapy. *33*, 172 (1987).
164 J. Mitsuyama, R. Hiruma, A. Yamaguchi and T. Sawai: Antimicrob. Ag. Chemother. *31*, 379 (1987).
165 S. A. Benson and A. Deloux: J. Bacteriol. *161*, 361 (1985).
166 T. I. Nicas and R. E. W. Hancock: J. Bacteriol. *153*, 281 (1983).
167 S. A. Lerner and J. P. Quinn: Chemioterapia. *4*, 95 (1985).
168 J. S. Bakken, C. C. Sanders and K. S. Thomson: J. Infect. Dis. *155*, 1220 (1987).
169 A. A. Madeiros, T. F. O'Brien, E. Y. Rosenberg and H. Nikaido: J. Infect. Dis. *156*, 751 (1987).
170 L. E. Bryan, K. O'Hara and S. Wong: Antimicrob. Ag. Chemother. *26*, 250 (1984).
171 L. Galbraith, S. G. Wilkinson, N. J. Legakis, V. Genimata, T. A. Katsorchis and E. T. Rietschel: Annals Microbiol. *135*, 121 (1984).
172 N. J. Legakis, A. Velonaki, E. M. Velonakis and T. Lymberopoulou: Chemioterapia *5*, 185 (1986).
173 A. S. Bayer, D. C. Norman and K. S. Kim: Antimicrob. Ag. Chemother. *31*, 70 (1987).
174 I. Chopra and S. J. Eccles: Biochem. Biophys. Res. Commun. *83*, 550 (1978).

175 L. F. Guymon, D. L. Walstadt and P. F. Sparling: J. Bacteriol. *136*, 391 (1978).
176 C. F. Beck, R. Mutzel, J. Barbe and W. Muller: J. Bacteriol. *150*, 633 (1982).
177 S. B. Levy. In: Antimicrobial Drug Resistance, p. 191. Ed. L. E. Bryan. Academic Press, New York (1984).
178 I. Chopra: J. Antimicrob. Chemother. *18*, Suppl. C 51 (1986).
179 L. McMurry, R. E. Petrucci and S. B. Levy: Proc. Natl Acad. Sci. USA *77*, 3974 (1980).
180 D. C. Coleman, I. Chopra, S. W. Shales, T. G. B. Howe and T. J. Foster: J. Bacteriol. *153*, 921 (1983).
181 D. F. Gaffney, E. Cundliffe and T. J. Foster: J. Gen. Microbiol. *125*, 113 (1981).
182 M. Kono and K. O'Hara: J. Antibiot. *29*, 176 (1976).
183 J. L. Burns, C. E. Rubens, P. M. Mendelman and A. L. Smith: Antimicrob. Ag. Chemother. *29*, 445 (1986).
184 J. L. Burns, P. M. Mendelman, J. Levy, T. L. Stull and A. L. Smith: Antimicrob. Ag. Chemother. *27*, 46 (1985).
185 L. Gutmann, R. Williamson, N. Moreau, M.-D. Kitzis, E. Collatz, J. Acar and F. W. Goldstein: J. Infect. Dis. *151*, 501 (1985).
186 K. Hirai, H. Aoyama, S. Suzue, T. Irikura, S. Iyobe and S. Mitsuhashi: Antimicrob. Ag. Chemother. *30*, 248 (1986).
187 D. C. Hooper, J. S. Wolfson, K. S. Souza, C. Tung, G. L. McHugh and M. N. Swartz: Antimicrob. Ag. Chemother. *29*, 639 (1986).
188 L. J. V. Piddock and R. Wise: J. Antimicrob. Chemother. *18*, 547 (1986).
189 K. Hirai, S. Suzue, T. Irikura, S. Iyobe and S. Mitsuhashi: Antimicrob. Ag. Chemother. *31*, 582 (1987).
190 L. J. V. Piddock, W. J. A. Wijnands and R. Wise: Lancet. *ii* (8564), 907 (1987).
191 L. J. V. Piddock. In: Microbial Resistance to Drugs, The Handbook of Experimental Pharmacology. Ed L. E. Bryan. Springer-Verlag, Berlin (in press, 1988).
192 R. E. Wooley, M. S. Jones and E. B. Shotts Jr: Vet. Microbiol. *10*, 57 (1984/85).
193 M. H. Miller, S. A. Feinstein and R. T. Chow: Antimicrob. Ag. Chemother. *31*, 108 (1987).
194 P. H. Plotz and B. D. Davis: Science *135*, 1067 (1962).
195 J. Nuesch and F. Knusel. In: Antibiotics, p. 499. Eds D. Gottlieb and P. D. Shaw. Springer-Verlag, New York (1967).
196 A. Hartmann, H.-P. Fiedler and V. Braun: Eur. J. Biochem. *99*, 517 (1979).
197 H. Zahner, H. Diddens, W. Keller-Schierlein and H.-U. Nageli: J. Antibiot. *30*, 5 (1977).
198 D. S. Plaha and H. J. Rogers: Biochim. biophys. Acta *760*, 246 (1983).
199 N.-A. Watanabe, T. Nagasu, K. Katsu and K. Kitoh: Antimicrob. Ag. Chemother. *31*, 497 (1987).
200 J. B. Neilands: Microbiol. Sci. *1*, 9 (1984).
201 J. B. Neilands, A. Bindereif and J. Z. Montgomerie: Curr. Top. Microbiol. Immunol. *118*, 179 (1985).
202 A. Bagg and J. B. Neilands: Microbiol. Rev. *51*, 509 (1987).
203 M. R. W. Brown and P. Williams: J. Antimicrob. Chemother. *15*, Suppl. A, 7 (1985).

Hypertension: Relating drug therapy to pathogenetic mechanisms

By David H.P. Streeten and Gunnar H. Anderson Jr.

State University of New York Health Science Center, Syracuse, N.Y. 13210, U.S.A.

Supported by a grant (RO1 AG03055) from the National Heart, Lung and Blood Institute, and a Clinical Research Center Grant (RR229) from Division of Research Facilities and Resources, U.S. Public Health Service.

1   **Introduction**

Hypertension is not a single disease. It is a disorder of cardiovascular physiology which may result from a large variety of causes – known and unknown – and may be the consequence of several distinct pathogenetic mechanisms. Since the causes and the pathogenesis of hypertension are not always the same, it is reasonable to believe that optimal therapy should not always be the same and that treatment should be aimed, as far as possible, at correcting the specific etiology or pathogenetic mechanism that is present in each individual. Our gradually developing understanding of the physiological derangements leading to hypertension, coupled with the availability of steadily increasing numbers of new drugs with known mechanisms of action, offers improved opportunities to match the drug used with the defect that is responsible for the hypertension.

One of the recently popular programs of antihypertensive therapy, "stepped care," comprised the automatic administration of medications, starting with a diuretic, and adding, if necessary, a drug that acted on the sympathetic nervous system, followed by a vasodilator (such as hydralazine) and finally guanethidine, until the blood pressure was restored to normal [1,2]. The rapid advent of a profusion of new, highly effective hypotensive agents led to modifications of "stepped care" [3,4] that deprived the program of its main asset, simplicity. The practicing physician, faced with a vast array of effective drugs, each highly touted by its producer, has few rational guidelines for choosing which drug to use first and what other drugs to add or substitute when the first proves to be ineffective. Apparently convincing claims for the advisability of beta-blockers [5,6], diuretics [2,7], angiotensin converting enzyme inhibitors [8], alpha-1-adrenergic antagonists [9] and alpha-2-adrenergic agonists [10,11] as first-step drugs abound in the recent literature. Disillusioned by the shortcomings of "stepped care" and confused by conflicting data and claims, the physician is badly in need of a rational therapeutic approach to patients with hypertension. The object of this chapter is to describe how many of the known pathogenetic types of hypertension may be recognized by (usually) outpatient procedures and how such recognition may be used to prescribe logical and frequently effective forms of treatment. To illustrate one way of detecting many of the known forms of "secondary" hypertension, we shall describe and utilize the results of a systematic study of the pathogenesis of hypertension which has been in progress in Syracuse, New York, for the past 15 years [12]. The results obtained will be compared with those reported in the literature, for patients with the specific pathogenetic types of hypertension that can be delineated and which are shown in Table 1. Since blood pressure is a function of cardiac output and peripheral resistance, the initial defect which leads to hypertension may be elevation of cardiac output and/or of peripheral resistance, although even those types of hypertension that are initially associated with increased cardiac output [13] are *eventually* most closely correlated with increased peripheral resistance presumably because of autoregulation [14].

2   Customary approach to hypertensive patients

It is recommended at present that physicians evaluate hypertensive patients with a careful medical history and physical examination, followed by laboratory tests restricted to an urinalysis and measurements of hematocrit and serum concentrations of electrolytes, creatinine, glucose and cholesterol [15]. When any of the findings elicited in these procedures suggest a specific form of hypertension, most physicians would seek more direct evidence of such specific types. However, the initial laboratory tests are deliberately held to a minimum in order to reduce costs and because of the belief that most "secondary" forms of hypertension can be excluded on the basis of the history and physical examination [15].

Although this is the generally recommended approach to hypertension, its appropriateness has been challenged by some [16]. Its greatest weakness is that it is based on inadequate evidence. Thus, to claim that primary aldosteronism has been excluded because the serum potassium was normal is simply not valid [17,18]. Similarly, as Kaplan has recognized [16, p. 268], in apparent conflict with his general recommendation cited above, other than the presence of an abdominal bruit in 12% of patients with renovascular hypertension, "one can expect little help from the history, physical examination or routine laboratory work in deciding which patients deserve a work-up for renovascular hypertension".

3   **Procedure for pathogenetic evaluation of hypertensives**

In spite of the undoubted need to contain the costs of health in disorders as prevalent as hypertension, we have investigated the yield and the potential usefulness of an 8-hour diagnostic procedure for patients who were refractory to conventional "stepped care" therapy. The

Table 1
Pathogenetic types of hypertension

A *Increased cardiac output* due to hypervolemia resulting from or associated with
  1 "Low renin hypertension" mechanism unknown
  2 Renal sodium and water retention due to
    (i)   reduced glomerular filtration resulting from intrinsic renal disorders of various types
    (ii)   excessive renal tubular reabsorption of sodium and chloride due to
        a. Primary aldosteronism
        b. Ingested or injected agents: licorice, fludrocortisone, deoxycorticosterone
        c. Adrenal enzymatic disorders with excessive production of deoxycorticosterone or other mineralocorticoids
    (iii) excessive sodium chloride intake in "salt-sensitive" individuals.

B *Increased peripheral resistance* may be caused by
  1 Excessive alpha-adrenergic activity associated with
    (i)   "neurogenic" hypertension, cause unknown
    (ii)   pheochromocytoma, adrenal medullary hyperplasia
    (iii)   anxiety, nervousness, pain and psychic tension
    (iv)   physical "stress"
    (v)   orthostasis
    (vi)   drugs which increase alpha-adrenergic activity
  2 Excessive angiotensin II blood levels, associated with
    (i)   renal or pre-renal arterial constriction or compression
    (ii)   intrinsic renal diseases of many types
    (iii)   renal outflow obstruction
    (iv)   extrinsic renal compression (e.g. by subcapsular hemorrhage)
    (v)   renal infarction and cholesterol embolization
    (vi)   renal tumors
    (vii) excessive renin substrate (estrogen action, Cushing's syndrome)
  3 Structural causes of decreased arterial and/or arteriolar compliance
    (i)   atherosclerosis
    (ii)   necrotizing arteriolitis
    (iii)   inflammatory vasculitides
    (iv)   fibrosis and calcification of major arteries

C *Hypertension caused by other (usually unknown) mechanisms*
  1 Toxemia of pregnancy
  2 Hyperparathyroidism
  3 Hypothyroidism and hyperthyroidism
  4 Acromegaly
  5 Increased intracranial pressure
  6 Autonomic insufficiency (exclusively recumbent hypertension)
  7 Baroreceptor disorder
  8 Normal renin essential hypertension

---

procedure was designed to investigate the presence of recognizable pathogenetic types of hypertension (Table 1) in a single, out-patient session. In most of the patients the studies were performed after their antihypertensive medications had been tapered and stopped for at least one week before the studies. This program, summarized in Table 2, has been utilized in a consecutive series of over 3000 patients. The results obtained in the first 1036 patients have been followed up, and analyzed, and will be summarized here.

Table 2
Syracuse out-patient procedure for evaluation of hypertension

0745-0759	Consent document read and signed by patient. Measurements of B.P. and heart rate in the recumbent and standing postures. Insertion of "butterfly" needle into forearm vein. Blood samples drawn for measurements of serum concentrations of electrolytes, creatinine, T_4 and TSH. I.v. injection of furosemide, 40 mg.
0800-0859	Recumbency for 2 h to promote natriuresis. History and physical examination.
0900-1059	Standing and walking for 2 hours.
1100	Blood drawn for measurements of plasma concentrations of renin activity, catecholamines and aldosterone.
1105-1214	Recumbent; automatic measurements of B.P. every minute.
1145-1214	Saralasin infusion at 0.05, 0.5, 5.0 and 10.0 µg/kg/min.
1215-1244	Luncheon
1215-1529	0.9% NaCl solution 2L by i.v. infusion.
1530	Blood drawn for plasma aldosterone and cortisol measurements.

The rationale of each procedure listed in Table 2 is straightforward. The initial measurements of B.P. and heart rate in the recumbent and the upright postures serve to confirm the presence of hypertension in recumbency, to reveal orthostatic hypotension and tachycardia – common manifestations of pheochromocytoma [19] – and to diagnose orthostatic hypertension [20,21]. Measurement of serum Ca concentration is a simple screening test for hyperparathyroidism which may cause hypertension [22,23]. Serum T_4 and TSH concentrations serve to indicate the presence of hyperthyroidism, which may elevate systolic B.P., or hypothyroidism, a common finding in our hypertensive patients and the apparent cause of the hypertension in over 1% of the hypertensive series [24]. The furosemide injection and subsequent orthostasis are potent stimuli to renin release [25]. Measurement of plasma renin activity after these stimuli facilitates the recognition of low-renin hypertension – a very common disorder – and is also a valuable adjuvant in the diagnosis of primary aldosteronism and angiotensinogenic hypertension. Plasma catecholamine measurements are used as screening tests for pheochromocytoma [26]. A hypertensive response to the angiotensin II antagonist, saralasin, provides rapid and useful confirmation of the presence of low renin hypertension [27], while a hypotensive response to saralasin, a fall of at least 10 mmHg systolic and 8 mmHg diastolic, indicates that the hypertension is, at least partially, angiotensinogenic in type [12,28]. An elevated plasma aldosterone concentration after the saline infusion and/or – in the presence of a low PRA – after furosemide and standing, provides strong evidence of the presence of primary aldosteronism [18, 29]. The plasma cortisol concentration is useful in suggesting or ruling out Cushing's syndrome [30].

4   Follow-up diagnostic procedures for specific types of hypertension

4.1   *Orthostatic hypo- or hypertension.* Orthostatic hypotension is present when the systolic B.P. falls by 20 mmHg or more and/or the diastolic pressure falls by at least 10 mmHg after the patient has been standing for at least 3 minutes with the arm on which the measurements are made, elevated and resting at the level of the atria [31].

Orthostatic hypotension in the presence of recumbent hypertension and in the absence of orthostatic tachycardia may result from autonomic insufficiency [32] or antecedent beta-blocker therapy. Failure of the plasma norepinephrine concentration to rise in the standing posture confirms the presence of autonomic insufficiency [33] which is usually associated with other clinical features (orthostatic lightheadedness or syncope, absent thermal sweating, impotence) and laboratory abnormalities [34].

When recumbent hypertension and orthostatic hypotension are associated with orthostatic tachycardia (a postural increase in heart rate of > 28 beats/min. or to above 108 beats/min. [31]), *pheochromocytoma* should be considered [19]. However, this combination may also occur in other forms of hypertension particularly when there has been excessive previous administration of diuretic therapy.

Orthostatic hypertension is diagnosed when the diastolic B.P. is below

90 mmHg in recumbency and is 98 mmHg or more in the standing posture [31]. This disorder is probably identical with *"borderline hypertension"*, judging from the observations of Hull et al. [35] and Safar et al. [36]. It seems likely that the inconsistent finding of diastolic hypertension in past designations of borderline hypertension might have resulted from measurements of the blood pressure in the recumbent and the sitting postures on different occasions.

4.2   *Hypercalcemia* may result from previous diuretic therapy particularly with thiazides. In the absence of such previous treatment, the finding of hypercalcemia should be followed by measurements of serum inorganic phosphate and parathyroid hormone concentrations and other procedures for the definitive diagnosis of primary hyperparathyroidism.

4.3   *Elevation of the serum creatinine concentration* may result from severe dehydration but commonly indicates bilateral renal disease with renal insufficiency. The possibility should be borne in mind, particularly in patients who have evidence of atherosclerosis, that bilateral renal arterial stenosis might be the cause of the renal insufficiency. Whether or not this possibility warrants renal angiography is sometimes a difficult decision, compounded by the dangers of a renal "shut-down" resulting from the administered dye load if the serum creatinine is above 3 mg/dl.

4.4   *Serum thyroxine (T_4) and thyrotropin (TSH) concentrations* are used to diagnose *hyperthyroidism*, which may elevate systolic B.P., and *hypothyroidism* which in some patients causes diastolic ($\pm$ systolic) hypertension [24].

4.5   *Serum electrolyte concentrations.* Hypokalemia in hypertensive patients most commonly results from recent diuretic therapy or from vomiting or diarrhea. In the absence of such previous treatment or symptoms, the presence of primary or secondary aldosteronism should be suspected. Although changes in serum sodium concentration are considerably less helpful, mild hypernatremia does occur occasionally in primary aldosteronism and hyponatremia in patients with severe angiotensinogenic hypertension [37].

4.6   *Stimulated plasma renin activity (PRA),* when it is above the normal range (1.7 – 8.5 ng · ml⁻¹ h⁻¹), provides confirmation of the presence of a hyperreninemic state. However, this does not reliably indicate that the hypertension results from the actions of angiotensin since high PRA levels may result from estrogen administration, cirrhosis, nephrosis and other conditions even in the absence of hypertension. Furthermore, our studies have shown that a hypotensive response to saralasin infusion is far more reliable than an elevated PRA level as an indicator of angiotensinogenic hypertension [28,38]. A suppressed PRA (below 1.7 ng · ml⁻¹, h⁻¹) is the hallmark of *low renin hypertension,* including primary aldosteronism, and probably indicates the presence of hypervolemia [39-41].

4.7   *Plasma catecholamine* concentrations are valuable screening tests for pheochromocytoma [26]. When plasma norepinephrine concentration exceeds 500 pg/ml in recumbency and/or 1200 pg/ml in the upright posture, there is a strong possibility that the patient has a pheochromocytoma. This possibility is enhanced, and an intra-adrenal pheochromocytoma is likely when plasma epinephrine concentration is above 150 pg/ml concomitantly. A computerized tomographic (CT) scan of the adrenals and of the para-aortic area, followed, if positive, by an MIBG (meta-iodo-benzyl guanidine) scan will usually provide conclusive evidence of pheochromocytoma.

4.8   *Plasma aldosterone concentrations* are elevated above 30 ng/dl despite a subnormal PRA after the furosemide injection and standing for 2 hours, and fail to fall below 8.5 ng/dl after the 2-liter saline infusion in patients with *primary aldosteronism* [18,29]. Similar findings may occur in patients who have angiotensinogenic hypertension and secondary aldosteronism but a high PRA and a hypotensive response to saralasin clearly distinguish secondary from primary aldosteronism. CT scan of the adrenals and adrenal vein catheterization for plasma aldosterone and cortisol determinations are necessary for the conclusive demonstration of an aldosterone-producing adenoma.

4.9   *Saralasin response.* A mean fall in B.P. of at least 10 mmHg (systolic) and 8 mmHg (diastolic) during saralasin infusion in a hypertensive patient, exceeds the response of healthy normotensive subjects and is a reliable indication that hypertension is largely angiotensinogenic [12,28,42-47]. Further studies are needed to demonstrate whether there is unilateral or bilateral renal arterial constriction (by digital subtraction angiography) and whether excessive renin production is occurring from one kidney (by renal vein PRA measurements).

4.10   *Plasma cortisol concentration* at the end of the evaluation procedure ($\pm$ 1530 h), is normally below 15 μg/dl. Higher values suggest Cushing's syndrome and require substantiation by measurements of urinary 17-hydroxycorticosteroid and creatinine excretion and the responses to dexamethasone suppression [30].

5   Results of pathogenetic studies in hypertensive patients

We have analyzed the results obtained in the first 1036 consecutive patients referred for evaluation of poor responses to "stepped care" therapy. Most had been off their previous antihypertensive therapy for 3–14 days. The following pathogenetic types of hypertension were distinguished among the patients:

5.1   *Low renin hypertension* was defined as hypertension associated with a peripheral PRA, which was below the range (1.7-8.5 ng ml⁻¹h⁻¹) found in 28 normal subjects, after receiving intravenous furosemide and standing for 2 hours. Of the 1036 patients studied, 30% had low renin hypertension by this definition. Most of these patients (93%) could be identified on the day of the testing by a rise in B.P. during the saralasin infusion. The saralasin-induced rise in B.P. was at least 10 mmHg systolic and 8 mmHg diastolic in 49% of the low-renin hypertensive patients [27]. Because it has been shown that patients with low renin hypertension usually experience an excellent hypotensive response to chronic diuretic therapy [48,49] referring physicians were advised to treat these patients with increasing doses of oral diuretics, adding spironolactone or K supplements, if necessary, to prevent hypokalemia. After 3-4 months of exclusively diuretic therapy the diastolic blood pressure in 115 randomly selected patients with low renin hypertension was found to have fallen below control levels in 95% and below 90 mmHg in 69% (79 patients). An analysis was made of the B.P. response in all of the 31 low renin hypertensives who happened to have been treated with hypotensive drugs other than diuretics before and were given only diuretic therapy after their referral for our studies. Diastolic B.P. fell to lower levels in response to diuretics than in response to their previous therapy in 29 (93%) of these patients, and fell below 90 mmHg on diuretic therapy alone in 22 (71%) of them. In these 31 low renin patients, as a group, the B.P. (mean $\pm$ SEM) fell during diuretic therapy from $180.4 \pm 5.6/111.0 \pm 2.3$ to $142.2 \pm 4.1/89.2 \pm 2.1$ (Fig. 1) which was highly significant ($p < 0.001$) for systolic and diastolic changes, by student's t-test.

Some authorities believe that low renin hypertension is not a pathogenetically distinct form of hypertension [50]. Whether they are correct or not, however, our data strongly confirm the overwhelming evidence that in patients with low stimulated PRA levels, the B.P. almost always

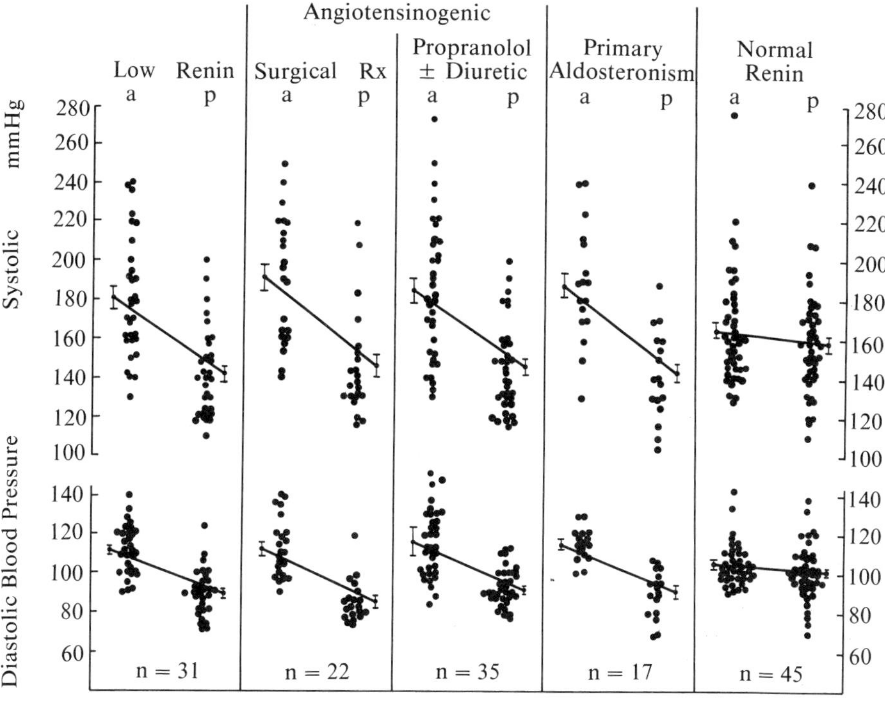

Fig.1

Blood pressures of hypertensive patients during therapy before (a) and after (p) the 8-hour study

Statistically significant reductions in mean systolic and diastolic blood pressures were found in 31 patients with low renin hypertension treated with diuretics exclusively, in 22 patients with surgically treated renovascular hypertension, in 35 patients with angiotensinogenic hypertension not due to renal arterial stenosis treated with propranolol with or without a diuretic, and in 17 patients with primary aldosteronism treated with surgical adrenalectomy (for adenoma) or Aldactone (for hyperplasia). No significant improvement occurred after therapeutic changes in 45 patients with normal renin essential hypertension.

falls, frequently into the normal range, in response to exclusively diuretic therapy [48,49]. In our experience, when diuretic drugs have failed to lower the B.P. in low renin hypertensive patients, it has usually been because the dosage or the type of diuretic used did not induce diuresis. When the dosage has been increased, or a longer-acting diuretic, such as chlorthalidone, has been substituted for a shorter-acting thiazide or loop diuretic in such patients, a more satisfactory reduction in B.P. has usually been accomplished. The use of this class of drugs provides simple, relatively inexpensive and very effective treatment in low renin hypertensives. Of course the efficacy of diuretic drugs may be determined empirically by starting all patients with hypertension on these drugs and continuing to use diuretics alone in those patients whose B.P. has responded adequately. In recent publications it has been shown that calcium channel blocking drugs are also particularly effective hypotensive agents in patients with low renin hypertension [51-53].

5.2   *Primary aldosteronism.* Among the 1036 patients studied there were 30 in whom either (i) plasma aldosterone concentration failed to fall below 8.5 ng/dl after the saline infusion and blood pressure did not fall during saralasin infusion, or (ii) a low stimulated PRA was associated with hypokalemia (serum $K < 3.5$ mEq/l) on admission. Twenty-seven of these 30 patients were admitted to the hospital for 3-day suppression tests, using deoxycorticosterone (DOC, 10 mg IM every 12 h) or fludrocortisone (0.2 mg every 12 h) [18]. In 22 of these 27 patients, urinary aldosterone excretion and/or plasma aldosterone concentration were not suppressed normally (below 8 μg/day and 8.5 ng/dl, respectively) by this mineralocorticoid administration and they were considered to have primary aldosteronism. A detailed discussion of their findings has been published elsewhere [18]. These patients were further studied for the presence of an unilateral adrenal adenoma with [131] I-iodocholesterol scans and/or computed tomography (CT scans) of the adrenals and/or adrenal vein catheterization for measurements of plasma aldosterone and cortisol concentrations in blood obtained from both adrenal veins and the inferior vena cava. Adrenalectomy was recommended in 10 patients who had a single adrenal adenoma. Seven of the 10 patients with evidence of a single adrenal adenoma were unilaterally adrenalectomized, one died of a cerebrovascular accident before surgery could be performed and two declined surgery. The 12 patients who were not operated upon were treated with spironolactone, 75-150 mg/day [54]. Surgery or spironolactone reduced B.P. in all except 1 of the 22 patients with primary aldosteronism, diastolic pressures falling below 90 mmHg in half of them. Mean change in B.P. from before to 6 months after the start of treatment in these patients was from $188.4 \pm 7.3/115.1 \pm 2.0$ to $142.2 \pm 5.6/81.1 \pm 2.9$ (Fig. 1) which was highly significant ($p < 0.001$) for systolic and diastolic changes.

The procedure used in our 8-hour studies to diagnose primary aldosteronism includes an adaptation of the saline infusion test first reported by Kem et al. [55]. However, if the plasma aldosterone response to saline infusion had been the sole method used to screen for primary aldosteronism, the sensitivity of the procedure would have been only 86% and its specificity 93%, since plasma aldosterone concentration was suppressed normally in some patients with aldosterone-producing adrenal adenomas while resistance to suppression was seen in 22% of angiotensinogenic hypertensives [18]. However, when unsuppressibility of plasma aldosterone level was combined with a low stimulated PRA, or with an agonistic response to saralasin, the sensitivity of these diagnostic tests reached 100%, and the specificity was 96% [29].

It is well known that in some patients with primary aldosteronism the blood pressure remains elevated after surgical removal of an aldosterone producing adenoma [56-58]. The cause of this phenomenon is unknown. Analysis of the individual responses to surgery in these patients has shown that all patients treated when they were less than 30 years of age were restored to normotension without need of medications but that as the age at time of surgery increased above 30, a progressively increasing number of patients required antihypertensive drugs postoperatively.

5.3   *Angiotensinogenic hypertension* was diagnosed in 105 patients by a fall in B.P. of $\geq 10/8$ mmHg during the saralasin infusion [12,28]. Renal vein catheterization, for measurements of PRA in both renal veins and in the inferior vena cava (below the renal veins), was performed in most of these patients, usually the next morning, after further diuresis with oral furosemide (80 mg) overnight. Blood samples for PRA measurements were obtained after the patients had been tilted, head-up, to 30° for 15 minutes. Isotopic renal scintigraphy, and/ or intravenous pyelography were performed on many of these patients in an attempt to define the nature of whatever renal lesion might be present. If the renal vein PRA ratio exceeded 1.5 or if the other studies suggested an unilateral renal arterial stenosis, renal arteriography was performed. This was required in a total of 51 patients (4.9% of the entire hypertensive group studied). A complete description of the results obtained has been published elsewhere [28]. Essentially, of 85 patients who could be re-evaluated, 22 were found to have a renal vein PRA ratio > 1.5, no significant renin production from the "better" kidney and arteriographic evidence of unilateral renal arterial stenosis or unilateral parenchymal renal disease. Unilateral renal arterial bypass surgery or nephrectomy lowered the diastolic B.P. considerably in all 22, and below 90 mmHg in 20 of these patients for more than 6 months without medical treatment. However, stenosis and hypertension recurred in 4 patients, who then required drug therapy. In this group of 22 patients after recurrence of the hypertension in 4, surgery lowered the mean B.P. from $191.0 \pm 6.9/111.9 \pm 3.5$ to $145.9 \pm 5.8/85.0 \pm 2.2$ mmHg (Fig. 1). The changes in systolic and diastolic B.P. resulting from surgery were highly significant ($p < 0.001$).

There are, of course, other methods of diagnosing renovascular hypertension. The consensus derived from published evidence [59] is, that, for the discovery of renovascular hypertension which is curable by angioplasty or by-pass renal arterial surgery, the sensitivity of the saralasin test (85.6%) is greater than that of the rapid-sequence intravenous urogram (76.9%) or the measurement of peripheral PRA (63.0%). Combination of the saralasin test with measurement of stimulated PRA should further increase the sensitivity of the screening for this form of hypertension. Moreover, renal arterial stenosis is a common finding in hypertensive patients without being the cause of the hypertension in more than about one third of the patients. In order to ensure that correction of a renal arterial stenosis will reduce blood pressure significantly, angiography should be supplemented by renin measurements and preferably by the saralasin test. Saralasin is no longer available for clinical use in the United States but can be purchased from the Röhm Pharma Company in Darmstadt, Germany.

Prior to the availability of captopril, the renin-lowering action of beta-adrenergic blockers [60] was utilized for the treatment of patients whose angiotensinogenic hypertension was associated with no demonstrable renal arterial stenosis or correctable renal lesion. Improvement in B.P. control was clearly evident (Fig. 1). More recently, captopril, enalapril, and other angiotensin converting enzyme (ACE) inhibitors have been used in this group of patients and have provided specific and highly effective therapy. It is of great importance in such patients, however, (a) to start with very small doses of ACE inhibitors to avoid excessive hypotensive responses to which these patients are prone, and (b) to be sure that serum creatinine concentration is below 2 mg/dl and does not rise with continued ACE therapy [61,62].

5.4   *Renal insufficiency.* Serum creatinine was above 1.5 mg/dl in 3.7% of the patients. Although it is recognized that impairment of renal excretory function may be a *result* of hypertension, there is strong evidence that reduced creatinine clearance is also an important *cause* of Na and water retention, with consequent hypertension [63]. In agreement with Blumberg et al. [63] we have recommended vigorous diuretic therapy and Na restriction as the best means of lowering the B.P. in such patients. Although the efficacy of this approach has been documented in several patients, its usefulness has not been systematically analyzed in the present series.

5.5   *Orthostatic hypertension* has been found in 10% of the patients studied. Subnormal orthostatic contractility of the veins with consequently excessive orthostatic venous pooling is the probable mechanism responsible for orthostatic hypertension [21,31]. Results of treatment in individuals of this type have shown a generally adverse effect of diuretic drugs. These sometimes aggravate the orthostatic rise in B.P., presumably by inducing hypovolemia and in this way worsening the reduction in venous return in the upright posture – the prime pathogenetic mechanism of orthostatic hypertension [21]. In other patients orthostatic hypertension has been converted by vigorous diuretic therapy into orthostatic hypotension associated with lightheadedness, "weakness", intolerance of the upright posture, syncope, and often inability to work [31].
It seems physiologically appropriate to administer alpha$_2$-adrenergic agonists in an attempt to increase venous contractility in these patients. A double-blind trial of the relative efficacy of clonidine and hydrochlorothiazide in patients with orthostatic hypertension, is in progress at present, in our Hypertension Unit. Until the results of this study are available, we believe it is important not to use diuretics in patients with orthostatic hypertension.

5.6   *Hypothyroid hypertension.* Measurements of blood pressure before and after the conversion of thyrotoxicosis to hypothyroidism by radioactive iodine therapy in 35 previously normotensive patients revealed a rise in diastolic B.P. to above 90 mmHg in 11 individuals [24]. In 9 of these 11 patients diastolic B.P. fell below 90 mmHg shortly after thyroxine therapy had restored euthyroidism. In a consecutive series of 688 hypertensive patients, hypothyroidism was found by serum T_4 and TSH measurements in 25 of them. When euthyroidism was restored by L-thyroxine administration, it was possible to maintain normotension in spite of stopping the use of conventional antihypertensive drugs in 8 of these patients. Thus, 1.2% of this series of hypertensive patients became normotensive on L-thyroxine replacement therapy without drugs of the conventional antihypertensive types [24].

5.7   *Rare types of hypertension.* Isolated instances of Cushing's syndrome (3 patients), pheochromocytoma (1 patient), and aortic coarctation (1 patient) have been encountered among the 1036 hypertensive patients reported here. Appropriate, specific forms of therapy have been administered, usually with good results.

5.8   *Normal renin essential hypertension* has been diagnosed by exclusion of the above 5 pathogenetic types of hypertension, and comprised 44% of the 1036 patients. Conventional therapy, usually of the "stepped care" type [1-3], has been recommended in these patients, and was not often very different after than before the 8-hour study. The mean of several B.P. recordings made before and during the 6 months after the 8-hour study were compared in 45 "normal renin" patients whose care before and after the study was undertaken in the outpatient department of our hospital. In these patients B.P. changed insignificantly from $165.1 \pm 4.1/105.3 \pm 2.1$ to $157.8 \pm 3.9/101.4 \pm 2.0$ mmHg (Fig. 1).
The blood pressure changes in these normal renin patients who continued to receive "stepped care" therapy, have been compared with the changes in the 3 groups for whom more specific forms of antihypertensive therapy were prescribed because of the findings of the 8-hour study. Before the study, the systolic and diastolic B.P. of the normal renin patients, $165.1 \pm 4.1/105.3 \pm 2.1$ was significantly lower than that of the other 3 goups as a whole: $185.2 \pm 3.0/113.1 \pm 1.4$ ($n = 105$, $p < 0.001$). After the study, when the normal renin patients were on

conventional "stepped care" treatment, they were found to have significantly *higher* systolic and diastolic blood pressures 157.8 ± 3.9/101.4 ± 2.0 mmHg than the 3 more specifically treated groups alluded to above: 144.2 ± 2.3/89.3 ± 1.1 mmHg (n = 105, p < 0.001). Thus, the reduction in B.P. observed after the 8-hour study was greater in the patients treated for low renin hypertension, angiotensinogenic hypertension and primary aldosteronism, -40.9 ± 2.6/-23.4 ± 2.6 than in the normal renin patients, -8.2 ± 3.5/-3.9 ± 1.9 (p > 0.001).

6   Therapeutic consequences of pathogenetic screening

It is calculated from the follow-up of the 105 patients in the various groups described above (excluding the normal renin group) that diastolic B.P. was reduced below the initial pressure by at least 10 mmHg in consequence of the recommended forms of therapy in 29 of 31 low renin hypertensives (93%), 46 of 57 surgically and medically treated angiotensinogenic hypertensives (81%), and 16 of the 17 patients with primary aldosteronism (94%), as well as some of the patients with renal insufficiency and the rarer causes of hypertension. Thus, among these groups of patients, it is estimated that 91 out of 105 patients (87%) who were treated as recommended, experienced improvement over their previous B.P. control. In contrast with these findings, only 14 of 45 patients (31%) with normal renin hypertension, whose treatment was not systematically changed after the 8-hour test, experienced a fall in diastolic blood pressure of more than 10 mmHg following the diagnostic procedure.

It is evident from the results obtained in these studies that improved control of hypertension was accomplished in almost 50% of patients by determining and treating the responsible pathogenetic mechanisms. These include the 30% of patients who have low renin hypertension, the 10% with angiotensinogenic hypertension and the 2% with primary aldosteronism, as well as smaller numbers of patients with hypothyroid hypertension, hyperparathyroidism, Cushing's syndrome, and pheochromocytoma. In addition to these patients, those with orthostatic hypertension constituted 10% of our referred patients and might make up a higher percentage of the entire hypertensive population, since the elevation of diastolic blood pressure is usually relatively mild in these patients. By recognizing patients with orthostatic hypertension and avoiding the use of diuretics in their therapy it is likely that side effects of treatment and consequent non-compliance will be reduced since undesirable side effects are an important cause of non-compliance. With the procedure described we have been able to find and cure 12 patients with pheochromocytoma in the past 4 years since catecholamine measurements by HPLC became available, compared with 4 patients in the preceding 23 years. The 8-hour program we have used has been modified and augmented over the course of the past 15 years. Though it is still imperfect, and requires a full-time nurse dedicated to the program, we believe it has been helpful in improving blood pressure control in a large number of patients, by attempting to fit the therapy to be pathogenetic mechanisms identified in individual patients.

7   Which hypertensives should be studied?

It is clear that patients with hypertension are not pathogenetically homogeneous and that many will be harmed if we ignore this fact. It is equally obvious that present attempts to match therapy with pathogenesis are too costly for universal application. What is badly needed is a simple and inexpensive diagnostic procedure which would guide physicians in their choice of therapy for individual patients. The procedure we have employed is only a first step in that direction, but has been beneficial to almost half of the patients we have studied. Until a more generally applicable diagnostic program evolves, we believe an approach similar to the one we have employed, might be beneficial and even cost-effective if used as follows:

1) A general physical examination should include 2 or 3 blood pressure determinations in the recumbent posture and 2 or 3 after the patient has been standing for about 3 minutes. This simple maneuver would diagnose orthostatic hypertension in about 10% of presently referred patients, in our experience, and perhaps a far higher percentage of hypertensives who are never referred for a second opinion. Whether patients with this disorder face the same risk of hypertensive complications as persistently hypertensive patients has never been studied. At least it is reasonably clear that they do not need expensive laboratory procedures for their further evaluation.

2) The vast majority of the remaining patients who have normal serum creatinine and potassium concentrations, no papilledema or fundal exudates or hemorrhages, no abdominal bruit, and no physical stigma-

ta of aortic coarctation, Cushing's syndrome, hypothyroidism or other clinically recognizable forms of "secondary" hypertension, should be started on diuretic therapy, as is the current practice in many clinics. If the blood pressure is satisfactorily controlled, as it will be in most low renin hypertensives, these patients (approximately 30%) will usually require no further studies unless hypokalemia becomes intractable, in which case primary and secondary aldosteronism will need to be excluded.

3) If effective diuretic therapy fails to lower blood pressure, the addition of a long-acting beta-blocker, or some other variant of the "stepped care" program [4] is worthy of trial.

4) Patients who have clinical or laboratory evidence of "secondary" hypertension, who have responded poorly to "stepped care", who are young, or who have malignant hypertension deserve further evaluation by procedures such as the one we have used, usually as out-patients or (for patients with severe or malignant hypertension) in the hospital.

References

1 M. Moser: New York State J. Med. *77*, 1753 (1977).
2 Joint National Committee on Detection, Evaluation and Treatment of High Blood Pressure: Arch. Intern. Med. *140*, 1280 (1980).
3 The 1984 Report of the Joint National Committee on Detection, Evaluation and Treatment of High Blood Pressure: Arch. Intern. Med. *144*, 1045 (1984).
4 A. Zanchetti: J. Hypertension *3* (Suppl 2), S57 (1985).
5 R. Turner, P. White and F. Benton: Practitioner *261*, 431 (1970).
6 F.R. Bühler, In F. Gross (Ed): The Cardioprotective Action of Beta Blockers. University Park Press, Baltimore (1976).
7 R.W. Gifford: Am. J. Med. *81* (Suppl 6C), 33 (1986).
8 Veterans Administration Cooperative Study Group on Antihypertensive Agents: Captopril. Clin. Sci. 63, *443s* (1982).
9 D.S. Bloom, C. Rosendorff and R. Kramer: Curr. Ther. Res. *18*, 144 (1975).
10 A.A. Lawson and M. Keston: Curr. Med. Res. Opin. *6*, 168 (1979).
11 M.P. Sambhi: Chest *83*, 427 (1983).
12 D.H.P. Streeten, G.H. Anderson, Jr., J.M. Freiberg and T.G. Dalakos: New Engl. J. Med. *292*, 657 (1975).
13 G.J. Wenting, A.J. Man in't Veld, R.P. Verhoeven, F.H.M. Derkx and M.A.D.H. Schalekamp: Circ. Res. *40* (Suppl 1), 163 (1977).
14 R.C. Tarazi, M.M. Ibrahim, E.L. Bravo and H.P. Dustan: New Engl. J. Med. *289*, 1330 (1973).
15 N.M. Kaplan: Clinical Hypertension, 3rd Edition, Williams & Wilkins, Baltimore (1983).
16 J.C. Melby and F.A. Finnerty: J. Am. Med. Assoc. *231*, 399 (1975).
17 M.H. Weinberger, C.E. Grim, J.W. Hollifield, D.C. Kem, A. Ganguly, N.J. Kramer, H.Y. Yune, H. Wellman and J.P. Donohue: Ann. Intern. Med. *90*, 386 (1979).

18   D.H.P. Streeten, N. Tomycz and G.H. Anderson, Jr: Am. J. Med. *67*, 403 (1979).
19   R.H. Smithwick, W.E.R. Greer, C.W. Robertson and R.W. Wilkins: New Engl. J. Med. 242, 252 (1950).
20   R.P. Sapru, P. Sleight, I.S. Anand, M.P. Sambhi, R. Lopez and P.N. Chuttani: Am. J. Med. *66*, 177 (1979).
21   D.H.P. Streeten, J.H. Auchincloss, Jr., G.H. Anderson, Jr., R.L. Robertson, F.D. Thomas and J.W. Miller: Hypertension *7*, 196 (1985).
22   T. Christensson, K. Hellström and B. Wengle: Eur. J. Clin. Invest. *7*, 109 (1977).
23   F.D. Rosenthal: Brit. Med. J. *4*, 396 (1972).
24   D.H.P. Streeten, G.H. Anderson, Jr., T. Howland, R. Chiang and H. Smulyan: Hypertension *11*, 78 (1988).
25   E.L. Cohen, J.W. Conn and D.R. Rovner: J. Clin. Invest. *46*, 418 (1967).
26   E.L. Bravo, R.C. Tarazi, R.W. Gifford and B.H. Stewart: New Engl. J. Med. *301*, 682 (1979).
27   G.H. Anderson, Jr., D.H.P. Streeten and T.G. Dalakos: Circ. Res. *40*, 243 (1977).
28   D.H.P. Streeten and G.H. Anderson, Jr: Kidney Internat. *15*, S44 (1979).
29   D.H.P. Streeten, G.H. Anderson, Jr. and J.M. Springer: Clin. Sci. *63*, 125s (1982).
30   D.H.P. Streeten, C.T. Stevenson, T.G. Dalakos, J.J. Nicholas, L.G. Dennick and H. Fellerman: J. Clin. Endocrin. Metab. *29*, 1191 (1969).
31   D.H.P. Streeten: Orthostatic Disorders of the Circulation. Plenum, New York 1987.
32   A.V. Chobanian, L. Volicer, C.P. Tifft, H. Gavras, C-S. Liang and D. Faxon: New Engl. J. Med. *301*, 68 (1979).
33   M.G. Ziegler, C.R. Lake and I.J. Kopin: New Engl. J. Med. *296*, 293 (1977).
34   R. Bannister: Autonomic Failure: A Textbook of Clinical Disorders of the Autonomic Nervous System. Oxford Univ. Press, Oxford 1983.
35   D.H. Hull, R.A. Wolthuis, T. Cortese, et al: Am. Heart J. *94*, 414 (1977).
36   M.E. Safar, Y.A. Weiss and J.A. Levenson: Am. J. Cardiol. *31*, 315 (1973).
37   J. Brown, D.I. Davis, A.F. Lever and J.I.S. Robertson: Brit. Med. J. *2*, 144 (1965).
38   D.H.P. Streeten, G.H. Anderson, Jr., T.G. Dalakos and J.M. Freiberg: Am. J. Med. *60*, 817 (1976).
39   P. Weidmann, D. Hirsch, C. Beretta-Piccoli, F.C. Reubi and W.H. Ziegler: Am. J. Med. *62*, 209 (1977).
40   A. Jose, J.R. Crout and N.M. Kaplan: Ann. Intern. Med. *72*, 9 (1970).
41   M. Esler, O. Randall and J. Bennett: Lancet 2, 115 (1976).
42   E.D. Vaughan, F.R. Bühler, J.H. Laragh, J.E. Sealey, L. Baer and R.H. Bard: Am. J. Med. *55*, 402 (1973).
43   L.S. Marks, M.H. Maxwell and J.J. Kaufman: Ann. Intern. Med. *87*, 176 (1977).
44   E.F. Poutasse, L. Gonzalez-Serva, J.R. Wendelken and J.P. Franz: J. Urol. *123*, 306 (1980).
45   L. Baer, J.Z. Parra-Carrillo, I. Radichevich and G.S. Williams: Ann. Intern. Med. *86*, 257 (1977).
46   H.M. Wilson, J.P. Wilson, P.E. Slaton, J.H. Foster, G.W. Liddle and J.W. Hollifield: Ann. Intern. Med. *87*, 36 (1977).
47   M.D. Lifchitz, M.A. Kirschenbaum, S.G. Rosenblatt and R. Gibney: Ann. Intern. Med. *88*, 23 (1978).
48   M.G. Crane and J.J. Harris: Am. J. Med. Sci. *260*, 311 (1970).
49   R.M. Carey, J.G. Douglas, J.R. Schweikert, Jr. and G.W. Liddle: Arch. Intern. Med. *130*, 849 (1972).

50 P.L. Padfield, D.G. Beevers, J.J. Brown, D.L. Davies, A.F. Lever, J.I.S. Robertson, M.A.D. Schalekamp and M. Tree: Lancet *1*, 548 (1975).
51 P. Erne, P. Bolli, O. Bertel, U.L. Hulthén, W. Kiowski, F. Müller and F.R. Bühler: Hypertension *5*, 11 (1983).
52 W. Kiowski, F.R. Bühler, M.O. Fadayomi, P. Erne, F.B. Muller, U.L. Hulthén and P. Bolli: Am. J. Cardiol. *56*, 81H (1985).
53 L.M. Resnick and J.H. Laragh: Am. J. Cardiol. *56*, 68H (1985).
54 J.J. Brown, D.L. Davies, J.B. Ferriss, R. Fraser, E. Haywood, A.F. Lever and J.I.S. Robertson: Brit. Med. J. *2*, 729 (1972).
55 D.C. Kem, M.H. Weinberger, D.M. Mayes, et al: Arch. Intern. Med. *128*, 380 (1971).
56 J.W. Conn: Harvey Lect. *62*, 257 (1968).
57 T.K. Hunt, M. Schambelan and E.R. Biglieri: Ann. Surg. *182*, 353 (1975).
58 J.B. Ferriss, D.G. Beevers, K. Boddy, J.J. Brown, D.L. Davis, R. Fraser, D. Kremer, A.F. Lever and J.I.S. Robertson: Am. Heart. J. *96*, 97 (1978).
59 D.H.P Streeten, G.H. Anderson, Jr., F.S. Sunderlin, J.S. Mallov and J. Springer, In Frontiers in Hypertension Research (Eds. J.H. Laragh, F.R. Bühler, D.W. Seldin). Springer-Verlag, New York 1981
60 F.R. Bühler, J.H. Laragh, E.D. Vaughan, H.R. Brunner, H. Gavras and L. Baer: Am. J. Cardiol. *32*, 511 (1973).
61 D.E. Hricik, P.J. Browning, R. Kopelman, W.E. Goorno, N.E. Madias and V.J. Dzau: New Engl. J. Med. *308*, 373 (1983).
62 J.J. Curtis, R.G. Luke, J.D. Whelchel, A.G. Diethelm, P. Jones and H.P. Dustan: New Engl. J. Med. *308*, 377 (1983).
63 A. Blumberg, H.B. Nelp, R.M. Hegstrom and B.H. Scribner: Lancet *2*, 69 (1967).

Adenosine receptors: Clinical implications and biochemical mechanisms

Vickram Ramkumar, George Pierson and Gary L. Stiles*
Departments of Medicine and Biochemistry, Duke University Medical
Center, Durham, NC 27710, USA

* G.L.S. is an Established Investigator of the American Heart Association, and is supported in part by National Heart, Lung, and Blood Institute Grant RO1HL35134, and Grant-in-aid 85612 from the American Heart Association, with funds contributed in part by the North Carolina Affiliate.

1   Introduction

Adenosine receptors comprise a group of cell surface receptors which mediate the physiological effects of adenosine. These receptors were originally classified by Burnstock as P_1 and P_2 purinergic receptors, depending on their preference for adenosine or adenine nucleotides [1]. The relative potencies for P_1 sites are adenosine $\geq$ AMP $\geq$ ADP $\geq$ ATP, while relative potencies for P_2 sites are ATP $\geq$ ADP $\geq$ AMP $\geq$ adenosine [1]. Further distinctions between P_1 and P_2 sites can be made by the use of selective antagonists. For example, methylxanthines, such as theophylline and 3-isobutyl-1-methylxanthine, are potent inhibitors of P_1 receptors, but are without effect on P_2 sites. These latter sites are selectively blocked by quinidine and high concentrations of 2-substituted imidazoline compounds such as antazoline and phentolamine, and 2'2-dipyridilisatogen.

The adenosine-sensitive P_1 sites have been further subdivided on the basis of their differential selectivity for a series of adenosine analogs (2,3). Thus, the potency series for A_1 adenosine receptor-mediated responses is R-phenylisopropyladenosine (R-PIA) = cyclohexyladenosine (CHA) > 5'-N-ethylcarboxamide adenosine (NECA) > S-PIA, while the A_2 receptors potency series is NECA > 2-chloroadenosine (2-CAdo) > R-PIA = CHA > S-PIA.

Methylxanthines are, in general, nonselective antagonists of these receptors. A number of substituted 8-phenylxanthine derivatives (4-7) have been shown to interact with A_1 receptors with high affinity in the nanomolar range. One such compound, 8-(4-[([[(2-aminoethyl)amino]carbonyl]methyl)oxy]phenyl]-1,8-diallylxanthine (XAC), appears relatively selective for A_1 adenosine receptors (7). Replacement of methyl groups at the 1- or 7-position of caffeine by propargyl or propyl groups increases selectivity at A_2 adenosine receptors (8).

The most extensively studied effector system coupled to adenosine receptors is the adenylate cyclase system. In most tissues, A_1 adenosine receptors are coupled to the inhibition of adenylate cyclase, while A_2 receptors mediate the activation of this enzyme (see Fig. 1). Adenosine receptors modulate adenylate cyclase activity via G proteins, G_s and G_i for A_2 and A_1 receptors, respectively. Optimal activation of these G proteins is dependent on guanine nucleotides such as GTP, Gpp(NH)p or GTPγS. In general, full activation of G_s is attained at GTP concentrations in the submicromolar range while activation of G_i

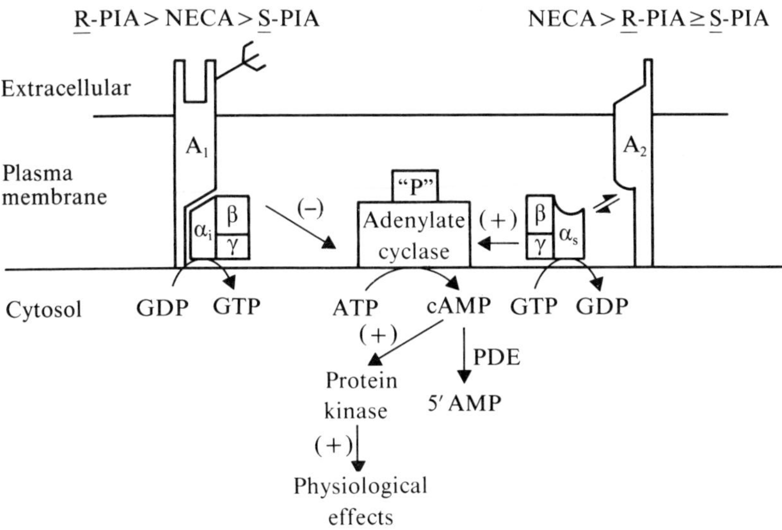

Fig. 1

Schematic representation of the components of the A_1- and A_2-adenosine receptor-adenylate cyclase system. Demonstrated at the top are the agonist potency series for both the A_1 and A_2 receptors. The A_1 receptor is shown as being tightly coupled to its guanine nucleotide regulatory protein, G_i, which is depicted as a complex of its three subunits ($\alpha_i\beta\gamma$). This complex inhibits the activity of the enzyme adenylate cyclase. The A_1 receptor is shown to contain an extracellular carbohydrate chain (ψ) as described in the text. The A_2 receptor is shown with its associated guanine nucleotide regulatory protein, G_s, which is depicted as a complex of three subunits ($\alpha_s\beta\gamma$). This complex activates adenylate cyclase. The 'P' site is a putative site on the catalytic unit through which ribose-modified adenosine analogs directly inhibit adenylate cyclase activity. Abbreviations are R-PIA, R-phenylisopropyladenosine; S-PIA, S-phenylisopropyladenosine, and NECA, 5′-N-ethylcarboxamide adenosine.

is only evident at nucleotide concentrations in the micromolar range (9,10). ADP ribosylation of G_i by pertussis toxin uncouples A_1 adenosine receptor from this protein and this results in a loss of R-PIA-mediated inhibition of fat cells (11). It has been suggested that presynaptic regulation of adenylate cyclase via A_1 and A_2 receptors might lead to inhibition and stimulation of neurotransmitter release, respectively (12).

Voltage sensitive Ca^{2+} channels appear to represent another possible site of adenosine's actions. The A_1 agonist, 2-chloroadenosine, inhibits K^+-dependent $^{45}Ca^{2+}$ uptake into cortical synaptosomal preparation (13). Furthermore, electrophysiological studies indicate that adenosine blocks voltage-sensitive Ca^{2+} channels in hippocampal neurons via a presynaptic A_1 receptors (14), while 2-chloroadenosine directly inhibits Ca^{2+} currents into rat dorsal root ganglion cells (12).

In addition to Ca^{2+} channels, activation of the 4-aminopyridine-sensitive K^+ channels by adenosine and its analogs has been demonstrated (14) in hippocampal neurons. A similar effect of adenosine has been described for cardiac K^+ conductance (15). In atrial myocytes, it appears that both adenosine and muscarinic acetylcholine receptors are coupled to the activation of the same K^+ conductance via a pertussis toxin sensitive GTP-binding protein (16). Activation of K^+ conductance by these agents account for their negative inotropic effects. Unlike the latter effects, inhibition of the Ca^{2+} channel by adenosine is associated with A_1 receptor-mediated decrease in adenylate cyclase activity (12, 13). In contrast, elevation of cyclic AMP presynaptically facilitates Ca^{2+} influx (13).

A third effector system activated by adenosine is the low K_m cyclic AMP phosphodiesterase. For example, adenosine enhanced the activity of this enzyme in homogenates of rat fat cells (17–19) and in brain (20). In fat, it appears that both insulin receptors and adenosine receptors are coupled to the activation of a common pool of the low K_m phosphodiesterase, which is independent of that coupled to β-adrenergic receptor activation (18). The identity of the adenosine receptor subtype responsible for the activation of this enzyme has not been reported.

Recent evidence has supported a fourth effector system activated by adenosine and its analogs, the particulate guanylate cyclase. In this respect, R-PIA is more potent than NECA, inferring coupling of guanylate cyclase to A_1-adenosine receptors. Activation of guanylate cyclase in smooth muscle might mediate, in part, the vasorelaxing effect of adenosine via A_1 receptors (21).

Several pieces of evidence have demonstrated that adenosine and its analogs can potentiate phosphotidylinositol hydrolysis stimulated by histamine in cerebral cortical slices of the guinea pig (22–24). This action of adenosine is specific for hydrolysis elicited by histamine, since no potentiation of noradrenaline or carbachol-mediated phosphotidylinositol hydrolysis was observed. Furthermore, augmentation of histamine stimulation by adenosine is likely mediated via A_1 adenosine receptors; the potency series of agonists in this respect is cyclopentyladenosine $>$ R-PIA $>$ NECA 2 Cl-Ado (24). However, unlike the high affinity interaction of agonists via A_1 adenosine receptors, the EC_{50} values for potentiating histamine-induced phosphotidylinositol hydrolysis were in the micromolar range (24). While the biochemical mechanism mediating the augmentation of adenosine on histamine response is not known, this action might derive from the interaction of adenosine at a novel and yet uncharacterized site.

In contrast, inhibition of phosphotidylinositol hydrolysis by adenosine and its analogs has been reported in peripheral tissues. In liver hepatocytes and vascular smooth muscle, this action is due to inhibition of phosphotidylinositol kinase, an enzyme which replenishes polyphosphoinositides following hormone activation of phosphotidylinositol turnover (25, 26). Inhibition of the kinase occurs at physiologically relevant concentrations of adenosine and might, therefore, contribute to the vascular smooth muscle relaxing property of adenosine. The addition of the adenosine receptor anatagonist IBMX failed to block this effect of adenosine, suggesting that this action is not receptor-mediated (25) and might be due to an intracellular site of action of adenosine and its analogs (26).

Adenosine receptors are widely distributed in both the central nervous system and peripheral tissues. The existence of adenosine receptors in most tissues has been supported by both radioligand binding assays and by functional studies (modulation of adenylate cyclase or guanylate cyclase activities). Detailed demonstration of A_1 adenosine receptors in rat brain has been provided using the agonist radioligand [^{3}H]cyclohexyladenosine (27, 28). High densities of A_1 receptors are localized to the cerebellum (molecular and granule cell layers), hippocampus, the cerebral cortex and certain thalamic nuclei (27, 28). In some of these regions, a presynaptic localization of A_1 – adenosine receptors has been suggested (14, 29–31), in accordance with a modulatory role of adenosine on neurotransmitter release. Both pre- and postsynaptic A_1 adenosine receptors have been demonstrated in the hippocampus (14). Furthermore, indirect evidence lends support for a presynaptic localization of A_2 adenosine receptors, activation of which facilitates neurotransmitter release (13, 32). The existence of A_1 binding sites (28, 33, 34) and more recently A_2 binding sites (35) in rat spinal cord coupled to the inhibition and stimulation of adenylate cyclase, respectively, has been demonstrated. In addition, the existence of presynaptic A_1 receptors on sympathetic nerve terminals has been suggested to account for the inhibition of evoked norepinephrine release (36).

The cardiac depressant effect of adenosine has been well documented (37–39) and is believed to be mediated by A_1 adenosine receptors (37,

40, 41). The nucleoside reverses both cardiac stimulation and elevation of cyclic AMP elicited by isoproterenol in the heart (42). Previous radioligand binding studies have failed to characterize in detail A_1 adenosine receptors in heart (43–45). However, a recent study (46) has demonstrated specific binding of ^{125}I-HPIA to rat ventricular membranes with pharmacology characteristics of the A_1 adenosine receptor subtype. In addition, several studies have suggested the presence of A_2 receptors (of low affinity) in the heart (47, 48). Adenosine is also a potent vasodilator, owing to its action to stimulate cyclic AMP via A_2 receptors in vascular smooth muscle cells (49–51). Recent evidence has also supported the presence of A_1 adenosine receptors in vascular smooth muscles, based on agonist potency series to activate particulate guanylate cyclase (21). Additionally, the A_1 adenosine receptor agonist R-PIA inhibited forskolin-stimulated cyclic AMP accumulation in cultured arterial smooth muscle cells (52).

Adipocytes contain both A_1 and A_2 adenosine receptors, negatively and positively coupled to adenylate cyclase, respectively (11). Under normal conditions, the inhibition of adenylate cyclase via A_1 adenosine receptors (3, 11, 53) predominates and this leads to the inhibition of lipolysis. Following pertussis toxin-mediated inhibition of G_i however, the stimulatory action of adenosine analogs on adenylate cyclase is manifested (11). A_1-adenosine receptors in adipocytes are also coupled to the activation of a low K_m cyclic AMP phosphodiesterase (17–19), providing another means by which cyclic AMP levels are reduced.

Other tissues containing adenosine receptors include the liver, which contains A_2-adenosine receptors (54), kidney (55), pancreas (56) smooth (21, 50) and skeletal muscle (57), blood cells and platelets (58, 59).

2 Physiological and clinical effects of adenosine receptor interactions
Nervous system

The role of adenosine as modulator of neurologic function was suggested by the initial work of Drury and Szent-Györgi [60]. It was not until 1970 when Sattin and Rall discovered the antagonism of the methylxanthines to the formation of adenosine 3′, 5′-monophosphate in intact brain slices that the magnitude of the adenosine's role in the central nervous system was appreciated [61]. They speculated that adenosine was a neurotransmitter substance filling a significant role in synaptic transmission (61). With the discovery that caffeine acts as an adenosine receptor antagonist it has become clear that adenosine receptors are among the greatest pharmacologically treated receptor groups. Since those initial observations adenosine has been found to be very active at the neuronal level. There are numerous reports of adenosine's influence on neurotransmitter release. Adenosine is known to effect the release of norepinephrine, dopamine, serotonin, acetylcholine, GABA, aspartate and glutamate (62–69). There is no clear unity as to the mechanism of how adenosine might participate in the role of neurotransmission. The mechanism that has generally been accepted is that adenosine acts through presynaptic modulation of neurotransmitter release.

It remains unclear, however, whether adenosine itself is a neurotransmitter or simply a potent modulator that acts on the release of other transmitters. There is unequivocal evidence which demonstrates the presence of extracellular high affinity binding sites within the central nervous system. These sites are known to be coupled to membrane bound adenylate cyclase through G proteins and act to regulate intracellular concentrations of cyclic AMP (70). Within the central nervous system there are regional differences in the number, density, and type of adenosine receptors. It is known the striatum contains relatively high numbers of A_1 and A_2 receptors while in the rest of the brain there are higher numbers of A_1 receptors (71). The sedative actions of adenosine are currently attributed to the activation of A_1 receptors with the consequent reduction of intracellular cyclic AMP levels (70).

Adenosine has been demonstrated to inhibit neurotransmitter release both peripherally and centrally. This release appears to be independent of calcium entry into the neuron. Adenosine and its analogs depress excitatory synaptic transmission more than inhibitory transmissions. It has been noted that in the presence of adenosine excitatory phase synaptic potentials are abolished. This may not be the singular action of adenosine in the nervous system, however. It has also been shown that neuronal spiking in calcium-free solution is abolished by the addition of adenosine. There is some evidence that adenosine causes hyperpolarization (mediated by activation of K^+ currents) in neuronal tissue. Numerous authors have noted the accumulation of cyclic AMP in neural tissue but its role has not been defined (72, 73). The role of cyclic AMP in modulating K$^+$ currents is likewise not clear. One of the best characterized actions of adenosine is its sedative effects. As early as 1954 Feldberg and Sherwood used intraventricular injections of adenosine to demonstrate the cessation of spontaneous motor activity in cats (74). The animals were not catatonic but were still capable of appropriate responses to various stimuli. These responses were amplified in the presence of adenosine uptake inhibitors. Adenosine also appears to exhibit facilitation of sleep time in the presence of hypnotic drugs, such as phenobarbital as well as lengthening natural sleep. Adenosine and its agonists do not produce sleep but probably play a role in the onset and maintenance of sleep (75–77). Hypothermia has been observed following injections with adenosine or adenosine analogs into intact animals. In high doses profound hypothermia has been observed with decreases up to six degrees centigrade reported in some cases. Furthermore, intracerebroventricular injections of R-PIA elicits marked hypotension and bradycardia with significant respiratory depression. These effects are believed to be central in origin, reflecting direct receptor activation in the CNS (70). Adenosine in high doses has been demonstrated to protect against seizure activity (78). These effects are believed to be receptor specific since they can be blocked by methylxanthines. Thus, some authors have even referred to adenosine as nature's endogenous anticonvulsant. Recently, there has been interest in determining if adenosine interacts with the central benzodiazepine receptor (79). It has been demonstrated that IBMX can interact with the benzodiazepine receptor in *in vitro* radioligand binding assays (79). While this supposition is interesting, considerable debate has been generated with no clear resolution as to the interaction of these receptors with adenosine (80). Dipyridamole has been demonstrated to inhibit adenosine A$_1$ receptor binding of benzodiazepine receptor agonists (81, 82). Conversely, however, benzodiazepine receptor antagonists do not inhibit the effect of dipyridamole on adenosine uptake (83).

Renal system

In the kidney there appears to be a biphasic release of renin in response to adenosine infusion. At low adenosine concentrations A$_1$ receptors are activated inhibiting renin release, while at high adenosine concentrations renin release is stimulated via activation of A_2 receptors (84). Adenosine is thought to be released from macula densa cells in response to an increased sodium load (85). Thus, at high sodium loads inhibition of renin release occurs, consequent to adenosine release (86). It has been speculated that increased adenosine release in the presence of an ischemic kidney may result in increased renin release, leading to hypertension in the Goldblatt kidney model (87).

In contrast to its usual vasodilator action in peripheral vasculature, adenosine is a potent vasoconstrictor in the kidney. Intrarenal infusions of adenosine result in a transient reduction in total renal blood flow due to cortical vasoconstriction (88). Adenosine-induced vasoconstriction is related to increased plasma renin activity and sodium loading (85). During prolonged adenosine infusion the flow through the cortical layers is normalized and in deep cortical layers it is enhanced (88). Both A_1 and A_2 receptors are found on afferent arterioles while efferent arterioles possess A_2 receptors exclusively (89). This implies that the predominant effect of adenosine on efferent arterioles is vasodilation (89). Consequently, adenosine has been shown to inhibit renin secretion and reduce glomerular filtration (90). In addition, dipyridamole decreases glomerular filtration rate (GFR) and total urine output (90). These effects are abolished by theophylline and augmented by indomethicin treatment (91). Thus, it is apparent that adenosine plays a significant role in the physiology of the kidney through its action on both vascular regulation and renin release.

Cardiovascular system

The initial investigations of the physiologic role of adenosine were centered on its cardiac effects. In the sentinel work of Drury and Szent Györgi in 1929, adenosine was found to play major roles in the regulation of cardiac rate and vascular smooth muscle contractility (92). Since those investigations, adenosine has been found to have a significant role in vascular smooth muscle dilation and inotropy, as well as the initially described chronotropic and dromotropic cardiac effects, and is currently being used as a pharmacologic agent for the treatment of a variety of cardiac dysrhythmias (90).

In their initial studies, Drury and Szent Györgi found a transient decrease in heart rate after adenosine infusion (92). They reported sinus slowing in intact rabbits, cats, dogs and guinea pigs. These findings served as the basis for extensive investigation into the electrophysiologic effects of adenosine. Adenosine appears to act directly on the SA and AV nodes and His bundle in experimental animals as well as in humans (93). The lower pacemakers, such as the His purkinje system, appear to be more sensitive to the effects of adenosine administration than the SA and AV nodes (94). These effects are blocked by methylxanthines and enhanced by adenosine uptake inhibitors such as dipyridamole (95). These actions are not affected by atropine but are antagonized by caffeine (96). The border of the nodal region and the cristae terminalis appears to be the anatomical site where adenosine exerts its effect on sinus exit block (97). As early as 1956 adenosine was found to cause an increase in the rate of atrial potential repolarization in guinea pig atrial fibers (98). Voltage clamp experiments indicated that membrane hyperpolarization seen with adenosine treatment is due to activation of K^+ conductance (99). Adenosine produces a dose-dependent inhibition of SA node automaticity in the guinea pig atria (100). It also increases the cycle length of the rabbit SA node and produces sinus slowing in the primary pacemaker cells in intact SA node preparations. Furthermore, adenosine increases maximum diastolic potential and decreases the rate of diastolic depolarization. The increase in maximum diastolic potential by adenosine correlates with the degree of cardiac slowing and is similar to that seen following vagal stimulation and acetylcholine exposure (101).

Adenosine's action on the AV node is restricted to the proximal portion. The primary action of adenosine on the conduction time is due to prolongation of the A-H interval without disturbing the H-V interval (97).

It is possible that adenosine and acetylcholine mediate their negative chronotropic and inotropic effects by activating the same K^+ channels through A_1 adenosine receptors and muscarinic receptors, respectively. Such a possibility is supported by recent findings (16, 97).

In humans, adenosine produced as a consequence of hypoxia and ischemia exerts pronounced effects on the SA node, causing bradycardia and various degrees of AV node block. There is growing evidence to suggest that adenosine is involved in bradycardias and conduction disturbances associated with myocardial ischemia and infarction (102). One report suggests that aminophylline treatment protects some patients from syncope by blocking the effects of adenosine on the conduction system (103). Thus, this treatment may prevent serious dysrhythmias in patients with myocardial infarctions.

In the purkinje system, adenosine partially antagonizes the action potential duration shortening produced by isoproterenol (104). The nucleoside also suppresses spontaneous depolarization in purkinje fibers, thereby eliminating their pacemaker capacities. This effect results from inhibition of phase four depolarization and is not observed in cells which were pretreated with beta-blocking agents such as propranolol (104, 105). Adenosine produces no effect on the resting potential of purkinje fibers or ventricular muscle fibers (105). In ventricular tissue, the greatest inhibitory effect of adenosine is exhibited when there is preexisting adrenergic tone (104). In this setting adenosine appears to blunt the effects of adrenergic stimulation.

A number of authors have noted the similarity of the action of acetylcholine and adenosine on cardiac tissues. This has led some to speculate that acetylcholine and adenosine act on different receptors but through a common pathway (106). Vagal induced bradycardia is enhanced in the presence of adenosine, while other authors report vagal blockade by adenosine. Thus, there still exists some controversy as to the exact role of adenosine in vagal stimulation.

Histamine produces a biphasic effect on the human heart. Positive inotropy is achieved by activation of histamine H-2 receptor while negative inotropy is mediated by H-1 receptor stimulation (107). Adenosine is known to antagonize the actions of H-2 receptors allowing the H-1 effects to occur unimpeded. Thus, histamine working in concert with adenosine produces more pronounced effects on the cardiovascular system than adenosine alone.

Adenosine has been used in the conversion of supraventricular tachycardias to normal sinus rhythm in humans with a great deal of success (108). It appears to be particularly effective in the treatment of tachycardias deriving from A-V nodal reentry as well as in slowing conduction of rapid atrial tachycardias (109). Recently, there have been reports that the administration of adenosine may represent another mode of treatment for certain ventricular tachycardias as well (110). Therapeutically, adenosine represents a rather safe agent, characterized by a short duration and rapid onset of action, only transient side effects and no reports of hypotension, as seen with calcium channel blockers (109). Its side effects are minimal; these include flushing, transient bradycardia, sinus arrest, various degrees of A-V block, malaise, headache and rarely bronchospasm (109). In view of its beneficial effects and limited side effects, adenosine represents a safe investigational drug for use in terminating various types of supraventricular tachycardias (111–116).

In 1963 Bern postulated that adenosine was a significant modulator of coronary blood flow (97). He and his co-workers demonstrated the presence of adenosine responsive vasodilation in coronary beds. Since those studies, investigators have been looking for the mechanisms of possible adenosine release. Hypoxia is believed to represent a signal by which adenosine is released into the blood. In addition to hypoxia, however, a number of other conditions are known to mediate adenosine release. These include aortic cross clamping, acidosis, isoproterenol infusion, rapid cardiac pacing, stellate ganglia stimulation and treadmill exercise in dogs. Furthermore, differences in adenosine concentrations during the cardiac cycle have been described (117).

Adenosine has long been known to dilate vascular beds (118) with the exception of the renal beds (85, 119) described above. The mechanisms by which the nucleoside dilates vascular smooth muscle is probably via activation of adenosine A_2 receptors on endothelial cells (117). As described earlier, activation of A_2 receptors leads to activation of G_s and consequently to stimulation of adenylate cyclase. Cyclic AMP is believed capable of promoting the release of endothelial-derived relaxing factor (EDRF) (119). EDRF then interacts with its specific receptors (EDRF receptors) on the smooth muscle cells to increase cyclic GMP production by activating the particulate fraction of guanylate cyclase. The increased cyclic GMP levels lead to vasodilation through relaxation of the vascular smooth muscle cell (117).

The existence of adenosine receptors on smooth muscle cells is generally accepted (118). These receptors are primarily of the A_2 subtype, but recently A_1 receptors have been identified as well (119). As early as 1985 Jonzon and coworkers (52) demonstrated the presence of A_1 receptors on vascular smooth muscle cells. In dog vascular tissue, adenosine is a much more powerful inhibitor of vasoconstriction than sympathetic nerve stimulation via β-receptor activation (120). In the adventitial layer of blood vessels there is evidence for the presence of presynaptic adenosine receptors which mediate the inhibition of neurotransmitter release (120). Fredholm speculated that by "interacting with these receptors adenosine could limit the influence of the sympathetic vasoconstrictor nerves. Thus, adenosine may cause vasodilatation by directly relaxing smooth muscle and indirectly by limiting the availability of vasoconstrictor transmitters" (117). The relationship between adenosine administration and hypotension is clearly dose-dependent. Furthermore, by blocking its reuptake, dipyridamole reduces the amount of adenosine needed to produce a given response (110). This hypotensive effect of adenosine is achieved primarily through reduction in systemic vascular resistance and not cardiac output (117).

Pulmonary system

Adenosine receptors play an important role in a variety of mechanisms in the pulmonary system. The methylxanthines (adenosine receptor antagonists) are among the most widely used drugs in clinical practice today. Theophylline has long been used to treat bronchospastic illnesses, such as asthma and chronic obstructive lung disease. The mechanism(s) by which the methylxanthines relieve bronchospasm have long been a topic of debate and at the present time no definitive answer is yet available. This topic will not be reviewed here in detail but recent monographs are available (121).

In normal subjects inhaled adenosine does not affect the bronchial tree but in asthmatics adenosine causes bronchoconstriction which is antagonized by theophylline (122, 123). The mechanism by which this is accomplished is thought to be through antagonism of adenosine at the receptor level. This antagonism is evident at plasma concentrations of theophylline which range from 50 to 100 micromolar (124). However, these data may not directly be applicable to non-adenosine induced asthma.

Theophylline has also been shown to inhibit adenosine potentiation of antigen-induced histamine release from guinea pig lung at therapeutic concentrations (125). Adenosine levels were found to be increased in asthmatic patients following antigen challenge (123). The nucleoside normally produces relaxation of stimulated guinea pig isolated smooth muscle (126), while contraction results when the muscle is exposed to adenosine in its resting state (127, 128). Napin and Temple suggested that A_2 receptors are responsible for these effects (129). They speculated that the bronchodilatory effect seen with theophylline is mediated by both its stabilizing effect on the mast cells of the pulmonary system as well as by decreasing the contraction of bronchial smooth muscle (129). Recently, A_2 receptors have been identified on rabbit alveolar macrophages where they mediate the expression of cell-associated procoagulant activity (130). There appears to be an inverse relationship between plasminogen activator release and expression of procoagulant activity in this tissue (130). The role of these mechanisms in the physiology of the bronchial tree remains to be defined.

Immune system

Adenosine deaminase deficiency has long been recognized as a source of profound disturbance of the immune system. The enzyme adenosine deaminase converts adenosine to inosine and its deficiency leads to high levels of endogenous adenosine. In 1970 adenosine was shown to inhibit the lectin-stimulated proliferative response of lymphocytes (131). It has recently been shown that adenosine inhibits interleukin-2 production by T-lymphocytes (132). Adenosine has also been found to induce phenotypic changes in helper-inducer T-lymphocyte subsets (133, 134). Inhibition of lymphocyte-mediated cell cytotoxicity has also been demonstrated in the presence of adenosine (135). Mast cells also display augmentation of IgE-mediated secretory granule release in the presence of adenosine (136). Adenosine can modify T-lymphocyte effector functions and inhibit or facilitate cytokine production (137). Adenosine has also been shown to alter the function of mononuclear phagocytes (138, 142). Suppression of superoxide and natural killer activity has been observed in neutrophils treated with adenosine (143). Alveolar macrophages produce tissue thromboplastin, a potent procoagulant of the extrinsic coagulation pathway. These cells also have the capacity to initiate fibrinolysis by producing a plasminogen activator. Adenosine has been shown to suppress this procoagulant activity while plasminogen activator activity was increased (130). Hasaday and Sitrin suggest that adenosine serves as an autoregulator in the maturation of mononuclear phagocytes (130). Some authors feel that adenosine may represent an autocoid similar to histamine and prostenoids (137) although direct evidence for this theory is not yet available. Clearly adenosine plays a powerful role in the immunologic system and all of its functions are yet to be defined.

Gastrointestinal tract

Caffeine has long been recognized as being able to augment the production of gastric ulcers associated with stressful situations, such as hypothermia or immobilization in rats (144, 145). Adenosine analogs have been used to induce gastric lesions as well (146). It is unclear whether this phenomenon is a local effect or an effect mediated through the central nervous system (144, 145). Adenosine receptors are present on the fundus of the stomach and appear to regulate gastric acid secretion at least to a certain extent (147). Ever since the suggestion that caffeine was a putative carcinogen in the pancreas the methylxanthines have undergone increasing scrutiny (148). Considerable controversy has been generated as to whether these findings were real or simply a result of sampling error (149, 150). Recently, Coffey and co-workers examined the role of caffeine in pancreatic exocrine function. These researchers found that caffeine increased pancreatic secretion when ingested during fasting and that chronic coffee drinkers had higher rates of secretion than non-coffee drinkers (151). Theophylline has been shown to release amylase from isolated pancreatic acinar cells (152). The exact role, if any, of adenosine in these responses is not known.

Platelets

In 1964 Born showed the anti-aggregatory effects of adenosine on platelets (153). This effect was associated with an increase in cyclic AMP concentration. Since that time platelets have been found to be rich in A_2 receptors (154). The anti-aggregatory effect is blocked by the methylxanthines and enhanced by the adenosine uptake inhibitor dipyridamole. Dipyridamole is used clinically in an attempt to prevent occlusion of venous grafts to coronary arteries following surgery (155). Adenosine is also used to prevent platelet destruction by extracorporeal circulation during open-heart surgery (156).

Breast

Caffeine has long been thought to be associated with fibrocystic disease of the breast and patients are told to abstain from caffeine consumption for 4 to 6 weeks in an effort to resolve breast lumps. Recently, a prospective single-blind study has shown there to be no association between caffeine consumption and change in breast lumps (157). There still remains considerable controversy of the role of caffeine in cystic breast disease (158–161).

3 Components of hormone sensitive adenylate cyclase systems

Hormone sensitive adenylate cyclase systems have been some of the most widely studied systems involved in receptor-mediated transmembrane signalling (162). The enzyme adenylate cyclase is regulated by a wide variety of hormones acting through specific receptors. These receptors can be divided into two general categories: those which stimulate its activity and those which inhibit enzyme activity. These systems are composed of three distinct protein components: (1) the receptor (A_1AR adenosine receptors) which recognize and bind ligands and under the influence of an agonist, such as adenosine, can interact with the second component of the system – a guanine nucleotide regulatory protein termed G or N. The currently preferred nomenclature is to use the abbreviation G rather than N.

The G protein can then, under the appropriate conditions, interact with the third component of the system – the enzyme adenylate cyclase (162). Thus, the binding of an agonist to a receptor promotes the interaction of the receptor with its G protein. This interaction releases GDP which is tightly bound to the G protein in its basal or inactive state. The G protein is then able to bind GTP (the active guanine nucleotide) and undergo dissociation of its component parts (see below) while simultaneously destabilizing the agonist-receptor-G protein complex so that the receptor is recycled back to begin the process again. The G protein liganded with GTP is then capable of regulating the activity of adenylate cyclase.

The G proteins represent an ever-increasing family of related proteins. Initially, G_s was described and was found to be the protein necessary for stimulation of adenylate cyclase (163). Next, G_i was postulated to exist (163) and then it was purified and characterized as the component associated with inhibition of adenylate cyclase. Subsequently, a variety of G proteins have been found including G_o, G_p, transducin and others (163). Recent molecular cloning studies have demonstrated that there are at least three "subclasses" of G_i (164). Studies on the biochemical purification of G_s have previously shown that there were two forms of G_s and recent studies have demonstrated that alternate splicing of the G_s transcript is responsible for these two protein forms (163). There are many similarities among all the G proteins including the fact that they are all heterotrimeric in nature having α subunits with molecular weight ranging from 39,000–53,000 daltons, β subunits of molecular weights 35,000–36,000 and a γ subunit of molecular weight ~ 10,000. Current dogma suggests that the β and γ subunits are interchangeable among all the G proteins and are functionally indistinguishable. Therefore, it is clear that G protein specificity in terms of receptor interactions and the regulation of adenylate cyclase or other effector activities must reside in the α subunits. Thus, the α subunit is the "business" component of the protein. It is now evident that the α subunit contains the guanine nucleotide binding site as well as the intrinsic "GTPase" activity which is responsible for deactivating the GTP-liganded α subunit by promoting the hydrolysis of GTP to GDP.

The α subunit is also the substrate for the various toxin-catalyzed ADP-ribosylations that occur. Thus, the α_s subunit of G_s is the substrate for cholera toxin-mediated ADP ribosylation which results in permanent activation of α_s and subsequent stimulation of adenylate cyclase. The α_i subunit of G_i is the substrate for pertussis toxin-catalyzed ADP-ribosylation. This covalent modification results in the uncoupling of G_i from inhibitory receptors and blocks the ability of GTP to promote inhibition of adenylate cyclase activity (10).

Finally, the α subunits are believed to contain the recognition sites for interaction with their appropriate receptors and effectors. These various domains of the α subunits are now beginning to be studied in detail (165).

Radioligand binding and adenylate cyclase studies have now provided some insight into how the various proteins, components and cofactors, such as guanine nucleotides and metal ions such as Mg^{++} and Na^+, interact in concert to regulate adenylate cyclase activity. These studies will be described in subsequent sections of the review.

4	**Radioligand binding studies**
4.1	Direct studies of adenosine receptors

Radioligand binding studies have only been attempted within the past 10 years and success has been of an even shorter duration. Initially, a tritiated form of the endogenous ligand adenosine was employed in an attempt to characterize adenosine receptors in a variety of tissues (166). In retrospect, the concept of using adenosine itself was fraught with problems and these manifested themselves fully in actual practice. The major difficulties were that adenosine has low affinity for the receptors (in the micromolar range), adenosine is susceptible to rapid degradation and binds to a variety of sites other than the receptors at the concentrations used in radioligand binding assays. Thus, binding was not to adenosine receptors in that binding was not saturable, did not display the appropriate pharmacological specificity, and did not have the characteristic expected of the physiologic relevant receptors. From these early studies it became obvious that in order to develop successful radioligand binding assays for adenosine receptors new types of adenosine analogs must be developed. The analog should be resistant to degradation, have high affinity for the receptor, display marked specificity for receptor sites and not bind to transport molecules or nonspecific adenosine sites. In addition, receptor subtype specificity (A_1 vs. A_2) would be useful since many tissues contain both subtypes. Antagonist radioligands would be preferable, but as will be seen, useful antagonists would only become available years after agonist radioligands.

The first successful radioligand binding studies were reported in 1980 by several different groups using a variety of radioligands (167–169). The radioligand binding displayed all the appropriate characteristics including saturability, reversibility, stereoselectivity and the correct pharmacological specificity expected of the physiologically relevant receptor.

Of the radioligands reported, three were agonists (adenosine analogs) including [³H]PIA, [³H]CHA and [³H]2-chloroadenosine and one was an antagonist [³H]DPX. Since 1980 a large variety of radioligands have been synthesized and utilized successfully (see Table 1). Only a brief discussion of some of the more useful radioligands will be presented here.

[³H]PIA has certainly fulfilled its promise as a useful A_1 selective agonist radioligand. Although initial studies suggested that [³H]PIA bound to A_1 receptors with a single affinity state and with all the appropriate pharmacological characteristics, most recent data suggest that [³H]PIA can in fact differentiate between high and low affinity states of the A_1 receptor (170). This finding is much more consistent with the paradigm established for agonist action in a variety of other model receptor systems (162). That is to say, the difference between an agonist, such as *R*-PIA, and an antagonist, such as isobutylmethylxanthine (IBMX), is that an agonist upon binding to the receptor induces a conformational change in the receptor allowing it to interact with a guanine nucleotide binding protein (such as G_i) which stabilizes a high affinity form of the

Table 1
Radioligands

	Radioligand	Isotope	Ag/Ant	Sp. Act.	K_D (nM)	Selectivity	Useful	Covalent labeling	References
1	Hydroxyphenylisopropyladenosine (HPIA)	125I	Agonist	2275	~ 1	A_1	Yes	Yes	172, 173, 188
2	*R*-phenylisopropyladenosine (*R*-PIA)	3H	Agonist	49.9	0.5,15	A_1	Yes	No	167, 170
3	Cyclohexyladenosine (CHA)	3H	Agonist	11.5	~ 1	A_1	Yes	No	168
4	Diethyl-8-phenylxanthine (DPX)	3H	Antagonist	13.4	60	A_1 and A_2	No	No	168
5	Aminobenzyladenosine (ABA)	125I	Agonist	2275	~ 2	A_1	Yes	Potential	229
6	2-Chloroadenosine (2ClAdo)	3H	Agonist	12	1,16	A_1	No	No	169
7	5'N-ethylcarboxyamido-adenosine (NECA)	3H	Agonist	10-30		A_1 and A_2	No	No	174
8	Xanthine amine congener (XAC)	3H	Antagonist	130	0.1-2	A_1	Yes	Potential	175
9	Paraaminophenylacetyl XAC (PAPAXAC)	125I	Antagonist	2275	0.1-1	A_1	Yes	Yes	176
10	Aminophenylethyladenosine (APNEA)	125I	Agonist	2275	1-2	A_1	Yes	Yes	184
11	Azidophenylethyladenosine (AZPNEA)	125I	Agonist	2275	1	A_1	Yes	Yes	185
12	2-Azido-N^6-hydroxyphenylisopropyl-adenosine (R-AHPIA)	125I	Agonist	2275	2	A_1	Yes	Yes	186
13	Azidobenyladenosine (AZBA)	125I	Agonist	2275	1	A_1	Yes	Yes	187

receptor. In contrast, an antagonist binds to the same receptor but does not induce the receptor-G protein interaction and thus binds with a single affinity state.

In addition, agonist binding is modulated by guanine nucleotides, such as GTP or Gpp(NH)p, which in turn interact specifically with the α subunit of the G protein. The G protein then dissociates into its subunits thereby perturbing R–G coupling and reducing the affinity of agonists for the receptor (see full description below).

Several important findings were made through the use of R-PIA and [^{3}H]R-PIA. First, Londos et al. (171) found that R-PIA was much more potent than NECA at A$_1$ receptors and that the reverse was true of A$_2$AR's. Second, early on it became clear that in order to maximize radioligand binding, pretreatment of membranes with adenosine deaminase had to take place. This pretreatment with adenosine deaminase removes endogenous adenosine which binds to the receptors precluding [^{3}H]R-PIA from interacting with the receptor. This phenomenon is quite general and can be documented with most tissues which contain adenosine receptors and with all the radioligands thus far developed. [^{3}H]PIA has been utilized to study A$_1$ receptors in a wide variety of tissues, such as rat cerebrum cerebellum, and adipose tissue. [^{3}H]CHA is strikingly similar in its characteristics to [^{3}H]PIA and for most radioligand binding situations they are quite interchangeable. [^{3}H]PIA has remained however, a more popular radioligand.

Another useful agonist radioligand has been [^{125}I]HPIA (172, 173). This ligand has several distinct advantages in that it is labeled to a much higher specific activity, thus allowing smaller quantities of tissue to be used in binding assays and allows the detection of receptors in tissues that have low receptor densities. In general, the nonspecific binding has been relatively low which is a great asset in terms of the technical aspects of radioligand binding and data analysis. Another advantage of an iodinated ligand is that the radioactivity can be detected without destroying the sample, i. e. no scintillation flour is required so the sample remains intact. This allows solubilized fractions to be counted and then retrieved for further analysis or manipulation. The use of iodinated ligands, however, makes it difficult to assess both the high and low affinity states since it would require very high amounts of radioactivity to be used in each tube to detect the low affinity form. Methods using dilution of radioligand with unlabeled ligand have been developed to overcome this technical problem but this technique

is fraught with difficulties and with the advent of good antagonist radioligands this is no longer necessary at all (vide infra).

[³H]NECA has been utilized as both a A_1- and A_2-AR radioligand
(174). There are several major drawbacks to the use of [³H]NECA.
These include the fact that it is not highly subtype selective and thus
binds to both A_1 and A_2 receptors, it has relatively low affinity for both
A_1 and A_2 receptors, and it is a full agonist at both A_1 and A_2 receptors.
There are now so many excellent agonist and antagonist radioligands
for A_1 receptors that [³H]NECA appears no longer to have a role in studying A_1 receptors. Therefore, its use can only be justified in tissues
containing only A_2 receptors, and it is hoped a more useful A_2 selective
ligand will soon be developed.

A recent development in the field of adenosine receptor radioligand
binding has been the synthesis of high affinity A_1 selective radioligands, such as [³H]XAC (xanthine amine congener). This radioligand
developed by Jacobson et al. offers several distinct advantages over
any previously available radioligand (175). First and foremost, it is an
antagonist so that it should recognize all A_1 receptors regardless of
their interaction with guanine nucleotide regulatory proteins making it
possible to obtain a true assessment of the quantity of A_1 receptors in a
given tissue. Second, it has very high affinity for the A_1 receptors (0.2-2
nM) depending on the tissue studied (175). Third, it is tritiated to high
specific activity (130 Ci/mmol). Fourth, it is structurally amenable to
alteration so that it can be made into a direct photoaffinity/photoaf-
finity crosslinking probe (176). This has now been successfully accomplished and recently published (176). Fifth, through agonist competition curves generated versus [³H]XAC, it will now be possible to study
agonist high and low affinity states and their modulation by guanine
nucleotides, mono- and divalent cation, pathophysiologic conditions
and following receptor solubilization. These studies should provide
dramatic new insights into the mechanism of how A_1AR interact with
G proteins and initiate transmembrane signalling. Very recently another high affinity antagonist (8-Cyclopentyl-1,3-dipropylxanthine
(DPCPX)) has been reported which should further add to our receptor
armamentarium (177).

Radioligands used primarily for photoaffinity crosslinking or photoaffinity probes will be discussed in a separate section.

4.2   Interpretation of radioligand binding studies

As pointed out previously, the hormone sensitive adenylate cyclase systems are composed of multiple components which must act in concert to either stimulate or inhibit the enzyme. Figure 1 depicts the various components of the adenosine receptor adenylate cyclase system. The primary components are the receptors which recognize and bind adenosine and its analogs and impart the pharmacological specificity to the system and the G proteins (G_s and G_i) which transduce the signal from the receptor to the third component, the enzyme adenylate cyclase. In addition to these protein components, specific ions such as Mg^{++} and cofactors such as guanine nucleotides (GTP) are required for full function of these systems. In addition to being required for transmembrane signalling, guanine nucleotides have interesting and important effects on agonist binding.

Radioligand binding studies have played an important role in delineating how the various components of the system interact to produce the biochemical and physiological effects noted in intact cells and in membranes derived therefrom. Much of the pioneering work has been performed in receptor systems where high affinity antagonist radioligands have been available for many years and it is only in the very recent past that the development of high affinity antagonists for the A_1AR has allowed this system to begin to be probed in detail. Future years will bring forth detailed information on the nuances of the A_1- and A_2-adenosine receptor systems, but for the present we can learn much by analogy from other receptor systems.

Figure 2 shows a competition binding curve for adenosine receptor antagonist isobutylmethylxanthine for occupancy of the A_1 adenosine receptors in soluble bovine brain membranes labeled with the A_1 receptor antagonist [^{3}H]XAC. This antagonist curve is steep, uniphasic and conforms to a model of one affinity state binding as predicted by the law of mass action. The curves drawn through these points are produced by a computer modeling system (178) capable of fitting data to a series of models and able to statistically determine the most appropriate model. These data, therefore, are consistent with a single ligand interacting with a single state of the receptor, i. e. ligand + receptor $\rightleftharpoons$ ligand · receptor. Attempts to fit such data to more complex models do not provide an improved description of the data.

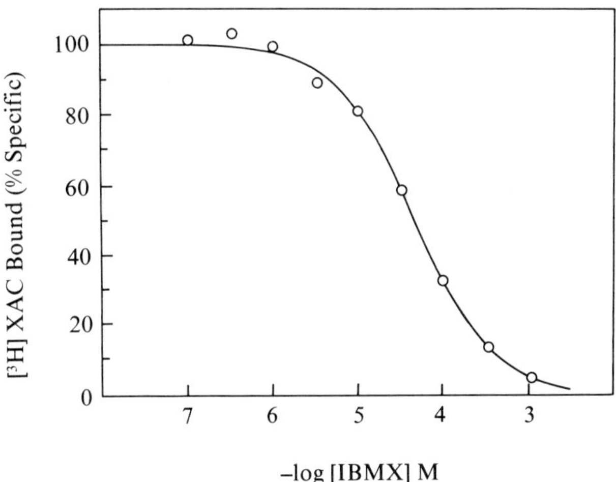

Fig. 2

Competition curve of IBMX using the radioligand [³H]XAC. Bovine brain membranes were prepared and solubilized using the detergent Chaps. [³H]XAC was present at a concentration of 0.3 nM with the indicated concentration of IBMX (antagonist). Binding was performed for 2 hrs. at 25⁰ C. The data points are experimentally determined while the line through the points are generated by a computer modeling system (see text).

In contrast to this result, agonist (*R*-PIA) competition curves display a quite different appearance (see Figure 3). In this case, the curve is shallow, complex and best described by a model for two states that have respectively high (K_H) and low (K_L) affinity for the agonist. In the presence of guanine nucleotides, such as Gpp(NH)p, the agonist curve is shifted to the right and steepened. Guanine nucleotides appear, therefore, to convert the equilibrium of high and low affinity states to a uniform low affinity state. This contrasts with the lack of effect of guanine nucleotides on antagonist competition curves. Thus, guanine nucleotides appear to specifically modulate agonist but not antagonist binding.

These findings are compatible with the notion that the unique property of agonists as compared to antagonists is that agonists uniquely induce, stabilize or recognize a high affinity form of the A_1AR. This stabilization of the high affinity form of the receptor in some way establishes conditions that favor the inhibition of adenylate cyclase. Furthermore, GTP which is required for inhibition of adenylate cyclase converts these high-affinity-state receptors to low-affinity-state receptors even as they are inhibiting the enzyme. This can be explained with a simple model.

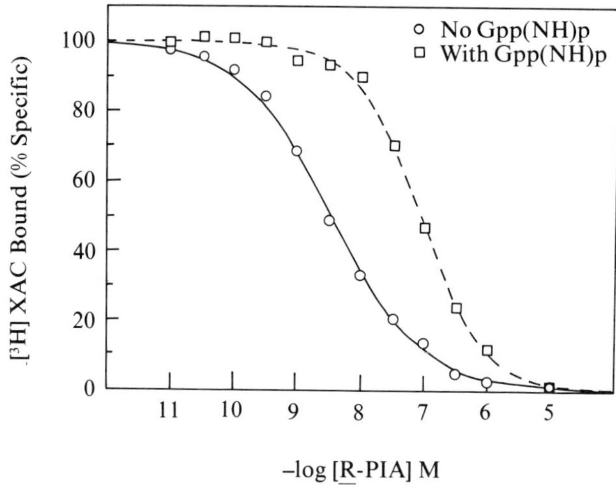

Fig. 3

Competition of *R*-PIA in the presence and absence of guanine nucleotides using the radioligand [³H]XAC. Bovine brain membranes were prepared and solubilized using the detergent Chaps. [³H]XAC was present at a concentration of 0.3 nM with the indicated concentration of *R*-PIA either in the absence or presence of Gpp(NH)p (10^{-4}M). Binding was performed at 25° C and data was analyzed using a computer modeling system as described in reference 178.

An agonist (Ag) interacts with the receptor (R) to form a low affinity complex Ag·R. The receptor can then undergo a conformational change allowing it to combine (or form a more tightly associated complex) with the G_i protein to form a high affinity ternary complex Ag·R·G$_i$. It is this complex which represents the active form of the receptor which subsequently modulates physiologic functions.

It can be predicted from this model that the extent to which an agonist stabilizes the high affinity state, the greater the intrinsic activity of an agonist. A second feature of the model is that guanine nucleotides destabilize the Ag·R·G$_i$ complex in association with the inhibition of adenylate cyclase.

It is now known that GTP binds directly to the α subunit of G$_i$. Upon binding to the α subunit, a series of events ensues during which the receptor and G$_i$ complex dissociate and the G$_i$ complex itself dissociates into α_i-GTP and $\beta\gamma$ subunits (163). This dissociated complex then mediates the inhibition of adenylate cyclase. The current proposed model for inhibition is that the released $\beta\gamma$ subunits from G$_i$ become available to bind up the active α_s and thus prevent it from activating the catalytic unit. In this manner, inhibitory receptors by promoting the activation of G_i can inhibit the enzyme adenylate cyclase (163). The activated G_i returns to the inactive form consequent to the hydrolysis of GTP to GDP by a GTPase activity intrinsic to the α subunit, favoring the reassociation of α_i with the $\beta\gamma$ subunits. This reassociation returns G_i to its basal state and permits it to reassociate with inhibitory receptors and begin the cycle again.

The essential feature of the current model is that each of the components of the adenylate cyclase system – the receptors, the G proteins and the catalytic moiety of the enzyme are capable of existing in active and inactive states through which they cycle. The pathways are linked with the cycle for G_i protein linking the receptor cycle to the cyclase pathway and the G_i cycle to the G_s cycle. The initial event is the binding of agonist to the receptor to form the low affinity complex (Ag $\cdot$ R). Under the influence of agonist binding this binary complex forms a ternary complex (Ag $\cdot$ R $\cdot$ G$_i$). *In vivo,* this intermediate complex is presumably transient and is acted upon by GTP. Upon formation of the Ag $\cdot$ R $\cdot$ G$_i$ complex, the rate limiting step in the subsequent reactions is the release of GDP from the α subunit of G_i. Upon the release of GDP, GTP then binds to α_i with the subsequent dissociation of Ag from R, R from G_i and α_i from $\beta\gamma$. This results in the reversion of high affinity binding to low affinity binding and the dissociation of the G protein into α_i-GTP and $\beta\gamma$. $\beta\gamma$ is then available to inhibit the activity of α_s. The GTPase activity of the α subunit then cleaves GTP to GDP and allows $\beta\gamma$ to rebind to α_i. If agonist remains in the system, then another cycle ensues and inhibition is maintained. A completely analogous series of reactions can explain the pathway for activation of adenylate cyclase by stimulatory receptors and G_s proteins (163).

It is appropriate to point out that much of the information obtained in other receptor systems by radioligand binding has been obtained using antagonist radioligands (162). This is important to note because much of the early information obtained on adenosine receptors was derived almost exclusively with agonist radioligands. As a consequence, it is difficult to assess and interpret agonist competition curves except when used to define pharmacological potency series. The reason for this is simple; agonists can, when used at appropriate concentrations, differentiate between the high and low affinity state of the receptor. Then when one competes with another agonist which differentiates two affinity states an extremely complex curve is generated which is difficult if not impossible to analyze quantitatively. Another important point to make is that a change in maximal binding capacity as determined with an agonist radioligand may not be able to differentiate between possibilities: one, is that receptor number is truly increased or decreased or two, that the increase or decrease in binding is consequent to increased or decreased coupling of receptor and G which may relate to quantitative or qualitative changes in G_i rather than receptor. With the advent of several high affinity antagonist radioligands, these dilemmas and difficulties should now be able to be resolved.

4.3   Regulation of ligand binding by ions and guanine nucleotides

As early as 1980, it was realized that agonist binding could be modulated by a variety of compounds and conditions (169). Since it was known that guanine nucleotides affected agonist binding in other hormone sensitive adenylate cyclase systems, studies on the effects of guanine nucleotides on agonist binding to A_1 adenosine receptors were undertaken. In preliminary studies, using single concentrations of radioligands – a practice which is fraught with error – it was determined that agonist binding decreased when guanine nucleotides, such as GTP, were included in the assay mixture (179). This finding was consonant with results previously published in a variety of receptor systems (162). Although GTP decreased agonist bindings, the use of a single concentration of radioligand does not allow one to determine whether the decreased binding is secondary to changes in maximal binding capacity or changes in agonist affinity. Therefore, full saturation curve analysis should always be undertaken.

More recent studies have documented that guanine nucleotides decrease the relative affinity of agonists for the A_1 receptor primarily by shifting the equilibrium between high and low affinity states to mostly low affinity state. Thus, the apparent decrease in agonist binding is really a change in affinity (K_D) with no significant change in B_{max}. This can be documented if radioligand binding is carried out with high enough concentrations of radioligand to saturate the low affinity state of the receptor. The shift in agonist affinity states can be documented by either full agonist saturation curves (170) or through the use of agonist competition curves constructed versus antagonist radioligands. In contrast to the dramatic effects of guanine nucleotides on agonist binding there appear to be little or no effects of GTP on antagonist binding (175).

Not all studies have made similar findings, however. For example, Yeung et al. reported that [³H]CHA binds to one affinity site in rat hippocampal membranes with a K_D of 1.8 nM and a B_{max} of 518 fmol/mg protein, while in the presence of Gpp(NH)p the K_D increases to ~ 13 nM (still a single affinity state was observed) and the B_{max} decreases to 272 fmol/mg protein. A third even lower affinity state could be discerned following N-ethylmalemide treatment of membranes by measuring the ability of agonists to compete for antagonist binding. They thus suggest that there are three agonist specific affinity states instead of the two (high and low) previously described in adenosine and other hormone sensitive receptor systems (180). In addition, this same group reported that guanine nucleotides increase the maximal binding capacity of [³H]DPX without affecting its K_D. This suggested to them that guanine nucleotides could reciprocally modulate agonist and antagonist binding.

In contrast to the above study, another group found that guanine nucleotides did not effect [³H]DPX binding in rat brain membranes (181). [³H]PIA was found to bind to a single class of high affinity states in the absence of GTP while in the presence of GTP two affinity states were discernable with the K_H being the same as the unique K_D in the absence of guanine nucleotides and a distinct low affinity state. In the presence of GTP 92 % of the receptors were in the low affinity state. They also assessed the effects of GTP on *R*-PIA competition curves and found two affinity states in the absence of GTP with similar K_H and $K_{L's}$ as determined from saturation curves with [³H]PIA. In the presence of GTP the *R*-PIA curve was shifted to the right and became uniphasic with only one low affinity state apparent. This was the same K_L found by [³H]PIA saturation curve analysis. No third very low affinity state was observed. In this same study antagonist competition curves (IBMX) were unaffected by GTP. This study, therefore, suggests two interconvertible agonist affinity states and that they can be assessed by either agonist saturation curves or by agonist competition curves.

The reason for the disparity in the findings between these two aforementioned studies is not obviously apparent, and further studies with the newer antagonist radioligands may be able to answer definitively how GTP acts on the A_1 adenosine receptor. Recent studies suggest the coupling of the A_1 receptor to the G_i protein may be very "tight" even in the absence of agonists (see below) and this may account for some of the unusual properties described in different studies (173).

Magnesium ions have long been known to be required for the effective functioning of adenylate cyclase systems. Mg^{++} appears to be an integral component of the ATP complex upon which the enzyme adenylate cyclase acts. In addition, Mg^{++} is necessary for the formation of the high affinity ternary complex of hormone-receptor and G protein. This is true for both stimulatory and inhibitory receptors. Magnesium appears to bind directly to the G proteins. Several studies have now documented that Mg^{++} is required to appreciate maximal agonist binding (182). The removal of the divalent ion by EDTA significantly decreases agonist binding. In contrast, Mg^{++} has no apparent effect on antagonist binding (182).

Sodium ions appear to play an important role in regulating or modulating the function of receptor systems coupled to adenylate cyclase systems in an inhibitory manner. Early studies in α_2-adrenergic receptor systems demonstrated that maximal inhibition of adenylate cyclase systems could not be obtained in the absence of Na^+ at concentrations ≥ 50 mM (183). There are some direct similarities in the phenomenology of the action of GTP and Na^+ ions. That is to say, both enhance the action of agonists on their effectors while at the same time decreasing the apparent affinity of agonists for the receptors at least as determined by radioligand binding studies.

As described earlier, it is fairly clear how GTP accomplishes this apparent paradox by interacting with G proteins to dissociate the high affinity $(Ag \cdot R \cdot G)$ complex while activating the α subunit of G proteins. The mechanism whereby Na^+ produces its effects remains much more elusive. Why it is required for the full manifestation of inhibitory agonists is still not understood.

Several studies on the effects of Na^+ ion on A_1-adenosine receptor binding have been carried out and no apparent concensus can thus far be reached. It has been reported that Na^+ decreases agonist but not antagonist binding in the brain but that Na^+ has very little effect on agonist or antagonist binding in adipocytes except under very special conditions (182). Much remains to be learned concerning the effects of sodium ions on the A_1 adenosine receptor system.

4.4   Photoaffinity radioligands

There are currently several approaches for the covalent labeling of membrane bound receptors. The basic requirements include a ligand which will bind with high affinity and the appropriate pharmacological specificity and stereoselectivity. In addition, the structure of the ligand must be amenable to alterations so as to include a radioactive label, such as ^{3}H or ^{125}I, and a reactive group through which the covalent linkage to the receptor can be made. It is axiomatic that following these structural manipulations the ligand must still interact with the receptor in the appropriate manner. A variety of different classes of covalent radiolabels have been synthesized for a wide range of membrane proteins und enzymes (162). At least three distinct types of compounds have been utilized. The first is a direct affinity radioligand in which a reactive chemical group, such as a bromacetyl moiety, is incorporated into the ligand which can directly interact with the protein to form a covalent bond (162). To our knowledge this approach has not as yet been successfully employed for adenosine receptors. A second type of probe is a radioligand which can be cross-linked into the receptor using a bifunctional cross-linking agent, such as SANPAH (N-succinimidyl-6-(4'-azido-2'-nitrophenylamino)hexanoate). An example of this is the radioligand [^{125}I]APNEA which can be utilized to label A$_1$ receptors (184).

For cross-linking radioligands the radioligand is first bound to the membrane receptor and then the cross-linking agent is added to the membranes. In the case of SANPAH and [^{125}I]APNEA, the succinimidyl group of SANPAH will react with the arylamine of the [^{125}I]APNEA forming a covalent bond. The membranes can then be exposed to ultraviolet light which then activates the arylazide in the SANPAH which in turn will then covalently incorporate into the receptor protein. This approach has been used for a wide range of receptors (162). A large number of different types of cross-linking agents (homobifunctional, heterobifunctional, cleavable) are commercially available which allow many radioligands to be incorporated into proteins.

The final approach that has been advocated is the synthesis of a direct photoaffinity probe (162). In this case the radioligand contains both a radiolabel and a photoactivatable group, such as an arylazide. An example of this type of probe is [^{125}I]AZPNEA (185).

Both of these latter types have proved to be very useful in the elucidation of the protein and glycoprotein structure of receptors including the A_1 adenosine receptor. The great advantage of these tools is that non-purified membrane bound receptors can be covalently labeled with a radioactive probe. The labeled receptors can then be solubilized and identified by SDS-PAGE and autoradiography. This permits the study of the protein nature of receptors to be undertaken without the requirement of purifying the receptor to homogeneity.

A number of agonist photoaffinity and photoaffinity cross-linking probes for the A_1 receptor have been synthesized in the last few years and very recently antagonists probes have become available. It is very gratifying that all the photoaffinity probes synthesized thus far appear to label the same A_1 adenosine receptor binding subunit (184–188). In addition, there appears to be very little species or tissue differences in the A_1 adenosine receptor. It should be mentioned that it is most appropriate to call the labeled peptide the binding subunit rather than the receptor since we will not know if the receptor is composed of subunits until the receptor has been purified to homogeneity and demonstrated to be the functionality intact moiety. That is to say, the purified receptor will bind ligands with the appropriate pharmacology and will interact with the G_i protein in a functional manner.

The first mammalian A_1 adenosine receptors were identified using [^{125}I]APNEA and SANPAH in rat brain and adipocyte membranes (184). This ligand binds to these aforementioned membranes with high affinity ($K_D \sim 2$ nM), reversibly and saturably. A_1 adenosine agonists will inhibit binding stereoselectivity and with the appropriate potency order, i. e. R-PIA > NECA > S-PIA.

SANPAH, the heterobifunctional cross-linking agent, covalently couples the radioligand to a protein of M_r 38,000 in both brain and adipocyte membranes as demonstrated by SDS-PAGE and autoradiography (see Figure 4). Inhibition of covalent labeling by adenosine analogs exhibited the same stereoselectivity and potency order typical of A_1 receptor ligands (184).

As would be expected from the fact that the radioligand is an agonist, guanine nucleotides, such as GTP or Gpp(NH)p, reduced both specific binding and covalent incorporation of the radioligand. This is demonstrated in Figure 4.

The A_1 receptor from a variety of tissues including rat, mouse, chicken and bovine brains, rat and mouse fat and bovine heart have been studied. All tissues appear to express A_1 receptor binding subunit as $\sim M_r$

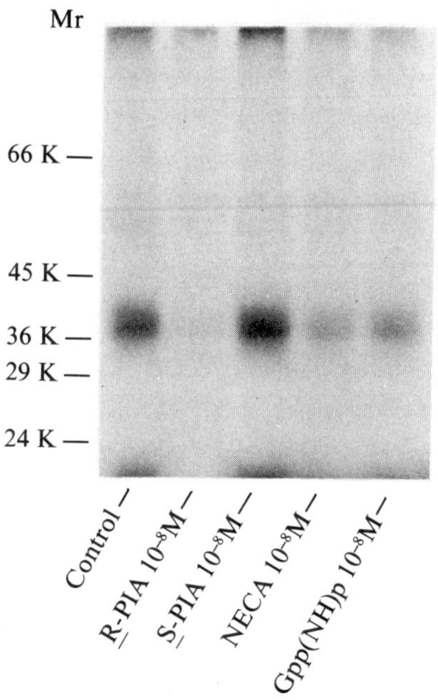

Fig. 4

Photoaffinity cross-linking of ^{125}I-APNEA into rat cerebral cortex membranes. Cerebral cortex membranes were photoaffinity cross-linked with ^{125}I-APNEA alone (control) or in the presence of the indicated concentration of competing ligand as described in reference 184. The samples were then solubilized and electrophoresed on a 10 % acrylamide gel. The relative molecular weight (M_r) scale was calibrated using iodinated protein standards.

38,000 peptides (176, 184, 189). The A_1 receptor is also present in substantial quantities in the rat testes and in this tissue the A_1 binding subunit appears to be slightly but significantly larger with an M_r 42,000 (190). The reason for this disparity remains unknown but it likely relates to differences in glycosylation rather than to differences in the peptide backbone. A description of the glycoprotein nature of the A_1 adenosine receptor will be provided in a later section.

[125I]AZPNEA is a direct photoaffinity probe which is also an A_1 selective agonist. This ligand can be directly incorporated in the A_1 receptor binding subunit without the use of a cross-linking agent. This probe also labels a M_r 38,000 peptide in a variety of tissues. Shown in Figure 5A is a gel autoradiograph of [125I]AZPNEA labeled A_1 receptor from rat brain and in Figure 5B a comparison of the A_1 receptor binding subunit from brain and adipocyte membranes.

The advantages of [^{125}I]AZPNEA compared to [^{125}I]APNEA is the relative ease with which incorporation can be achieved and the fact that the efficiency of incorporation is 10–20 times greater for the arylazide compared to the cross-linking procedure. This makes detection of the receptor binding subunit much easier for tissues without high concentrations of receptor and permits the autoradiographic films to be developed much more quickly.

Very recently we have developed an antagonist photoaffinity cross-linking radioligand based on the xanthine amine congener approach as described by Jacobson et al. (176). The radioligand [^{125}I]PAPAXAC was found to have very high affinity ($K_D \cong 0.1$ nM) for the A_1 adenosine receptor of bovine brain membranes (176). Using the same approach as that described for [^{125}I]APNEA, we were able to covalently incorporate [^{125}I]PAPAXAC in a M_r 38,000–40,000 peptide using SAN-PAH.

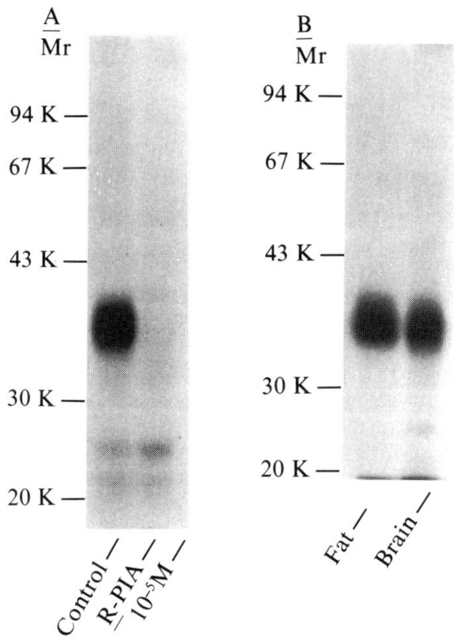

Fig. 5

Photoaffinity labeling of the A_1 adenosine receptor. *A*: Photoaffinity labeling of A_1 adenosine receptors in rat brain with [^{125}]AZPNEA. Aliquots of brain membrane were incubated with [^{125}I]AZPNEA and photolabeled as described in reference 185. The samples were solubilized and electrophoresed on a 11 % acrylamide gel. The molecular weight standards were determined with radioiodinated premixed electrophoresis standards. The M_r 38,000 labeled protein is the only labeled band demonstrating specific binding. *B*: Gel autoradiograph of [^{125}I]AZPNEA-labeled A_1 receptors from rat cerebral cortex and rat fat. The conditions were as in A.

The binding of the radioligand, as well as the covalent labeling, displayed all the appropriate stereoselectivity and potency order typical for bovine brain A_1 receptors (176).

As demonstrated in Figure 6, the peptide labeled with [^{125}I]PAPAXAC is strikingly similar to that labeled with the agonist probes described above. This is not surprising and, in fact, suggests that agonist and antagonists bind to exactly the same subunit.

One striking difference between the agonist and antagonist photoaffinity probes is that in contrast to the agonist radioligand [^{125}I]APNEA neither [^{125}I]PAPAXAC binding nor covalent labeling is affected by guanine nucleotides (compare Fig. 4 Lane 5 with Fig. 6 Lane 5). This would be anticipated from the model of agonist-mediated activation of G_i protein described previously. This characteristic of [^{125}I]PAPAX-AC should make it very useful for studying A_1 receptors since the binding of [^{125}I]PAPAXAC is not dependent on A_1AR-G_i interactions for its

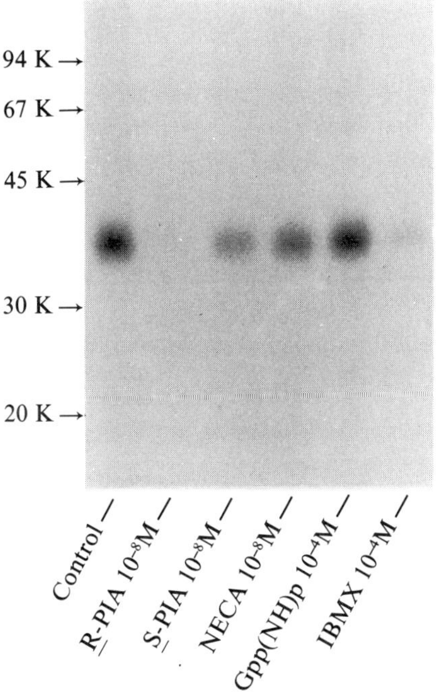

Fig. 6
Photoaffinity cross-linking of [^{125}I]PAPAXAC into bovine cerebral cortex membranes. Cerebral cortex membranes were photoaffinity cross-linked with [^{125}I]PAPAXAC alone (control) or in the presence of the indicated concentration of competing ligands as described in reference 176. The samples were then solubilized and electrophoresed on a 12 % acrylamide gel. The relative molecular weight (M_r scale) was calibrated using iodinated protein standards.

binding nor is its binding perturbed by guanine nucleotides.

These photoaffinity probes, as well as those described by others, should be very useful for studying potential alterations in receptor structure following pathophysiological perturbations or for use during the purification of the A_1 receptor. At the time of this writing there are, to our knowledge, no known photoaffinity probes for the A_2 adenosine receptor and nothing is known concerning its structure.

5   **Biochemical characterization of A_1 adenosine receptors**
5.1   Receptor-guanine nucleotide regulatory protein coupling

As noted in the previous sections, guanine nucleotides, such as GTP, play a central role in the transmembrane signalling process by interacting with and ultimately activating the G protein. The traditional litany has been that it is only under the influence of agonists that the receptor and G protein (α subunit) come together to form a tightly associated complex (see references 162, 163 for review). We and others have recently demonstrated that this simplistic view of R-G coupling does not hold true for the A_1 adenosine receptor and may not be applicable to a variety of receptors which have in common the ability to inhibit the enzyme adenylate cyclase [173].

The inital observation that suggested the A_1 adenosine receptor might not follow the same paradigm established for stimulatory receptors was that guanine nucleotides did not totally eliminate agonist binding. For example, Figure 7 demonstrates an [125I]HPIA saturation curve in rat cerebral cortex membranes. As can be seen the agonist binding is of high affinity ($K_D = 0.7$ nM) and is saturable. When the same curve is constructed in the presence of the nonhydrolyzable form of GTP (Gpp(NH)p) several interesting changes occur. First, total binding is decreased by 50 %; second, the affinity of [125I]HPIA for the receptor remains unchanged. The fact that binding is only decreased by 50 % and not completely ablated is in marked contradistinction to the findings in stimulatory receptor systems wherein high affinity agonist binding is totally eliminated and only low affinity agonist binding remains upon exposure to guanine nucleotides. It should be noted that under the conditions utilized for the agonist binding shown in Figure 7, we are only detecting high affinity binding. If we were to carry out the saturation curves to [125I]HPIA concentrations of 35–50 nM we could detect the low affinity state but practical considerations, such as

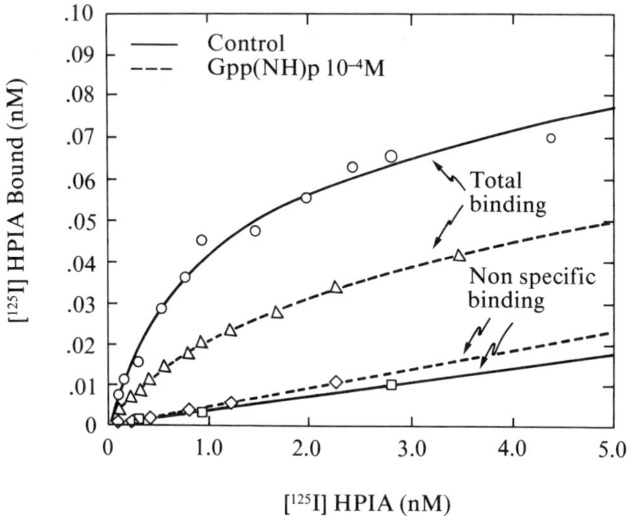

Fig. 7

[^{125}I]HPIA saturation curves with and without Gpp(NH)p (10^{-4}M) in rat cerebral cortex membranes. Membranes were prepared and radioligand binding was performed as described in reference 173. [^{125}I]HPIA was added at a final concentration shown on the abscissa. The concentration of [^{125}I]HPIA bound is on the ordinate. Nonspecific binding was defined with 10^{-6}M unlabeled R-PIA. The data points are means of duplicate determinations. This experiment is representative of 3 similar experiments. The curves were drawn with the aid of a computer modeling program based on the law of mass action (178).

the amount of radioactivity needed, preclude such experiments. In addition, the detection of the low affinity state per se is not germane to our current discussion.

The loss of only 50 % percent of high affinity agonist binding suggested to us that perhaps there was something unusual concerning R-G coupling in this model system. To further probe this question, we devised a method to solubilize the receptor while retaining its ability to recognize and bind ligands. What we found was that following solubilization in the detergent digitonin, the receptor was still able to bind agonists with the same high affinity as that found in intact membrane binding experiments. This is in spite of the fact that all endogenous adenosine was removed from membranes with adenosine deaminase thus precluding the possibility that we had formed a ternary complex of adenosine -R-G$_i$ prior to solubilization. Thus, if agonist high affinity binding is dependent on R-G$_i$ coupling, then the R-G$_i$ complex might be solubilized as an intact unit and may, therefore, demonstrate high affinity binding as evidenced by R-G$_i$ coupling with agonist occupancy.

This was indeed found to be the case since agonist binding was still modulated by guanine nucleotides. When Gpp(NH)p was added to soluble binding assays, all agonist binding was eliminated and no specific [^{125}I]HPIA could be detected. This suggests that in order for the A_1-G_i complex to totally dissociate it requires first solubilization and then treatment with Gpp(NH)p.

To further evaluate the molecular properties of the solubilized guanine nucleotide sensitive A_1 receptor complex, sucrose density gradient centrifugation was performed on A_1 adenosine receptors which had been labeled either prior to solubilization or following solubilization with [^{125}I]HPIA. If prelabeling of receptor in membranes with agonists leads to a stable agonist-A_1-G_i complex which is solubilized intact but solubilization of non-prelabeled A_1 receptor leads to a receptor which still binds agonists with high affinity and is modulated by guanine nucleotides but is not associated with G_i, then these two species should migrate differently in sucrose gradients (173). If, in fact, they comigrate then this is sound evidence that the same A_1-G_i complex is solubilized regardless of agonist occupancy. In fact, the complexes do comigrate on sucrose gradients to exactly the same location and both species (prelabeled and postlabeled) remain sensitive to guanine nucleotide modulation as manifested by total loss of specific binding. Thus, the A_1 receptor must be tightly associated with the G_i protein in the membranes and this association is stable upon solubilization.

There are now available data in other inhibiting receptor systems to suggest that this "tight coupling" is not unique to A_1 adenosine receptors but rather is characteristic to some degree of all inhibitory receptor-G_i protein interactions (193, 194). The biochemical mechanisms underlying the tight association between A_1 and G_i remain to be determined and should provide new insights into the structural domains of the A_1 receptor and the G_i protein which directly interact.

The advent of molecular cloning techniques has recently made the definition of what the G_i protein is much less certain. Although we have glibly used the term G_i to refer to the G protein which is the mediator of the inhibition of adenylate cyclase, it is now clear that there are at least three clones which represent a family of G_i proteins which have been termed α_{i1}, α_{i2}, and α_{i3}. At the present time it is unclear which of these, if any, represents what has traditionally been termed α_i.

5.2   Glycoprotein characteristics of A_1 adenosine receptor

With the advent of photoaffinity labeling techniques it has become possible to assess some of the detailed structure of the peptide which represents the A_1 receptor binding subunit.

An initial approach our laboratory has taken has been to directly compare the A_1 receptor binding subunit from different tissues and species. This is of interest because different tissues appear to display slight differences in pharmacological agonist potency series in relation to physiological responses. This finding has lead to the suggestion that the receptor expressed in each tissue might be slightly different and that the A_1 receptor may actually represent a family of receptors. As mentioned in the section on photoaffinity radioligands, this hypothesis appears not to be true at least at the level of the overall molecular weight of the receptor binding subunit. Thus, by SDS-PAGE/autoradiography the A_1 receptor binding subunit from a wide variety of tissues and species migrate as M_r 38,000–40,000 proteins (184, 185, 190).

To explore the detailed structure of the A_1 receptor, we have utilized the technique of two dimensional gel electrophoresis and peptide mapping to directly compare the peptide maps generated from the photoaffinity labeled A_1 receptor binding subunits from different tissues (191, 192). These techniques provide a very powerful method to determine whether two proteins are homologous in terms of amino acid sequence or if they differ significantly.

When directly comparing the A_1 receptor from cerebral cortex with that from adipocytes, it was found that the series of peptide fragments generated from each receptor was exactly analogous regardless of the proteinase used to digest the proteins. For example, when staphylococcus aureus V8 proteinase was used, 5 distinct peptides were generated from each receptor and there was a one to one correspondence in the molecular weight of peptides generated. If a proteinase of markedly different specificity, such as papain was used, six different peptides were generated compared to those found with staphylococcus V8 but again there was one to one correspondence between the fragments generated from the A_1 receptors from each tissue (see reference 191 for details).

These data provide very strong evidence that the A_1 receptor expressed in these two tissues are very homologous if not identical. The final proof of the exact amino acid sequence homology will only come about through the molecular cloning of the A_1 adenosine receptor. Most, if not all, membrane bound receptors are glycoproteins and contain either N-linked carbohydrate chains or O-linked chains or both (192). Recent cloning studies have confirmed the presence of consensus N-linked glycosylation sites for several membrane receptors and it is reassuring that the number of sites for glycosylation predicted from endoglycosidase experiments, using photoaffinity labeled receptor, for the β-adrenergic receptor have proven to be correct.

We have recently used a variety of endo- and exoglycosidases to study the carbohydrate nature of the A_1 adenosine receptor. These studies are possible because under the appropriate conditions these enzymes which display marked specificity for carbohydrates can act upon photoaffinity labeled binding subunits, thus, precluding the necessity of using purified receptor.

The enzyme endoglycosidase F under the appropriate conditions will cleave all N-linked carbohydrates from glycoproteins. This treatment then can totally deglycosylate the receptor in terms of N-linked carbohydrate chains (191, 192). As we have previously reported in other receptor systems, the carbohydrate chains may be removed sequentially by performing a time course experiment (192). For example, the mammalian β-adrenergic receptors contain two N-linked chains. The removal of carbohydrate chains from proteins alters the ability of SDS to bind to the protein and thus affects the receptors mobility on SDS-PAGE out of proportion to the molecular mass of glycan removed. Using SDS-PAGE/autoradiography one can then determine if there are carbohydrate chains present and how many chains are on each receptor molecule.

When these techniques are applied to the photoaffinity labeled A_1 receptor binding subunit, it can be determined that the A_1 receptor from both brain and adipocyte contain a single N-linked carbohydrate chain (see Fig. 8). Following deglycosylation the receptor binding subunit now migrates as a M_r 32,000 peptide. This is true for both the adipocyte and the cerebral cortex A_1 receptor.

To further define what type of carbohydrate chain is present, there are at least three distinct types termed complex, high mannose or hybrid. By utilizing exoglycosidases which have specificity for the terminal sugar residue present one can determine which type of chains are present. For example, complex chains terminate with sialic acid (neuraminic acid) and hence are sensitive to neuraminadase but are resistant to the enzyme α-mannosidase. In contrast, high mannose chains have terminal α-mannose and are thus sensitive to α-mannosidase but are resistant to neuraminadase. By using these enzymes, we have been able to demonstrate that the carbohydrate chains associated with the A_1 receptor of both adipocytes and cerebral cortex are likely of the complex type and contain no high mannose type chains.

Although much has been learned concerning the detailed structure of the A_1 receptor, much more remains to be learned concerning the functional domains of the receptor and why there is the "tight" association with its G protein. Information on the A_2 receptor lags far behind the A_1 receptor. Hopefully, this disparity will be overcome in the next several years.

6   Regulation

Regulation of cell surface receptors represents an important mechanism by which a cell can modulate the effects of drugs and hormones

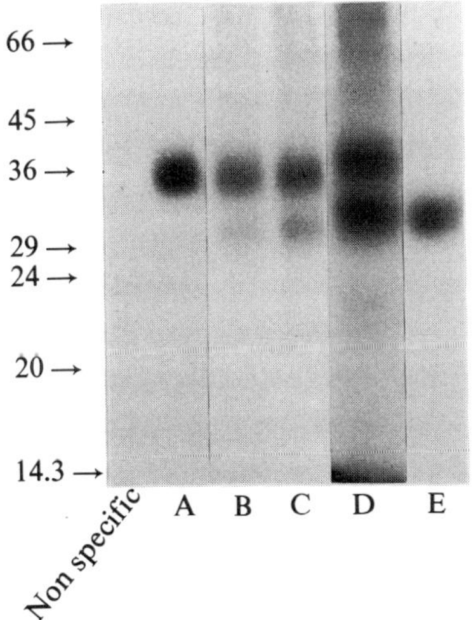

Fig. 8

Effect of endoglycosidase F on the A_1 receptor-binding subunit of rat fat. Membranes were labeled with [125I]APNEA alone or in the presence of 10^{-5}M R-phenylisopropyladenosine and solubilized as described in reference. Aliquots were treated with no Endo F (A, control), 5 units/ml for 5 h (B), 10 units/ml for 8 h (C), 10 units for 16 h (D), and 10 units for 26 h at 37° C (E). The samples were then subjected to SDS-polyacrylamide gel electrophoresis on a 12.5 % acrylamide gel. The molecular weight markers are shown X10^{-3}. This gel is a compilation of several experiments, each repeated several times.

acting via these receptors. Generally, exposure to agonists leads to uncoupling of the receptor-effector system and/or subsequent internalization of receptor from the cell surface (down regulation); the entire process results in desensitization of the response elicited by agonists (195). Activation of a receptor by an agonist might lead to desensitization of the response elicited via activation of that specific receptor (homologous desensitization) or by activation of other receptor systems similarly coupled to a common effector (heterologous desensitization) (196). In contrast, exposure to antagonists can lead to increases in receptor number and/or increased coupling of receptors to effector systems, leading to a sensitization of the systems to agonists (170).

6.1   Desensitization

Desensitization or tachyphylaxis refers to a state of refractoriness of a cell to normally activating stimuli following chronic exposure of cells to agonists (196). This biological phenomenon has only recently been characterized biochemically in the adenosine receptor-adenylate cyclase system. However, it is known that chronic administration of the agonist R-PIA to mice is characterized by pharmacological tolerance to subsequent effects of this drug (197). In this study, no concomitant changes in A_1 adenosine receptors in the brain were observed.

Evidence from our laboratory has demonstrated changes in multiple components of the adenosine receptor-adenylate cyclase system following chronic infusion of R-PIA to rats (195). This treatment results in a decrease in the number of A_1-adenosine receptors (as measured by [^{3}H]R-PIA binding), together with decreased inhibitory and enhanced stimulatory actions on adenylate cyclase. Furthermore, regulation of the levels of the G proteins was observed, G_i (α_i) being decreased while G_s (α_s) was increased. Thus, not only are A_1 adenosine receptors decreased but the levels of the G proteins (G_i and G_s) are also regulated reciprocally by R-PIA. This is the first described regulation of both α_s and α_i by a hormone acting via its membrane receptor. A recent report (198) has confirmed these findings using primary cultures of rat adipocytes. The loss of A_1 adenosine receptors induced by *in vitro* exposure to R-PIA was both dose and time-dependent, half maximal effects being observed using 16 nM R-PIA and following 24-h treatment. Similarly, both dose and time-dependent loss in the inhibitory guanine nucleotide protein (G_i) were observed by R-PIA treatment. These changes were accompanied by desensitization of the antilipolytic effect of R-PIA, together with an attenuation of the effect of insulin on glucose transport, the latter reflecting a heterologous mode of desensitization. Recent evidence from our laboratory indicate that desensitization to R-PIA results in a true loss in A_1 adenosine receptors (as determined by the binding of the antagonist [^{3}H]XAC) and a diminished ability of R-PIA to compete for radioligand binding (Longabaugh and Stiles, unpublished observations).

Desensitization of the A_2-adenosine receptor-adenylate cyclase system has also been described. Treatment of normal rat kidney (NRK) fibroblasts with an adenosine analog, for example, leads to a selective diminution of the stimulatory effect of adenosine on adenylate cyclase (homologous desensitization), with the stimulatory effects of epinephrine and sodium fluoride remaining intact. Studies using the neuronal clonal cells (NG 108–15), which contain both A_2-adenosine receptors and prostaglandin E (PGE_1) receptors coupled to the activation of adenylate cyclase, suggest a more complex mode of regulation of the receptor-adenylate cylase system. Treatment of these cells with PGE reduces the effects of both PGE_1 and adenosine to activate adenylate cyclase (heterologous desensitization). In contrast, homologous desensitization to 2-chloroadenosine was observed in this cell line (199).

6.2   Sensitization

Sensitization of the A_1 adenosine receptor-adenylate cyclase complex has been demonstrated following prolonged *in vivo* administration of certain methylxanthines, which are nonselective antagonists of adenosine receptors. Representative members of this class of drugs include caffeine, theophylline and aminophylline. Acute administration of caffeine is associated with arrhythmias (200), increased heart rate (201), stroke volume (202) and blood pressure (203), in addition to central stimulant actions (204). It is generally believed that these effects of caffeine are mediated through blockade of A_1 adenosine receptors (205, 206).

Prolonged administration of caffeine, followed by abrupt cessation of drug intake, leads to a number of symptoms such as headaches, myalgias, fatigue and anxiety (207–209), collectively termed "caffeine withdrawal syndrome". In addition, tolerance develops rapidly to both the peripheral and central effects of caffeine in man (210) and rodents

(211). While the basis for these changes are not clear, it is likely linked to concommitant changes in the A_1-receptor-adenylate cyclase complex induced by prolonged caffeine intake. For example, caffeine increases A_1 adenosine receptors in the brain (197, 172, 212–214). Increases in these receptors were observed using both agonist (172, 212–214) and antagonist (197, 230) radioligands, indicating that caffeine treatment elevates both the agonist high affinity state and the total receptor population. In accordance with increases in agonist high affinity state of the A_1-adenosine receptors, concurrent enhancement of R-PIA-mediated inhibition of cerebal cortical adenylate cyclase activity was also observed following similar treatment conditions (172). In this latter study, the sensitivity of [^{3}H]R-PIA binding to the guanine nucleotide Gpp(NH)p was increased substantially in the caffeine-treated rats, suggesting possible regulation of G_i induced by this treatment. The existence of such a possibility has recently been demonstrated in our laboratory (230). Following chronic caffeine ingestion, the level of G_i in the cortex was increased by 33 % over control level (230). Thus, chronic caffeine ingestion appears to regulate multiple components of the A_1 adenosine receptor-adenylate cyclase system in rat cerebral cortex. Taken together, these changes could account for sensitization of the A_1 receptor inhibitory system to endogenous adenosine induced by the methylxanthine. These different means of adaptation might account for the development to tolerance to the central actions of caffeine, along with the central nervous system manifestations of caffeine withdrawal syndrome.

Up-regulation of rat cerebral cortical adenosine receptors has also been demonstrated following chronic administration of theophylline for 21 days. In this study, a significant increase in [^{3}H]CHA binding sites was observed in the cortex, with no significant change in the hippocampus in the treated rats (215). Furthermore, substantial increases in A_1 adenosine receptors in mouse brain have been demonstrated following chronic morphine administration (216). Whether increases in A_1 adenosine receptors in these two studies reflect increased receptor-G_i coupling or a true increase in receptor number (derived from antagonist radioligand) is not clear.

6.3   Thyroid status

Adenosine, interacting with A_1 adenosine receptors in adipose tissue, is a potent antilipolytic agent. Since these receptors are negatively coupled to adenylate cyclase, it is believed that inhibition of lipolysis in adipose tissue by adenosine results from reduction in the level of cyclic AMP. Conversely, catecholamine-mediated activation of adenylate cyclase enhances lipolysis. In the hypothyroid state, the ability of adrenaline to increase lipolysis is severely diminished while the antilipolytic effect of R-PIA is enhanced (217). This change does not appear to be due to changes at the A_1 adenosine receptor, since [^{3}H]cyclohexyladenosine binding sites in the hypothyroid rats were not altered compared to controls (218). However, both R-PIA and GTP-mediated inhibition of adenylate cyclase were exaggerated in adipocyte membranes prepared from hypothyroid rats, inferring a post-receptor amplification of this inhibitory system. Subsequent studies have supported this contention by demonstrating a 2–3 fold increase in the level of the inhibitory G_i protein in the hypothyroid state (218). This increase in G_i will necessarily facilitate coupling of inhibitory receptors to the catalytic subunit of adenylate cyclase, thus explaining the augmentation of the inhibitory effects of R-PIA. Furthermore, such an explanation might also account for the reduction in catecholamine-mediated stimulation of adenylate cyclase in hypothyroid rats.

In contrast to hypothyroidism, the hyperthyroid state is characterized by enhanced lipolytic activity and cyclic AMP accumulation in adipocytes (219). This likely relates to a loss in the inhibitory tone mediated via A_1 receptors due to a 35 % reduction in [^{3}H]cyclohexyladenosine binding sites (220). Thus, in presence of adenosine (in absence of added adenosine deaminase), inhibition of adenylate cyclase via a lower than normal complement of A_1-adenosine receptors is decreased, leading to potentiation of the effects of stimulatory agonists on cyclic AMP accumulation. In the presence of exogenous adenosine deaminase (absence of adenosine), cyclic AMP accumulation in adipocytes stimulated by adrenaline and forskolin was significantly reduced. This finding, together with those showing no changes in the levels of G_i, G_o and G_s induced by hyperthyroidism, point to deficiency in the catalytic subunit of adenylate cyclase in order to explain the diminished response to adrenaline and forskolin (220).

6.4   Pregnancy and lactation

The rate of lipolysis in adipose tissue is greatly accelerated during early lactation in order to maintain adequate milk production. In lactating sheep, for example, the lipolytic effect of noradrenaline is increased over control, unmated sheep (221). However, R-PIA-mediated inhibition of lipolysis was also increased during lactation (221). The reason for these two opposing changes is not clear but might reflect a compensatory measure to prevent significant depletion of fat stores during lactation.

In rats, lactation is associated with depressed noradrenaline stimulated lipolysis in adipocytes together with an increased sensitivity to the anti-lypolytic activity of adenosine (222). Thus, there appears to be species differences in the effects of lactation on hormone-stimulated lipolysis. With the availability of suitable ligands for the A_1 adenosine receptors and methods for the quantitation of G proteins, studies dealing with the regulation of these systems during lactation might soon be forthcoming.

6.5   Starvation

In man, starvation impairs the lipolytic activity of the catecholamines, noradrenaline and adrenaline, compared to the fed state (223). However, basal activity of lipolysis in adipocytes is significantly higher during starvation. The reason for the diminished stimulatory effect of the catecholamines is not clear but apparently involves adenosine. In the fed state and at normal levels of adenosine the preferred target of these catecholamines is the β-adrenergic receptors in adipose tissue, the activation of which increases lipolysis. However, during starvation the levels of adenosine are decreased and the preferred target of the catecholamines appears to be α_2 adrenergic receptors which mediate the inhibition of lipolysis. This latter effect is mimicked by adenosine deaminase which metabolizes adenosine to inosine (223). Thus, starvation dictates which effector system is activated by catecholamines via regulation of adenosine levels in man. The situation appears to be reversed in rats, where starvation increases the lipolytic activity of epinephrine in adipose tissue (224). While the reason for this is unclear, this finding points to species differences in the biochemical responses to starvation.

6.6   Obesity

Adenosine is a local hormone which mediates its antilipolytic activity either directly via inhibitory A_1 adenosine receptors or by potentiating the effects of insulin. Alterations in the inhibitory system are observed in physiological states characterized by altered insulin sensitivity, inferring a close and physiologically relevant association between these two receptor systems coupled to the inhibition of lipolysis. Diminution of the antilipolytic effects of *R*-PIA in adipocytes has been reported in obesity (225), a state also characterized by reduced responsiveness to insulin. The effectiveness of this adenosine analog in this respect bears an inverse correlation with body weight (225). Furthermore, the ability of *R*-PIA to inhibit cyclic AMP accumulation in subcutaneous abdominal fat cells was substantially reduced in obese subjects (225). Taken together, these findings demonstrate co-regulation of the A_1 adenosine receptor and insulin receptor systems in obesity.

6.7   Aging

Aging is associated with diminished responsiveness of several tissues to stimulatory hormones. For example, the effects of catecholamines on the cardiovascular system are reduced with age (226, 227), while catecholamine-stimulated lipolysis in rat adipocytes is also diminished (228). Since lipolysis is under inhibitory control by A_1 adenosine receptors stimulated by endogenously released adenosine, the possibility that this inhibitory system is amplified with aging was tested (228). It was observed that isoproterenol-stimulated glycerol release (measure of lipolysis) in adipocytes was considerably reduced or abolished upon addition of adenosine. Furthermore, following adenosine deaminase treatment, subsequent addition of *R*-PIA reinstates the blunted response of isoproterenol observed in adipocytes prepared from the older rats. Thus, it appears that the enhanced inhibitory tone in adipose tissue which accompanies aging limits the lipolytic effect of catecholamine. Further, studies are needed to localize the exact component(s) of the A_1 adenosine receptor-adenylate cyclase complex responsible for this enhanced inhibitory tone.

7 Summary

The past 60 years has produced a slow but progressive increase in our knowledge of the physiological effects of adenosine on the various organ systems of the body. The last 20 years has brought forth an understanding of the mechanisms of action of adenosine at the organ and cellular level and ushered in an appreciation that many of the effects of adenosine are mediated via specific membrane receptors. Recently, within the past 5–10 years, the biochemical mechanisms underlying receptor-effector coupling which serve to transduce a transmembrane signal have begun to be elucidated. Several important findings have recently become clear including the fact that both A_1 and A_2 adenosine receptors are coupled to effector systems other than adenylate cyclase. For example, there is good evidence that A_1 adenosine receptors are coupled in an inhibitory manner to adenylate cyclase, in a stimulatory manner to K^+ channels, phosphodiesterases and likely to guanylate cyclase in a stimulatory manner. It is of interest that activation of these different effector systems displays similar agonist potency series, suggesting it is indeed the same receptor which produces these different effects. In some cells two different effector systems seem to be activated by the same receptor suggesting the specificity of the receptor for the effector must reside in the coupling or G proteins. The elucidation of how this comes about should be a fruitful area of research over the next several years.

The structure of the adenosine receptor has now begun to be probed at the molecular level and future research should provide insight into how the receptor is synthesized and how this synthesis is regulated under normal and pathophysiological conditions. In addition, through the tools of molecular biology it is hoped that the receptor could be regulated or altered for therapeutic benefit.

Although there have been great strides made in our understanding of adenosine and adenosine receptor actions, the next 5–10 years should produce new and hitherto undreamed of insights into how adenosine regulates physiological function.

References

1 G. Burnstock, in: Cell Membrane Receptors for Drugs and Hormones: A Multidisciplinary Approach, p. 107–118. Eds. L. Bolis and R. W. Straub. Raven Press, New York (1978).
2 D. Van Calker, M. Miller and B. Hamprecht: J. Neurochem. *33*, 999 (1979).
3 C. Londos, D. M. F. Cooper and J. Wolff: Proc. Natl. Acad. Sci. USA *77*, 2551 (1980).
4 R. F. Bruns, J. W. Daly and S. H. Snyder: Proc. Natl. Acad. Sci. USA *80*, 2077 (1988).
5 K. A. Jacobson, K. L. Kirk, W. L. Padgett and J. W. Daly: J. Med. Chem. *28*, 1334 (1985).
6 H. W. Hamilton, D. F. Ortwine, D. F. Worth, E. W. Badger, J. A. Bristol, R. F. Bruns, S. J. Haleen and R. P. Steffen: J. Med. Chem. *28*, 1071 (1985).
7 K. A. Jacobson, D. Ukena, K. L. Kirk and J. W. Daly: Proc. Natl. Acad. Sci. USA *83*, 4089 (1986).
8 J. W. Daly, K. A. Jacobson and D. Ukena in: Cardiac Electrophysiology ans Pharmacology of Adenosine and ATP: Basic and Clinical Aspects, pp. 41–63. Alan R. Liss, Inc., New York (1987).
9 D. M. F. Cooper, W. Schlegel, M. C. Lin and M. Rodbell: J. Biol. Chem. *254*, 8927 (1979).
10 T. Murayama and M. Ui: J. Biol. Chem. *258*, 3319 (1983).
11 J. A. Garcia-Sainz and M. L. Torner: Biochem. J. *232*, 439 (1985).
12 A. C. Dolphin, S. A. Prestwich and S. R. Forda in: Adenosine: Receptors and Modulation of Cell Function, p. 107. Alan R. Liss, New York, 1985.
13 Y. Kuroda in: Adenosine : Receptor and Mudulation of Cell Funktion, pp. 233–239. Eds. V. Stafanovich, K. Rudolphi and P. Schubert. IRL Press Limited, Oxford, England (1985).
14 P. Schubert in: Adenosine: Receptor and Modulation of Cell Function, pp. 117–129. Eds. V. Stefanovich, K. Rudolphi and P. Schubert. IRL Press Limited, Oxford, England (1985).
15 M. Bohm, R. Bruckner, J. Neumann, W. Schmitz, H. Scholz and J. Starbutty: Naunyn-Schmiedebergs Arch. Pharmacol. *332*, 407 (1986).
16 Y. Kurachi, T. Nakajima and T. Sugimoto: Pflugers Arch. *407*, 264 (1986).
17 T. S. Teu, S. Ooi and E. H. A. Wong: FEBS Lett. *128*, 75 (1981).
18 E. H. A. Wong, S. Ooi, E. G. Loten and J. G. T. Sneyd: Biochem. J. *227*, 815 (1985).
19 P. De Mazancourt and Y. Guidicelli: FEBS Lett. *173*, 385 (1984).
20 P. De Mazancourt and Y. Guidicelli: Brain Res. *300*, 211 (1984).
21 A. Kurtz: J. Biol. Chem. *262*, 6296 (1987).
22 E. B. Hollingsworth, A. De La Cruz and J. W. Daly: Eur. J. Pharmacol. *122*, 45 (1986).
23 S. J. Hill and D. A. Kendall: Br. J. Pharmacol. *89*, 771P (1986).
24 S. J. Hill and D. A. Kendall: Br. J. Pharmacol. *91*, 661 (1987).
25 J. J. O'Shea, C. A. Suarez-Quian, R. A. Swank and R. D. Klausner: Biochem. Biophys. Res. Commun. *146*, 561 (1987).
26 S. R. Doctrow and J. M. Lowenstein: Biochem. Pharmacol. *36*, 2255 (1987).
27 M. E. Lewis, J. Patel, S. Moon Edley and P. J. Marangos: Eur. J. Pharmacol. *73*, 109 (1981).
28 R.R. Goodman and S. H. Snyder: J. Neurosci. *2*, 1230 (1982).
29 J. W. Phillis and P. H. Wu: Prog. Neurobiol. *16*, 187 (1981).
30 W. J. Wojcik and N. Neff: J. Neurochem. *41*, 759 (1983).
31 R. R. Goodman, M. J. Kuhar, L. Hester and S. H. Snyder: Science *220*, 967 (1983).
32 A. C. Dolphin and P. Greengard: J. Neurosci. *1*, 192 (1981).
33 T. F. Murray and D. L. Cheney: Neuropharmacology *21*, 575 (1982).

34   J. D. Geiger, F. S. Labella and J. I. Nagy: J. Neurosci. *4*, 2303 (1984).
35   J. I. Choca, H. K. Proudfit and R. D. Green: J. Pharmacol. Exp. Ther. *242*, 905 (1987).
36   P. Hedquist and B. B. Fredholm: Naunyn-Schmiedeberg's Arch. Pharmacol. *293*, 217 (1976).
37   D. B. Evans, J. A. Schenden and J. A. Bristol: Life Sci. *31*, 2425 (1982).
38   L. Belardinelli, F. L. Belloni, R. Rubio and R. M. Berne: Circ. Res. *47*, 684 (1980).
39   J. G. Dobson, Jr., R. A. Fenton and F. D. Romano in: Cardiac Electrophysiology and Pharmacology of Adenosine and ATP: Basic and Clinical Aspects, p. 331. Alan R. Liss, Inc., New York, 1987.
40   R. W. von Borstel, R. J. Wurtman and L. A. Conlay: Life Sci. *32*, 1151 (1983).
41   D. N. Paton: J. Auton. Pharmacol. *1*, 287 (1981).
42   L. Belardinelli and G. Isenberg: Circ. Res. *53*, 287 (1983).
43   V. Hausleithner, M. Freissmuth and W. Schultz: J. Recept. Res. *6*, 311 (1986).
44   M. J. Loshe, D. Ukena and U. Schwabe: Naunyn-Schmiedeberg's Arch. Pharmacol. *328*, 310 (1985).
45   J. Linden, C. E. Hollen and A. Patel: Circ. Res. *56*, 728 (1985).
46   D. Martens, M. J. Lohse, B. Rauch and U. Schwabe: Naunyn-Schmiedeberg's Arch. Pharmacol. *336*, 342 (1987).
47   M. Huang and G. I. Drummond: Biochem. Pharmacol. *25*, 1713 (1979).
48   E. Leung, C. I. Johnston and E. A. Woodcock: Biochem. Biophys. Res. Commun. *110*, 208 (1983).
49   S. Kusachi, R. D. Thompson and R. A. Olsson: J. Pharmacol. Exp. Ther. *227*, 316 (1983).
50   M. B. Anand-Srivastava, D. J. Franks, M. Cantin and J. Genest: Biochem. Biophys. Res. Commun. *108*, 213 (1982).
51   M. G. Collis and C. Brown: Eur. J. Pharmacol. *96*, 61 (1983).
52   B. Jonzon, J. Nilsson and B. B. Fredholm: J. Cell. Physiol. *124*, 451 (1985).
53   C. Londos, D. M. F. Cooper, W. Schlegel and M. Rodbell: Proc. Natl. Acad. Sci. USA *75*, 5362 (1978).
54   D. M. F. Cooper and C. Londos: J. Cyclic Nucleotide Res. *5*, 289 (1979).
55   P. C. Churchill: J. Pharmacol. Exp. Ther. *222*, 319 (1982).
56   I. L. Campbell and K. W. Taylor: Biochem. J. *204*, 689 (1982).
57   J. A. Ribeiro and A. M. Sebastiao: Br. J. Pharmacol. *84*, 911 (1985).
58   E. Huttemann, D. Ukena, V. Lenschow and U. Schwabe: Naunyn-Schmiedeberg's Arch. Pharmacol. *325*, 226 (1984).
59   G. Marone, M. Plaut and L. M. Lichtenstein: J. Immul. *121*, 2153 (1978).
60   A. N. Drury and A. Szent-Györgi: Br. J. Pharmacol. *77*, 347p (1929).
61   A. Sattin and T. W. Rall: Mol. Pharmacol. *6*, 13 (1970).
62   T. W. Stone: Neuroscience *6*, 523 (1981).
63   G. Spignoli, F. Pedata and G. Pepeu: Eur. J. Pharmacol. *97*, 341 (1984).
64   R. P. Ebstein and J. W. Daly: Cell Mol. Neurobiol. *2*, 205 (1982).
65   A. Carlsson: Pre- and Postsynaptic Receptors, pp. 49–67. Eds. E. Usdin and W. E. Bunney, Jr., New York (1975).
66   Y. Kuroda in: Physiology and Pharmacology of Adenosine Derivaties, pp. 245–256. Eds. J. W. Daly, Y. Kuroda, J. W. Phillis, H. Shimizu and M. Ui, Raven Press, New York (1983).
67   J. A. Riberio: Ann. N. Y. Acad. Sci. *377*, 874 (1981).
68   H. H. Harms, G. Wardeh and A. H. Mulder: Neuropharmacology *18*, 577 (1979).
69   S. Govoni, V. Perkov, O. Montefusco, C. Missale, F. Battaini, et al.: J. Pharm. *36*, 458 (1984).
70   T. V. Dunwiddie: Int. Rev. Neurobiol. *27*, 63 (1985).

71 C. Ehersolt, J. Brenunt, A. Prochantz, M. Perez and J. Bucknert: Brain Res. *267*, 123 (1983).
72 J. W. Phillis and R. A. Barraco in: Advances in Cyclic Nucleotide and Protein Phoshorylation Research, Vol. 19, pp. 243–257. Eds. D.M.F. Cooper and K.B. Seamon. Raven Press, New York (1985).
73 T. W. Dunwiddie and B. B. Fredholm: Naunyn-Schmiedeberg's Arch. Pharmacol. *326*, 294 (1984).
74 W. Feldberg and S. L. Sherwood: J. Physiol. (Lond.) *123*, 148 (1954).
75 M. Radulovacki and R. M. Virus: Pharmacol. Exp. Ther. *228*, 268 (1983).
76 M. Radulovacki, R. M. Virus and M. Djunivic-Nedelsen: *271*, 391 (1983).
77 G. Yank and M. Radulovack: Brain Res. *401*, 362 (1987).
78 M. Dragunow, G. V. Goddard and R. Laverty: Epilepsia *26*, 486 (1985).
79 M. Williams: J. Med. Chem. *26*, 619 (1983).
80 M. Williams, E. A. Risley and J. R. Huff: Can J. Physiol. Pharmacol. *59*, 897 (1981).
81 L. P. Davies, S. C. Chow, and G. A. R. Johnston: Eur. J. Pharmacol. *97*, 325 (1984).
82 L. P. Davies, A. F. Cook, A. M. Poonian and K. M. Taylor: Life Sci. *26*, 1089 (1980).
83 P. F. Morgan, H. G. Lloyd and T. W. Stone: Neurosci. Lett. *41*, 183 (1983).
84 S. Intoh, O. A. Carretero and R. D. Murray: J. Clin. Invest. *76*, 1412 (1985).
85 M. A. Lang, A. S. Preston, J. S. Handler and J. N. Forrest: Am. J. Physiol. *249*, C330 (1985).
86 J. E. Hall and J. P. Granger: Am. J. Physiol. *250*, F32 (1986).
87 R. E. Katholi, G. R. Hagman, P. L. Whitlow and W. T. Woods: Hypertension *5*, I149 (1983).
88 J. F. Macias, M. Fiksten-Olsen, J. C. Romero and F. G. King: Am. J. Physiol. *244*, H138 (1983).
89 M. A. Lang, A. S. Preston, J. S. Hander and J. N. Forrest: Am. J. Physiol. *249*, C330 (1985).
90 B. B. Fredholm and A. Solleir: Clin. Physiol. *6*, 1 (1986).
91 W. S. Spielman and C. I. Thompson: Am. J. Physiol. *242*, F423 (1982).
92 A. N. Drury and A. Szent-Gyorgi: J. Physiol. (Lond.) *68*, 213 (1929).
93 A. Pellig: Amer. Heart J. *110*, 688 (1985).
94 L. Belardinelli, F. Bellini, R. Rubio and R. M. Bern: Circ. Res. *47*, 684 (1980).
95 L. Belardinelli, E. C. Maltos and R. M. Bern: J. Clin. Invest. *68*, 195 (1981).
96 R. M. Bern, J. P. DiMarco and L. Belardinelli: Circ. *69*, 1195 (1984).
97 L. Belardinelli: Cardiac Electrophysiology and Pharmacology of Adenosine ATP Basic and Clinical Aspects, pp. 109–118. Alan R. Liss Inc. (1987).
98 E. A. Johnson and M. C. McKinnon: Nature *178*, 1174 (1956).
99 J. Bailey and D. Rardon: Cardiac Electrophysiology and Pharmacology of Adenosine ATP Basic and Clinical Aspects, pp. 119–133. Alan R. Liss Inc. (1987).
100 G. A. West: Cardiac Electrophysiology and Pharmacology of Adenosine ATP Basic and Clinical Aspects, pp. 97–108. Alan R. Liss Inc. (1987).
101 T. Mazzalov, L. S. Dreifus, E. L. Michelson and A. Pellig: Cardiac Electrophysiology and Pharmacology of Adenosine ATP Basic and Clinical Aspects, pp. 195–219. Alan R. Liss Inc. (1987).
102 A. Pellig, T. Mitsuoka, T. Mazzalov and E. Michelson: Cardiac Electrophysiology and Pharmacology of Adenosine ATP Basic and Clinical Aspects, pp. 375–384. Alan R. Liss, Inc. (1987).
103 A. Pellig, T. Mitsuoka, T. Mazzalov, and E. Michelson: Cardiac Electrophysiology and Pharmacology of Adenosine ATP Basic and Clinical Aspects, pp. 384–395. Alan R. Liss, Inc. (1987).

104	J. Dobson, R. A. Fenton and F. D. Romano: Cardiac Electrophysiology and Pharmacology of Adenosine ATP Basic and Clinical Aspects, pp. 331–343. Alan R. Liss, Inc. (1987).

105	A. D. Sharma and G. J. Klein: Cardiac Electrophysiology and Pharmacology of Adenosine ATP Basic and Clinical Aspects, pp. 315–318. Alan R. Liss, Inc. (1987).

106	T. Mazzalov, L. Dreifus, E. L. Michelson and A. Pellig: Cardiac Electrophysiology and Pharmacology of Adenosine ATP Basic and Clinical Aspects, pp. 195–219. Alan R. Liss, Inc. (1987).

107	A. Genovese and R. Levi: Cardiac Electrophysiology and Pharmacology of Adenosine ATP Basic and Clinical Aspects, pp. 345–360. Alan R. Liss, Inc. (1987).

108	T. D. Sellers, J. B. Kirchhoffer and T. A. Modesto: Cardiac Electrophysiology and Pharmacology of Adenosine ATP Basic and Clinical Aspects, pp. 283–299 Alan R. Liss, Inc. (1987).

109	D. Belhassen and A. Pellig: Am. J. Cardiol. *4*, 414 (1984).

110	B. B. Lerman and L. Belardinelli: Cardiac Electrophysiology and Pharmacology of Adenosine ATP Basic and Clinical Aspects, pp. 301–314 Alan R. Liss, Inc. (1987).

111	D. Belhassen and A. Pellig: Am. J. Cardiol. *54*, 225 (1984).

112	B. Belhassen, A. Pellig: D. Shoshani, B. Geva and S. Lamado: Circ. *68*, 827 (1983).

113	J. P. DiMarco, T. D. Sellers, R. M. Bern, G. A. West and L. Belandinelli: Circ. *68*, 1254 (1983).

114	J. P. DiMarco, T. D. Sellers, B. B. Lerman, M. L. Greenberg, R. M. Bern and L. Belardinelli: J. Am. Coll. Cardiol. *6*, 417 (1985).

115	R. Greco, B. Musto, V. Arienzo, A. Alborino, S. Garofalo and F. Marsico: Circ. *66*, 504, (1982).

116	A. Pellig, B. Belhassen, R. Ilia and S. Lanardo: Am. J. Cardiol. *55*, 571 (1985).

117	R. A. Olsson and R. Boyer: Progress in Cardiovas. Dis. *XXIX*, 369 (1987).

118	Y. O. Li and B. B. Fredholm: Acta. Physiol. Scan. *124*, 253 (1985).

119	J. F. Macias, M. Fiksen-Olsen, J. C. Ramero and J. G. Knox: Am. J. Physiol. *244*, H138 (1983).

120	A. Solevi and B. B. Fredholm: Acta. Physiol. Scan. *119*, 15 (1983).

121	C. G. A. Persson, K. E. Anderson and G. Kjellin: Life Sci. *38*, 1057 (1986).

122	M. J. Cushey, A. E. Tatterfield and S. T. Holgate: Am. Rev. Respir. Dis. 40, 380 (1984).

123	M. J. Cushey and S. T. Holgate: J. Allergy Clin. Immunol. *74*, 272 (1985).

124	B. B. Fredholm: Acta. Med. Scand. *217*, 149 (1985).

125	A. F. Welton and B. A. Simko: Biochem. Pharmacol. *29*, 1085 (1980).

126	R. A. Coleman: Science *57*, 51 (1976).

127	B. B. Fredholm, K. Brodkin and K. Straudberg: Acta. Pharmacol. Toxicol. *45*, 336 (1979).

128	C. Adveneir, D. Bidet, A. Floch-St.-Aubin and R. Reiner: Br. J. Pharmacol. *77*, 39 (1982).

129	N. Napin and D. M. Temple: J. Pharm. Pharmacol. Toxicol. *45*, 336 (1979).

130	J. Hasaday and R. G. Sitrin: J. Lab. Clin. Med. *110*, 264 (1987).

131	R. Hurschhornn, J. Grossman and G. Weissman: Proc. Soc. Exp. Biol. Med. *13*, 1361 (1970).

132	L. E. Averill and G. M. Kanner: Clin. Res. *33*, 839A (1985).

133	R. E. Birch and S. H. Polmar: Clin. Exp. Immunol. *48*, 218 (1982).

134	R. E. Birch, A. K. Rosenthal and S. H. Polmar: Clin. Exp. Immunol. *48*, 231 (1982).

135	T. P. Zimmermann, G. Wolberg and G. S. Duncan in: R. M. Bern, T. M. Rall and R. Rubio Reds Regulatory Function of Adenosine , pp. 261–273. Martinns Nijhoff, Boston (1983).

136 S. T. Holgate, R. A. Lewis and A. F. Austen: Proc. Natl. Acad. Sci. USA *77*, 6800 (1980)
137 G. M. Kanner: J. Lab. Clin. Med. *110*, 255 (1987).
138 G. Marone, R. Petracca and S. Vigorita: Int. Arch. Allergy Appl. Immnol. *77*, 259 (1985).
139 I. Garcia-Casto, J. M. Mato, G. Vasnthakumar, W. P. Wasmann, E. Schiffman and P. K. Chay: J. Biol. Chem. *258*, 4345 (1983).
140 B. B. Fredholm, M. Jondal, F. Lanefelt and J. Ng: Scand. J. Immunol. *801*, 511 (1984).
141 D. Lappin and K. Whaley: Clin. Exp. Immunol. *57*, 454 (1984).
142 R. R. Askanitt, P. S. Backlund, Jr. and G. L. Cantini: J. Biol. Chem. *258*, 20 (1983).
143 B. N. Cronstein, E. D. Rosenstein, S. B. Kramer, G. Weisman and G. Hirschhorn: J. Immunol. *135*, 1366 (1985).
144 J. D. Geiger and G. B. Glavin: Eur. J. Pharmacol. *115*, 195 (1985).
145 A. H. Watt, D. J. Lewis and J. J. Horne: British Med. J. *294*, 10 (1987).
146 I. Ushijina, Y. Mizuki and M. Yamada: Brain Res. *339*, 351 (1985).
147 J. G. Gerber, A. S. Nies and N. A. Payne: J. Pharm. Exp. Ther. *233*, 623 (1985).
148 B. MacMahon, S. Yen, D. Trichopoulos, K. Warren and G. Nardi: N. Engl. J. Med. *304*, 630 (1981).
149 A. R. Feinstein, R. I. Horwitz, M. D. Spitzer and R. M. Bottista: JAMA *246*, 957 (1981).
150 H. R. Goldstein: N. Engl. J. Med. *306*, 947 (1982).
151 R. J. Coffey, W. G. Yagheary, A. R. Zimmerstein and E. P. DiMayno: Pancreas *1*, 55 (1986).
152 L. Y. Korman, M. D. Walker and J. D. Garder: Am. J. Physiol. *239*, G324 (1980).
153 G. Born: Nature *202*, 95 (1964).
154 N. J. Cusack and S. M. O. Hourani: Br. J. Pharmacol. *72*, 443 (1981).
155 B. B. Fredholm and A. Uwi: Clin. Pharmacol. *6*, 1 (1986).
156 A. Solber, L. Thorsell, B. B. Fredholm, G. Settergren and M. Blomhack: Scand. J. Thorac. Cardiovasc. Surg. *19*, 155 (1985).
157 S. Allen and D. Froberg: Surgery *101*, 720 (1987).
158 J. P. Minton, M. K. Foecking, D. J. T. Webster and R. H. Mathews: Surgery *86*, 105 (1979).
159 J. R. Minton, H. Ahou-Issa, N. Reichus and J. M. Roseman: Surgery *90*, 299 (1981).
160 P. G. Brooks, S. Gart, A. J. Heldford, M. L. Margolia and M. S. Allen: J. Rep. Med. *26*, 279 (1981).
161 V. L. Ernester, L. Mason, W. H. Goodsen, et al.: Surgery *91*, 263 (1982).
162 G. L. Stiles, M. G. Caron and R. J. Lefkowitz: Physiol. Rev. *84*, 661 (1984).
163 A. G. Gilman: Ann. Rev. Biochem. *56*, 615 (1987).
164 D. T. Jones and R. R. Reed: J. Biol. Chem. *262*, 14241 (1987).
165 S. B. Masters, R. M. Stroud and H. R. Bourne: Protein Enging. *1*, 47 (1986).
166 C. C. Malbon, R. C. Hart and J. N. Fain: J. Biol. Chem. *253*, 3114 (1978).
167 T. Trost and U. Schwabe: Mol. Pharmacol. *19*, 228 (1981).
168 R. F. Bruns, J. W. Daly and S. H. Snyder: Proc. Natl. Acad. Sci. USA *77*, 5547 (1980).
169 M. Williams and E. A. Risley: Proc. Natl. Acad. Sci. USA *77*, 6892 (1980).
170 R. Green and G. L. Stiles: J. Clin. Invest. *77*, 222 (1986).
171 C. Londos, D. M. F. Cooper and J. Wolff: Proc. Natl. Acad. Sci. USA *77*, 2551 (1980).
172 U. Schwabe, V. Lenschow, D. Ukenas, D. R. Ferry and H. Glossman: Naunyn-Schmiedebergs Arch. Pharmacol. *321*, 84 (1982).
173 G. L. Stiles: J. Biol. Chem. *260*, 6728 (1985).

174 W. Schutz, E. Tuisl and O. Kraupp: Naunyn-Schmiedebergs Arch. Pharmacol. *319*, 34 (1982).
175 K. A. Jacobson, D. Ukena, K. L. Kirk and J. W. Daly: Proc. Natl. Acad. Sci. USA *83*, 4089 (1986).
176 G. L. Stiles and K. A. Jacobson: Mol. Pharmacol. *32*, 184 (1987).
177 R. F. Bruns, J. H. Fergas, E. W. Badger, J. A. Bristol, L. A. Santay, J. Hartman, S. J. Hays and C. C. Huang: Naunyn-Schmiedebergs Arch. Pharmacol. *335*, 59 (1987).
178 R. S. Kent, A. DeLean and R. J. Lefkowitz: Mol. Pharmacol. *17*, 14 (1980).
179 M. Gavish, R. R. Goodman and S. H. Snyder: Science *215*, 1633 (1982).
180 S. Yeung and R. D. Green: J. Biol. Chem. *258*, 2334, (1983).
181 M. J. Lohse, V. Lenschow and U. Schwabe: Mol. Pharmacol. *26*, 1 (1984).
182· U. Ukena, E. Poeschla and U. Schwabe: Naunyn-Schmiedebergs Arch. Pharmacol. *326*, 241 (1984).
183 K. H. Jacobs, W. Saur and G. Schultz: FEBS Lett. *85*, 167 (1978).
184 G. L. Stiles, D. T. Daly and R. A. Olsson: J. Biol. Chem. *260*, 10806 (1985).
185 G. L. Stiles, D. T. Daly and R. A. Olsson: J. Neurochem. *47*, 1020 (1986).
186 K. N. Klotz, G. Cristalli, M. Grifantini, S. Vittari and M. J. Lohse: J. Biol. Chem. *260*, 14659 (1985).
187 J. I. Choca, M. M. Kwatra, M. M. Hosey and R. D. Green: Biochem. Biophys. Res. Commun. *131*, 115 (1985).
188 A. Green, C. A. Stuart, R. A. Pietrzyk and M. Partin: FEBS Lett. *206*, 130 (1986).
189 K. Klotz and M. J. Lohse: Biochem. Biophys. Res. Commun. *140*, 406 (1986)
190 G. L. Stiles, G. Pierson, S. Sunay and W. J. Parsons: Endocrinology *119*, 1845 (1986).
191 G. L. Stiles: J. Biol. Chem. *261*, 10839 (1986).
192 G. L. Stiles, J. L. Benovic, M. G. Caron and R. J. Lefkowitz: J. Biol. Chem. *259*, 8655 (1984).
193 R. A. Cerione, J. W. Regan, H. Nakata, J. Codina, J. C. Benovic, P. Gierschik, R. L. Saners, A. M. Spiegel, L. Birnbaumer, R. J. Lefkowitz and M. G. Coran: J. Biol. Chem. *261*, 3901 (1986).
194 C. P. Berric, N. J. M. Birdsall, E. C. Hulme, M. Keen and J. M. Stockton: Br. J. Pharm. *82*, 853 (1984).
195 W. J. Parsons and G. L. Stiles: J. Biol. Chem. *262*, 841 (1987).
196 D. R. Sibley and R. J. Lefkowitz: Nature *317*, 124 (1985).
197 M. K. Ahlijanian and A. E. Takemori: J. Pharmacol. Exp. Ther. *236*, 615 (1986).
198 A. Green: J. Biol. Chem. *262*, 15702 (1987).
199 J. G. Kenimer and M. Nirenberg: Mol. Pharmacol. *20*, 585 (1981).
200 D. J. Dobmeyer, R. A. Stine, C. V. Leier, R. Greenberg and S. F. Schaal: N. Engl. J. Med. *308*, 814 (1983).
201 H. P. T. Ammon and C. J. Estler: Med. Exp. *19*, 161 (1969).
202 C. Pilcher, C. P. Wilson and T. R. Harrison: Am. Heart J. *2*, 618 (1927).
203 K. A. Conrad, J. Blanchard and J. M. Trang: J. Am. Geriat. Soc. *30*, 267.
204 S. H. Snyder, J. J. Katims, Z. Annau, R. F. Braun and J. W. Daly: Proc. Natl. Acad. Sci. USA *78*, 3260 (1981).
205 B. B. Fredholm: Trends Pharmacol. Sci. *1*, 129 (1980).
206 G. L. Stiles: Trends Pharmacol. Sci. *7*, 486 (1986).
207 J. Greden, B. Victor, P. Fontaine and M. Lubetsky: Psychosomatics *21*, 411 (1980).
208 B. White, C. Lincoln, N. Pearce, R. Reeb and C. Vauda: Science (Wash. DC) *209*, 1547 (1980).
209 L. Roller: Med. J. Aust. *20*, 146 (1981).

210 D. Robertson, D. Wade, R. Workman, R. L. Woosley and J. A. Oates: J. Clin. Invest. *67*, 1111 (1981).
211 D. T. Chou, S. Khan, J. Forde and R. Hirshk: Life Sci. *36*, 2347 (1985).
212 B. B. Fredholm: Acta. Physiol. Scand. *115*, 283 (1982).
213 J. P. Boulenger, J. Patel, R. M. Post, A. M. Parma and P. J. Marangos: Life Sci. *32*, 1135 (1983).
214 R. Corradetti, F. Pedata, G. Pepeu and M. G. Vannucchi: Br. J. Pharmacol. *88*, 671 (1986).
215 T. Murray: Eur. J. Pharmacol. *82*, 113 (1982).
216 M. K. Ahlijanian and A. E. Takemori: J. Pharmacol. Exp. Ther. *236*, 615 (1986).
217 J. J. Ohisalo and J. E. Stouffer: Biochem. J. *178*, 249 (1979).
218. C. C. Malbon, P. J. Rapiejko and T. J. Marigano: J Biol. Chem. *260*, 2558 (1985).
219 C.C. Malbon, F. J. Moreno, R. J. Cabelli and J. N. Fain: J. Biol. Chem. *253*, 671 (1978).
220 P. J. Rapiejko and C. C. Malbon: Biochem. J. *241*, 765 (1987).
221 R. G. Vernon and E. Finley: Biochem. J. *230*, 651 (1985).
222 R. G. Vernon, E. Finley and E. Taylor: Biochem. J. *216*, 121 (1983).
223 H. Kather, E. Wieland, B. Fischer, A. Wirth and G. Schlierf: Eur. J. Clin. Invest. *15*, 30 (1985).
224 J. Zapf, M. Waldvogel and E. R. Froesch: FEBS Lett. *76*, 135 (1977).
225 J. J. Ohisalo, S. Ranta and I. T. Huhtaniemi: Metab. Clin. Exp. *35*, 143 (1986).
226 J. H. Fleisch: Trends Pharmacol. Sci. *2*, 337–339.
227 I. B. Abrass, J. L. Davis and P. J. Scarpace: J. Gerontol. *37*, 156 (1982).
228 B. B. Hoffman, H. Chang, Z. Farahbakhsh and G. Reaven: J. Clin. Invest. *74*, 1750 (1984).
229 J. Linden, A. Patel and S. Sadek: Circ. Res. *56*, 299 (1985).
230 V. Ramkumar, J. R. Bumgarner, K. A. Jacobson and G. L. Stiles: J. Clin. Invest. (1988) in press.

Stereoselective drug metabolism and its significance in drug research

By Bernard Testa and Joachim M. Mayer
Ecole de Pharmacie, Université de Lausanne,
CH–1005 Lausanne, Switzerland

Dedicated to the memory of Professor Peter J. Meffin, whose untimely death in November 1987 deprives the scientific community of a prominent figure in the field of stereoselective drug metabolism.

1   Introduction

Far from being a curiosity, *chirality* is a ubiquitous phenomenon in nature, ranging from the level of quanta and the parity violation in all weak interactions (e.g. the intrinsic left-handedness of the electron and its neutrino, and the right-handedness of the positron and its antineutrino) to an apparent excess of left-handed galaxies [1-5]. At the levels of material organisation explored by chemists, biochemists and biologists, namely the levels of molecules, biomolecules and living systems, chirality is of common occurrence and has since long been the object of intense interest.

To the pharmacologist (in the broadest sense of the word), the field of study are not drugs per se, but their interactions with biological systems. Schematically, the interactions between drugs and biosystems have three faces, namely the pharmacodynamic effects, the pharmacokinetic effects, and complex intertwined interactions whereby both types of effects cannot be separated [6]. A vast majority of drugs of natural origin, and a significant proportion of synthetic drugs, exhibit chirality, i.e. they exist as enantiomers. But while chiral drugs of natural origin usually contain only one of two possible enantiomers, synthetic chiral drugs are marketed either as racemic mixtures or as single isomers. Of capital significance in the present context is the fact that enantiomers interact differently with biological systems, which are themselves chiral and made of chiral constituents. This phenomenon of chiral recognition (see Section 3) explains why enantiomers are expected to produce distinct pharmacodynamic responses and to differ in their pharmacokinetic behaviour.

The specific problems posed by chiral drugs, namely the mechanisms and consequences of their enantioselective interactions with biological systems, are receiving an ever-increasing attention in all fields of drug research. This was clearly recognized by Robert L. Smith in his message as the retiring president of the ISSX (International Society for the Study of Xenobiotics) [7], one of the major issues of current interest being the compared therapeutic benefits of racemic mixtures versus single enantiomers. Enantioselectivity is frequently observed in pharmacodynamic effects, and according to Pfeiffer's rule [8] the enantiomeric potency ratio increases with potency. In modern nomenclature [9, 10], the more active and the less active enantiomer of a chiral drug are termed the *eutomer* and the *distomer*, respectively, while their potency ratio or receptor affinity ratio is known as the eudismic ratio. These pharmacodynamic aspects, despite their obvious importance, will only be considered here inasmuch as they are essential to place drug metabolism in the broader perspective of drug research and pharmacotherapy.

Pharmacokinetic events in the broadest meaning of the term comprise a number of phenomena such as absorption, distribution, binding, excretion and metabolism (biotransformation). While the present chapter is limited to metabolic processes, it may be useful to mention that enantioselectivity is not an exception as far as the other pharmacokinetic events are concerned [11–14].

2 Essential notions of stereochemistry for the absent-minded pharmacologist

2.1 Introduction

Not all readers of this review will claim to be fully conversant with organic stereochemistry. Furthermore, medicinal chemists not uncommonly observe fanciful and capricious uses of stereochemical nomenclature in even reputable pharmacology journals. This section thus offers an ultrashort reminder of a number of notions essential for a proper understanding of the pages to follow. The reader willing to pursue the study of stereochemistry is invited to consult the two works [15, 16] from which the following material is taken.

A few definitions are a necessary starting point. *Isomers* can be defined as molecules which closely resemble each other, but fail to be identical due to one difference in their chemical structure. When they share the same molecular formula (i. e. the same atomic composition) but differ in their constitution (i. e. in the connectivity of their atoms), they are called *constitutional isomers* (e. g. 1-propanol and 2-propanol). When structural isomers have identical composition but differ in the spatial arrangement of their atoms, they are designated as *stereoisomers*.

Stereoisomers can be subdivided according to two distinct criteria, namely geometry and energy criteria. The geometry criterion classifies stereoisomers into *enantiomers* and *diastereomers*. Either two stereoisomers are related to each other as object and nonsuperimposable mirror image, or they are not. In the former case, they share an enantiomeric relationship. This implies that the molecules are dissymmetric (chiral), and the property of *chirality* is the necessary and sufficient condition for the existence of enantiomeric objects. Enantiomers are also referred to as *optical isomers* since they show optical rotations of opposite sign and ideally of identical amplitude. Note however that this optical rotation may be too small to be detected, and that both amplitude and also sign are condition-dependent (e. g. wavelength, solvent, concentration, temperature). Stereoisomers which are not enantiomers are diastereomers, implying that by definition enantiomeric and diastereomeric relationships are mutually exclusive.

Independently from the geometry-based classification, stereoisomers can be discriminated according to the energy necessary to convert one isomeric form into another. Here, the energy barrier becomes the criterion of classification. In qualitative terms, a „high"-energy barrier separates *configurational isomers,* while a „low"-energy barrier separates *conformational isomers (conformers).* In contrast to the sharp enantiomerism-diastereomerism classification, the configuration-conformation classification of stereoisomers lacks a well-defined borderline. In the continuum of existing energy values, intermediate cases exist whose classification is difficult, but in the authors' opinion the boundary should be viewed as a broad energy range encompassing the value of 80 kJ/mol (ca. 20 kcal/mol) which is the limit of fair stability under ambient conditions. The classification of stereoisomers according

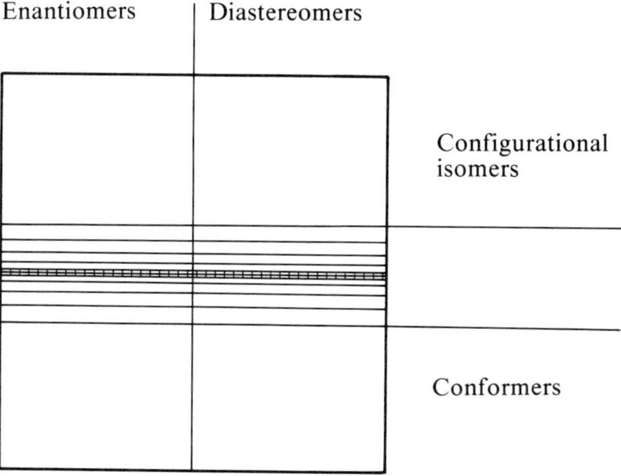

Figure 1

A classification of stereoisomers [15] (reproduced with the permission of Marcel Dekker Inc., New York).

to the two independent criteria of geometry and energy is presented graphically in Figure 1.

The presentation to follow is restricted to configurational isomerism.

2.2   Enantiomerism

Chirality, as explained above, is the property displayed by chemical compounds (and for that matter by any object) to exist as two enantiomers. To be chiral, a molecule must contain one or more *elements of chirality,* i.e. centres of chirality, axes of chirality, or planes of chirality. Axes and planes of chirality are seldomly encountered in medicinal chemistry and will not be discussed here.

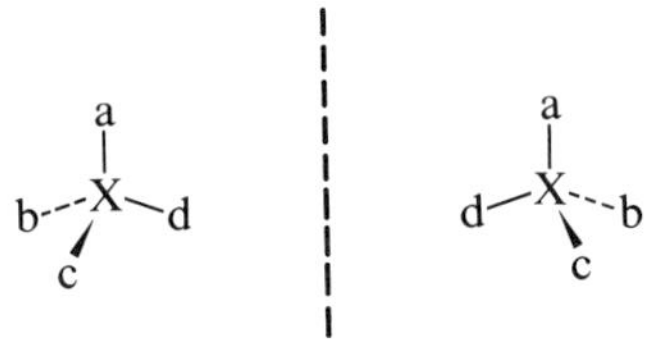

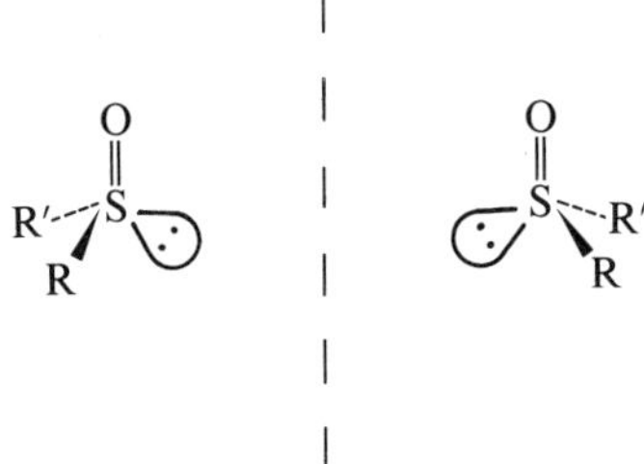

Figure 2
Enantiomerism due to the presence of a chiral tetracoordinate centre.

By far the most frequently encountered centre of chirality (also called asymmetric centre) is a carbon atom bearing four different substituents. Other chiral tetracoordinate centres include the quaternary nitrogen, silicon derivatives and phosphonium salts, the general case being depicted in Figure 2. A few configurationally stable tricoordinate centres are also known in medicinal chemistry, e.g. sulfoxides (Fig. 3) and sulfonium salts. Tricoordinate derivatives of nitrogen and other

Figure 3
Configurationally stable, chiral sulfoxides. The lone pair of electrons is formally treated as the fourth ligand.

first-row elements undergo fast inversion and are thus configurationally unstable, e. g. Figure 4.

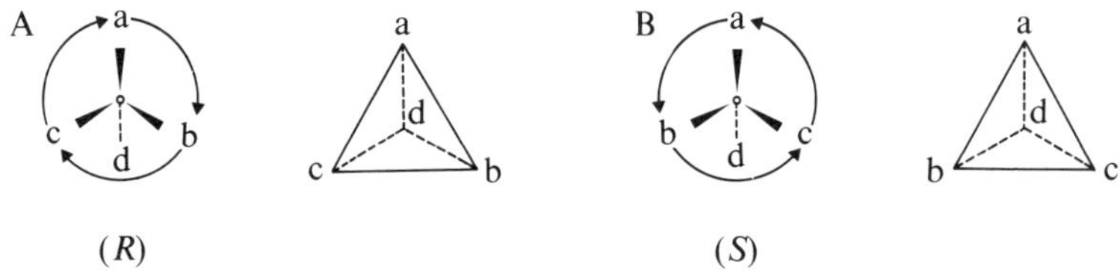

Figure 4
Configurational instability of tertiary amines.

The unambiguous designation of enantiomers is a topic of capital significance in pharmacology as much as in chemistry, but the awareness
of existing rules is unfortunately quite different among pharmacologists and chemists. *Absolute configuration,* when known, is by far the
best way to designate enantiomers provided an adequate notation is
used. The *D* and *L* nomenclature is a most useful one for amino acids,
sugars and analogs, but its extension to various chemical classes meets
with difficulties and has become rare. In contrast, the *R* and *S* nomenclature of Cahn, Ingold and Prelog [17] is applicable universally and
with relative ease. Briefly, the essential part of this nomenclature is the
sequence rule, i. e. a set of consistent but arbitrary rules which allow
hierarchical assignment of the four different substituents
(a > b > c > d). A presentation of these rules will not be undertaken
here. By convention, the chiral centre is viewed with a, b and c pointing
towards the observer, and d pointing away. The path from a to b to c to
a can then be either clockwise (Fig. 5 A) in which case the configuration ist designated (*R*) (rectus), or counterclockwise (Fig. 5 B) which
means an (*S*) configuration (sinister).
When the absolute configuration is not known, there is no option left
but to use the *sign of optical rotation,* i. e. (+) and (−), to label the two

enantiomers, bearing in mind that this sign may be solvent-dependent. Designating the enantiomers with the prefixes *dextro* and *levo* is also correct albeit cumbersome. In contrast, the lower-case prefixes *d* and *l* should never be used since they are so easily confused with, or changed into, the upper-case prefixes *D* and *L* which as discussed above describe absolute configuration (a structural property) and not optical rotation (a physical property), the two properties not being related at all. That this gross error occurs at all in even the best pharmacology journals is a frustrating sight for the medicinal chemist.

Quite customary, and in fact useful, is the combination of two prefixes (sign of rotation and absolute configuration) to designate enantiomers. The strict application of chemical nomenclature demands that a physical property precedes a structural property, e.g. $(-)-(R)-$adrenaline, and not vice versa, although the frequent non-respect of this rule is harmless.

When a *racemate* is meant, it should be explicitly designated as such to avoid potentially dangerous confusions. The prefixes $(\pm)$ or (d, l) (the latter to be avoided) are sometimes used. Simonyi has recently made the very useful proposal of systematically employing the prefix *rac* [18].

2.3   Diastereomerism

The differences between enantiomers and diastereomers are not only geometric, but cover a number of properties. Thus, while a given molecule can have one and only one enantiomer, it may have several diastereomers. Also, not one but a number of structural features can generate diastereomerism, and diastereomers can be either chiral or achiral. Finally, while two enantiomers have identical physicochemical properties and can be discriminated only with a chiral handle (chiral reagent, left- and right handed circularly polarized components of plane-polarized light, etc), diastereomers differ in every physicochemical property, however minute the difference, and can be discriminated by nonchiral tools.

Stereoisomerism about double bonds is best designated as π-*diastereomerism* rather than the usual and ambiguous labels of *geometric isomerism* or *cis-trans*-isomerism. The designation of two such diastereomers as *cis* and *trans* (see Fig. 6) is usually quite convenient, although the *Z* and *E* nomenclature is more rigorous and is now of

256   Bernard Testa and Joachim M. Mayer

a\C=C/c   a\C=C/H
H/   \H   H/   \c

cis, Z   *trans, E*

Figure 6

π-Diastereomerism (commonly called geometric isomerism)

common use. Briefly, the two substituents at one carbon are ranked according to the sequence rule, as are the two substituents on the other carbon. If the two sequence rule-preferred substituents are on the same side of the double bonds (i. e. „zusammen"), the isomer is labeled Z; if they are opposite (i. e „entgegen"), the isomer is E (Fig. 6). Note that Z and E usually, but not always, correspond to *cis* and *trans*. Further note that the same nomenclature also applies to, e. g., carbon-nitrogen and nitrogen-nitrogen double bonds.

Another kind of diastereomerism arises in *acyclic molecules containing n (n $\geq$ 2) centres of chirality*. The number of possible stereoisomers in such cases is 2^n or less depending on various molecular features. When 2^n stereoisomers exist, they can be subdivided into $2^{(n-1)}$ diastereomeric pairs of enantiomers. Thus, $(-)$-$(1R;2S)$-ephedrine has one enantiomer, namely $(+)$-$(1S;2R)$-ephedrine, and two diastereomers, namely the $(-)(1R;2R)$- and $(+)$-$(1S;2S)$-isomers. Diastereomers which differ in the configuration of a single chiral centre (i. e. which have identical configuration on n-1 chiral centres) are called *epimers*.

Configurational isomerism is also encountered in *cyclic molecules*, more accurately in bi- and polysubstituted cyclic compounds as well as in fused ring systems. In the case of bisubstituted monocyclic systems, *cis-trans* isomerism exists if the two substitutents are not geminal. Thus 1,2-, 1,3- and 1,4- disubstituted cyclohexane derivatives show this *cis-trans* diastereomerism. In addition, some of these derivatives are chiral (*trans*-1,2 and *trans*-1,3), some are achiral (*cis*- and *trans*-1,4), while others (*cis*-1,2 and *cis*-1,3) are achiral or chiral depending on whether the two substituents are identical or different, respectively.

2.4   Prostereoisomerism

Prostereoisomerism results from some singular relationships existing within intact molecules between groups or atoms of identical constitu-

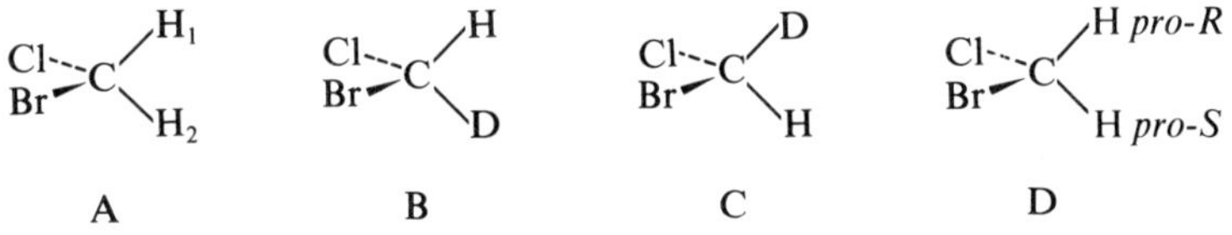

Figure 7
Bromochloromethane and its two enantiotopic hydrogens.

tion. Such intramolecular relationships are of fundamental importance in understanding the stereochemistry of various metabolic reactions, and as such deserve the reader's attention. Enantiotopic and diastereotopic relationship will be discussed in turn.

In a simple model molecule such as bromochloromethane (Fig. 7A), the two hydrogen atoms are not equivalent. Indeed, the sequence of the three atoms Br-Cl-H is clockwise when viewed from H_1, and counterclockwise from H_2. The molecular environments of H_1 and H_2 are thus enantiomeric. Also, substituting in turn H_2 and H_1 with for example deuterium yields two enantiomeric compounds (Fig. 7B and 7C, respectively). Because of their non-equivalence, the two hydrogen atoms in bromochloromethane are designated as *enantiotopic*. If H_1 is arbitrarily preferred over H_2, an *R* configuration is obtained (e.g. Fig. 7C); H_1 is therefore designated *pro-R*, while H_2 is *pro-S* (see Fig. 7D). As regards bromochloromethane itself, it is not chiral since it contains a plane of symmetry. But because it bears two enantiotopic groups it is said to be prochiral. The concept of *prochirality* has its origin in biological studies showing that in molecules like citric acid (Fig. 8) two chemically identical groups such as CH_2COOH are biochemically quite distinct.

Molecules such as bromochloromethane and citric acid contain a centre of prochirality, or prochiral centre. It is a serious error to confuse enantiotopic groups or atoms and prochiral centres. Enantiotopic groups belong to prochiral centres; they are not themselves prochiral. Just as unfortunate is the designation of molecules as enantiotopic,

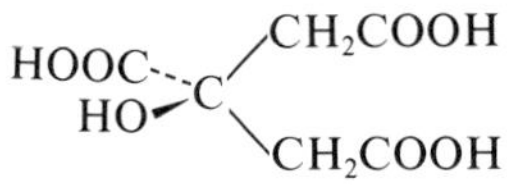

Figure 8
Citric acid and its two enantiotopic CH_2COOH groups.

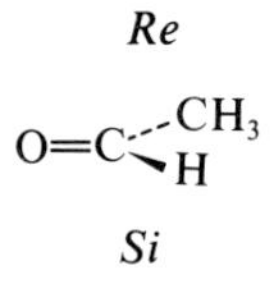

$$Re$$

$$O{=}C{\overset{\textit{--}CH_3}{\diagdown H}}$$

$$Si$$

Figure 9
Acetaldehyde and its two enantiotopic faces.

since enantiotopism is a property displayed only by groups *within* a molecule.

The presence of a centre of prochirality is not an obligatory condition for a molecule to be prochiral. Indeed, other elements of prochirality exist, namely axes and planes of prochirality. The concept of prochirality can also be applied to trigonal centres, i. e. to *faces* of suitable molecules. In acetaldehyde (Fig. 9), the two faces of the molecule are not equivalent but enantiotopic. Indeed, the groups O-Me-H define a clockwise path when the molecule is viewed from above, and a counterclockwise path when viewed from below. The two faces are designated as *Re* and *Si*, respectively.

Diastereotopic groups reside in diastereomeric environments, cannot be interchanged by symmetry operations, and upon substitution by chiral or prochiral groups generate diastereomeric structures. Diastereotopic groups in a molecule imply prostereoisomerism as an element of prochirality or of *proachirality*. For example, chloroethylene (Fig. 10) contains two geminal hydrogen atoms which are diastereotopic. But no element of prochirality exists in this molecule, and the carbon atoms carrying the two hydrogens build a proachiral centre.

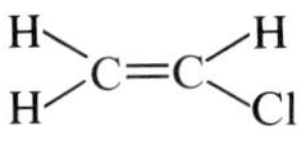

$$\underset{H}{\overset{H}{\diagdown}}C{=}C\underset{Cl}{\overset{H}{\diagup}}$$

Figure 10
Chloroethylene and its two diastereotopic hydrogens on carbon-2.

In contrast, 1,2-propanediol (Fig. 11) contains a centre of chirality (C-2) and is therefore chiral. Carbon-1 on the other hand is prochiral; in turn replacement of one of the two adjacent hydrogens generates diastereomeric products. As a consequence, the two hydrogens at C-1 are diastereotopic groups adjacent to a prochiral centre.

Diastereotopic faces also exist, as seen in the achiral 4-methyl-

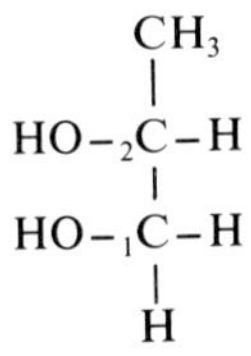

$$
\begin{array}{c}
CH_3 \\
| \\
HO-_2C-H \\
| \\
HO-_1C-H \\
| \\
H
\end{array}
$$

Figure 11
1,2-Propanediol and its two diastereotopic hydrogens on carbon-1.

cyclohexanone and in the chiral 2-methylcyclohexanone (Fig. 12).

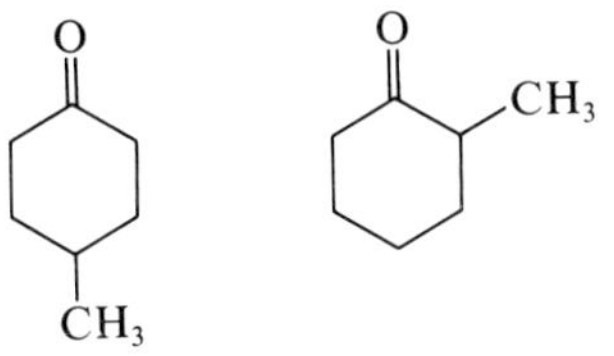

Figure 12
4-Methyl- and 2-methylcyclohexanone each have two diastereotopic faces.

The biochemical discrimination of stereoheterotopic (enantiotopic or diastereotopic) groups leads to the formation of stereoisomeric metabolites, as discussed below. A sufficient grasp of the above rules is thus a prerequisite for understanding and interpreting the mechanism and stereoselectivity of a number of major metabolic reactions.

3   General concepts in stereoselective drug metabolism
3.1   Stereoselective processes in drug metabolism
             and disposition

One of the fundamental principles of pharmacodynamics, as formulated by Paul Ehrlich, is that compounds do not act unless bound *(corpora not agunt nisi fixata)*. However, this rule is of such significance and scope that, far from being restricted to pharmacodynamic events, it can also be transcribed to pharmacokinetic events. Indeed, many processes in drug metabolism and disposition involve binding to biological macromolecules or macromolecular assemblies. As discussed elsewhere [6, 19], such binding processes cover a continuum of energies ranging from a few kcal/mol (weak reversible bonds) to approximately 100 kcal/mol and more (covalent bonds). The only pharmacokinetic events not involving biological binding *stricto sensu* are reactions of non-enzymatic biotransformation [20] and processes of passive diffusion (transport, excretion), although the passive partitioning into membranes is based on the same weak interactions as reversible binding and is thus conceptually comparable.

These molecular processes are necessary for a proper understanding of the fact that the enantiomers of a chiral drug display differences in metabolism, protein binding, storage, distribution and excretion, as abundantly documented [11-14, 21]. Indeed, their binding to a chiral biomolecule such as a macromolecule or an enzyme results in diastereomeric complexes and, at least in principle, in *chiral recognition*. The latter phenomenon is of fundamental significance in biology and is particularly marked when covalent binding leads to diastereomeric transition states displaying a relatively large energy difference (kinetic control) [15, 16]. In contrast to enantiomers, diastereomers differ in their physicochemical properties and can be discriminated without a chiral tool; but because the tools of the biosphere (enzymes, macromolecules, membranes, …) are always chiral, such a theoretical distinction between enantiomers and diastereomers is of little pharmacokinetic relevance in practice. For these reasons, enantiomers are receiving most of the attention in the present review. However, the arguments developed here for enantiomers often apply to diastereomers also; when this in the case, the term of *stereoselectivity* will be preferred over that of *enantioselectivity*.

As far as stereoselectivity is concerned, pharmacokinetic events must be classified into two groups:

– To the processes of absorption, distribution, binding and excretion corresponds only one type of stereoselectivity, namely the differential disposition displayed by stereoisomers. This situation is comparable to that found in pharmacodynamic events.

– In contrast, metabolism (i.e, biotransformation) displays two basic types of stereoselectivity, namely substrate stereoselectivity and product stereoselectivity. Note that stereochemical choice extends far beyond xenobiotic metabolism and is encountered in innumerable enzymatic reactions, as aptly illustrated by Overton [22].

Prelog [23], studying the enzymatic reduction of ketones, was the first to distinguish the two types of stereoselectivity. When the two enantiomers of a chiral substrate were metabolized at different rates he used the concept of substrate stereoselectivity. When an asymmetric centre was created during the reduction reaction, he observed that the two possible stereoisomeric products were formed at different rates, thus leading to the concept of product stereoselectivity. Such a classification brought considerable clarification to the very large body of data generated by Prelog's studies. In 1973, these concepts were extended and for the first time applied to drug metabolism, proving their broad applicability and their power in extracting valuable additional information from experimental studies [24]. Our current understanding of these concepts [6, 21, 25–28] is discussed and illustrated in the pages to follow.

3.2   Substrate stereoselectivity

Substrate stereoselectivity, which corresponds to the situation encountered with the other pharmacokinetic events, is seen when stereoisomers are metabolized:
– differently (in quantitative and/or qualitative terms),
– by the „same" biological system, and under identical conditions.
It is a trivial statement that the metabolism of stereoisomers can be validly compared only when all biological and experimental conditions are identical. In addition, we shall see in Section 7 that the separate or common incubation or administration of stereoisomers may not yield identical results due to metabolic interactions.
Substrate stereoselectivity, and principally enantioselectivity, is abundantly documented in the literature, be it under in vivo [29] or in vitro conditions. An interesting and thoroughly studied case is provided by the uricosuric-diuretic agent indacrinone [30]. In the rhesus monkey, its two enantiomers (Fig. 13, upper panel) are metabolized by *para*-hydroxylation, and the (*R*)-isomer is visibly eliminated much faster than the (*S*)-isomer (Fig. 13, lower panel). A detailed pharmacokinetic analysis quantitated these differences (Table 1), which appear to be accounted for by the very large difference (40-fold) in the metabolic clearance of the enantiomers. Indacrinone thus provides a telling example of a drug metabolized by an apparently single route with very high substrate enantioselectivity.
When several metabolic routes are operative, the situation may become more difficult to analyze. Indeed, competitive reactions may not remain without effect on their relative selectivities, implying that only apparent substrate enantioselectivities will be obtained. Morphine of-

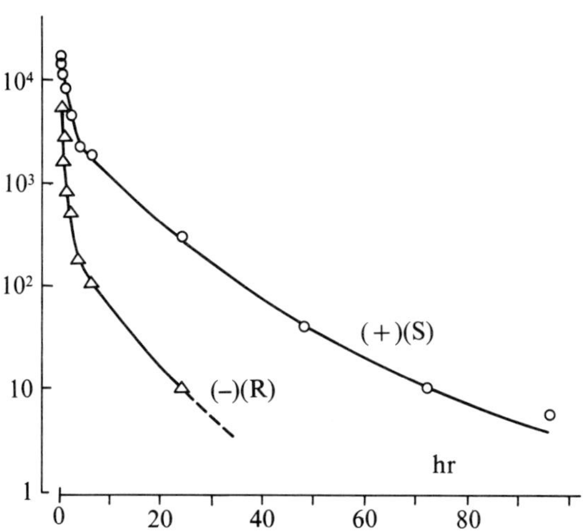

Figure 13
upper panel: The (–)-(*R*)- and (+)-(*S*)-enantiomers of indacrinone (left and right, respectively).
lower panel: Plasma concentrations of indacrinone in the rhesus monkey following separate i.v. injection of the enantiomers [30] (reproduced with the permission of Williams and Wilkins Co, Baltimore).

fers a good illustration of a drug metabolized by a number of routes displaying quantitative and qualitative differences in their substrate enantioselectivity. In rat liver microsomes, the natural (–)-enantiomer of morphine was found to undergo 3-O-glucuronidation and N-demethylation as major and minor routes, respectively [31]. The 3-O-glucuronidation and N-demethylation of the unnatural (+)-enantiomer are 7–8 times and 2 times slower, respectively. In addition, the latter enantiomer also undergoes 6-O-glucuronidation as its major route, but the overall glucuronidation of the unnatural enantiomer remains ca 2 times slower than of the natural enantiomer (Fig. 14) [31]. It

Table 1
Pharmacokinetics of indacrinone in the rhesus monkey [30]

	$(+)$-(S)-I	$(-)$-(R)-I	$(+)$-(S)-pOH	$(-)$-(R)-pOH
AUC$_\infty$ (μg·hr/ml)	58.4±9.4	8.08±2.4	46.3±5.1	1.7±0.4
V$_D$ (steady state) (l)	0.60±0.17	1.99±0.85	0.55±0.08	7.7±3.2
Plasma clearance (ml/min)	1.25±0.27	9.45±4.13	1.62±0.23	43.2±9.2
Renal clearance (ml/min)	0.88±0.18	4.7±2.9	1.10±0.18	25.9±4.5
Metabolic clearance (ml/min)	0.05±0.02	2.0±0.8	–	–

I: indacrinone; pOH: its para-hydroxy metabolite.

can be postulated that the N-demethylation of ($-$)-morphine in rat liver microsomes would be faster, and the substrate enantioselectivity of this reaction would be higher, were it not decreased by the predominant 3-O-glucuronidation.

The reactions of glucuronidation and glutathione conjugation are of particular interest in the present context because they introduce additional centres of chirality into a chiral substrate, yielding metabolites which are diastereomers and no longer enantiomers. This fact has consequences for the stereoanalysis of the metabolites, but we simply wish to stress here that such cases pertain to substrate enantioselectivity and not to product enantioselectivity to be discussed below. Indeed, (D)-glucuronic acid and (L, L)-glutathione exist in the body as single enantiomers (i.e. they are homochiral), and each enantiomer of the substrate can lead to only one stereoisomer of the conjugate. Examples in-

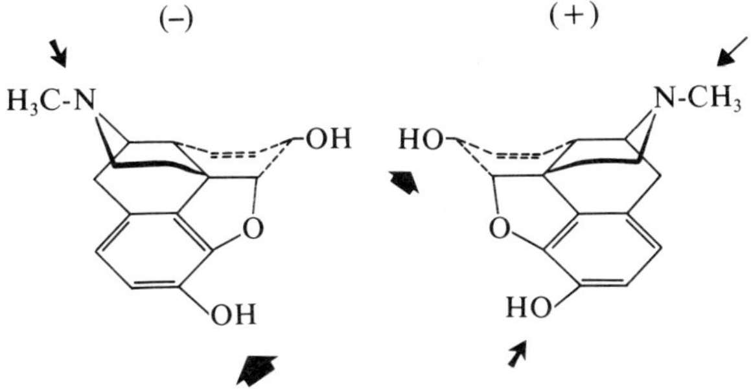

Figure 14
Sites of enantioselective glucuronidation and N-demethylation of morphine enantiomers in rat liver microsomes. The thickness of the arrows indicates the relative rates of metabolism [31] (reproduced with the permission of Williams and Wilkins Co, Baltimore).

clude the glucuronidation of the ($+$)-(S)- and ($-$)-(R)-enantiomers of oxazepam [32] and oxaprotiline [33].

3.3   Product stereoselectivity

Product stereoselectivity, which by definition is only encountered as the result of a (bio)chemical reaction, occurs when stereoisomeric metabolites are generated:
– differentially (in quantitative or qualitative terms),
– from a single prochiral or proachiral substrate.
A given product stereoselectivity applies only when the substrate is examined under a given set of biological and experimental conditions. Should any of the conditions, in particular the biological system, be changed, a different product stereoselectivity will be found.
All documented cases of product enantioselectivity, and most cases of product diastereoselectivity, result from the metabolic generation of a centre of chirality. A necessary condition for such a situation to occur is the presence in the substrate molecule of a suitable prochiral centre or face (see Section 2.4). The most representative and common reactions leading to the metabolic creation of a centre of chirality are shown in Figure 15. A number of these cases will be illustrated in the present and later sections.
The selective hydroxylation of a prochiral methylene group of the type R, CH_2, R' (Fig. 15C) is certainly the most frequently encountered and discussed case of product stereoselectivity. Of similar interest but rarer occurrence is the case shown in Figure 15D, as illustrated by phenytoin. This achiral drug has C-3 as a prochiral centre carrying two enantiotopic phenyl rings. In all species investigated except the dog, 5- (4-hydroxyphenyl)-5-phenylhydantoin is the major phase I metabolite, with an approximately 10- to 20-fold excess of the ($-$)-(S)-enantiomer (Figure 16) [e. g. 34–36]. Under these biological conditions, the oxidation of phenytoin thus occurs with a high degree of selective recognition of its *pro-S* ring.
Reactions of the type shown in Figure 15B are exemplified by the selective reduction in humans of the C(20)-C(22) double bond of digoxin to (20R)-dihydrodigoxin [37]. Metabolic reduction of carbonyl groups (Fig. 15A) is more frequently observed than carbon-carbon double bond reduction. Simple aryl alkyl ketones as studied by Coutts and colleagues [38] are reduced with a high degree of product enantioselec-

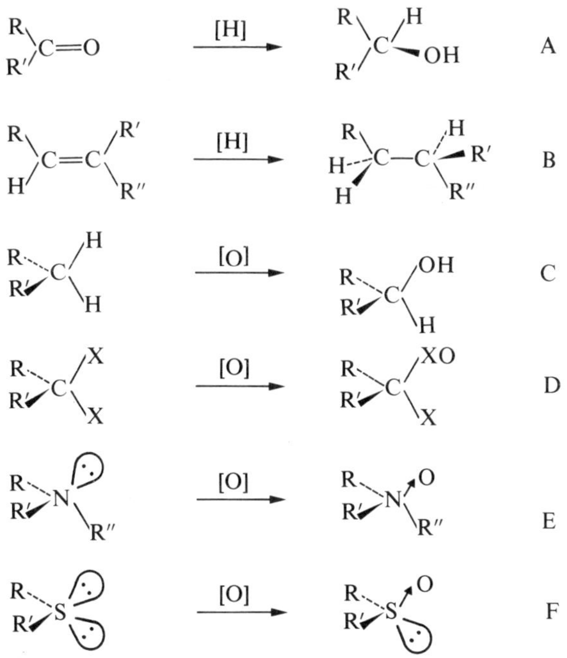

Figure 15
Some reductive and oxidative pathways producing new centres of chirality in substrate molecules.

A: ketone reduction;
B: reduction of carbon-carbon double bonds;
C: hydroxylation of a prochiral methylene group;
D: oxygenation of an enantiotopic moiety;
E: oxygenation of a configurationally labile tertiary amine to an N-oxide;
F: oxygenation of a sulfide to a configurationally stable sulfoxide.

The configurations depicted are arbitrary (modified from ref. 147 with the permission of Elsevier Science Publ., Amsterdam).

tivity in rats and rabbits under in vivo and in vitro conditions (Fig. 17A). Interestingly, these results and others reviewed elsewhere [24] can be rationalized in terms of the model proposed by Baumann and Prelog (Fig. 17B) to account for the microbiological reduction of numerous ketones [39]. This suggests that closely related carbonyl reductases are involved.

All above examples pertain to product enantioselectivity. Cases of product diastereoselectivity resulting from the creation of an additional centre of chirality will be considered in the next section, but it should be noted here that product diastereoselectivity can also involve achiral metabolites. Thus, metabolic hydroxylation of substituted acetophenone imines leads to the formation of isomeric oximes, and the

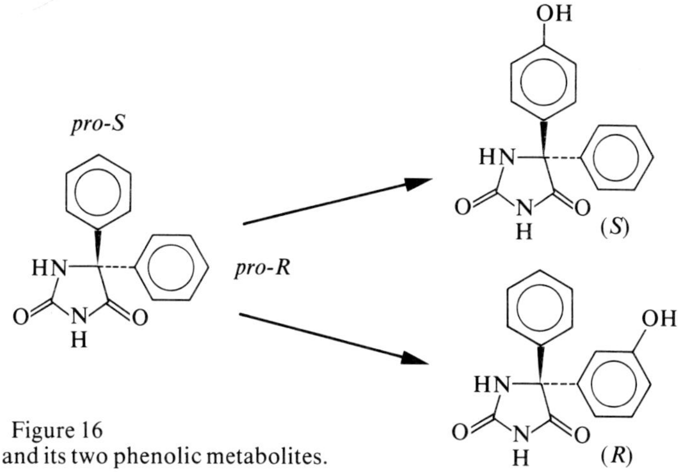

Figure 16
Phenytoin and its two phenolic metabolites.

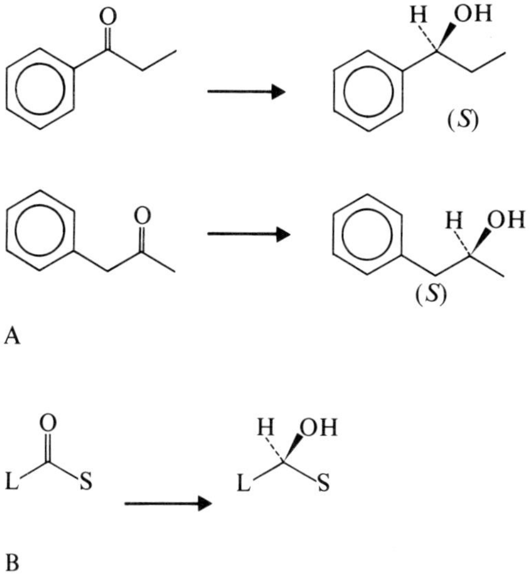

Figure 17
A: Stereoselective reduction of ketones [38].
B: General model proposed by Baumann and Prelog (L = large; S = small) [39].

(*E*)-isomers were found to predominate largely over the (*Z*)-isomers
[40].
There is no *a priori* reason why conjugation reactions should not be
comparable to functionalization reactions in terms of their product
stereoselectivity. At present however, relevant evidence is rather

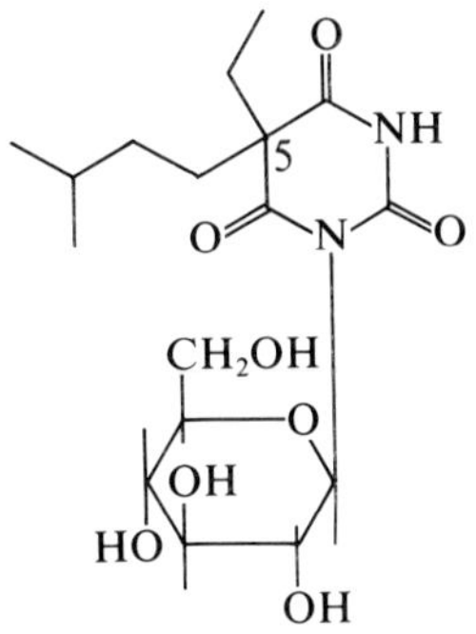

Figure 18
The structure of N-glucosylamobarbital.

scarce, rendering the few available data all the more interesting. This is the case of an intriguing preliminary report on the N-glucosylation of amobarbital in humans [41]. The reaction destroys the plane of symmetry in the molecule and transforms C-5 from a prochiral centre into a chiral centre. Glucose being chiral, the conjugate (Fig. 18) exists as two diastereomers which were isolated from urine. One diastereomer was found to be in very large excess over the other, and we can look forward to the publication of their absolute configuration.

3.4   Substrate-product stereoselectivity and biological factors

The question may be asked as to the influence of an existing centre of chirality of the metabolic generation of a second such centre in a given molecule. The phenomenon is known in synthetic chemistry as *asymmetric induction* and implies the preferential formation of one diastereomeric product from a chiral reactant. In asymmetric induction there is only one chiral tool, namely the preexisting chiral centre. This contrasts with xenobiotic metabolism where an enzyme is most frequently operative as a second chiral tool. Whatever the influence of the enzyme's chirality, it cannot be neglected, meaning that the concept of asymmetric induction must take a special meaning when applied to enzymatic reactions. Yet it is a common observation that product stereoselectivity can be different for two enantiomeric substrates, in other words that product stereoselectivity is itself substrate-enantioselective. The term „substrate-product stereoselectivity" has been proposed a number of years ago to designate such situations [24] and it appears to have found acceptance among specialists in the field of stereoselective metabolism.

A highly illustrative example of substrate-product stereoselectivity can be found in the 3'-hydroxylation of pentobarbital (Fig 19). The stereochemical aspects of the reaction have been studied in the dog upon administration of $(+)$-(R)-, $(-)$-(S)- and $(\pm)$-pentobarbital [42, 43]. The two enantiomers yielded comparable amounts $(35\% \pm 3\%)$ of 3'-hydroxylated pentobarbital, and substrate enantioselectivity regarding overall 3'-hydroxylation is non-existent. However, the diastereomeric ratio varied considerably with the substrate; while the (R)-enantiomer yielded equal amounts of the $(1'R;3'S)$- and $(1'R;3'R)$-diastereomers, (S)-pentobarbital yielded the $(1'S; 3'R)$- and $(1'S; 3'S)$-diastereomers in a 1/5 ratio. Thus, no product enantioselectivity exists for (R)-pentobarbital, as opposed to (S)-pentobarbital. The in vivo results for pentobarbital hydroxylation much resemble the in vitro results obtained using rat liver microsomes (Table 2) [44]. The K_m values for the $(1'R; 3'S)$- and $(1'S; 3'R)$-enantiomers are the same as are the K_m values for the $(1'S; 3'S)$- and $(1'R; 3'R)$-enantiomers. These two pairs of values, however, clearly differ, perhaps suggesting that each pair of enantiomeric products is generated by a distinct isozyme. It is also apparent from Table 2 that the V_{max} values parallel the in vivo product stereoselectivity.

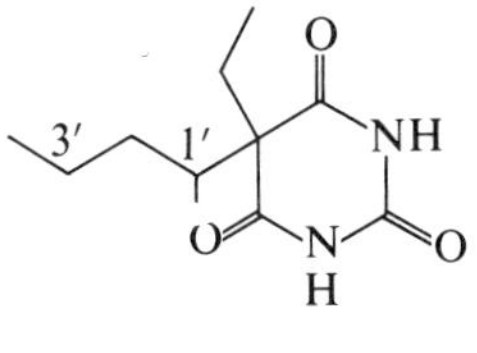

Figure 19
The structure of pentobarbital.

Table 2
Substrate-product stereoselective 3'-hydroxylation of pentobarbital [42–44]

Substrate	Metabolite	Percentage of dose in urine of dog	Rat liver microsomes	
			K_m mM	V_{max} nmol (mg prot)$^{-1}$ min^{-1}
$(+)$-(R)	$(1'R;3'S)$	16.6	0.195	0.581
	$(1'R;3'R)$	16.2	0.087	0.654
$(-)$-(S)	$(1'S;3'R)$	6.2	0.198	0.296
	$(1'S;3'S)$	31.6	0.097	1.017

The influence of biological factors on stereoselective drug metabolism is now classical knowledge and has been amply reviewed [e. g., 21, 24, 27]. Three examples are given here, merely documenting species variation in substrate enantioselectivity, product enantioselectivity, and substrate-product stereoselectivity, respectively. A closer examination of this phenomenon at the enzymatic level will be attempted in the next section.

The in vitro sulfate conjugation of 4′-hydroxypropranolol shows species-dependent substrate enantioselectivity, the $(-)/(+)$ metabolic ratio being 0.62 ± 0.02 in the hamster, 0.73 ± 0.01 in the dog, 0.95 ± 0.04 in the cat, and 1.07 ± 0.02 in the rat [45]. These differences, while not marked, are nevertheless significant in statistical terms and are representative of a rather general phenomenon. A more spectacular example, which pertains to product enantioselectivity, becomes apparent when the story of the aromatic hydroxylation of phenytoin is told in greater detail. In Section 3.3, we discussed the metabolism of phentytoin in humans and most animal species and mentioned the dog to be an exception. In this species, the *para*-hydroxylated metabolite is produced in a $(S)/(R)$ ratio of only 2/1, indicating a smaller product enantioselectivity than in most species. But the dog has the pecularity of also producing 5-(3-hydroxyphenyl)-5-phenylhydantoin as the major phase I metabolite, a product present only in trace amounts in human urine. The absolute configuration of this metabolite is (R) (Fig. 16), i. e. the configuration opposite to that of the *para*-hydroxylated metabolite [46]. Furthermore, the *meta*-hydroxylated metabolite excreted by the dog is believed to be enantiomerically pure, i. e. to apparently result from a completely stereoselective (stereospecific) enzymatic process.

An example aptly illustrating species variation in substrate-product stereoselectivity is that of the 1′-N-oxidation of nicotine. This chiral molecule exists as the $(+)$-(R)- and $(-)$-(S)-enantiomer, the latter being the natural form. The N-1′ centre is not configurationally stable, but becomes a stable centre of chirality upon N-oxidation. In hepatic preparations from five different species, (S)-nicotine was preferentially oxidized to the *cis*-$(1′R;2′S)$-isomer, while (R) in all investigated species except the rabbit gave rise preferentially to the *trans*-$(1′R; 2′R)$-isomer [47, 48] (Fig. 20). The species variations (Fig. 21) can be rationalized in terms of the reaction showing marked and comparable product stereoselectivity in the guinea pig, hamster and mouse, leading to a high predominance of the $(1′R)$-configurated metabolite independ-

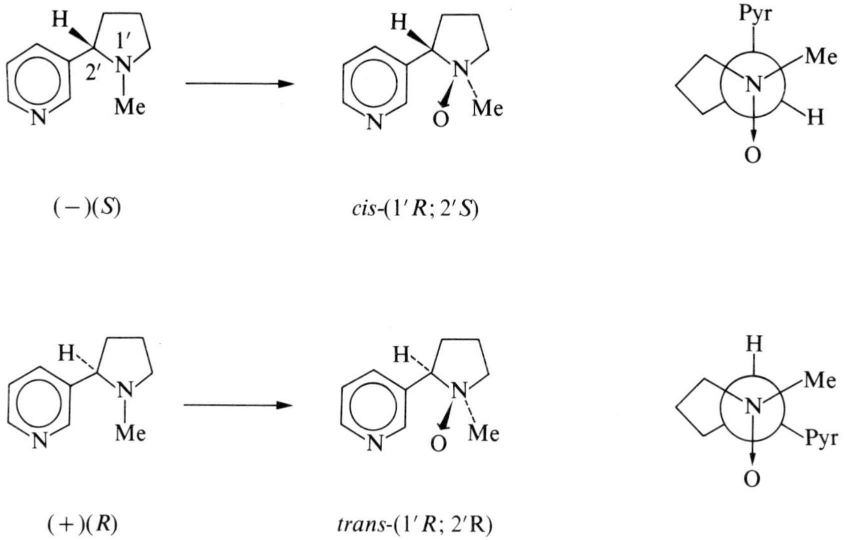

$(-)(S)$ cis-$(1'R; 2'S)$

$(+)(R)$ $trans$-$(1'R; 2'R)$

Figure 20

The predominant stereoisomers of metabolically produced nicotine 1'-N-oxide, shown in classical representation and Newman projection [26] (reproduced with the permission of Elsevier Science Publ., Amsterdam).

ently of the configuration at C-2'. Product stereoselectivity is comparatively modest in the rabbit (*cis/trans* ratios smaller than 2), while the rat shows product stereoselectivity for (R)-nicotine only.

While the concepts of substrate, product, and substrate-product stereoselectivity are shown above to be indeed useful in ordering and clarifying many observations, they do not add to our understanding of underlying molecular mechanisms. Here other approaches are necessary, as discussed in the following section.

4   Mechanisms of stereoselective drug metabolism
4.1   Enzymatic factors

All biological processes can, at least conceptually, be subdivided into three steps. There is first penetration, then recognition, and finally activation [19]. In xenobiotic metabolism, recognition can be equated with binding to the enzyme, and activation is the catalytic step. Thus, the phenomenon of chiral recognition (Section 3.1) can be postulated to occur at the *binding step* and/or at the *catalytic step,* resulting in different affinities and/or reactivities of the two enantiomers. In other

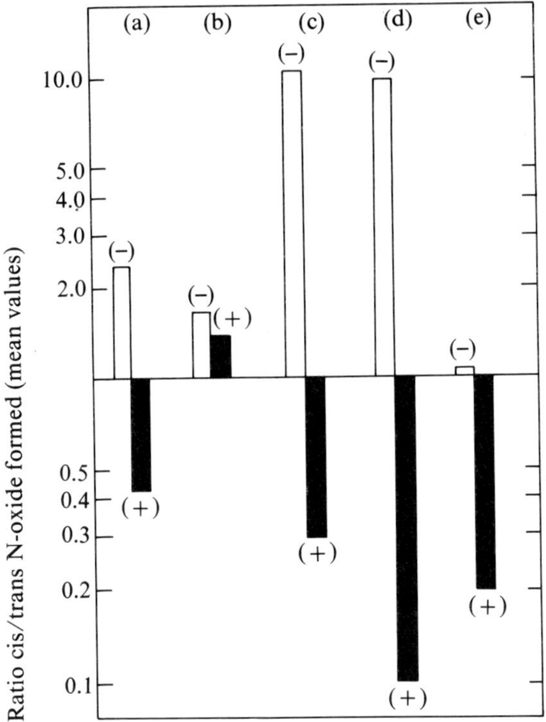

Figure 21

Species variation in diastereomeric 1′-N-oxide ratio produced from (+)-(*R*)- and (–)--(*S*)-nicotine: (a) guinea pig; (b) rabbit; (c) hamster; (d) mouse; (e) rat [47] (reproduced with the permission of Taylor and Francis, Basingstoke).

words, both binding and reactivity factors can be expected *a priori* to contribute to substrate enantioselectivity and to product enantioselectivity (Fig. 22). It is the aim of this section to examine whether and to what extent these two contributions can be disentangled to reveal some aspects of the mechanisms of enantioselective drug metabolism. Before so doing, however, enzymatic factors *per se* must be acknowledged, i.e. the discrimination of enzymes and isozymes by stereoisomers.

The question of the origin of species variation in stereoselective drug metabolism (Section 3.4) was for years an object of debate. In particular, these differences were explained in terms of distinct enzymatic sites, or of distinct modes of binding to the same site. Other poorly explained factors were enzyme induction and inhibition and their influence on stereoselective drug metabolism. For example, the substrate enantioselective in vitro glucuronidation of the enantiomers of oxaze-

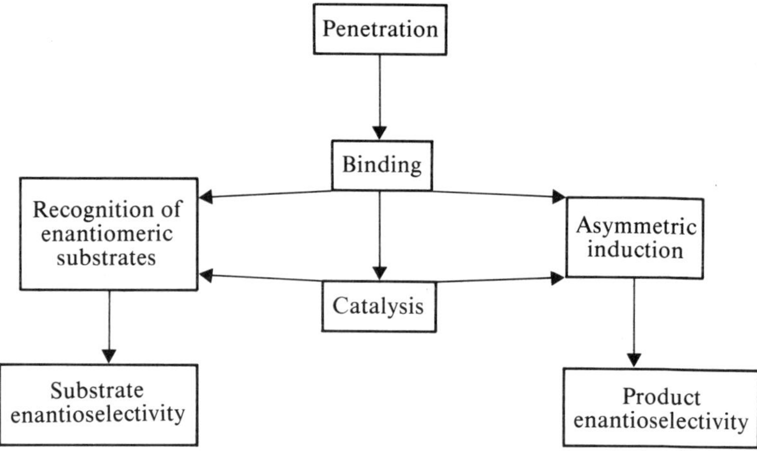

Figure 22

Possible contribution of the binding and catalysis steps to the phenomena of substrate enantioselectivity (which results from the recognition of enantiomeric substrates) and product enantioselectivity (which results from asymmetric induction) [146] (reproduced with the permission of Pergamon Press Ltd, Oxford).

pam is markedly influenced by a variety of inducers, the R/S ratio ranging from 0.76 to 1.25 [49]. Similarly, the cytochrome P450 mediated, product enantioselective epoxidation of styrene (a prochiral compound) to styrene oxide (a chiral metabolite) is altered by pretreatment of the animals with selective inhibitors [50].

Current evidence now indicates that *enzyme multiplicity* is the primary factor (but perhaps not the only one) accounting for substrate and product selectivity, in particular regio- and stereoselectivity, and substrate selectivity of analogues. One of the earliest and most convincing indications to this effect was the elegant work of Pohl et al. on the rat liver microsomal oxidation of warfarin (Fig. 23) [51]. By examining the sub-

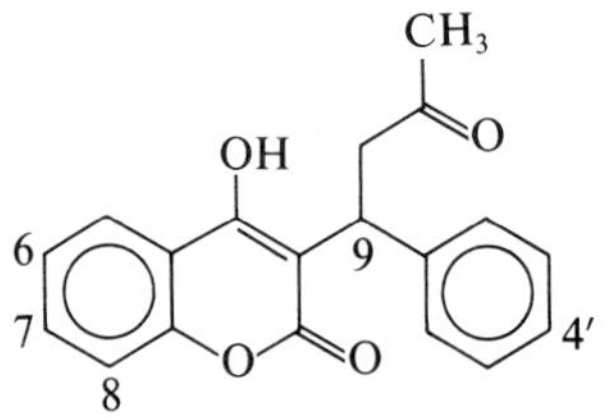

Figure 23

The chemical structure of warfarin showing the chiral centre (C-9) and the major sites of hydroxylation.

strate enantioselectivity and the product regioselectivity of hydroxylation at the 9- (benzylic), 6-, 7-, 8-, and 4'-position, they found marked differences between liver microsomes from untreated, phenobarbital-induced, and 3-methylcholanthrene-induced rats. Thus, the former appear particularly active in 7-hydroxylating (R)-warfarin, while the latter are even more active in 8-hydroxylating (R)-warfarin. From these results, Pohl et al. were able to deduce the involvement of at least four enzymatic forms which they designated as A, B, C and D [51]. Using the same substrate, Kaminsky at al. then reported one of the first uses of purified forms of rat liver microsomal cytochrome P450 in stereoselective metabolism studies [52]. Considerable differences in regio- and stereoselectivity, as well as in overall metabolism, were found between one isozyme isolated from untreated animals, another isozyme isolated from untreated and phenobarbital-induced animals, at least one (but possibly three) additional form(s) isolated from phenobarbital-induced rats, and one distinct form from 3-methylcholanthrene-treated rats [52].

Examples of this type have now accumulated. They make up an impressive body of evidence pointing to the central role of enzyme multiplicity in accounting for biological variability in xenobiotic metabolism and connected selectivites. As an additional example, the epoxidation of styrene mentioned above was found to occur with a R/S ratio ranging from 0.9 to 2.0 depending on the purified isocytochrome P450 being used [53].

Interestingly, examples exist of *different enzymes* catalyzing the same reaction but with different enantioselectivites. This is the case of sulfoxidation as catalyzed by cytochromes P450 and by FAD-monooxygenases. Thus, the formation of the chiral sulfoxide from the prochiral 4-tolyl ethyl sulfide exhibited R/S ratios of 20/1 and 1/4, respectively, when purified FAD-monooxygenases and isocytochromes P450 were used [54]. Such examples indeed prove that substrate and product stereoselectivities vary with isozymes/enzymes or with isozyme populations, a fact of major significance when pharmacokinetic or pharmacodynamic consequences of stereoselective metabolism are considered (see later). In addition, this biological factor renders more difficult any mechanistic interpretation of stereoselectivity in xenobiotic metabolism.

4.2   Substrate enantioselectivity in binding and catalysis

The contribution of *binding* to substrate enantioselectivity, in other words the differential binding of enantiomers to an enzyme, is due to the chirality of the substrate binding site, usually a hydrophobic pocket in the enzyme protein. But is this binding enantioselectivity sufficient to account for the substrate enantioselectivity resulting from enzymatic reactions? Examination of a number of publications suggests that probably a large proportion of workers believe, either consciously or subconsciously, this to be true.

Here, we wish to prove that the *catalytic step* may indeed, in an number of cases, contribute to substrate enantioselectivity. Before discussing direct or indirect evidence, some facts must be considered. First, it should be noted that substrate selectivity can occur without binding; this is documented for chemical models of enzymes. For example, a flavin hydroperoxide was used as a model of the flavoprotein monooxygenase; it did oxidize arylamines efficiently and with marked substrate selectivity, the reaction rate being correlated with the substrates' nucleophilicity [55]. Secondly, the chirality of the binding site should not push into oblivion the chirality of the active site, as exemplified by the chiral orientation of the prosthetic heme in cytochrome P450 [56]. The importance of the chirality of the catalytic site can be deduced from the fact that enzyme reactivity rests on the proper positioning and alignment of the catalytic and target groups [57, 58]; only so can the reaction proceed with adequate rapidity and stereoelectronic control. A good application of these principles can be found in the reaction mechanism of NAD(P)/NAD(P)H dehydrogenases/reductases [59].

In the remainder of this section, three arguments are used to unravel the relative contributions of binding and catalysis to substrate enantioselectivity. These arguments are enzyme kinetics, structure-metabolism considerations, and metabolic interactions between enantiomers. Despite its limitations, *Michaelis-Menten analysis* offers a potentially interesting but insufficiently explored approach to evaluate the binding and catalytic components of substrate enantioselectivity. A compilation of kinetic parameters taken from the recent literature is quite revealing. In some of the examples reported, the two enantiomers have identical or closely resembling affinities, but the velocity or rate constant of the reaction differ. This is the case of the hydroxylation of

3-PPP [60], of the oxidation of the two higher homologues of phenylaminoethanes [61], of the hydrolysis of *para*-nitrophenyl alpha-methoxyphenyl acetate [62], of the methylation of DOPA [65], and of the glutathione conjugation of benz[a]anthracene and benzo[a]pyrene oxide [68]. Other cases show the opposite behaviour, namely analogous rates but markedly different affinities, e. g. the hydration of naphthalene and anthracene oxide [64]. One example in the list shows no meaningful substrate enantioselectivity, namely the methylation of 3-(3-4-dihydroxyphenyl)-N-propylpiperidine [66]. And finally, the list also reports cases of substrate enantioselectivity caused by differences in both affinity and rate, namely glucuronidation of 9,10-dihydroxy-9, 10-dihydrophenanthrene (DDP) and 4,5-dihydroxy-4,5-dihydrobenzo[a]pyrene (DDBP) [67], and the hydrolysis of phenyloxirane [63].

Table 3 thus indicates that the molecular origin of substrate enantioselectivity may lie either in different affinities of the two enantiomers, or in different rate constants or velocities. The latter case is the one most frequently encountered in Table 3, a possibly fortuitous observation. Alternatively, substrate enantioselectivity may be caused by differences in both affinity and velocity. In other words, Michaelis-Menten analysis suggests that the molecular mechanism of substrate enantioselectivity can be found in the binding step (different affinities), in the catalytic step (different reactivities), or in both steps. Note that Table 2 (Section 3.4) indicates that Michaelis-Menten analysis may also contribute to a better understanding of substrate-product stereoselectivity.

Analysis of structure-metabolism relationships can also be expected to help explain mechanisms of substrate enantioselectivity. This approach is adequately illustrated by the selective hydroxylation of some coumarin anticoagulants. Thus, liver microsomes of rats pretreated with beta-naphthoflavone or 3-methylcholanthrene hydroxylate warfarin and phenprocoumon in the 6- and 8-position with high product regioselectivity, and with high but opposed substrate enantioselectivity, the preferred substrates being (+)-(*R*)-warfarin and (−)-(*S*)-phenprocoumon. This opposed enantioselectivity remained unexplained for years until Heimark and Trager [69, 70] showed that (*R*)-warfarin binds to cytochrome P450 as the cyclic hemiketal tautomer which renders it topographically equivalent to (*S*)-phenprocoumon despite opposed absolute configurations (Fig. 24). This 3-dimensional congru-

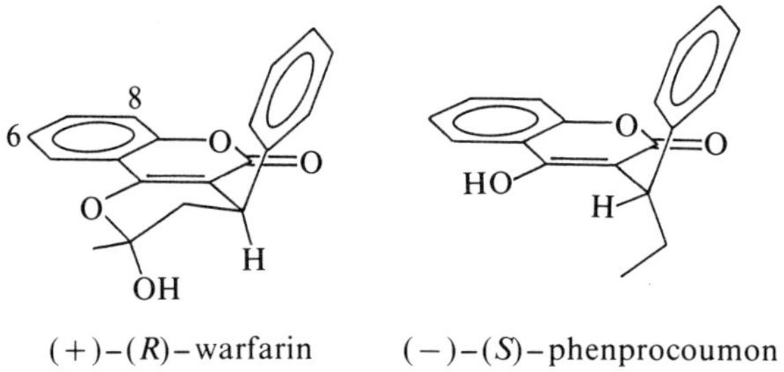

$(+)-(R)$-warfarin $(-)-(S)$-phenprocoumon

Figure 24
The topographically equivalent conformations of (R)-warfarin and (S)-phenprocoumon as they bind to the active site of cytochrome P450 [69].

ence brings evidence for a substrate enantioselectivity controlled by the binding step, but the argument is not devoid of ambiguity.

Metabolic inhibitions between enantiomers can also be interpreted in terms of mechanism of substrate enantioselectivity. Indeed, competitive metabolic interactions between enantiomers are to be expected when both have sufficient affinity for the same enzymes/isozymes. When this is the case, substrate stereoselectivity occurs, which as discussed above results from stereoisomer recognition at the binding and/or catalysis step. Conceivably, enantiomers can be substrates and/or inhibitors, leading to a number of possible situations. For example, both enantiomers can be substrates of the same enzyme/isozyme while simultaneously acting as mutual competitive inhibitors. Another possibility is for only one enantiomer to be substrate, and for the other to be inhibitor. Both metabolic situations are documented in the literature and will be discussed here. Note that the present Section is restricted to in vitro inhibitions, while in vivo metabolic interactions form the subject of Section 6.

The first situation is aptly illustrated by the rabbit liver microsomal metabolism of *para*-chloroamphetamine [71]. When incubated alone, the $(-)-(R)$- and $(+)-(S)$-enantiomers were oxidized at rates of 0.329 ± 0.024 and $0.258 \pm 0,029$ mmol (mg protein)$^{-1}$min^{-1}, respectively, indicating a modest substrate enantioselectivity. However, when the racemate was incubated and the enantiomers monitored separately by an enantiospecific analytical method, the rates of metabolism of the two forms were 0.037 ± 0.008 and 0.145 ± 0.20 mmol (mg protein) $^{-1}$ min $^{-1}$, respectively. In other words, the (S)-substrate had a nine-fold inhibitory effect on the oxidation rate of the (R)-substrate, while the

Table 3
Michaelis-Menten analyses of enantioselective metabolic reactions

	K_m (μM)	rate	rate/K_m	ref.
4-Hydroxylation of 3-(3-hydroxyphenyl)-N-propylpiperidine (3-PPP)				
(+)-(R)	1.0 ± 0.1	1.85 ± 0.05[a]	1.85	[60]
(−)-(S)	1.4 ± 0.2	2.33 ± 0.18	1.66	
Cytochrome P455 nm complex formation with 2-X-1-phenyl-2-aminoethanes				
X = methyl (R)	ca. 270	ca. 1.5[b]	ca. 0.006	[61]
(S)	ca. 280	ca. 1.6	ca. 0.006	
X = ethyl (R)	23 ± 5	3.2 ± 0.5	0.14	
(S)	30 ± 3	4.0 ± 0.5	0.13	
X = n-propyl (R)	28 ± 6	5.1 ± 0.4	0.18	
(S)	32 ± 4	9.1 ± 0.5	0.28	
X = n-butyl (R)	16 ± 3	5.7 ± 0.1	0.36	
(S)	21 ± 5	10.5 ± 0.3	0.50	
Albumin hydrolysis of *para*-nitrophenyl alpha-methoxyphenyl acetate				
(D)	75	37.0[c]	0.49	[62]
(L)	59	9.6	0.16	
Hydrolysis of phenyloxirane catalyzed by epoxide hydrolase				
(+)-(R)	29	11.6[a]	0.40	[63]
(−)-(S)	155	44.1	0.28	
Hydrolysis of naphthalene oxide (NO) and anthracene oxide (AO) catalyzed by epoxide hydrolase				
(+)-(1R;2S)-NO	1	2.2[d]	2.2	[64]
(−)-(1S;2R)-NO	12	3.6	0.3	
(+)-(1R;2S)-AO	1.0 ± 0.3	29 ± 5	29	
(−)-(1S;2R)-AO	7.1 ± 1.6	17 ± 0.6	2.4	
Catechol-O-methyltransferase (COMT) catalyzed methylation of DOPA				
D-DOPA	2.05 ± 0.02[e]	195 ± 3[f]	95	[65]
L-DOPA	1.70 ± 0.02	292 ± 5	172	
COMT catalyzed methylation of 3-(3,4-dihydroxyphenyl)-N-propylpiperidines				
(*R*)	0.12 ± 0.02[e]	(42)[g]	346 ± 67	[66]
(*S*)	0.08 ± 0.01	(37)	485 ± 74	
Glucuronidation of 9,10-dihydroxy-9,10-dihydrophenanthrene (DDP) and 4,5-dihydroxy-4,5-dihydrobenzo[a]pyrene (DDBP)				
(9R;10R)-DDP	16.1 ± 0.8	0.070 ± 0.001[h]	0.0044	[67]
(9S;10S)-DDP	1280 ± 70	1.40 ± 0.03	0.0011	
(4S;5S)-DDBP	172 ± 47	4.1 ± 0.7	0.024	
(4R;5R)-DDBP	1620 ± 450	0.37 ± 0.15	0.00023	
(4S;5R)-DDBP	73 ± 15	0.23 ± 0.02	0.0032	
(4R;5S)-DDBP			< 0.00005	
Activity of glutathione S-transferase C towards benz[a]anthracene oxide (BAO) and benzo[a]pyrene oxide (BPO)				
(−)-(5R;6S)-BAO	2.0 ± 0.5	16.0 ± 0.8[d]	8.0	[68]
(+)-(5S;6R)-BAO	1.6 ± 0.2	5.0 ± 0.2	3.1	
(−)-(4R;5S)-BPO	4.6 ± 0.7	12.0 ± 0.6	2.6	
(+)-(4S;5R)-BPO	2.9 ± 0.5	4.1 ± 0.2	1.4	

a) V_{max} nmol (mg prot)$^{-1}$ min^{-1}; b) V_{max} ΔA cm^{-1} min^{-1} (mM P450)$^{-1}$; c) $k_2 \times 10^3$ sec^{-1}; d) k_{cat} min^{-1}; e) mM; f) mU (mg prot)$^{-1}$; g) calculated from the V_{max}/K_m data, no units given; h) k_c sec^{-1}.

latter had a ca twofold inhibitory effect on the oxidation rate of the former. The results of this study can be summarized in terms of (S)-*para*-chloroamphetamine being a slightly poorer substrate, but a distinctly better ligand, of monooxygenases than its (R)-enantiomer. The observed substrate enantioselectivity may thus be ascribed predominantly to the catalytic step, with the binding of the higher-affinity ligand being comparatively unproductive.

Data are missing to conclude whether the findings discussed above are but an isolated case, or whether they exemplify a general phenomenon. However, the 4-hydroxylation of 3-PPP (see Table 3) is worth mentioning again, since the V_{max} value for the racemate was found to be 1.75 nmol (mg prot)$^{-1}$ min $^{-1}$, i. e. slightly smaller than that of either enantiomer tested separately [60]. This result certainly suggests weak mutual inhibition of the two enantiomers.

The second case above is an extreme one, namely one enantiomer only being the substrate, and the other the inhibitor. To the best of our knowledge no examples have been published which would document this situation for monooxygenase-mediated reactions. In contrast, an example of particular clarity can be found in the reaction of nicotine with azaheterocycle N-methyltransferases. Here, full enantioselectivity is seen in that only (+)-(R)-nicotine, which is present in tobacco smoke, undergoes N-methylation (K_m 14.2 μM with guinea pig lung aromatic azaheterocycle N-methyltransferase) [72, 73]. In contrast, (−)-(S)- nicotine is totally unreactive but acts as a strong competitive inhibitor of (R)-nicotine N-methylation (K_i 62.5 μM). The complete lack of reactivity of (S)-nicotine despite its high affinity must therefore be of catalytic origin, arising from an unproductive mode of binding which positions the target pyridyl nitrogen out of reach of S-adenosylmethionine.

4.3   Mechanisms of product enantioselectivity

Schematically, product stereoselectivity may be due to the action of two or more distinct isozymes to each of which a substrate binds by a single mode, and/or it may result from different modes of binding of the substrate molecule to the active site of a single isozyme. The involvement of *distinct isozymes* relates to the differential affinity and binding of a substrate, and more generally to the molecular mechanisms by which a given substrate „discriminates" enzymes or isozymes.

A case in point is that of acetanilide [74]. The compound, which is hydroxylated in the aromatic ring by cytochrome $P450_{LM4}$, binds to this 3-methylcholanthrene-inducible isozyme with the phenyl ring in close proximity to the iron atom. In contrast, acetanilide is not ring-hydroxylated by the phenobarbital-inducible cytochrome $P450_{LM2}$, an isozyme to which it binds with the phenyl ring orientated away from the iron atom. Differences in binding thus result in contrasting formation of metabolite(s).

The second schematic situation involves *different modes of binding* to a single isozyme, although caution is necessary here; this the „single" isozyme of today may well be awaiting technical refinements for its multiplicity to be uncovered. The Aristotelian dichotomy between binding to distinct isozymes and different modes of binding to a single isozyme should thus be qualified carefully. When a single isozyme is assumed, each *productive* binding mode will bring a different target group in the vicinity of the catalytic site. Assuming optimal positioning, the ratio of products will depend on the relative probability of the various binding modes in case of product stereoselectivity, as well as on the relative reactivity of the target groups in case of product regioselectivity. The above example of acetanilide can also be interpreted in this context. Consider also the oxidation of *D*-camphor, adamantanone and adamantane by cytochrome $P450_{cam}$ [75]. In each case, reaction with this enzyme led to one product only, namely the 5-exo-, 5- and 1-hydroxylated derivative, respectively. These three positions are topographically congruent and indicate a tight enzyme-substrate complex. In contrast, the same substrates each yielded two or three metabolites when hydroxylated by cytochrome $P450_{LM2}$. The rations of products were approximately proportional to the chemical reactivity of the abstractable hydrogen atoms, indicating moderate steric constraint in the binding site and considerable movement of the bound substrate.

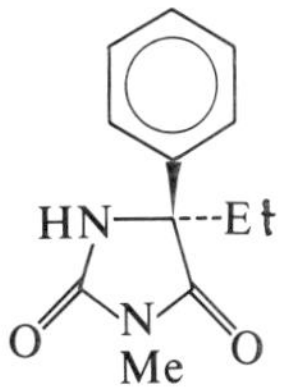

$(+)-(S)-$mephenytoin

Figure 25
Substrate enantioselective *para*-hydroxylation of $(+)$-(S)-mephenytoin.

A classical example of product enantioselectivity as discussed in Sections 3.3 is that of phenytoin, a prochiral compound displaying highly selective para-hydroxylation of the *pro*-S ring. This example can be gainfully compared with that of mephenytoin, a closely related yet chiral compound whose *para*-hydroxylation is highly selective for the (+)-(*S*)-enantiomer (Figure 25) [e. g. 76] (see Section 7.2). Interestingly, the predominantly *para*-hydroxylated phenyl rings in mephenytoin and phentytoin are topographically equivalent (compare Fig. 16 and 25), suggesting that the two compounds might share the same mode of binding to the common isocytochrome(s) P450 mediating both reactions. However, there is now evidence to prove that the *para*-hydroxylation of (*S*)-mephenytoin and the *pro*-S *para*-hydroxylation of phenytoin are mediated by distinct isozymes. Indeed, phenytoin is a poor in vitro inhibitor of mephenytoin hydroxylation [77]. Even more important is the fact that poor (*S*)-mephenytoin hydroxylators (see Section 7.2) are deficient not in *pro*-S but in *pro*-R phenytoin hydroxylation [78], indicating that (*S*)-mephenytoin and *pro*-R phenytoin *para*-hydroxylation are mediated by the same isocytochrome P450, while *pro*-R and *pro*-S *para*-hydroxylation of phentytoin is mediated by (at least) two distinct isozymes. Thus, product enantioselectivity in phenytoin hydroxylation appears to be controlled by the relative affinities of the prochiral substrate to the two isozymes. For each of the isozymes, a single productive binding mode is postulated, each bringing another of the two enantiotopic phenyl rings in the proximity of the catalytic site.

Recent studies of rare elegance bring final proof that product enantioselectivity in cytochrome P450 mediated reactions, and hence the enzymatic recognition of enantiotopic groups, is not an inherent feature of the reaction mechanism itself, but is primarily a function of isozyme structure and the chirality of the catalytic site [79]. These studies have employed prochiral substrates rendered chiral by the stereospecific labelling of the enantiotopic target groups and have investigated kinetic aspects of the enzymatic reaction of hydroxylation. Thus, the selective hydroxylation of the enantiotopic methyl groups in 2-phenylpropane (cumene) was found to occur with a small degree of incomplete equilibrium of the two target groups in the enzyme-substrate complex [80]. In contrast, methoxychlor contains two bulky enantiotopic anisyl groups which interchanged only slowly in the enzyme-substrate complex, resulting in a large degree of incomplete equilibrium

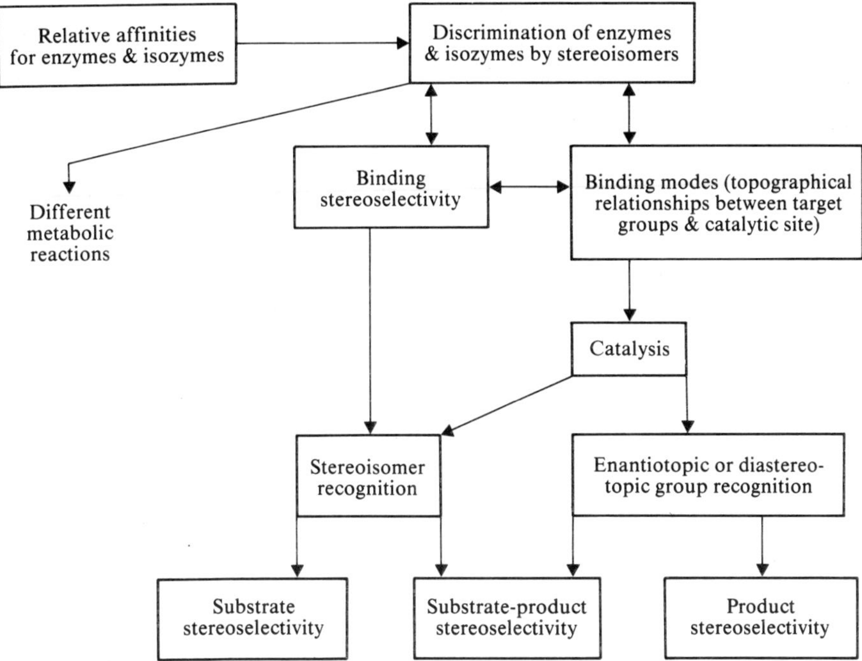

Figure 26

Enzymatic and mechanistic factors underlying substrate, product, and substrate-product stereoselectivity in drug metabolism [146] (reproduced with the permission of Pergamon Press Ltd, Oxford).

[81]. What these studies have confirmed, in other words, is the critical role of the catalytically productive orientation of the target groups, the relative importance of which is governed by the equilibrium constants between the various binding modes.

The importance of the spatial position of target groups is also documented by some cases of conformer selectivity in xenobiotic metabolism. Thus, UDP-glucuronyltransferase has been shown to be specific for equatorial hydroxyl groups in dihydrodiols of polycyclic aromatic hydrocarbons [82].

4.4   Mechanistic outlook

To summarise, the above discussion indicates that *substrate enantioselectivity (or stereoselectivity)* is the result of *enantiomer (or stereoisomer) recognition* at the binding and/or catalytic step, the relative contribution of the factors varying from case to case. *Product enantioselectivity (or stereoselectivity)* on the other hand results from *enantiotopic (or diastereotopic) group recognition* (proximity of target group and catalytic centre) originating in the binding mode(s) of the substrate. The combined existence of stereoisomer recognition and stereotopic group recognition leads to complex cases of *substrate-product stereoselectivity* [21, 24, 26, 28]. These facts, together with structural differences in binding sites, offer a rationale for the variations in substrate and product stereoselectivity seen among isozymes. Figure 26 presents a logical scheme summarising such an outlook.

For a number of enzymes, considerable information on stereoselectivity and structure-metabolism relationship (SMR) has been summarised in the form of *topographical models of binding/catalytic sites*. Thus, many data on the regioselectivity and product enantioselectivity of cytochrome P450c mediated epoxidation of polycyclic aromatic hydrocarbons have been rationalized in terms of the steric model shown in Figure 27 [e.g. 83–87]. Some limitations of this model have been uncovered [88]. Another case in point is the model of epoxide hydrolase (Fig. 28) as derived from the substrate-product stereoselectivity in the hydration of various epoxycyclohexanes [89]. Other examples include UDP-glucuronyltransferase [90] and glutathione transferase [68, 91]. Such models, despite their limitations and reductionistic character, contribute significantly to a better understanding of SMR and shall lead to the use of computer graphics for improved metabolic predictions.

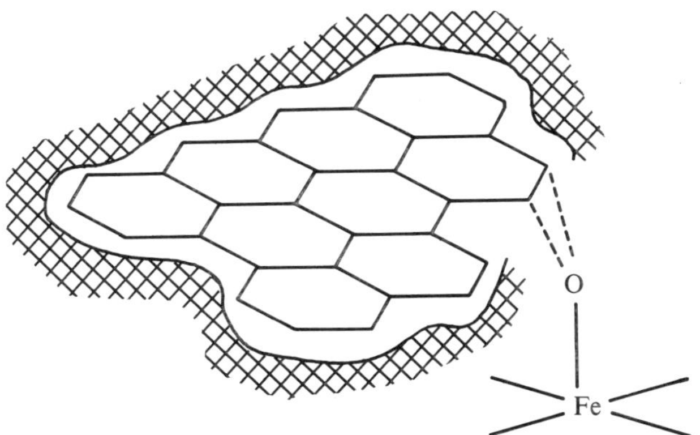

Figure 27
Steric model of the binding-catalytic site of cytochrome P450c [83–87].

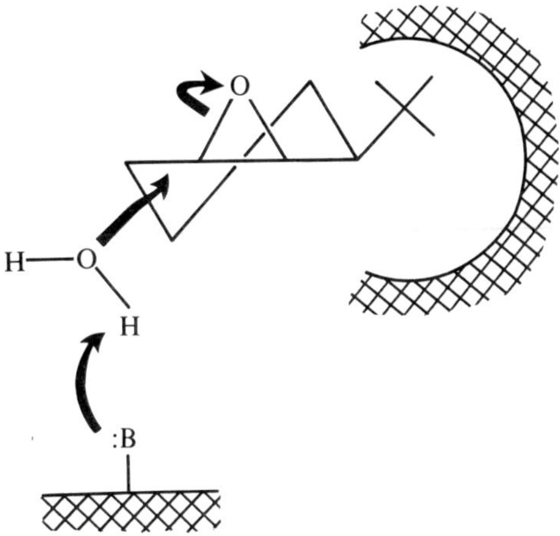

Figure 28
Steric and mechanistic model of the active site of epoxide hydrolase [89].

5   Enzymatic enantioselectivity in synthetic chemistry

In this section, and before considering stereoselective drug metabolism in a more clinical perspective (Sections 6–9), a brief glimpse on biosynthetic chemistry is offered. For many years, chemists have used a variety of micro-organisms to carry out key regio- and stereospecific steps in the synthesis of complex chemicals such as hormones and drugs. Commercially available enzyme preparations now offer additional possibilities for reactions of this type. Aside from a small number of books, two most valuable reviews have been published recently, one by Whitesides and Wong featuring over 300 references [92], and the other by Jones containing almost 400 references [93]. French readers will also enjoy a nice little overview by Sicsic [94].
The general principle of these biosynthetic methods is aptly illustrated by the work of Cambou and Klibanov [95], who used esterase-catalyzed transesterifications in biphasic media to carry out the preparative resolution of chiral alcohols (Fig. 29). The enzymatic reaction displays very high substrate enantioselectivity under optimized conditions, and the unreacted enantiomeric alcohol is separated from the ester produced, which is then hydrolyzed chemically to yield the other enantiomer. The method allows, for example, the preparation of a chiral synthon of vitamin E.

284   Bernard Testa and Joachim M. Mayer

$$
CH_3CH_2-\overset{\overset{\displaystyle O}{\|}}{C}-OCH_3 + HOCH_2CH_2CHRR' \xrightarrow{\text{esterase}}
$$

$$
HOCH_2CH_2-C\overset{R}{\underset{R'}{\cdots H}} \quad + \quad CH_3CH_2-\overset{\overset{\displaystyle O}{\|}}{C}-OCH_2CH_2-C\overset{H}{\underset{R'}{\cdots R}} \quad + \quad CH_3OH
$$

$$
\Big\downarrow HO^{\ominus}
$$

$$
CH_3CH_2COOH + HOCH_2CH_2-C\overset{H}{\underset{R'}{\cdots R}}
$$

Figure 29

Use of esterase-catalyzed transesterification for the preparative resolution of enantiomeric alcohols [taken from 95].

As pointed out by Jones [93], more than 2000 enzymes are known, several hundred of which are commercially available, including a significant number in immobilized form. The most useful in synthesis are oxidoreductases, transferases, hydrolases and lyases. Many current studies employ esterases and lipases of various origins (e. g. mammalian or bacterial), and a small selection of papers published in 1986 and 1987 is given [96–108].

While pH is a parameter of obvious importance in influencing enzymatic reactions, it does not appear to have a marked impact on enantioselectivity. In contrast, significant improvements have been obtained in mixtures of water and an organic solvent. Thus the progressive addition of dimethylsulfoxide decreased the reaction rate, but markedly improved the substrate enantioselectivity, in the pig liver esterase catalyzed hydrolysis of 2,2-dialkylated malonic acid dimethyl esters [109]. Similarly, optimal methanol concentration, pH and temperature allowed practically complete substrate enantioselectivity to be achieved in the pig liver esterase catalyzed hydrolysis of 3,3-dialkylated glutaric acid dimethyl esters [110]. From a large set of substrates, it was concluded that the hydrolysis is *pro*-S selective for the diesters with small C-3 substituents and reverses to *pro*-R preference when C-3 substituents are large [110]. Based on these studies, an active site model was proposed (Fig. 30) which has later been found to be compatible with a variety of other substrates [111].

Whatever the potential of traditional enzymes in synthetic chemistry, novel promises are in store. Indeed, site-directed mutagenesis has

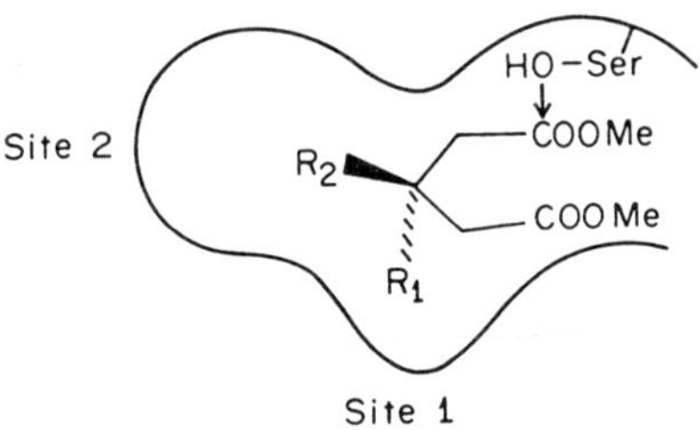

Figure 30
Orientation of chiral glutarate esters in the active site of liver esterase. The target ester group must locate close to the nucleophilic serine function. Site 1 binds the larger group until its steric dimensions are exceeded [110] (reproduced with the permission of the American Chemical Society, Washington).

been recently used to produce enzymes with rationally modified surface charge and altered selectivity [112]. The use of monoclonal antibodies as enzymatic catalysts is another breakthrough [113, 114], all the more so now that they have demonstrated stereoselective control [115].

6   Enantioselectivity in metabolic interactions

Sections 3–5 have presented various molecular and mechanistic aspects of stereoselective xenobiotic metabolism. Sections 6–9 now take up a more biological and clinical perspective, starting with enantioselective metabolic interactions already discussed in a mechanistic context in Section 4.2. Two situations can be distinguished, namely metabolic interactions between two enantiomers, and differential interactions of two enantiomers with another drug. These two types of interactions have been briefly reviewed by Trager and Testa [28] and by Walle and Walle [29]. Yet despite their potential or established significance in fundamental and clinical pharmacology, they are still receiving a relatively modest attention. It should be the conclusion of this section that the phenomenon may be more frequent than currently recognized.

6.1   Metabolic interactions between enantiomers

An early observation in the human [116] and in the rat [117] showed the half-life of (R)-propranolol to be shorter when given alone than when administered in the racemate. The half-life of the two enantiomers when administered as a racemic mixture could not be predicted based on studies with the individual enantiomers, suggesting metabolic interaction. A classical example showing the therapeutic potential of such interactions has been provided by Cooper and Anders [118], who examined the effect of dextromethorphan, a non-analgesic compound, on the analgesic activity of the active enantiomer levomethorphan. They found that the co-administration of dextromethorphan and levomethorphan led to a significantly enhanced and prolonged analgesic response relative to the administration of a corresponding dose of levomethorphan alone. Similarly, the inactive (–)-propoxyphene was found to markedly intensify the analgesic effect produced by the active enantiomer, (+)-propoxyphene, but without a prolongation in activity [118]. The latter result is somewhat surprising, since a prolongation in activity should be expected as a consequence of metabolic inhibition. In fact, it was later demonstrated that (–)-propoxyphene does not interfere with the metabolism of its active enantiomer but causes the release of the latter from liver binding sites, thereby increasing its systemic concentration [119]. This example serves to stress the fact that a combination of techniques is needed to unambiguously assess the origin of interactions observed in vivo.

It should be noted that metabolic interaction between enantiomers may not be a necessary occurrence. Examples documenting the absence of a metabolic interaction are known, but authors seldomly stress the fact. Thus, (–)-(R)- and (+)-(S)-hexobarbital exhibited a highly different plasma clearance in the rat when administered separately (226 ± 36 versus 1620 ± 186 ml min^{-1} kg^{-1}, respectively) [120]. These values remained unchanged when the enantiomers were monitored following administration of the racemate (243 ± 50 versus 1512 ± 381 ml min^{-1} kg^{-1}, respectively). At present, the number of studies specifically searching for interactions between enantiomers is relatively limited. Documented interactions may be of non-metabolic origin, being due for example to displacement from binding sites in tissues [e.g. 119] or plasma proteins [e.g. 121]. From the evidence presented in Sections 4.2 and 6.1, it can provisionally be concluded that metabolic interactions are to be expected for enantiomers subjected to a common metabolic route. However, many more studies are needed before the significance of the phenomenon can be better assessed. Specifically, the publication of positive and negative results should be equally encouraged.

6.2   Metabolic interactions between enantiomers and another drug

When in addition to the two enantiomers of a chiral drug, another (achiral) compound is administered, the situation becomes considerably more complex. Three structurally distinct xenobiotics are now present, each of which may interact with the two others.

One possibility may be for the metabolism of the achiral drug to be differently influenced by the two enantiomers. To be detected, such an enantioselective phenomenon requires administration of the two enantiomers on separate occasions. This effect was seen following administration of the enantiomers of *trans*-stilbene oxide to rats [122]. Indeed, the (+)-enantiomer proved a better inducer of cytochrome P450, aminopyrine N-demethylase and aldrin epoxidase than the (–)-enantiomer. Similarly, the former induced cytosolic epoxide hydrolase while the latter was inactive. However, these differences in inducing activity may not be intrinsic but may be due to the fact that (–)-*trans*-stilbene oxide is metabolized more rapidly than the (+)-form [122]. Another study has shown that lauric acid 12-hydroxylase is induced more potently by (+)-(S)- than by (–)-(R)-2-phenylpropionic acid [123]; but because the (S)-enantiomer is efficiently converted in vivo into the (R)-enantiomer (see Section 8), the potency of induction of the two isomers cannot be measured with confidence.

The interaction most clearly detected in vivo is when an achiral drug differently influences the metabolism or disposition of the two enantiomers of a chiral drug. In such cases, therapeutic consequences can be expected if the two enantiomers differ in their pharmacological potency. A number of relevant examples can be found in the literature, many of which bring evidence for an enantioselective metabolic inhibition caused by the achiral drug. Particular attention has been paid to the anticoagulant drugs warfarin and phenprocoumon, the reasons presumably being the diversity of their substrate-product stereoselective and regioselective metabolism, and the therapeutic import of its enantioselective inhibition [28]. Illustrative evidence includes the inhibition by cimetidine of (+)-(R)- but not (–)-(S)-warfarin metabolism in humans. Indeed, the AUC (area under the plasma concentration-time curve) and plasma half-live of (R)-warfarin were increased by 31 % and 21 %, respectively, while the total plasma clearance was decreased by 27 %; the pharmacokinetic parameters of (S)-warfarin were not influenced by cimetidine [124]. Similarly, the chemotherapeutic agent enoxacin enantio- and regioselectively inhibited the metabolism of warfarin in humans: only (R)-warfarin 6-hydroxylation was decreased (by 50 % under the conditions of the study) [125]. Other drugs enantioselectively influencing the metabolism of warfarin and/or phenprocoumon include phenylbutazone [126, 127], sulfinpyrazone [128], metronidazole [129], chloramphenicol [130] and amiodarone [131]. These drugs are known to be inhibitors of xenobiotic-metabolizing enzymes, and their enantioselective effects are but one facet of the enzyme/isozyme selectivity commonly displayed by substrates, inducers and inhibitors.

The above examples conclusively show that stereoselective interactions at the metabolic level occur under a variety of conditions and with a number of drugs. The phenomenon is obviously one element among many in the fate of drugs and their intertwined pharmacokinetic and pharmacodynamic components. Furthermore, stereoselective metabolic interactions are of importance not only in vivo but also in vitro and shed some light on drug-enzyme interactions. While no generalization can be offered at present which would render the phenomenon and its consequences partly predictable, this frustrating situation may change with a better understanding of underlying phenomena.

7   The significance of enantioselectivity in polymorphic drug metabolism

For some metabolic reactions affecting a number of drugs, human populations exhibit bimodal polymorphism, i.e. some subjects are poor metabolizers (PM) while others are extensive metabolizers (EM). The N-acetylation of isoniazide and other amines is a well-known example. Inasmuch as metabolism is a major factor influencing the pharmacological and toxicological activity of many drugs, polymorphism in xenobiotic metabolism may have far-reaching pharmacodynamic implications [132].

The cytochrome P450 mediated oxidation of several drugs is also under genetic control, and a number of distinct phenotypes are known to date:

– The debrisoquine 4-hydroxylation phenotype, which also includes sparteine oxidation, bufuralol 1′-hydroxylation and encainide O-demethylation;

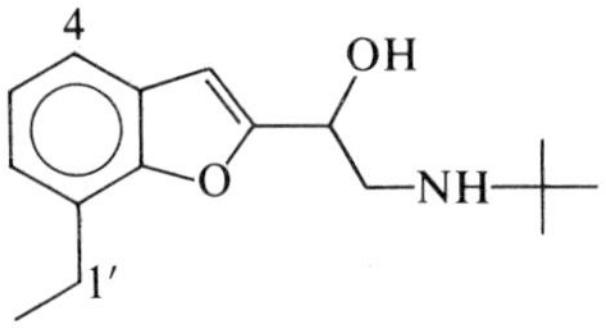

Figure 31
The structure of bufuralol.

– The (*S*)-mephenytoin *para*-hydroxylation phenotype, which also
 includes tolbutamide hydroxylation;
– The phenacetine O-deethylation phenotype;
– The nifedipide oxidation phenotype, which also includes the oxidation of other dihydropyridines [133].

The first and second of these phenotypes display a fascinating stereochemical dimension in that they involve reactions the substrate or product enantioselectivity of which is strongly affected in the slow metabolizers.

7.1 Bufuralol 1′-hydroxylation and debrisoquine 4-hydroxylation

The β-blocker bufuralol (Fig. 31) undergoes 4- and 1′-hydroxylation,
and the two reactions are under the same genetic control [134]. The
stereochemical aspects of the reaction of 1′-hydroxylation (carbinol
formation) have been particularly investigated, showing for example
that poor metabolizers excreted less of the carbinol metabolite than extensive metabolizers. The latter individuals excreted larger amounts of
the carbinol from the inactive (+)-(*R*)- than from the active (–)-(*S*)-bu-
furalol. In contrast, poor metabolizers excreted about the same

Table 4
Kinetic parameters of bufuralol 1′-hydroxylation in human liver microsomes from extensive metabolizers (EM) and poor metabolizers (PM), and in reconstituted isocytochromes P450 buf I and buf II [134, 135].

| | K_m (μM) | | V_{max} [a] | | Ratio |
	(*S*)	(*R*)	(*S*)	(*R*)	(*S*)/(*R*)
EM	35.3	47.3	4.1	8.6	0.48
PM	233.5	182.5	1.5	1.9	0.77
P450 buf I	50.8	60.3	8.9	53.9	0.16
P450 buf II	304.0	245.0	36.1	36.4	0.99

a) nmol (nmol P450)$^{-1}$ (15 min)$^{-1}$.

amounts of the carbinol from both enantiomers, showing that the poly-
morphic pathway is selective for (R)-bufuralol.

In vitro investigations using human liver microsomes have significantly contributed to our understanding of the molecular mechanisms involved [134, 135]. As shown in Table 4, there is only negligible enantioselectivity in the binding of bufuralol, which however has a much
smaller affinity for PM microsomes and cytochrome P450 buf II than
for EM microsomes and cytochrome P450 buf I, respectively. The
reaction is substrate enantioselective for (R)-bufuralol in microsomes
from EM and with cytochrome P450 buf I, but the enantioselectivity is
fully lost in microsomes from PM and with cytochrome P450 buf II.
These results are explained by the fact that cytochrome P450 buf I is
missing in poor metabolizers, where bufuralol 1'-hydroxylation is mediated by cytochrome P450 buf II, an isozyme with little affinity and
no substrate enantioselectivity for bufuralol [136, 137].

In addition to bufuralol, good evidence indicates differential substrate
enantioselective metabolism of metoprolol in extensive and poor debrisoquine metabolizers [138, 139]. This suggests that the phenomenon
may affect other β-blockers the metabolism of which is also under
genetic control.

In the example of bufuralol 1'-hydroxylation discussed above, only the
substrate enantioselectivity of the reaction has been uncovered. The
1'-carbon atom is a prochiral centre, containing two enantiotopic
hydrogen atoms, but to date the substrate-product enantioselectivity
of 1'-hydroxylation does not appear to have been reported. Product enantioselectivity in the debrisoquine/bufuralol phenotype therefore remained a mere possibility until Eichelbaum and collaborators [140,
141] recently elucidated the stereochemistry of debrisoquine 4-hydrox-
ylation. This drug is achiral but it contains three prochiral methylene
groups, one of which is the site of metabolic hydroxylation, i.e. C-4
(Fig. 32). In all extensive metabolizers, a very high to complete enan-

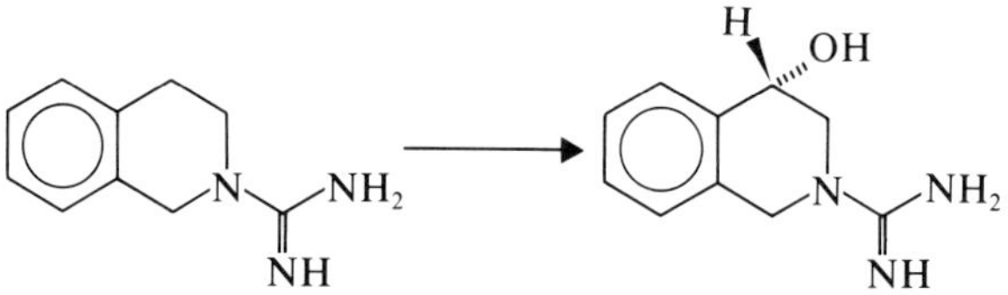

Figure 32
Debrisoquine and its major metabolite, ($+$)-(S)-4-hydroxydebrisoquine.

tioselectivity was found, (+)-(*S*)-4-hydroxydebrisoquine being produced with an enantiomeric excess ranging from ca 98 to 100 %. Poor metabolizers in contrast were characterized not only by a marked decrease in total 4-hydroxylation, but also by a loss in product enantioselectivity. The percentage of the (–)-(*R*)-enantiomer produced by these subjects ranged from 5 to 36, i.e. the enantiomeric excess of (+)-(*S*)-4-hydroxydebrisoquine ranged from 90 to 28. A significant negative correlation was found between the urinary metabolic ratio and the (+)-(*S*) enantiomeric excess, indicating a relationship between the product enantioselectivity and the metabolic capacity of the reaction of 4-hydroxylation. The molecular mechanism accounting for this situation remains to be elucidated but may well prove to be comparable to that described for bufuralol (see above).

7.2 (*S*)-Mephenytoin *para*-hydroxylation

The anticonvulsant mephenytoin (Fig. 25) is another telling example of a drug the metabolism of which is under genetic control in humans. While (*S*)-mephenytoin (t½ ≈ 1 h) is rapidly *para*-hydroxylated, its (*R*)-enantiomer (t½ ≈ 70 h) is slowly N-demethylated to 5-phenyl-5-ethylhydantoin (PEH), an active metabolite which tends to accumulate due to very slow elimination (t½ ≈ 150–200 h) [142]. In approximately 5 % of a Caucasian population, a genetic defect exists which is not related to the debrisoquine phenotype [77, 143]. In such slow metabolizers, (*S*)-mephenytoin is not significantly *para*-hydroxylated, has a t½ of ca 70 h similar to that of its enantiomer, and, like the latter, is now available for N-demethylation to PEH, which is thus formed in amounts approximately threefold higher than in extensive hydroxylators [142].

In vitro studies have shed additional light on the phenomenon. Indeed, liver microsomes from extensive metabolizers *para*-hydroxylated (*S*)-mephenytoin ten times faster than its (*R*)-enantiomer (3.74 ± 1.81 vs 0.37 ± 0.13 nmol (mg protein)⁻¹ h⁻¹). In liver microsomes from poor metabolizers, only (*S*)-mephenytoin hydroxylation was affected, its rate being reduced to that of (*R*)-mephenytoin (0.49–0.52 vs 0.37–0.57, same units) [144].

A drug closely related to mephenytoin is mephobarbital. But while *para*-hydroxylation of the former is substrate enantioselective for the (*S*)--enantiomer, *para*-hydroxylation of the latter is substrate enantioselec-

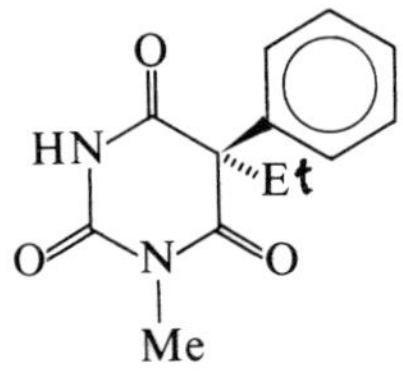

Figure 33
The structure of (*R*)-mephobarbital.

tive for the (*R*)-enantiomer (Fig. 33). Of particular interest in the present context is the fact that the genetic defect in (*S*)-mephenytoin *para*-hydroxylation cosegregates with poor (*R*)-mephobarbital *para*-hydroxylation [145]. Comparing Figures 25 and 33 shows the close three-dimensional similarity of the two enantiomeric substrates and suggests an identical mode of binding to the isocytochrome P450 absent or defective in mephenytoin poor metabolizers.

8   Pharmacodynamic consequences of enantioselective drug metabolism

The pharmacodynamic (pharmacological and toxicological) consequences of drug metabolism are documented in countless publications. Thus, a drug may be eliminated mainly by biotransformation into inactive metabolites, or it may yield active metabolites responsible for some or all therapeutic effects. Similarly, toxic metabolites may be generated and detoxified by further biotransformation. As amply documented in this review, stereoselectivity in drug metabolism is the rule and not the exception, and it is only logical that stereoisomeric factors should contribute an additional dimension in the already complex interplay between pharmacokinetic and pharmacodynamic events. Indeed, beyond rather straightforward consequences such as stereoselective activation or inactivation, metabolic interactions between stereoisomers may also operate with beneficial or detrimental therapeutic consequences.

A particularly complex and as yet partly assessed example is that of propranolol, whose *β*-adrenoceptor antagonistic activity resides in the (−)-(*S*)-enantiomer. Some impacts of its substrate enantioselective biotransformation on activity have been briefly reviewed [146]. Another example combining extensive biotransformation and significant sub-

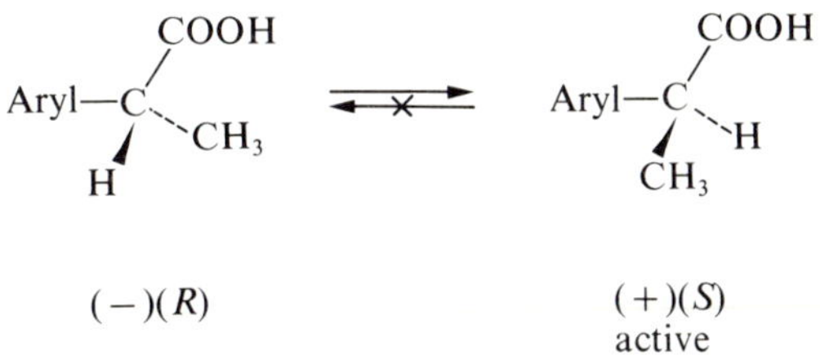

$$\text{Aryl}-\underset{\underset{\text{H}}{|}}{\overset{\overset{\text{COOH}}{|}}{\text{C}}}-\text{CH}_3 \quad \xrightleftharpoons{\;\times\;} \quad \text{Aryl}-\underset{\underset{\text{CH}_3}{|}}{\overset{\overset{\text{COOH}}{|}}{\text{C}}}-\text{H}$$

$(-)(R)$ $(+)(S)$
active

Figure 34

The unidirectional chiral inversion of profens.

strate-product stereoselectivity with the production of active metabolites is that of the anorectic drug amfepramone (diethylpropion) [147]. In the present Section, another class of drugs will be discussed, namely the anti-inflammatory 2-arylpropionic acids or profens.

8.1   The case of 2-arylpropionic acids

Profens (2-arylpropionic acids) are a major group of antiinflammatory drugs. These compounds are chiral due to the presence of an asymmetric carbon atom alpha to the carboxyl function. Their in vitro activity resides practically exclusively in the $(+)$-(S)-enantiomers [148], yet all profens except naproxen are marketed as racemates. These drugs also exhibit enantioselective pharmacokinetic behaviour [149–152]. Thus, 2-phenylpropionic acid (hydratropic acid) [153] and ketoprofen [154] bind stereoselectively to serum albumin, while ester glucuronide formation occurs stereoselectively for, e.g., 2-phenylpropionic acid [155] and ibuprofen [156].

The most intriguing aspect in the biotransformation of profens is their chiral inversion of configuration from the inactive $(-)$-(R)- to the active $(+)$-(S)-isomer (Fig. 34). The reaction is unidirectional in that no (R)-enantiomer is formed from the (S)-enantiomer; a few recent findings however suggest that this unidirectionality may not be absolute [152]. Thus, (S)-2-phenylpropionic acid yields small but genuine amounts of its (R)-enantiomer in rats [157]. The reaction has been documented in vivo for a number of profen analogues in a number of animal species, as comprehensively reviewed by Hutt and Caldwell [149]. A fit illustration of the reaction is provided by benoxaprofen, a drug withdrawn from the market due to its suspected toxicity. As shown in Figure 35, there is a fast enrichment of the (S)-form in rats, reaching al-

$(+)(S)$ in plasma as % of total

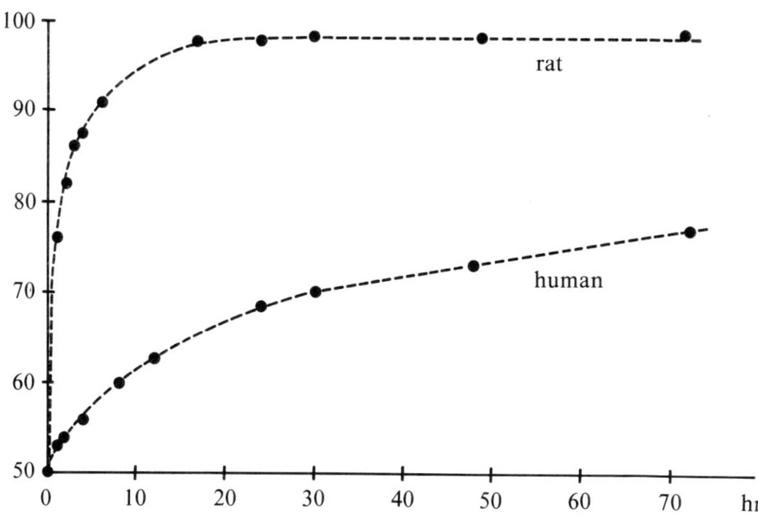

Figure 35
Plasma levels of $(+)$-(S)-benoxaprofen in rats and humans, as percentage of total, following administration of the racemate [158].

most 100 % of the total drug in plasma after 1 day; the reaction is much slower but nevertheless significant in humans [158]. More recent in vivo evidence has been reported in a number of studies [153, 155–157, 159–162]. While the rate of unidirectional inversion appears to depend on both the animal species [157] and the nature of the substrate, no clear generalisations emerge from available data. Of interest in this context is the fact that no inversion occurs for 2-phenylpropionic in the mouse [157] and for tiaprofenic acid in humans [163].
The reaction of chiral inversion of profens is a clear case of metabolic activation, i.e. the $(-)$-(R)-form is a prodrug of its enantiomer. Thus, for the profens administered as racemates, both intensity and duration of the therapeutic effect should depend not only on the total clearance of the active enantiomer, but also on the rate of chiral inversion.
The enantiospecific inversion of profens may have yet other pharmacodynamic effects which appear to result from the molecular mechanism of the reaction. A growing body of converging evidence [149, 152] indicates that the most likely mechanism involves activation of the 2-arylpropionic acid by formation of the acyl-coenzyme A thioester, as originally suggested by Wechter et al. [164] and Nakamura et al. [165].

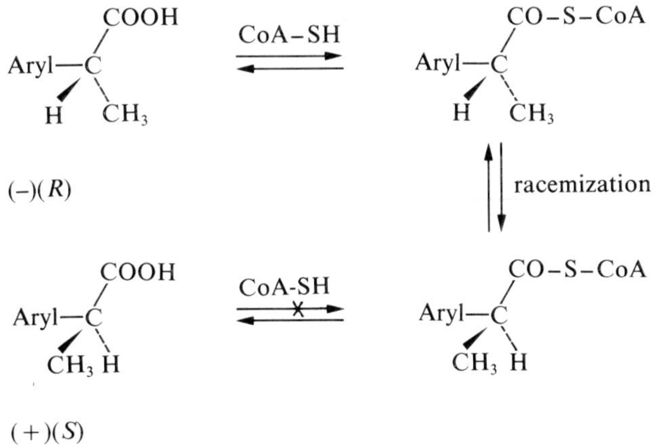

Figure 36
The postulated mechanism of the unidirectional chiral inversion of profens.

The unidirectional chiral conversion is explained in terms of the acyl-CoA formation being substrate stereospecific for the $(-)$-(R)-enantiomer, while the $(+)$-(S)-enantiomer is not a substrate for the acyl-CoA synthetase(s) catalyzing this reaction. The (R)-thioester may then racemize either enzymatically due to the action of an isomerase, or more likely chemically [166] due to the increased acidity of the methine proton. The last step in this pathway is the hydrolysis (enzymatic and/or non-enzymatic) of the thioester resulting in an enrichment of the $(+)$-(S)-enantiomer. Figure 36 summarizes the postulated mechanism. The fact that profens, like a variety of xenobiotic acids [169, 170], can enter metabolic routes followed by fatty acids may not be devoid of biological effects. Thus, pirprofen has recently been found to exhibit a concentration-dependent inhibition of mitochondrial β-oxidation of fatty acids, a finding postulated to account for the microvesicular steatosis produced by this drug [167]. Ibuprofen and tiaprofenic acid similarly inhibited β-oxidation, but it is significant that (S)-naproxen displayed no such effect [167].

Another cause for concern comes from the elegant work of Williams et al. [168]. Following administration of (R)-ibuprofen to rats, these authors found in adipose tissues both the (R)- and (S)-ibuprofenoyl moieties incorporated in hybrid glycerides; the administration of (S)-ibuprofen did not lead to the formation of such conjugates. These two sets of findings appear of clear toxicological significance, suggesting that the coenzyme A conjugates of (R)-profens represent a pivotal

point not only in their pharmacological activation, but also in toxica-
tion pathways.

8.2   Should only eutomers be used in therapy?

The above section has illustrated some pharmacological and toxicological consequences of stereoselective drug metabolism. As for sections 4.2 and 6.1, they have demonstrated that due to various interactions the pharmacokinetic behaviour of mixtures of stereoisomers (e.g.
racemates) is not always the simple addition of the behaviour of individual stereoisomers [12, 29].
We believe that these facts must be fully acknowledged when attempting to answer the question "Should only eutomers be used in therapy?". Indeed, many drugs are marketed as racemates while others are
pure enantiomers, and it is fashionable in some circles to consider
racemates as drugs containing 50 % impurity [171, 172]. A blunt statement of this sort is more caricatural than scientific, and some nuance is
certainly in order. Enforcing compulsory resolution of all chiral drugs

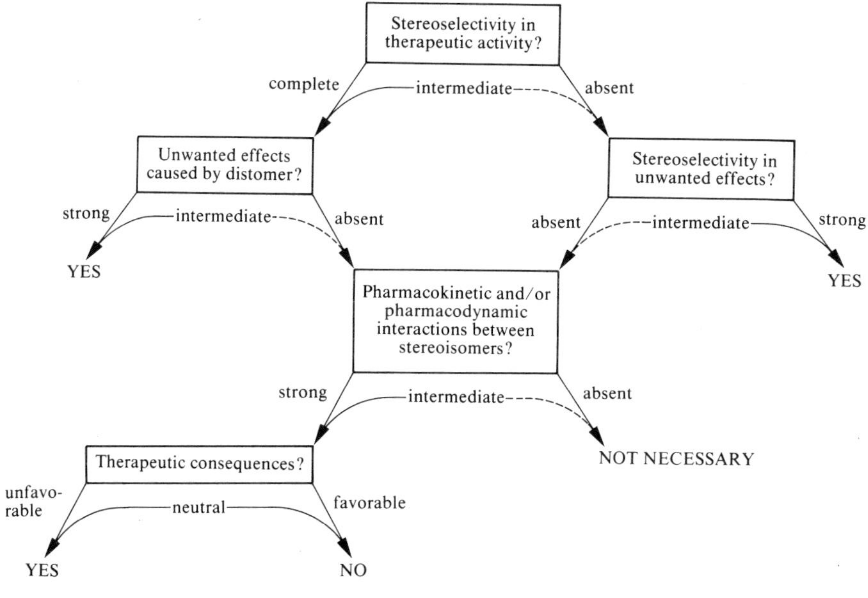

Figure 37
Proposed logical scheme addressing the question "Should chiral drugs (and more generally drugs existing in stereoisomeric forms) be used in therapy as the pure eutomer?"
[146] (reproduced with the permission of Pergamon Press Ltd, Oxford).

would increase their costs severalfold. While certainly desirable in many cases such as profens, a rigid legislation of this sort may also be useless or even detrimental in other cases. In an effort to rationally approach this important problem, a logical scheme has been proposed addressing the above question (Fig. 37) [146, 147]. A number of questions should be considered in turn, in particular unwanted effects and interactions (pharmacokinetic and pharmacodynamic) between the stereoisomers. One difficulty of such an approach lies with the limitations of bivalued (Aristotelian) logic, which ignores intermediate cases. Many factors, implicit and explicit, may have to be taken into account before reaching "yes" or "no" answers. Such factors must be primarily scientific, and they require a wealth of pharmacokinetic and pharmacodynamic data which is seldomly available. This is a second, and sometimes prohibitive, difficulty of the approach. It follows that in a number of cases, only preliminary or arbitrary decisions can be taken.

9   Conclusion

Enantiomers may be seen as identical or different depending on the probe used to study them. When enantiomers interact with a probe or an environment that is achiral, they behave identically and are seen as identical. When observed with a chiral probe or in a chiral environment, their behaviour is different as a matter of principle. Biological environments are made of various chiral constituents, and their differential interactions with chiral (and more generally stereoisomeric) xenobiotics are the key to biological stereoselectivity.

The main goal of this review has been to illustrate, clarify and interpret stereoselectivity in drug metabolism. At the biochemical level, stereoselectivity offers an insight into the enzymatic steps of recognition and catalysis. Thus, the mechanisms by which enzymes recognize stereoisomeric substrates and stereoheterotopic groups lead to general conclusions regarding the binding of ligands and substrates to enzymes. Similarly, essential aspects of reaction mechanisms have been unveiled, such that stereoelectronic arguments have entered the conceptual framework of biochemists and enzymologists.

At the pharmacological and clinical levels, pharmacokinetic and pharmacodynamic events have long since been recognized as essential and interdependent determinants of the therapeutic response, and for each stereoselectivity is the rule. In this context, stereoselective drug metabolism ceased many years ago to be regarded as an academic curiosity. Increasing attention is now being paid to the differential pharmacokinetic behaviour of stereoisomers, which clinical pharmacologists, following in the steps of biochemists and medicinal chemists, are learning to regard as distinct compounds.

We conclude that stereoselectivity in general, and stereoselective drug metabolism in particular, have gained significance at all levels of drug research, the implicit guiding thread of the present review. Advances in drug research and in drug metabolism are and will remain intimately linked, and we can look forward to fascinating factual findings and conceptual breakthroughs.

References

1   S. F. Mason: Trends Pharmac. Sci. *7*, 20 (1986).
2   A. J. MacDermott: Nature *323*, 16 (1986).
3   S. F. Mason: Nouv. J. Chim. *10*, 739 (1986).
4   M. Gardner: The Ambidextrous Universe. Penguin Books, New York 1982.
5   H. R. Pagels: The Cosmic Code. Penguin Books, New York 1986.
6   B. Testa: Trends Pharmac. Sci. *8*, 381 (1987).
7   R. L. Smith: ISSX Newsletter *5*, 2 (1986).
8   C. C. Pfeiffer: Science *124*, 29 (1956).
9   P. A. Lehmann, J. F. Rodrigues de Miranda and E. J. Ariëns: in Progress in Drug Research, vol. 20 (E. Jucker, ed.), Birkhäuser, Basel 1978, p. 101 and 126.
10   P. A. Lehmann: Trends Pharmac. Sci. *7*, 281 (1986).
11   M. Simonyi: Med. Res. Rev. *4*, 359 (1984).
12   K. Williams and E. Lee: Drugs *30*, 333 (1985).
13   D. E. Drayer: in Drug-Protein Binding (M. M. Reidenberg and S. Erill, eds), Praeger, New York 1986, p. 332.
14   M. Simonyi, I. Fitos and J. Visy: Trends Pharmac. Sci. *7*, 112 (1986).
15   B. Testa: Principles of Organic Stereochemistry, Dekker, New York 1979.
16   B. Testa: in Stereochemistry (Ch. Tamm, ed.), Elsevier, Amsterdam 1982, p. 1.
17   R. S. Cahn, C. K. Ingold and V. Prelog: Experientia *12*, 81 (1956). Angew. Chem. Int. Ed. *5*, 385 (1966).
18   M. Simonyi: in Pharmacology (M. J. Rand and C. Raper, eds). Elsevier, Amsterdam 1987, p. 807.
19   B. Testa: in Advances in Drug Research, vol. 13 (B. Testa, ed.), Academic Press London, 1984, p. 1.
20   B. Testa: Drug Metab. Rev. *13*, 25 (1982).
21   B. Testa and P. Jenner: in Concepts in Drug Metabolism, part A (P. Jenner and B. Testa, Eds), Dekker, New York 1980, p. 53.
22   K. H. Overton: Chem. Soc. Rev. *8*, 447 (1979).
23   V. Prelog: Ind. Chem. Belge *11*, 1309 (1962).
24   P. Jenner and B. Testa: Drug Metab. Rev. *2*, 117 (1973).
25   L. K. Low and N. Castagnoli Jr.: in Annual Reports in Medicinal Chemistry (F. H. Clarke, ed.), Academic Press, New York 1978, p. 304.

26 B. Testa and W. F. Trager: in Topics in Pharmaceutical Sciences 1983 (D. D. Breimer and P. Speiser, eds), Elsevier, Amsterdam 1983, p. 99.

27 N. P. E. Vermeulen and D. D. Breimer: in Stereochemistry and Biological Activity of Drugs (E. J. Ariëns, W. Soudijn and P. B. M. W. M. Timmermans, eds), Blackwell, Oxford 1983, p. 33.

28 W. F. Trager and B. Testa: in Drug Metabolism and Disposition: Considerations in Clinical Pharmacology (G. R. Wilkinson and M. D. Rawlins, eds), MTP Press, Lancaster 1985, p. 35.

29 Th. Walle and U. K. Walle: Trends Pharmac. Sci. *7*, 155 (1986).

30 A. G. Zacchei, M. R. Dobrinska, T. I. Wishousky, K. C. Kwan and S. D. White: Drug Metab. Disp. *10*, (1982).

31 A. Rane, B. Gawronska-Szklarz and J. O. Svensson: J. Pharmac. exp. Ther. *234*, 761 (1985).

32 S. F. Sisenwine, C. O. Tio, F. V. Hadley, A. L. Liu, H. B. Kimmel and H. W. Ruelius: Drug Metab. Disp. *10*, 605 (1982).

33 W. Dieterle, J. W. Faigle, H. P. Kriemler and T. Winkler: Xenobiotica *14*, 311 (1984).

34 J. H. Poupaert, R. Cavalier, M. H. Claessen and P. A. Dumont: J. med. Chem. *18*, 1268 (1975).

35 T. C. Butler, K. H. Dudley, D. Johnson and S. B. Roberts: J. Pharmac. exp. Ther. *199*, 82 (1976).

36 J. H. Maguire and J. S. McClanahan: in Biological Reactive Intermediates III (J. J. Kocsis, D. J. Jollow, C. M. Witmer, J. O. Nelson and R. Snyder, eds), Plenum, New York 1986, p. 897.

37 R. H. Reuning, T. A. Shepard, B. E. Morrison and H. N. Bockbrader: Drug Metab. Disp. *13*, 51 (1985).

38 D. B. Prelusky, R. T. Coutts and F. M. Pasutto: J. Pharm. Sci. *71*, 1390 (1982).

39 P. Baumann and V. Prelog: Helv. Chim. Acta *41*, 3262 (1958).

40 J. W. Gorrod and M. Christou: Xenobiotica *16*, 575 (1986).

41 W. H. Soine, P. J. Soine, B. W. Overton and L. K. Garrettson: Drug Metab. Disp. *14*, 619 (1986).

42 K. H. Palmer, M. S. Fowler, M. E. Wall, L. S. Rhodes, W. J. Waddell and B. Baggett: J. Pharmac. exp. Ther. *170*, 355 (1969).

43 K. H. Palmer, M. S. Fowler and M. E. Wall: J. Pharmac. exp. Ther. *175*, 38 (1970).

44 J. L. Holtzman and J. A. Thompson: Drug Metab. Disp. *3*, 113 (1975).

45 D. D. Christ and T. Walle: Drug Metab. Disp. *13*, 380 (1985).

46 J. H. Maguire, T. C. Butler and K. H. Dudley: J. med. Chem. *21*, 1294 (1978).

47 P. Jenner, J. W. Gorrod and A. H. Beckett: Xenobiotica *3*, 563 & 573 (1973).

48 B. Testa, P. Jenner, A. H. Beckett and J. W. Gorrod: Xenobiotica *6*, 553 (1976).

49 G. S. Yost and B. L. Finley: Drug Metab. Disp. *13*, 5 (1985).

50 J. M. Mathews and J. R. Bend: Molec. Pharmac. *30*, 25 (1986).

51 L. R. Pohl, W. R. Porter, W. F. Trager, M. J. Fasco and J. W. Fenton II: Biochem. Pharmac. *26*, 109 (1977).

52 L. S. Kaminsky, M. J. Fasco and F. P. Guengerich: J. biol. Chem. *255*, 85 (1980).

53 C. Harris, R. M. Philpot, O. Hernandez and J. R. Bend: J. Pharmac. exp. Ther. *236*, 144 (1986).

54 D. J. Waxman, D. R. Light and C. Walsh: Biochemistry *21*, 2499 (1982).

55 D. R. Doerge and M. D. Corbett: Biochem. Pharmac. *33*, 3615 (1984).

56 P. R. Ortiz de Montellano, K. L. Kunze and H. S. Beilan: J. biol. Chem. *258*, 45 (1983).

57 F. M. Menger: Acc. chem. Res. *18*, 128 (1985).

58   D. G. Gorenstein: Chem. Rev. *87*, 1047 (1987).
59   K. P. Nambiar, D. M. Stauffer, P. A. Kolodziej and S. A. Benner: J. Am. chem. Soc. *105*, 5886 (1983).
60   H. Rollema, D. Masterbroek, H. Wikström and A. S. Horn: J. Pharm. Pharmac. *37*, 314 (1985).
61   U. B. Paulsen-Sörman, K. H. Jönsson and B. G. A. Lindeke: J. med. Chem. *27*, 342 (1984).
62   Y. Kurono, T. Kondo and K. Ikeda: Archs Biochem. Biophys. *227*, 339 (1983).
63   T. Watabe, N. Ozawa and A. Hiratsuka: Biochem. Pharmac. *32*, 777 (1983).
64   P. J. van Bladeren, J. M. Sayer, D. E. Ryan, P. E. Thomas, W. Lewin and D. M. Jerina: J. biol. Chem. *260*, 10226 (1985).
65   R. H. Gordonsmith, M. J. Raxworthy and P. A. Gulliver: Biochem. Pharmac. *31*, 433 (1982).
66   H. Rollema, D. Mastebroek, H. Wikström, K. Svensson, A. Carlsson and S. Sundell: J. med. Chem. *29*, 1889 (1986).
67   D. A. Lewis and R. N. Armstrong: Biochemistry *22*, 6297 (1983).
68   D. Cobb, C. Boehlert, D. Lewis and R. N. Armstrong: Biochemistry *22*, 805 (1983).
69   L. D. Heimark and W. F. Trager: J. med. Chem. *27*, 1092 (1984).
70   L. D. Heimark and W. F. Trager: J. med. Chem. *28*, 503 (1985).
71   M. M. Ames and S. K. Frank: Biochem. Pharmac. *31*, 5 (1982).
72   K. C. Cundy, C. S. Godin and P. A. Crooks: Biochem. Pharmac. *34*, 281 (1985).
73   K. C. Cundy, P. A. Crooks and C. S. Godin: Biochem. biophys. Res. Comm. *128*, 312 (1985).
74   R. F. Novak and K. P. Vatsis: Molec. Pharmac. *21*, 701 (1982).
75   R. E. White, M. B. McCarthy, K. D. Egeberg and S. G. Sligar: Archs Biochem. Biophys. *228*, 493 (1984).
76   A. Küpfer, P. V. Desmond, S. Schenker and R. A. Branch: J. Pharmac. exp. Ther. *221*, 590 (1982).
77   W. Kalow: Xenobiotica *16*, 379 (1986).
78   S. Fritz, W. Lindner, I. Roots, B. M. Frey and A. Küpfer: J. Pharmac. exp. Ther. *241*, 615 (1987).
79   R. E. White, J. P. Miller, L. V. Favreau and A. Bhattacharyya: J. Am. chem. Soc. *108*, 6024 (1986).
80   K. Sugiyama and W. F. Trager: Biochemistry *25*, 7336 (1986).
81   R. Ichinose and N. Kurihara: Biochem. Pharmac. *36*, 3751 (1987).
82   R. N. Armstrong, D. A. Lewis, H. L. Ammon and S. M. Prasad: J. Am. chem. Soc. *107*, 1057 (1985).
83   H. Yagi and D. M. Jerina: J. Am. chem. Soc. *104*, 4026 (1982).
84   P. J. van Bladeren, R. N. Armstrong, D. Cobb, D. R. Thakker, D. E. Ryan, P. E. Thomas, N. D. Sharma, D. R. Boyd, W. Levin and D. M. Jerina: Biochem. biophys. Res. Comm. *106*, 602 (1982).
85   K. P. Vyas, P. J. van Bladeren, D. R. Thakker, H. Yagi, J. M. Sayer, W. Levin and D. M. Jerina: Molec. Pharmac. *24*, 115 (1983).
86   P. J. van Bladeren, K. P. Vyas, J. M. Sayer, D. E. Ryan, P. E. Thomas, W. Levin and D. M. Jerina: J. biol. Chem. *259*, 8966 (1984).
87   S. K. Yang, M. Mushtaq and P. L. Chiu: in Polycyclic Hydrocarbons and Carcinogenesis (R. G. Harvey, ed.), American Chemical Society, Washington, DC, 1985, p. 19.
88   S. K. Wang and Z.-P. Bao: Molec. Pharmac. *32*, 73 (1987).
89   G. Belluci, G. Berti, R. Bianchini, P. Cetera and E. Mastrorilli: J. org. Chem. *47*, 3105 (1982).
90   H. Van de Waterbeemd, B. Testa and J. Caldwell: J. Pharm. Pharmac. *38*, 14 (1986).

91 L. A. Dostal, A. Aitio, C. Harris, A. V. Bhatia, O. Hernandez and J. Bend: Drug Metab. Disp. *14*, 303 (1986).
92 G. M. Whitesides and C. H. Wong: Angew. Chem. Int. Ed. *24*, 617 (1985).
93 J. B. Jones: Tetrahedron *42*, 3351 (1986).
94 S. Sicsic: La Recherche *18*, 626 (1987).
95 B. Cambou and A. M. Klibanov: J. Am. chem. Soc. *106*, 2687 (1984).
96 K. Adachi, S. Kobayashi and M. Ohno: Chimia *40*, 311 (1986).
97 Q.-M. Gu, C.-S. Chen and C. J. Sih: Tetrahedron Lett. *27*, 1763 (1986).
98 D. Seebach and M. Eberle: Chimia *40*, 315 (1986).
99 J. K. Whittesell and R. M. Lawrence: Chimia *40*, 318 (1986).
100 C.-S. Chen, S.-H. Wu, G. Girdaukas and C. J. Sih: J. Am. chem. Soc. *109*, 2812 (1987).
101 R. Dernoncour and R. Azerad: Tetrahedron Lett. *28*, 4661 (1987).
102 G. Gil, E. Ferre, A. Meou, J. Le Petit and C. Triantaphylides: Tetrahedron Lett. *28*, 1647 (1987).
103 V. Kerscher and W. Kreiser: Tetrahedron Lett. *28*, 531 (1987).
104 M. Luyten, S. Müller, B. Herzog and R. Keese: Helv. Chim. Acta *70*, 1250 (1987).
105 A. L. Margolin, J. Y. Crenne and A. M. Klibanov: Tetrahedron Lett. *28*, 1607 (1987).
106 P. Mohr, L. Rösslein and C. Tamm: Helv. Chim. Acta *70*, 142 (1987).
107 Y. Morimoto, Y. Terao and K. Achiwa: Chem. Pharm. Bull. *35*, 2266 (1987).
108 Th. Oberhauser, M. Bodenteich, K. Faber, G. Penn and H. Griengl: Tetrahedron *43*, 3931 (1987).
109 F. Björkling, J. Boutelje, S. Gatenbeck, K. Hult, T. Norin and P. Szmulik: Bioorg. Chem. *14*, 176 (1986).
110 L. K. P. Lam, R. A. H. F. Hui and J. B. Jones: J. Org. Chem. *51*, 2047 (1986).
111 G. Sabbioni and J. B. Jones: J. Org. Chem. *52*, 4565 (1987).
112 A. J. Russell and A. R. Fersht: Nature *328*, 496 (1987).
113 A. Tramontano, K. D. Janda and R. A. Lerner: Science *234*, 1566 (1986).
114 S. J. Pollack, J. W. Jacobs and P. G. Schultz: Science *234*, 1570 (1986).
115 A. D. Napper, S. J. Benkovic, A. Tramontano and R. A. Lerner: Science *237*, 1041 (1987).
116 C. F. George, T. Fenyvesi, M. E. Conolly and C. T. Dollery: Eur. J. clin. Pharmac. *4*, 74 (1972).
117 K. Kawashima, A. Levy and S. Spector: J. Pharmac. exp. Ther. *196*, 517 (1976).
118 M. J. Cooper and M. W. Anders: Life Sci. *15*, 1665 (1974).
119 P. J. Murphy, R. C. Nickander, G. M. Bellamy and W. L. Kurz: J. Pharmac. exp. Ther. *199*, 415 (1976).
120 M. van der Graaff, P. H. Hofman, D. D. Breimer, N. P. E. Vermeulen, J. Knabe and L. Schamber: Biomed. Mass Spectrom. *12*, 464 (1985).
121 K. M. Giacomini, W. L. Nelson, R. A. Pershe, L. Valdivieso, K. Turner-Tamiyasu and T. F. Blaschke: J. Pharmacokinet. Biopharm. *14*, 335 (1986).
122 P. Rauch, M. Püttmann, F. Oesch, Y. Okamoto and L. W. Robertson: Biochem. Pharmac. *36*, 4355 (1987).
123 S. Fournel, J. Caldwell, J. Magdalou and G. Siest: Biochim. Biophys. Acta *882*, 469 (1986).
124 I. A. Choonara, S. Cholerton, B. P. Haynes, A. M. Breckenridge and B. K. Park: Br. J. clin. Pharmac. *21*, 271 (1986).
125 S. Toon, K. J. Hopkins, F. M. Garstang, L. Aarons, A. Sedman and M. Rowland: Clin. Pharmac. Ther. *42*, 33 (1987).
126 R. J. Lewis, W. F. Trager, K. K. Chan, A. Breckenridge, M. L'Orme, M. Rowland and W. Schary: F. clin. Invest. *53*, 1607 (1974).
127 R. A. O'Reilly, W. F. Trager, C. H. Motely and W. Howald: J. clin. Invest. *65*, 746 (1980).

128   L. D. Heimark, S. Toon, M. Gibaldi, W. F. Trager, R. A. O'Reilly and D. A. Goulart: Clin. Pharmac. Ther. *42*, 312 (1987).

129   R. A. O'Reilly: N. Engl. J. Med. *295*, 354 (1976).

130   J. Halpert, C. Balfour, N. E. Miller, E. T. Morgan, D. Dunbar and L. S. Kaminsky: Molec. Pharmac. *28*, 290 (1985).

131   R. A. O'Reilly, W. F. Trager, A. E. Rettie and D. A. Goulart: Clin. Pharmac. Ther. *42*, 290 (1987).

132   R. L. Smith: Arch. Toxic. *Suppl. 9*, 138 (1986).

133   F. P. Guengerich, D. R. Umbenhauer, P. F. Churchill, P. H. Beaune, R. Böcker, R. G. Knodell, M. V. Martin and R. S. Lloyd: Xenobiotica *17*, 311 (1987).

134   U. A. Meyer, J. Gut, T. Kronbach, C. Skoda, U. T. Meier and T. Catin: Xenobiotica *16*, 449 (1986).

135   J. Gut, T. Catin, P. Dayer, T. Kronbach, U. Zanger and U. A. Meyer: J. biol. Chem. *261*, 11734 (1986).

136   P. Dayer, T. Kronbach, M. Eichelbaum and U. A. Meyer: Biochem. Pharmac. *36*, 4145 (1987).

137   U. M. Zanger, F. Vilbois and U. A. Meyer: Abstracts of 10th International Congress of Pharmacology, Sydney, 23–28 August 1987, Abstr. P95.

138   M. S. Lennard, G. T. Tucker, J. H. Silas, S. Freestone, L. E. Ramsay and H. F. Woods: Clin. Pharmac. Ther. *34*, 732 (1983).

139   M. S. Lennard, G. T. Tucker, J. H. Silas and H. F. Woods: Xenobiotica *16*, 435 (1986).

140   C. O. Meese and M. Eichelbaum: N. S. Arch. Pharmac. Suppl. *332*, R95 (1986).

141   M. Eichelbaum: Biochem. Pharmac. *37*, 93 (1988).

142   P. J. Wedlund, W. S. Aslanian, E. Jacqz, C. B. McAllister, R. A. Branch and G. R. Wilkinson: J. Pharmac. exp. Ther. *234*, 662 (1985).

143   A. Küpfer, R. Patwardhan, S. Ward, S. Schenker, R. Preisig and R. A. Branch: J. Pharmac. exp. Ther. *230*, 28 (1984).

144   U. T. Meier, P. Dayer, P. J. Malè, T. Kronbach and U. A. Meyer: Clin. Pharmac. Ther. *38*, 488 (1985).

145   A. Küpfer and R. A. Branch: Clin. Pharmac. Ther. *38*, 414 (1985).

146   B. Testa: Biochem. Pharmac. *37*, 85 (1988).

147   B. Testa: Trends Pharmac. Sci. *7*, 60 (1986).

148   T. Y. Shen: in Burger's Medicinal Chemistry, 4th Edn, Part III (M. E. Wolff, ed.), Wiley, New York 1981, p. 1205.

149   A. J. Hutt and J. Caldwell: J. Pharm. Pharmac. *35*, 693 (1983).

150   A. J. Hutt and J. Caldwell: Clin. Pharmacokin. *9*, 371 (1984).

151   A. J. Hutt and J. Caldwell: J. Pharm. Pharmac. *37*, 288 (1985).

152   J. Caldwell, A. J. Hutt and S. Fournel-Gigleux: Biochem. Pharmac. *37*, 105 (1988).

153   M. E. Jones, B. C. Sallustio, Y. J. Purdie and P. J. Meffin: J. Pharmac. exp. Ther. *238*, 288 (1986).

154   S. Rendic, T. Alebic-Kolbah, F. Kajfez and V. Sunjic: Farmaco Ed. Sci. *35*, 51 (1980).

155   T. Yamaguchi and Y. Nakamura: Drug Metab. Disp. *13*, 614 (1985).

156   E. J. D. Lee, K. Williams, R. Day, G. Graham and D. Champion: Br. J. clin. Pharmac. *19*, 669 (1985).

157   S. Fournel and J. Caldwell: Biochem. Pharmacol. *35*, 4153 (1986).

158   R. G. Simmonds, T. J. Woodage, S. M. Duff and J. N. Green: Eur. J. Drug Metab. Pharmacokin. *5*, 169 (1980).

159   A. Rubin, M. P. Knadler, P. P. K. Ho, L. D. Bechtol and R. L. Wolen: J. Pharm. Sci. *74*, 82 (1985).

160   P. J. Meffin, B. C. Sallustio, Y. J. Purdie and M. E. Jones: J. Pharmac. exp. Ther. *238*, 280 (1986).

161   A. Abas and P. J. Meffin: J. Pharmac. exp. Ther. *240*, 637 (1987).
162   P. J. Hayball and P. J. Meffin: J. Pharmac. exp. Ther. *240*, 631 (1987).
163   N. N. Singh, F. Jamali, F. M. Pasutto, A. S. Russell, R. T. Coutts and K. S. Drader: J. Pharm. Sci. *75*, 439 (1986).
164   W. J. Wechter, D. G. Loughhead, R. J. Reischer, G. J. Vangiessen and D. G. Kaiser: Biochem. biophys. Res. Comm. *61*, 833 (1974).
165   Y. Nakamura, T. Yamaguchi, S. Takahashi, S. Hashimoto, K. Iwatani and Y. Nakagawa: J. Pharmacobio-Dyn. *4*, s-1 (1981).
166   J. M. Mayer, C. Bartolucci, J. M. Maître and B. Testa: Xenobiotica, *18*, 533 (1988).
167   J. Geneve, B. Hayat-Bonan, G. Labbe, C. Degott, P. Letteron, E. Freneaux, T. Le Dinh, D. Larrey and D. Pessayre: J. Pharmac. exp. Ther. *242*, 1133 (1987).
168   K. Williams, R. Day, R. Knihinicki and A. Duffield: Biochem. Pharmac. *35*, 3403 (1986).
169   J. Caldwell and M. V. Marsh: Biochem. Pharmac. *32*, 1667 (1983).
170   J. Caldwell: Biochem. Soc. Trans. *12*, 9 (1984).
171   E. J. Ariëns: Eur. J. clin. Pharmac. *26*, 663 (1984).
172   E. J. Ariëns: Med. Res. Rev. *6*, 451 (1986).

Immunostimulation with peptidoglycan or its synthetic derivatives

By Duncan E. S. Stewart-Tull

Department of Microbiology, Alexander Stone Building, University of Glasgow, Garscube Estate, Bearsden, Glasgow G 61 1 QH, Scotland

1   Introduction

For a time during the 1960s the emphasis shifted from prevention to treatment of infectious diseases. The resurgence of interest in vaccine development arose because of the realization that continued antibiotic therapy brought its own problems. For example, a leprosy patient treated with rifampicin for long periods gradually developed a secondary resistance to the drug. This problem was exacerbated when a close relative contracted a leprosy infection which did not respond to rifampicin treatment; an example of primary drug resistance [1]. Subsequently, conventional living attenuated or crude, heat-killed, whole-cell vaccines became a liability because of the risks of litigation, due to the reported harmful effects of vaccine whether unfounded or not. Much work has been done to improve the existing range of vaccines with particular importance placed on the search for candidate antigens required for the stimulation of a protective immune response. The advent of the current era of vaccine research occurred following the publication of the results of the two most significant biological studies since the discovery of DNA structure, namely, those which led to the production of monoclonal antibodies and the techniques of genetic manipulation. The potential exists to sequence genes coding for protective antigens and to produce peptide antigens. A disadvantage is that these synthetic peptides are poor antigens and must either be conjugated to a higher molecular weight carrier or be used in conjunction with an immunostimulatory agent (adjuvant). Vane and Cuatrecasas [2] pointed out that progress in the development of single protein or synthetic peptide vaccine relies on the continued search for improved adjuvants (immunopotentiators). They also indicated that a greater understanding of the basis of adjuvant action was required.

2   Arguments for and against the use of adjuvants

Differing opinions about the merits or disadvantages of immunopotentiators are frequently expressed, these are often diametrically opposed so confusion is generated. For instance, Paquet, Raines and Brownback [3] concluded "there is certainly a need for the development of more immunopotentiators which could be added to our arsenal of weapons used in the battle against disease". On the other hand, some workers refer to adverse reactions as contra-indications to the

use of immunopotentiators; some patients given oil-adjuvanted cholera or tetanus vaccines experienced reactions at the injection sites and
some of these required surgical removal [4].
In a WHO tetanus field trial in New Guinea good protection was stimulated against neo-natal tetanus after the injection of a Drakeol oil-
adjuvanted vaccine. Similarly, in the Philippines cholera field trials
with an oil-adjuvanted vaccine some reactions were observed in 19/56
patients, these were judged to be acceptable especially as high titres of
vibriocidal antibodies persisted for at least fifteen months after immunization [5,6]. Nevertheless, these field trials are still cited as the definitive results which indicate the inadvisability of using oil-based adjuvanted vaccines in human beings.
These studies were made long before the minimal active component of
the classical Freund's complete adjuvant (FCA) had been elucidated
[7] or the importance of using mineral oils containing long-chain (C_{15}
to C_{19}) hydrocarbons was described [8]. Considerable efforts have been
made during the last decade to construct chemically a range of derivatives of the minimal adjuvant component, muramyl dipeptide (I) but
there is still little evidence of the oil chemists demonstrating such activity in the search for biodegradable oils.
In the scheme of testing FCA or similar mycobacterial adjuvants it was
believed that the formation of an epithelioid-macrophage granuloma
at the site of injection was a prerequisite of good immunopotentiating
activity [9]. Similarly, muramyl dipeptide stimulated the formation of
such a granuloma if it was injected in a water-in-mineral oil emulsion

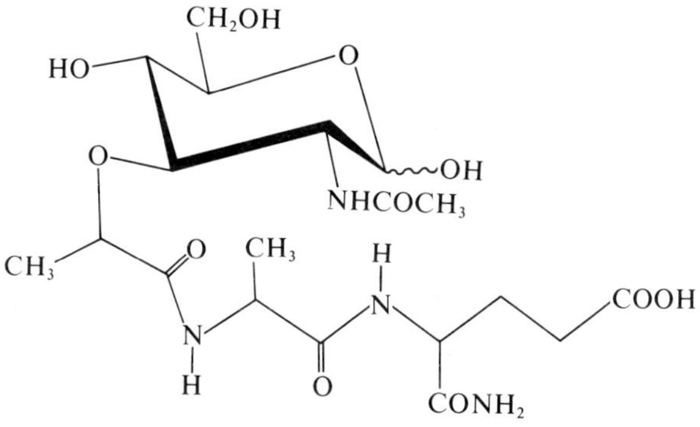

I N-acetyl-muramyl-L-alanyl-D-isoglutamine (M.D.P.)

but not in phosphate-buffered saline [10,11]. However, the 6-0-acyl derivatives (B30–MDP, B46–MDP and BH48–MDP), with the branched acyl group conjugated to muramic acid, stimulated extensive granuloma formation in phosphate-buffered saline whereas the linear acyl derivative (L30–MDP) did not [12]. This study confirmed that the mineral oil of the water-in-oil emulsion or the α-branched fatty acids were responsible for the formation of granulomas by mycobacterial cells or peptidoglycolipid in CFA; this can now be avoided.

It is worthwhile reiterating that no adjuvant researcher would countenance the use of Freund's complete adjuvant in the human being. However, this does not preclude the possibility that Freund's incomplete adjuvant (FIA) might be used in the future if a non-toxic, non-granuloma-inducing oil base could be found.

3   The optimum dose of immunopotentiator

In the laboratory there is little doubt that Freund's adjuvant can be used in a sensible way to stimulate the production of high levels of antibody in animals [13]. It is still surprising how many workers inject an arbitrary dose of adjuvant, subcutaneously and repetitively into animals. In any animal, my own experience is that only *one* injection of FCA should be administered and this by the deep intramuscular (i.m.) route; subcutaneous (s.c.) and cutaneous injections cause the formation of slow-healing ulcers.

One reason for the lack of efficacy associated with some whole cell vaccines is due to "over-adjuvanting". The cholera vaccine contains lipopolysaccharide (LPS), peptidoglycan (PG) and enterotoxin; 2.0 μg cholera enterotoxin was adjuvant-active in mice [14]. Subsequently, it was shown that as little as 1.0–2.0 μg of cholera enterotoxin stimulated an adjuvant response against sheep red blood cells when measured at five days [15]. Theophylline (50.0 μg), an inhibitor of cAMP phosphodiesterase increased the adjuvant activity of 0.01 μg of cholera enterotoxin but the same amount decreased the activity of 0.05 μg of the toxin. Similarly, the whooping cough vaccine contains LPS, PG and pertussis toxin. The latter was adjuvant-active at a dose of 0.1 μg (Dr. A. Robinson, personal communication).

For the inexperienced worker interested in using adjuvants it is useful to include the details for the preparation of adjuvant mixtures. The usual injection mixture contains equal parts of physiological saline containing the protein antigen and the mineral oil/emulsifier (eg. Bayol F 80.0 parts: Arlacel A 20.0 parts by volume) containing dried, heat-killed *Mycobacterium tuberculosis* cells. Care should be exercised when handling the mycobacterial cells because the dried organisms can cause a delayed hypersensitivity reaction in the lungs if the organisms are inhaled! Similarly, there has been a report of accidental inoculation of the laboratory worker with the formation of a local granuloma in the finger [16]. In practice the doses of antigen and immunopotentiator for experimental use are:

Animal species	Protein antigen	*M. tuberculosis* dried organisms	Volume of mixture	Injected dose
	(mg)	(mg)	(ml)	(ml)
Sheep	20.0	2.0	10.0	10.0
Rabbit	10.0	1.0	4.0	2.0
Guinea-pig	10.0	1.0	1.0	0.2

One month after the initial primary injection, if a booster dose is required, this is usually administered by the intraperitoneal (i.p.) route with the same dose of the experimental mixture in FIA, i.e. without the mycobacterial cells. The dose of immunopotentiator to be used is considerably less when water-soluble components or synthetic muramyl dipeptide derivatives are used.

4   Peptidoglycan immunopotentiators in human vaccines

Although one often hears that microbial immunopotentiators cannot be used in human vaccines, in reality this has been done unwittingly for many years [17]. Nearly sixty years have elapsed since the immunogenic properties of diphtheria toxoid precipitated with alum were discovered but "the mineral gels remain to this day the only adjuvants employed in human vaccines" [18]. However, this is not exactly true, the *M. bovis* BCG vaccine was produced in 1921 in France and introduced into the U. K. in the 1950s. The vaccine in current use contains 8.0–26.0 million living and the same number of dead *M. bovis* organisms per millilitre (0.3 mg dry weight). The mycobacterial cell wall accounts for 20.0–35.0 % of the total dry weight and of this amount 50.0–80.0 % is peptidoglycan. Therefore, a conservative estimate of the weight of peptidoglycan in the vaccine would be 30.0–48.0 μg ml^{-1}. Similarly, 34.0–40.0 % of the cell is composed of the mycolic acid-containing wax, peptidoglycolipid; approximately 102.0–120.0 μg ml^{-1}. This means that even at the lower of these two values 3.0–5.0 μg peptidoglycan and 10.0–12.0 μg peptidoglycolipid would be administered in the 0.1 ml dose of BCG vaccine.

Both *M. leprae* and *M. bovis* BCG were injected into patients in an experimental leprosy vaccine [19]; the calculated minimal weights of peptidoglycan and peptidoglycolipid would be 30.0–134 μg and 25.0–165.0 μg, respectively.

Davies [20] reported that BCG caused a low incidence of adverse effects, for example lymphadenitis and a lesion at the site of the injection. His argument about the adverse effects of immunopotentiators in man was endorsed by reference to a very singular study involving the administration to cancer patients of 75.0 mg BCG (1000 times the normal dose) [21]. These patients were given an incredible dose containing approximately 7,500 μg of BCG peptidoglycan! It is hardly surprising that these authors reported an increase in temperature, hepatic dysfunction, hepatic granuloma, adenopathy, splenomegaly and pruritis at the site of injection. The results of this particular study are not relevant to the meaningful debate about the inclusion of immunopotentiators in human vaccines, rather they indicate a totally unacceptable dosage level.

These examples indicate how the impassive attitudes and aversions to the use of immunopotentiators have developed. There are still some criticisms which must be answered before the resistance will begin to disappear, meanwhile adjuvant researchers must continue to build on the experiences gained from the use of accepted vaccines.

5   Criteria for acceptable immunopotentiators

The number and variety of biological responses in the mammal following the injection of an immunopotentiator is considerable. Of these effects some are physiological, pharmacological or immunological. Many conclusions are based on *in vitro* tests so it has become a difficult task to specify the particular biological activities associated with the *in vivo* adjuvant response. At present, there are no guidelines for the testing of suitable candidate immunopotentiators. Until these have been formulated it is my belief that they must initially be tested in parallel with the original FCA or, with aluminium hydroxide in view of the statement by Bomford [18]. There is a wealth of information about these to enable useful comparisons to be made; for FCA see Stewart-Tull [22] and for Al(OH)$_2$ see Bomford [23]. Thus it will be possible to monitor the relative merits of immuno-potentiators as stimulatory agents of a) a humoral response b) a cell-mediated response (delayed-type hypersensitivity) and c) a non-specific resistance to microbial infection. It is important to target one or more of these by parenteral injection or oral administration. For many years it was believed that stimulation of (a) and (b) were inseparable because of the results obtained with FCA in the guinea-pig [24,25] however, it is now known that it is possible to stimulate one without the other. Similarly, the choice of the suspending medium for the antigen and immunopotentiator can be varied (oil or saline) or both can be incorporated into artificial lipid bilayers (liposomes) in order to potentiate the immunogenicity of the vaccine antigen. Investigation of these parameters with natural microbial products will continue to assist the chemists in their search for the ideal synthetic derivative.

Some criteria for the definition of a safe immunopotentiator have been proposed [17, 26–30]. These aims may or may not be attainable but we must address ourselves to the problems in order to convince the various agencies to licence the use of immunopotentiators.

5.1   The substance should be biodegradable

Difficult to achieve with natural microbial products because the adjuvant response is associated with the initial activation of macrophages due to the persistence of the immunopotentiator in the tissues. The D-amino acids and diaminopimelic acid (meso-A$_2$pm) of peptidoglycan are not easily catabolized in the mammalian system, although the liver does contain a D-amino-acid oxidase, [31]. Parant *et al.* [32] showed that intact synthetic ^{14}C-labelled muramyl dipeptide was rapidly excreted in the urine after parenteral injection. Other workers provided evidence that some muramic acid was retained in the tissues. For instance, after the injection of 20.0 mg of streptococcal cell walls into a rat it was possible to detect 30-200 μg muramic acid per gram of spleen tissue for a week after injection; none could be detected in the tissues of normal animals with the alditol acetate derivatives examined by the gas liquid chromatographic – mass spectrophotometric method used [33]. Subsequently, muramic acid was extracted from the brain, liver and kidney tissue of normal rats with 8 % (w/v) trichloro-acetic acid

(TCA) and extraction of the TCA extract with ether [34]. The dried samples were hydrolysed with 6M HCl and muramic acid was purified by cation-exchange chromatography. Thin-layer chromatography of fluorescamine derivatives of muramic acid significantly increased the limit of detection, for example 100–150 pmol muramic acid gm^{-1} normal liver.

The lack of detection methods may explain the reports that MDP is not cleaved by the serum enzyme, N-acetylmuramyl –L–alanine amidase [35]. However, Harrison and Fox [36] used normal serum from Lewis rats to demonstrate by thin layer chromatography that it contained N-acetylmuramyl- L-alanine amidase activity. N-acetylmuramic acid and L-alanine – D-isoglutamine were released, the latter was broken down to constituent amino acids by a peptidase. This process could be important in terminating inflammatory events but it is conceivable that it could be the homeostatic mechanism required to control the levels of peptidoglycan residues in the mammalian system. This work indicates that such synthetic peptidoglycan derivatives are therefore biodegradable.

5.2   The immunopotentiator should be neither toxic nor pyrogenic

A number of natural and synthetic products have been eliminated from the list of possible immunopotentiators because of excessive levels of toxicity or pyrogenicity. Robbins [37] expressed the opinion that "we are prepared to accept no toxicity; any toxicity that we accept is a compromise".* The fear of litigation arising from the use of a new vaccine is apparent but we must continue to search for the adjuvant compromise. There is reasonable agreement that the macrophage is the main target for the immunopotentiator in the initiation of the adjuvant response. A totally innocuous immunopotentiator may lack the irritant nature required to activate these cells. Will the Robbins' compromise depend ultimately on the skill of the chemist so that a particular grouping will leave a *soupçon* of these activities in the synthetic immunopotentiator to stimulate cellular activation?

---

* In Japan, highly skilled chefs are employed to cook *Fugu* (the globe fish) because the liver contains a toxin lethal to man. The art of preparing the dish is to leave a trace of tetrodo toxin sufficient to cause a slight numbing sensation in the mouth – the acceptable compromise?

5.3   An immunopotentiator should not induce a hypersensitivity
           reaction to host tissue nor to the immunopotentiator
5.3.1   Adjuvant arthritis

The injection of FCA, *M. bovis* and *M. smegmatis* cell walls induced a polyarthritis in rats [38–40]. A detailed review of the early studies of the polyarthritogenic activity of FCA is to be found in Stewart-Tull [22]. A mycobacterial glyco-peptide, containing peptidoglycan residues, lacked this activity and we suggested that the length of the glycan (muramic acid – N-acetylglucosamine) chain was important in the induction of this response [41] but see p. 316. This study also showed that an immunopotentiator could be obtained which lacked the polyarthritogenic activity. Koga *et al.* [42] compared these responses with enzyme digests of *Lactobacillus plantarum* and *Staphylococcus epidermidis* cell walls. Peptidoglycan components with more than five disaccharide repeating units induced polyarthritis whereas those with two to three disaccharide units did not. Following intraperitoneal injection of streptococcal cell walls an erosive synovitis of the small joints of the fore and hindlegs was observed [43]. The clinical course of this polyarthritis was similar to that seen in human rheumatoid arthritis – "the degree of clinical arthritis in an individual joint correlated with the amount of cell wall material deposited". Much of the cell wall material was catabolized within three days and only some 10.0 % remained in the tissues examined. During the peak rate of excretion between days 1 and 2 less than 1.0 μg of cell wall material day^{-1} was detected. Cell wall residues persisted in the tissues for at least 65 days especially in the liver (with antigen localization in Kupffer cells and small granuloma formation without necrosis which did not affect liver function) and spleen and lymph nodes (focal necrosis and haemorrhage with infiltration of macrophages containing cell wall antigens). As mentioned previously, by a GLC - mass spectrophotometric method it was possible to detect 0.2–1.0 μg gm^{-1} of tissue from an arthritic joint whereas tissue from control rats had no detectable muramic acid [33]. These studies are relevant to the interpretation of the previous results, namely, that the synthetic N-acetyl-muramyl-L-alanyl-D-isoglutamine (MDP) was inactive in PVG/c strain rats [42]. However, when 0.4 mg MDP was mixed with 0.8 mg type II collagen from human costal cartilage in FIA and injected into one hind footpad of 25 PVG/c rats a severe form of collagen-induced arthritis was stimulated in all of the animals whereas all of the 12 PVG/c rats given 0.4 mg MDP in FIA, containing Arlacel A and liquid paraffin oil (1.5:8.5, v/v), showed no arthritis. Once again it is apparent that the particular oil chosen may be a critical factor since Arlacel A and Bayol F (1.5:8.5, v/v) were used to induce arthritis in WKA rats with MDP [44]. This strain is also less susceptible to poly-arthritis than the PVG/c strain. A further example of variation in the experimental design which can lead to the recording of conflicting results. The use of a single strain or species of animal can also be mis-leading; an aqueous suspension of cell walls or peptidoglycan from a Group A streptococcus and *Lactobacillus casei* was injected into mice and induced acute joint lesions in BALB/c, DBA/1J, (BALB/c × DBA/1J)$_{F1}$ and C3H/He strains but not in C57BL/6, DBA/2 and AKR strains [45]. Cell walls or peptidoglycan from *Nocardia coeliaca, Nocardia canicruria* ATCC 17896, *Nocardia corynebacteroides* ATCC 14898, *L. plantarum* ATCC 8014, *S. epidermidis* ATCC 155 and *Streptococcus mutans* BHT were all effective in BALB/c mice. Another factor to consider in the interpretation of these experimental models is the role of complement, since a C5 deficiency has been reported in DBA/2 and AKR mice [46].

Recent work from Holland and Israel indicates the role of antigenic cross-reactivity between the microbial component and joint cartilage in adjuvant arthritis. It is worth remembering in this context that non-arthritogenic peptidoglycan fragments with (disaccharide)$_{2-3}$ and MDP are haptens. First, an active adjuvant arthritis was induced in Lewis rats with FCA, four weeks later 20µg *M. tuberculosis* antigen or 20 µg chondroitin sulphate or 0.1 ml of a 1/20 dilution of synovial fluid from a human osteoarthritic joint were injected into the rat ear [47]. The percentage increase in ear thickness, as a measure of delayed-type hypersensitivity was *M. tuberculosis* 86.0, chondroitin sulphate 16.0 and synovial fluid 33.0. An arthritogenic T-cell line, designated A2b, was injected into irradiated (750R) Lewis rats and these were also tested for delayed-type hypersensitivity to the same three antigens; the percentage increase in ear thickness was 113.0, 54.0 and 70.0 respectively [48,49]. The antigenic mimicry between *M. tuberculosis* and joint cartilage is not unique, as there may be sharing of antigens between *M. tuberculosis* and neoplastic cells [50].

As mentioned previously, FCA is injected only once to stimulate a good humoral response in animals. Repeated injections of *M. tuberculosis* depress the response and care must be exercised to ensure that different vaccines, containing the same immunopotentiator, administered to one individual do not also cause severe contraindications. For example, the induction of arthritis in cancer patients treated with repeated BCG injections has been reported [51].

The questions posed by the cross-reactivity of *M. tuberculosis* epitopes with other antigens requires further examination. The interaction between mycobacterial peptidoglycan component (glycopeptide) and guinea pig antiovalbumin IgG_1 and IgG_2 immunoglobulins has interested us for many years. After treatment with the cell wall glycopeptide, a potent immunopotentiator, there was an altered electrophoretic mobility of the IgG_2 immunoglobulin associated with an increase in carbohydrate content; 1.3 % to 38.0–40.0 % (w/w). The Fab fragment of the IgG_2 contained 12.25 % and the Fc fragment 25.15 % carbohydrate. This change in the glycosylation of the IgG_2 was reproducible and the bond between immunoglobulin and glycopeptide survived ion-exchange chromatography and gel filtration [52,53]. Immune complexes consisting exclusively of immunoglobulin are found in rheumatoid arthritis [54]; it is interesting that the IgG_2 of FCA-stimulated guinea pigs appeared as a highly aggregated complex under the electron microscope [54]. Although we suggested that the complex formation with the IgG_2 was a consequence of FCA activity a link with immune complex deposition was not apparent. However, the recent association between rheumatoid arthritis, primary osteoarthritis and changes in the glycosylation pattern of total human serum IgG [55] may suggest a link between our results on the glycosylation of IgG and those on FCA-stimulated polyarthritis. Parekh *et al.*'s [55] elegant study, conducted in parallel in Oxford and Tokyo showed that the extent of galactosylation of the IgG decreased from 75 % in normal individuals to 50 % in rheumatoid arthritis and to 65 % in osteoarthritic patients.

An exciting advance has come from this work as the lines of arthritogenic anti-*M.tuberculosis* T-cells from adjuvant arthritic Lewis rats have been used to screen mycobacterial antigens expressed in *Escherichia coli*, genetically engineered truncated proteins and synthetic peptides. Both T-cell clones A2b and A2c showed proliferation in the presence of $10\mu g$ ml^{-1} of the purified 65K *M.bovis* BCG protein prepared from the 65K over-producing *E.coli* host strain 1046. A synthetic nonapeptide 180–188 of the 65K protein (1.0 μg ml^{-1}) was recognized by the arthritogenic clone A2b and the protective clone A2c. Although the 65K protein in oil did not induce adjuvant arthritis in Lewis rats, it protected these animals against subsequent challenge with *M. tuberculosis* organisms in oil [56].

These results raise questions about some of the earlier experiments with enzymic digests of bacterial cell walls. Did they contain contaminating proteins, with crossreactive epitopes, primarily responsible for the adjuvant arthritis in the presence of a potent peptidoglycan immunopotentiator? It is possible that the length of the glycan chain [41] was not the important factor but the degree of preparation purity; hence the decreasing inductive activity from *M.tuberculosis* → cell wall enzymic digest → peptidoglycan → to no activity with glycopeptides or MDP. The same argument may be true for other cell wall preparations because the *M.bovis* BCG 65K protein may be related to an antigen shared with numerous organisms which have been involved in arthritis. Nevertheless, these studies clearly show that an epitope of the *M.tuberculosis* 65K protein is responsible for arthritis induction and not the dissimilar tetrapeptide of the peptidoglycan or the dipeptide of MDP.

5.3.2   Induction of allergic responses to the injected antigen or to a competing antigen

The increased attention being given to the oral administration of experimental vaccines demands some care since other antigens in the gut eg. food proteins could become associated with the adjuvant response. In many experiments with mycobacterial adjuvants we have been unable to detect antibodies directed against the immunopotentiator itself. Similarly, RIBOMUNYL (a commercial preparation containing 375 μg *Klebsiella pneumoniae* proteoglycan mixed with ribosomes from *K. pneumoniae, Strep. pneumoniae, Strep. pyogenes* and *Haemophilus influenzae* (35:30:30:5 by parts)) failed to stimulate IgE antibodies against the proteoglycan after oral administration, although IgG anti-proteoglycan antibodies were detected (Dr. D A Levy, personal communication).

Ham l/CR mice and Hartley strain guinea-pigs were injected with or fed ovalbumin, lactalbumin, soyabean or gluten. Intraperitoneal injections were given in saline, FIA or FCA mixtures containing different mycobacterial preparations. Animals were fed the proteins for 24 hr before and after the oral administration of similar protein: adjuvant mixtures. The mice did not show any IgG- or IgE-mediated allergic responses after oral administration and the few IgE responses after i.p. injection showed no consistent pattern. Similarly, in the guinea-pigs there were no IgE allergic responses to lactalbumin or gluten but some reactions were obtained with FCA plus soyabean and ovalbumin; atypical proteins in the animals' diet. IgG antibodies were obtained with all protein adjuvant mixtures after feeding or i.p. injection. We concluded that allergy induction was not a serious problem provided the diet remained constant.

5.3.3   Induction of autoimmune diseases

Early workers induced a demyelinating encephalomyelitis by repeated injections of homologous brain tissue or by one injection in FCA. Subsequently, a variety of autoimmune diseases could be induced with mycobacterial peptidoglycolipid extracts containing peptidoglycan or some water-soluble extracts [22]. The injection of synthetic encephalitogen, another nona-peptide, with FCA containing 100 μg *M. tuberculosis* or 1.0 μg MDP in FIA also stimulated encephalomyelitis in guinea-pigs [57,58]. An experimental orchiepididymitis was induced in guinea-pigs with 500 μg MDP in FIA but at a dose of 50 μg MDP was inactive [59].

In B 10.BR (H-2^k) high-responder mice thyroiditis was induced in 2/3 animals given thyroglobulin in CFA. By comparison, MDP, N-acetyl-muramyl – L-alanyl –D-isoglutamine – L-alanyl – D-glycerol mycolate [MDP – L-Ala – Glyc – Myc], N-acetylmuramyl – L-alanyl – D-glutamyl – (decyl)methyl ester [MDP (decyl)methyl] and N-acetylmuramyl – L-alanyl – D-glutamine – n-butyl ester [MDP –(Gln)-OnBu or Murabutide II,] were tested for their activity in FIA or saline. In CBA (H-2^k) high responders FIA with 20 μg thyroglobulin and 100 μg MDP, [MDP – L-Ala – Glyc – Myc] or [MDP (decyl)methyl] or with 150 μg thyroglobulin and 100 μg Murabutide induced thyroiditis in 5/6, 0/6, 6/6 and 2/4 mice respectively. However, in saline with 50 μg thyroglobulin the comparative incidences of thyroiditis were 0/5, 1/5, 0/6 and 0/6. Allergic thyroiditis was not detected in low responder mice, BALB/c (H-2^d) or B10.D2 (H-2^d). More importantly, MDP or Murabutide injected alone some forty times to high responder mice did not stimulate autoantibody formation [60].

In rabbits, MDP, N-acetyl – L-α-aminobutyryl – D-isoglutamine [(Abu)^{1}MDP] and 6-0-stearoyl (Abu)^1 MDP caused a reversible uveitis whereas Murabutide did not [61].

II   Murabutide

These experimental autoimmune diseases are very artificial because in each instance the particular tissue antigen is included in the injection mixture. Again it must be stated that many whole cell vaccines have been used in millions of doses and there has not been an associated high incidence of autoimmune phenomena. In addition, I asked a leading clinician and mycobacteriologist if there was any evidence of an increased incidence of autoimmune disease in tuberculosis patients or in BCG recipients. He believed that if there had been a high incidence data would have been collated but there were none on record. Others believe there is excessive anxiety concerning this possibility since "autoimmune diseases can only be produced if there preexist defects in normal autoregulation mechanisms. Otherwise the use of a large number of vaccines would have produced a very high incidence of autoimmune reactions" [62].

6   **Beneficial aspects of natural and synthetic adjuvants**
6.1   Stimulation of non-specific host resistance to infection

It was noticed in 1977 that various synthetic immunopotentiators would protect animals from a subsequent lethal bacterial challenge [63]. Numerous examples have been recorded of protection against a variety of bacterial, fungal and protozoal infections.
Tanaka [64] injected ICR mice with a water-in-oil emulsion containing 100 μg MDP or with the water-in-oil emulsion alone. Four days later the animals were challenged intraperitoneally or intravenously with

$CH_3(CH_2)_{16}CO.OCH_2$

III   6-O-stearoyl-MDP (L18 MDP)

4×10^5 c.f.u. *L. monocytogenes;* 70 % of the animals were still alive 11 days later. Curiously, the peptidoglycan (100 μg) of *L.monocytogenes* from the rabbit strain EGD protected mice after i. p. injection from subsequent challenge with 5×10^8 *Candida albicans* (16/20 survived), although the preparation from the human strain ISC was less protective (8/20 survived) [3]. Such antifungal activity could be beneficial in compromised patients but it would be important to know the duration and extent of the non-specific protection after the injection of immunopotentiator.

In Japan, there has been interest in some of the acyl-MDP derivatives, especially 6-O-stearoyl-MDP (L18 –MDP) III and Nα – MDP– N-stearoyl – L-lysine (MDP – Lys (L18)) IV. Both of these were effective in stimulating non-specific resistance against infection with *Escherichia coli, C. kutscheri, Staphylococcus aureus, Pseudomonas aeruginosa* and *C. albicans* but not against *Salmonella typhimurium* or *L. monocytogenes* [65–69]. Nor-MDP when administered in an oil emulsion or in liposomes protects against *P. aeruginosa* and *C. albicans* infections [70–71].

6.2 Practical use of immunopotentiators in experimental vaccines

The gradual process of eliminating the criticisms and aversions to the use of immunopotentiators has increased the awareness of their poten-

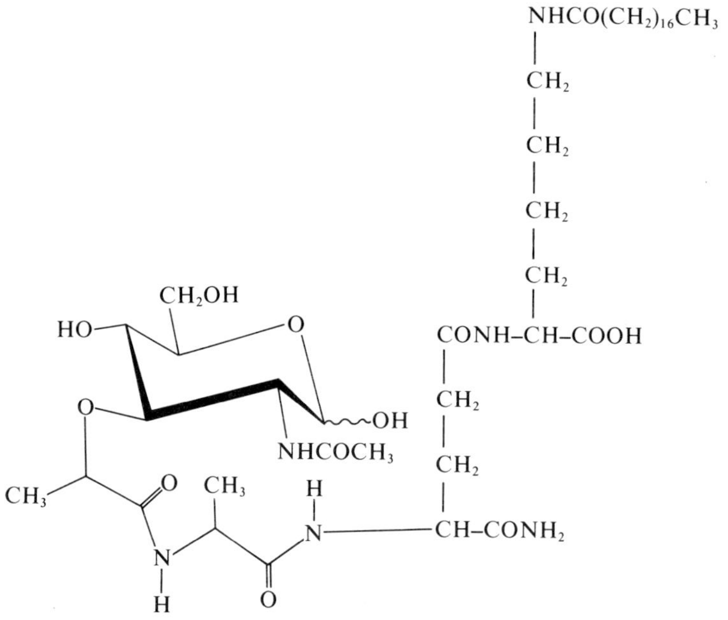

IV   MDP-Lys (L18)

tial benefits; some 2000 publications last year in comparison to a score
in the 1960s. Consequently, in this short review mention can be made
of only a few examples which indicate the successful uses of immunopotentiators and where further investigation is warranted.

Organic chemists have synthesized a great range of derivatives from
the parent MDP. Some of these retain harmful activities and cannot be
considered for use in human vaccines. In general, the opinion seems to
be that the muramyl residue should be left essentially unmodified, the
L-alanyl residue can be substituted by another aliphatic L-amino acid
but not by another D-amino acid. The D-glutamyl residue is essential
for the immunopotentiating activity and should not be removed [72].
The position of the substituting residue may also be significant: a cell-
mediated hypersensitivity response to azobenzenearsonate – N-acetyl
L-tyrosine was induced in guinea-pigs with 6-amino – 6-deoxy – N-ace-
tylmuramylpeptides or their 6-acylamino –MDP derivatives, however,
6-acylamino – 6-deoxy – N-(acyl)MDP derivatives were inactive [73].
Some of the many derivatives are proving to be acceptable for vaccine
inclusion. For example, Murabutide which is non-pyrogenic in rabbits

[74], does not produce a synergistic toxicity with LPS [74], does not cause haemorrhagic necrosis in guinea-pigs [75], induction of an auto-immune disease in rabbits [61], transient leucopaenia in rabbits [74] or guinea-pig distress syndrome [76].

It is important to test different vaccine formulations under varying experimental conditions before making the optimal selection. In an attempt to stimulate the production of IgA antibodies at the mucosal surface rats were immunized s.c. with 10 μg of the major outer membrane protein of *Neisseria gonorrhoeae* (PI) or with 50 μg directly into the proximal duodenum (i.i.) with or without the adjuvants $AlPO_4$ (PI 1.0:$AlPO_4$ 100 w/v) or MDP (PI 1.0: MDP 1.0 w/w). The levels of anti-PI antibodies was greatest with $AlPO_4$ in the vaccine given by the s.c. route and with MDP in the vaccine given by the i.i.route [77].

The purified hepatitis B virus surface antigen was used in conjunction with 100 μg murabutide or 40 μg Alugel 50 to immunize BALB/c mice. After a single injection the mean antibody titres against the surface antigen were 850 for murabutide and 3,500 for alum after 4 months. If the mice were given a booster injection one month after the first inoculation the titres were 15,000 for both adjuvants. A most interesting result was the synergism between the two adjuvants, when administered together, in mice boosted after 30 days; the antibody titre to the hepatitis B antigen was as high as 524,288 twenty days later and was 32,768 even after 4 months. IgE titres in these animals, determined by passive cutaneous anaphylaxis on day 50 after immunization was only 160 [78]. Antibody titres for a single injection regimen were not presented for

V Threonyl – MDP

the 4 month bleed but nevertheless these results merit further consideration of the combined utilization of immunopotentiators in particular vaccine formulations. As mentioned before there is also the problem that the combined use of immunopotentiators may 'over-adjuvant' the vaccine [17].

Similarly, threonyl-MDP (NAcMur –L-thr –D–isoglu) V lacks many of the harmful effects and would seem to be a potentially beneficial immunopotentiator [30,61]. In one 'adjuvant formulation' threonyl-MDP or murabutide plus antigen were mixed with a pluronic polymer L121 VI [79] squalane and Tween 80. This mixture was effective in potentiating both the humoral and cell-mediated immune responses [31]. The non-ionic block copolymer surfactant L121 is composed of hydrophilic polyoxyethylene (POE) and hydrophobic polyoxypropylene (POP).

$$\left[\mathrm{HO(CH_2-CH_2-O)_a} \quad - \quad \underset{\underset{CH_3}{|}}{\mathrm{(CH-CH_2-O)_b}} \quad - \quad \mathrm{(CH_2-CH_2-O)_a} \right]$$

POE M_r 200   — POP M_r 4000   — POE M_r 200

VI L 121

By itself, L121 had an immunopotentiating effect on both the humoral and cell-mediated responses in either saline or 1% oil-in-water (Drakeol or Eicosane) [79]. It seems likely that the amphipathic L121 may interact with the threonyl-MDP in a manner similar to that described earlier with mycobacterial glycopeptide and Gram-negative LPS adjuvants [17,80].

Table 1
Adjuvant properties of Gram-negative LPS, mycobacterial glycopeptide and L121

Biological activity	Mycobacterial glycopeptide	Amphiphiles LPS	L121
Mitogenicity	−	+ + +	+ +
Pyrogenicity	−	+ + +	−
Toxicity	−	+ + +	−
Adjuvant-activity	+ + +	+ + +	+ + +

The existence of a conjugated complex between MDP and the amphipathic trehalose-dimycolate (TDM) and between TDM and the hydrocarbon Squalane has also been proposed. Such a complex may assist in the movement of MDP into the cell membrane and in the expression of the biological function of the MDP [81]. This interaction between amphiphiles of different classes of immunopotentiator is another area worthy of closer attention in the search for a range of useful adjuvants. Some of the experimental synthetic vaccines which are being assessed require that a synthetic peptide should be bound to a suitable carrier. The carrier is often one of the highly immunogenic toxoids from existing vaccines eg. diphtheria and tetanus toxoids. The possibility that natural microbial antigens might be replaced in vaccines by synthetic antigens containing unique epitopes has been of interest for many years [82–87]. Initially, a synthetic polypeptide antigen was covalently linked to MDP, 10 μg conjugate contained 1.0 μg MDP. Specific antibodies were stimulated in mice by this conjugate not only in a water-in-oil emulsion but also in saline [88]. With the development of genetic manipulation techniques, nucleotide sequencing and peptide synthesis this characteristic of the MDP-type adjuvants will be of great significance.

Two distinct fragments of the diphtheria toxin can be isolated: the enzymatic fragment A responsible for the toxicity and the B fragment which binds receptors on cells. The A and B subunits are hinged together by a loop of 16 amino acids between residues 186–201. This octadecapeptide was synthesized and conjugated to MDP- poly DL-alanine –poly L-lysine via its SH and COOH groups with glutaraldehyde. This totally synthetic vaccine in saline protected guinea-pigs against the dermonecrotic effect of diphtheria toxin [89–91]. Peptides corresponding to amino acid sequences within the cholera toxin B.subunit have also been synthesized and conjugated to tetanus toxoid, 1.0 mg of conjugate in 0.5 ml phosphate buffered saline when emulsified with FCA stimulated significant levels of anti-cholera toxin antibodies in rabbits [92].

A synthetic peptide fragment of the hepatitis B virus surface antigen conjugated to diphtheria toxoid was injected into mice with 100 μg murabutide or 40 μg Alugel 50. The anti-peptide antibody titres were 25,000 and 40,000 respectively for the primary response and 380,000 and 703,000 respectively for the secondary response. The titres of anti-diphtheria toxoid antibodies were not stated [78].

The use of these antigen: adjuvant conjugates has been used in non-infectious models. Immunological castration of outbred Swiss mice was achieved with a totally synthetic vaccine composed of 8.6 μg of the decapeptide of the luteinizing hormone-releasing hormone conjugated to 4.3 μg MDP-lys in saline [93]. After 50 days spermatogenesis was absent and there was a marked atrophy in 9/10 animals. The authors doubted whether this vaccine could be used in man but considered it an alternative to castration in the veterinary field. Since the early 1970s the World Health Organization has sponsored the development of an anti-fertility vaccine based on the carboxy-terminal polypeptide of the β-subunit of human chorionic gonadotrophin (hCG), a protein hormone produced by the early trophoblast. One of the functions of hCG is the maintenance of progesterone production by the corpus luteum. It was proposed that prevention of hCG activity by immunization would result in physiological luteolysis and menstruation.

For anti-fertility studies in baboons an experimental vaccine was prepared with diphtheria toxoid as carrier, synthetic peptide of β-hCG residues 109–145 as antigen and the adjuvant MDP in saline. This was emulsified with squalene in the presence of Arlacel A (mannide mono-oleate which would be unacceptable for use in man). Baboons injected with the vaccine or with diphtheria toxoid alone showed 4.6 % and 70.0 % pregnancies respectively [94]. However, the utilization of carriers does have limitations such as dosage (5.0 μg tetanus toxoid will only conjugate 1.0 μg peptide), sensitization and specific epitopic suppression [95]. Fatty acids coupled to peptides or carbohydrates may prove to be acceptable non-immunogenic carriers [30]. Recently, a polyvalent synthetic vaccine was constructed without a carrier [96]. Four synthetic peptides, copies of two from bacterial antigens *(Streptococcus pyogenes* M protein (S34 peptide) and *C. diphtheriae* toxin (residues 186–201), one from the hepatitis B surface antigen (residues 99–121) and one from the circumsporozoite protein of *Plasmodium knowlesi* (peptide PK 26) were conjugated together with glutaraldehyde. The antibody titres against these individual peptides were low in the primary response when the peptide construct was mixed with murabutide but they were considerably higher with FCA. On the other hand, the titres in the secondary response were much greater with both murabutide and FCA. It seemed from the titres that the magnitude of the responses varied from one peptide to another. Thus with FCA the maximum titre was against the protozoan peptide PK 26 whereas with murabutide it was against the S34 peptide. This study creates the exciting prospect of manufacturing synthetic vaccines against a variety of bacterial and viral antigens provided that one of the epitopes does not become immunodominant. It will also be essential to determine the stability of the glutaraldehyde conjugation.

There is much work to be done in the initial search for antigens which will stimulate the production of protective immunity against virulent microorganisms. It is essential to understand the basis of pathogenesis and of host protection against these organisms before one moves from the natural products to synthetic constructs. Similarly, it will be necessary to examine a variety of immunopotentiators eg. alum, FCA, MDP or derivatives, TDM or saponin to ensure the maximum stimulation of the immune response. A mood of optimism prevails and the future use of adjuvanted synthetic vaccines now seems to be a realistic goal.

References

1 M.F.R. Waters, A.B.G. Laing and R.J.W. Rees: Lepr. Rev. *49*, 127 (1978).
2 J. Vane and P. Cuatrecasas: Nature *312*, 303 (1984).
3 A. Paquet, K.M. Raines and P. C. Brownback: Infect. Immun. *54*, 170 (1986).
4 R. MacLennan, F. D. Schofield, M. Pittman, M. C. Hardegree and M. F. Barile: Bull. WHO *32*, 683 (1965).
5 J. C. Azurin, A. Cruz, T. P. Pesigan, M. Alvero, T. Camena, R. Suplido, L. Ledesma and C. Z. Gomez: Bull. WHO *37*, 703 (1967).
6 H. Ogonuki, S. O. Hashizume and B. Takashashi: Bull. WHO *37*, 729 (1967).
7 F. Ellouz, A. Adam, R. Ciorbaru and E. Lederer: Biochem. Biophys. Res. Commun. *59*, 1317 (1974).
8 D. E. S. Stewart-Tull, T. Shimono, S. Kotani and B. A. Knights: Int. Archs. Allergy App. Immunol. *52*, 118 (1976).
9 R. G. White: Ann. Rev. Microbiol. *30*, 579 (1976).
10 K. Emori and A. Tanaka: Infect. Immun. *19*, 613 (1978).
11 A. Tanaka and K. Emori: Am J. Pathol. *98*, 733 (1980).
12 K. Emori, S. Nagao, N. Shigematsu, S. Kotani, M. Tsujimoto, T. Shiba, S. Kusamoto and A. Tanaka: Infect. Immun. *49*, 244 (1985).
13 D. E. S. Stewart-Tull and R. E. C. Rowe: J. Immunol. Meths. *8*, 37 (1975).
14 R. S. Northrup and A. S. Fauci: J. Inf. Dis. *125*, 672 (1972).
15 F. V. Chisari, R. S. Northrup and L. C. Chen: J. Immunol. *113*, 729 (1974).
16 P. B. Stones: Brit. Met. J. *1*, 1627 (1979).
17 D. E. S. Stewart-Tull: Immunology of the Bacterial Cell Envelope. Ed. D. E. S. Stewart-Tull and M. Davies. J. Wiley & Sons, Chichester p60 (1985).
18 R. Bomford: Progress Towards Better Vaccines. Eds. R. Bell and G. Torrigiani. Oxford University Press, Oxford. p177 (1986).
19 J. Convit, N. Aranzazu, M. Pinardi and M. Ulrich: Clin. Exp. Immunol. *36*, 214 (1979).
20 I. ab I. Davies: Adv. Drug React. Ac. Pois Rev. *1*, 1. (1986).
21 L. Schwarzenberg, M. C. Simmler and J. L. Pico: Cancer Immunol. Immunother. *1*, 69 (1976).

22   D. E. S. Stewart-Tull: The Biology of the Mycobacteria Vol. 2. Eds. C. Rat-
     ledge and J. Stanford. Academic Press, London p3 (1983).
23   R. Bomford: Animal Cell Biotechnology. Vol. 2. Eds. R. E. Spier and J. B.
     Griffiths. Academic Press, London p235 (1985).
24   R. G. White, L. Bernstock, R. G. S. Johns and E. Lederer: Immunology *1*, 54
     (1958).
25   R. G. White, P. Jollès, D. Samour and E. Lederer: Immunology *7*, 1958
     (1964).
26   World Health Organization: Technical Report Series No. 595. WHO, Gen-
     eva p3 (1976).
27   R. H. Gisler, F. M. Dietrich, G. Baschang, A. Brownbill, G. Schumann, F.
     G. Staber, L. Tarcsay, E. D. Wachsmuth and P. Dukor: Drugs and Immune
     Responsiveness. Eds. J. L. Turk and D. Parker. Macmillan Press, London
     p133 (1979).
28   R. Edelman: Revs. Infect. Dis. *2*, 370 (1980).
29   D. E. S. Stewart-Tull: Ann. Immunol. hung. *26*, 197 (1986).
30   A. C. Allison, N. E. Byars and R. V. Waters: Advances in Carriers and Adju-
     vants for Veterinary Biologics. Eds. R. M. Nervig, P. M. Gough, M. L. Kae-
     berle and C. A. Whetstone. Iowa State University Press Ames, Iowa p91
     (1986).
31   D. E. S. Stewart-Tull: Ann. Rev. Microbiol. *34*, 311 (1980).
32   M. Parant, F. Parant, L. Chedid, A. Yapo, J. F. Petit and E. Lederer: Int. J.
     Immunopharmacol. *1*, 35 (1979).
33   A. Fox, J. H. Schwab and T. Cochran: Infect. Immun. *29*, 526 (1980).
34   Z. Sen and M. L. Karnovsky: Infect. Immun. *43*, 937 (1984).
35   B. Ladesic, J. Tomasic, S. Kveder and I. Hrsak: Biochem. Biophys. Acta,
     678, 12 (1981).
36   J. Harrison and A. Fox: Infect. Immun. *50*, 320 (1985).
37   J. B. Robbins: New Developments with Human and Veterinary Vaccines.
     Eds. A. Mizrahi, I. Hertman, M. A. Klingberg and A. Kohn. Alan R. Liss
     Inc, New York p333 (1980).
38   C. M. Pearson: Proc. Soc. Exptl. Biol. and Med. *91*, 95 (1956).
39   I. Azuma, F. Kanetsuna, Y. Kada, T. Takashima and Y. Yamamura: Jap. J.
     Microbiol. *16*, 333 (1972).
40   F. Audibert, M. Parant, J. F. Petit and A. Adam: Compt. rendu. Acad. Sci.
     277, 2097 (1973).
41   D. E. S. Stewart-Tull, T. Shimono, S. Kotani, M. Kato, Y. Ogawa, Y. Ya-
     mamura, T. Koga and C. M. Pearson: Immunology *29*, 1 (1975).
42   T. Koga, K. Maeda, K. Onoue, K. Kato and S. Kotani: Mol. Immunol. *16*,
     153 (1979).
43   R. Eisenberg, A. Fox, J. J. Greenblatt, S. K. Anderle, W. J. Cromartie and J.
     H. Schwab: Infect. Immun. *38*, 127 (1982).
44   S. Nagao and A. Tanaka: Infect. Immun. *28*, 624 (1980).
45   T. Koga, K. Kakimoto, T. Hirofuji, S. Kotani, H. Ohkuni, K. Watanabe, N.
     Okada, H. Okada, A. Sumiyoshi and K. Saisho: Infect. Immun. *50*, 27
     (1985).
46   S. H. Ohanian, J. H. Schwab and W. J. Cromartie: J. exp. Med. *129*, 37
     (1969).
47   W. Van Eden, J. Holoshitz, Z. Nevo, A. Frenkel, A. Klajman and I. R. Co-
     hen: Proc. Natl. Acad. Sci. USA *82*, 5117 (1985).
48   J. Holoshitz, Y. Naparstek, A. Ben-Nun and I. R. Cohen: Science *219*, 56
     (1983).
49   J. Holoshitz, A. Matitiau and I. R. Cohen: J. clin. Invest. *73*, 211 (1984).
50   P. Minden, J. K. McClatchy, M. Wainberg and D. W. Weiss: J. Nat. Can.
     Inst. *53*, 1325 (1974).

51   M. Tovisu, T. Miyahara, N. Shinohara, K. Ohsato and H. Sonozahi: Cancer Immunol. Immunother. *5*, 77 (1978).
52   D. E. S. Stewart-Tull, P. C. Wilkinson and R. G. White: Immunology *9*, 151 (1965).
53   D. E. S. Stewart-Tull and P. C. Wilkinson: Immunology. *24*, 205 (1973).
54   D. E. S. Stewart-Tull, J. P. Arbuthnott and J. H. Freer: Immunochemistry *12*, 941 (1975).
55   R. B. Parekh, R. A. Dwek, B. J. Sutton, D. L. Fernandes, A. Leung, D. Stanworth, T. W. Rademacher, T. Mizuochi, T. Taniguchi, K. Matsuta, F. Takeuchi, Y. Nagano, T. Miyamoto and A. Kobata: Nature *316*, 452 (1985).
56   A. Noordzij, J. D. A. van Embden, E. J. Hensen and I. R. Cohen: Nature *331*, 171 (1988).
57. Y. Nagai, K. Akiyama, K. Suzuki, S. Kotani, Y. Watanabe, T. Shimono, T. Shiba, S. Kusumoto, F. Ikuta and S. Takeda: Cell Immunol. *35*, 158 (1978).
58   Y. Nagai, K. Akiyama, S. Kotani, Y. Watanabe, T. Shimono, T. Shiba and S. Kusumoto: Cell Immunol. *35*, 168 (1978).
59   F. Toullet, F. Audibert, G. A. Voisin and L. Chedid: Ann. d'Immunol. *128*, 267 (1977).
60   Y. M. Kong, F. Audibert, A. A. Giraldo, N. R. Rose and L. Chedid: Infect. Immun. *49*, 40 (1985).
61   R. V. Waters, T. G. Terrell and G. H. Jones: Infect. Immun. *51*, 816 (1986).
62   L. Chedid: Ann. Inst. Past. Immunol: *136D*, 283 (1985).
63   L. Chedid, M. Parant, F. Parant, P. Lefrancier, J. Choay and E. Lederer: Proc. Natl. Acad. Sci. *74*, 2089 (1977).
64   A. Tanaka: Immunomodulation by Microbial Products and related synthetic compounds. Eds. Y. Yamamura, S. Kotani, I. Azuma, A. Koda, T. Shiba. Excerpta Medica Amsterdam p72 (1982).
65   C. Ishihara, N. Hamada, K. Yamamoto, J. Iida, I. Azuma and Y. Yamamura: Vaccine *3*, 370 (1985).
66   C. Ishihara, K. Yamamoto, N. Mikami and I. Azuma: Vaccine *2*, 261 (1984).
67   K. Matsumoto, H. Ogawa, T. Kusama, O. Nagase, N. Sawaki, M. Inage, S. Kusumoto, T. Shiba and I. Azuma: Infect. Immun. *32*, 748 (1981).
68   K. Matsumoto, T. Otani, T. Une, Y. Osada, H. Ogawa and I. Azuma: Infect. Immun. *39*, 1029 (1983).
69   K. Matsumoto, Y. Osada, T. Une, T. Otani, H. Ogawa and I. Azuma: Immunostimulants: Now and Tomorrow. Ed. I. Azuma and G. Jollès p79 (1987).
70   P. A. Morozumi, E. Brummer and D. A. Stevens: Mycopathologia *81*, 35 (1983).
71   E. B. Fraser-Smith and T. R. Matthews: Infect. Immun. *34*, 676 (1981).
72   A. Adam and E. Lederer: Progress Towards Better Vaccines. Eds. R. Bell and G. Torrigiani. Oxford Univ. Press, Oxford (1986).
73   I. Azuma, H. Okumara, I. Saiki, Y. Tanis, M. Kiso, A. Hasegawa and Y. Yamamura: Infect. Immun. *32*, 1305 (1981).
74   L. A. Chedid, M. A. Parant, F. M. Audibert, G. J. Riveau, F. J. Parant, E. Lederer, J. P. Choay and P. L. Lefrancier: Infect. Immun. *35*, 417 (1982).
75   S. Kotani, I. Azuma, H. Takada, M. Tsujimoto and Y. Yamamura: Adv. Exp. Med. Biol. *166*, 117 (1983).
76   N. E. Byars: Infect. Immun. *44*, 344 (1984).
77   S. H. M. Jeurissen, T. Sminia and E. C. Beuvery: Infect. Immun. *55*, 253 (1987).
78   F. M. Audibert, G. Przewlocki, C. D. Leclerc, M. E. Jolivet, H. S. Gras-Masse, A. L. Tartar and L. Chedid: Infect. Immun. *45*, 261 (1984).
79   R. L. Hunter and B. Bennett: Advances in Carriers and Adjuvants for Veterinary Biologics. Eds. R. M. Nervig, P. M. Gough, M. L. Kaeberle and C. A. Whetstone. Iowa State Univ. Press, Ames. p61 (1986).

80   M. Davies and D. E. S. Stewart-Tull: Immunology of the Bacterial Cell Envelope. Eds. D. E. S. Stewart-Tull and M. Davies. John Wiley and Sons, Chichester. p239 (1985).

81   E. Lederer: Immunomodulation by microbial products and related synthetic compounds. Eds. Y. Yamamura, S. Kotani, I. Azuma, A. Koda and T. Shiba. Excerpta Medica. p3 (1982).

82   R. Arnon: Immunity of Viral and Rickettsial Diseases. Eds. A. Kohn and A. M. Klinberg. Plenum Press, New York. p209 (1972).

83   R. Arnon: Pharmac Ther. *6*, 275 (1979).

84   R. Arnon: Behring Inst. Mitt. *69*, 19 (1981).

85   M. Sela: Science *166*, 1365 (1969).

86   M. Sela: Antiviral Mechanisms. Perspectives in Virology IX. Ed. M. Pollard. Academic Press New York. p91 (1975).

87   R. Arnon and M. Sela: Ann. Inst. Pasteur. Immunol. *136D*, 271 (1985).

88   E. Mozes, M. Sela and L. Chedid: Proc. Natl. Acad. Sci. USA *77*, 4933 (1980).

89   F. Audibert, M. Jolivet, L. Chedid, J. E. Alouf, P. Boquet, P. Rivaille and O. Siffers: Nature *289*, 593 (1981).

90   F. Audibert, M. Jolivet, L. Chedid, R. Arnon and M. Sela: Proc. Natl. Acad. Sci. USA. *79*, 5042 (1982).

91   F. Audibert, M. Jolivet and C. Carelli: Adv. Immunopharmacol. *2*, 429 (1983).

92   C. O. Jacob, M. Sela and R. Arnon: Proc. Natl. Acad. Sci. USA *80*, 7611 (1983).

93   C. Carelli, F. Audibert, J. Gaillard and L. Chedid: Proc. Natl. Acad. Sci. USA. *79*, 5392 (1982).

94   V. C. Stevens, J. E. Powell, A. C. Lee and D. Griffin: Fertility and Sterility *36*, 98 (1981).

95   F. Audibert and L. Chedid: Modern Approaches to Vaccines. Eds. R. M. Chanock and R. A. Lerner. Cold Spring Harbor, New York. p397 (1984).

96   M. E. Jolivet, F. M. Audibert, H. Gras-Masse, A. L. Tartar, D. H. Schlesinger, R. Wirtz and L. A. Chedid: Infect. Immun. *55*, 1498 (1987).

Organizing for drug discovery

By Michael Williams and Gary L. Neil*

Research Department, Pharmaceuticals Division, CIBA-GEIGY,
Summit, NJ 07901, USA, and *Biotechnology and Basic Research Support,
The Upjohn Company, Kalamazoo, MI 49001, USA

1   Introduction

The role of the pharmaceutical industry has been defined as the application of "basic knowledge in biology and chemistry to the discovery of new medicines" [1]. While it is generally acknowledged that the industry has made significant contributions to the health and financial well-being of society in this role [2], there is, by and large, no single formula by which success in the discovery process can be guaranteed. The choice of a research philosophy, the identification of realistic targets based on unmet medical needs, hiring of skilled, goal-oriented personnel and the tactical plan by which to implement chosen strategies are all problems that research management has traditionally faced [3]. As the cost of research increases faster than company revenues [4] and the cost of introduction of new therapeutic entities, which can take from 10–15 years, now exceeds the $ 100 million mark [5, 6], the hope that the process of drug discovery will lend itself to being a quantifiable science with rules, guidelines and formulae for success is eagerly anticipated.

The definition of what sequence of events constitute the process of drug discovery, as opposed to development, is an important one in evaluating the strategy, cost and potential for success of a research organization. Drug discovery is a team-based effort comprising; the chemists synthesizing compounds; the biologists performing the test procedures that sequentially identify the activity, efficacy and selectivity of action of a compound; the pharmacokineticists who measure bioavailability; the pharmaceutical chemists who provide formulation expertise; the batch chemists who provide the kilogram quantities necessary for formulation and toxicological workup; toxicologists; and more recently, drug delivery experts. Beyond the endeavors of these individuals lies the expertise of production chemists and the individual skills and experience that comprise development and marketing (Fig. 1). The process of drug discovery therefore encompasses far more than the evaluation of compounds at the bench level but less than the corporate effort required to put a product on the pharmacist's shelf.

Behind the creativity associated with the innovative idea there is a good deal of less glamorous effort in other scientific disciplines to move a compound from being a new chemical entity (NCE) to a drug candidate. The expertise of the drug formulator or the drug delivery

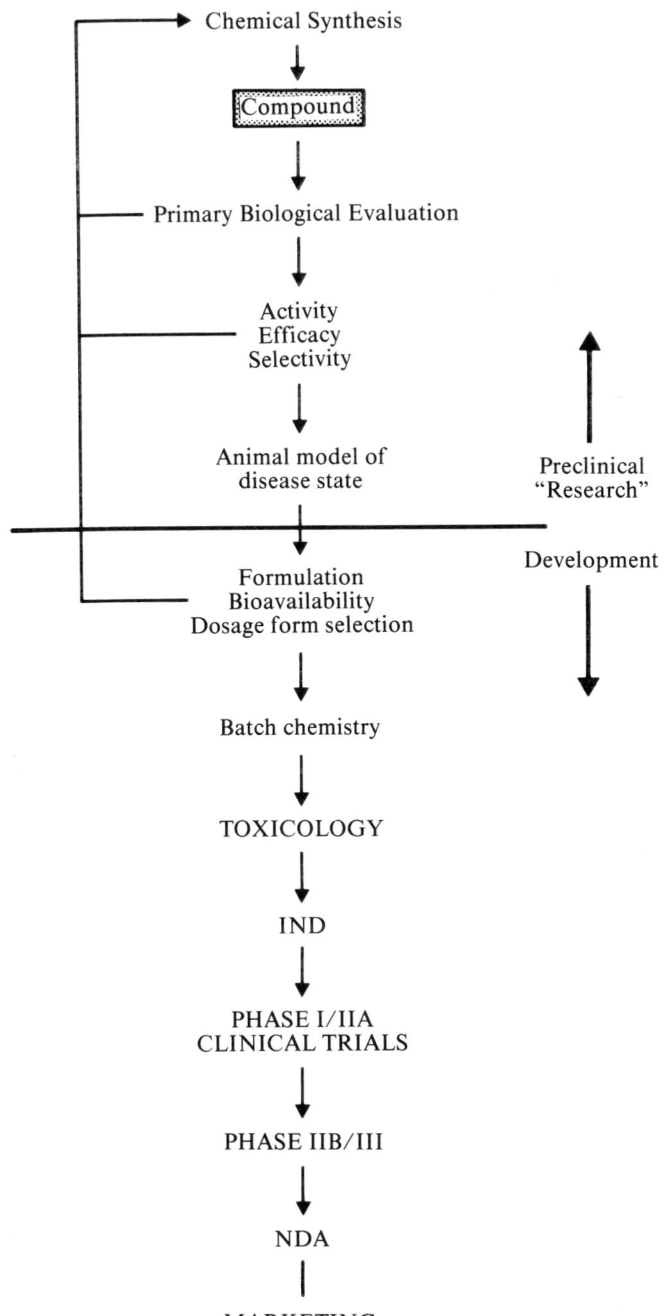

Fig. 1
The drug discovery process – a schematic

expert in getting a compound to its site of action in sufficient quantities for efficacy can often be as important as the initial idea, yet such individuals are more often than not the unsung heroes of the drug discovery process.

In organizing for drug discovery, the extent to which the Research Director has responsibility for the various elements described above will dictate, in part, the type of organization that can be formed. If this responsibility extends to clinical trials, then the organization would be clearly different from that where only preclinical technical areas are part of the reporting scheme. Given that these areas of reporting responsibility and of therapeutic area targets are well defined, by both research *and* corporate management, and assuming that an adequate budget is available, the choice of the type of research approach is probably the next major factor in the design of the drug discovery organization.

The rational, or mechanistic approach that epitomized the discovery of cimetidine [7] and captopril [8], while leading to the not unreasonable conclusion that a receptor/enzyme targeted approach is *the* key to successful drug discovery, has not always yielded compounds in other areas. For example, since the discovery of the enkephalins and endorphins in the mid 1970s, the search for novel opiate-type analgesics based on structure activity relationships at μ, ∂ and κ opiate receptors has been both singularly expensive and unsuccessful. This approach, is however, by its very nature, unpredictable. Whether or not a compound proves to be active in a targeted disease can provide important knowledge as to the role of the given target in the given disease. At the very least therefore, the mechanistic approach is a positive learning experience. The more classical, serendipitous approach has also been successful in drug discovery as evidenced, not only by entities such as the antipsychotic, chlorpromazine [9], but more recently, the anxiolytics, diazepam [10] and buspirone [11] and also by that major exception to the laws of rational drug design, morphine [12].

The ability to balance critical resource mass together with a necessary degree of flexibility to take advantage of unforeseen scientific breakthroughs or of unexpected findings with experimental compounds, the opportunistic approach, are problems probably unique to the pharmaceutical industry that do not readily lend themselves to analysis in the same way that the demands related to the design and production of a new automobile or computer can [13]. Similarly, the need to balance the incentives likely to ensure the growth and commitment of individual scientists to the discovery of new drugs with the uncertainty of the pharmaceutical business environment is an especial challenge. Indeed, the application of the traditional "management-by-objectives" approach to the pharmaceutical industry has been censured [4, 5] with the suggestion that once having directed a research effort toward a particular target, "the best thing management can do is stay out of the way" [4]. de Stevens has similarly concluded [5] that drug-related research can neither be managed by a rule book nor subject to manipulation in response to a short-term business cycle which inevitably leads to the production of uninnovative, "me-too"-type compounds.

Such viewpoints are hardly reassuring for those charged with managing research in an industry where success has most often been historically associated with a methodical, directed approach rather than entrepreneurship. In addition, the current dearth of new products has been attributed to industry bureaucracy and an inability to respond to a changing environment rather than the often cited increase in governmental constraints [14].

The genesis of the present article was a symposium entitled "Organizing for Drug Discovery" held in September, 1987, under the auspices of the Drug Discovery Management Group of the Research and Development Steering Committee of the Pharmaceutical Manufacturer's Association. Drawing on the presentations made at this meeting and other sources, the current state of various approaches to the process of drug discovery, both theoretical and organizational are reviewed within the context of the various scientific, regulatory and marketing, both commercial and social, changes with which the industry has found itself faced in the past decade. The reader will note that the authors, much to their chagrin, are unable to offer an organizational chart to ensure success. Neither are they able to document the necessary prerequisities for the ideal individual to lead the industry into the 21st century. Rather, it is hoped that they have been able to present some guidelines and thoughts from which the reader can draw his or her own conclusions.

2.   Historical perspective

2.1   The golden age

Among the high technology industries, the pharmaceutical industry is perhaps the oldest in continuous existence, having its origins in the German chemical companies of the 18th and 19th centuries. The industry was well entrenched by the 1940s with the successes of various antibiotics. It reached a zenith in the 1950s and 1960s when psychotropics such as chlorpromazine and diazepam, as well as evolving therapies for cardiovascular diseases, such as the β-blockers, began to produce major improvements in the quality of life and life-span. In addition, such drugs altered the economics of medical treatment providing relief to patients previously denied effective treatment. Many of the drugs that form the armamentarium of the Western physician have been discovered since 1930, and it is of interest that 95 % of all documented drug research has been carried out in this century [15].

Many pharmaceutical companies, in tracing their origins back to the 19th century, have a deep sense of tradition in the way that they conduct business. They often enjoy high recognition in the medical and academic communities. Accordingly, the company image has been as much a factor in the success and continued viability of marketed products as the quality of the products themselves. This "golden age of pharmaceuticals" [16], with its high return on investment ostensibly ended in 1962 [17, 18], with the amendments made to the Food, Drug and Cosmetic Act in the U.S.A. as a result of the thalidomide tragedy. Since then, the time and cost of developing a new drug has almost tripled with an attendant reduction in the effective patent life, from 17 to 8 years [1]. The number of new drugs being introduced per year decreased from 93 in 1961 to 48 in 1980 [18] although this number increased to 58 in 1987 [19]. At the same time, patents on older products expired making these open to generic competition where cost, rather than patent exclusivity, is the determining factor in drug usage [4]. Of 1,948 IND requests in 1987, 709 were approved for ANDAs. It is noteworthy that while many industrial enterprises return approximately 2–8 % of their income to research, the pharmaceutical industry on the whole returns about 15 %. Furthermore, from a product perspective, less than 4 % of new drugs reach sales of greater than $ 50 M a year while a median selling drug can take in excess of 9 years to recoup its development costs.

2.2   Health cost containment

The worldwide movement to contain health care costs has further changed the economic climate of drug marketing. While lists of government approved drugs are relatively common in Europe and Japan, it is only recently that the trend to reduce health costs has been manifest in the U.S. The role that the individual physician has in prescribing medication is thus becoming subordinated to the third party health care organizations (HMOs, socialized medical plans, insurance companies) that actually cover these costs. In North America, state and provincial government also provide lists of approved drugs in the effort to contain costs. In Canada, such controls and the resultant boost in generics has made the investment in innovative drug research in that country particularly unattractive, an issue that has only recently been resolved in the Canadian Parliament. In an increasing number of instances therefore, the industry is either faced with only being able to sell certain patented drugs or with selling in bulk to highly cost conscious organizations that specify to their patients which medications they will and will not pay for, rather than convincing the individual physician to prescribe optimal therapy where there may be minimal concern regarding cost.

In reviewing this trend, Schwartz [20] has commented on the increased involvement of the U.S. Federal Government in the pharmaceutical industry evidenced by the Medicare catastrophic medical insurance bill. The possibility that socialized medicine in the U.S. may be inevitable has further led this commentator on the drug industry to suggest that because of the free market in the U.S., America has, in the past, subsidized the rest of the world when drug costs were government mandated. A diminution of profit margins may therefore be anticipated to reduce the amount of money available to identify new drugs through research efforts.

2.3   The educated public

In addition to being willing participants in seeking to contain health care costs, the public has also become far more educated in regard to the inherent value of health care. Various governmental and public organizations are thus focussing heavily on the ethical issues related to drug usage, availability and need [21]. These concerns relate not only to the overall quality of health care [22, 23], the prolongation of life at the expense of its quality, but also the apparent inability of the industry to find "cures" for cancer and more recently, AIDS [24] while devoting research resources to third and fourth generation "me-too" therapeutics for which there is often no perceived need. This is compounded by a "zero-risk" orientation in terms of drug safety. In view of the dramatic role that the industry has had in society through the discovery of drugs such as antibiotics, birth control pills, psychotropic agents for the treatment of anxiety, schizophrenia and depression, and various vaccines, the increasing interdependence occurring between the various branches of the health care industry requires a clearer recognition of the commitment of the industry to health care, as both an integral element and as a consumer concerned in containing costs. While drug costs may increase, they inevitably are cheaper than the alternatives of chronic hospitalization or surgery.

2.4   Wall St focus

Concommitant with this paradox of increased costs and decreased revenues have come the first "billion dollar a year" drugs – the histamine H_2 blockers, angiotensin converting enzyme (ACE) inhibitors and, very likely, the HMG CoA reductase inhibitors. The apparently endless opportunity for profit has attracted a considerable amount of venture capital and investment interest. It is accordingly, a rare week that the industry is not the focus of a pundit from one Wall Street brokerage house or another; indeed, many of these individuals often appear to know more about a company's research plans than do the company's employees! The increased interest of the financial community in pharmaceuticals has also been fueled by the advent of the biotechnology revolution [25, 26], with molecular biology serving as the impetus for the proliferation of a new breed of biotechnology based pharmaceutical companies (e.g. Genentech, Cetus, Chiron, Collagen Corp., Genetics Institute, Genex, California Biotechnology, Collaborative Research, Hybridtech, Biogen, ImmunoGenetics, Praxis, Centocor and Amgen). In nearly every major pharmaceuticial house, based on the somewhat unrealistic optimism of biotechnology management as to how "easy" and quick it is to get their products approved, biotechnology in one form or another has been welcomed enthusiastically as a magic formula by which new products will be readily discovered and also as the stimulus for the revitalization of a status quo that is deemed unproductive. Apart from the relative ease in producing large amounts of naturally occurring substances such as insulin or tissue plasminogen activator (tPA), this viewpoint has yet to be substantiated and it appears probable that biotechnology will be an additional, albeit important, tool to add to those already in use in drug discovery, rather than an end unto itself [27]. As evidenced by the problems encountered by Genetech with the FDA in obtaining approval for tPA, biotechnology-derived drugs are no easier to bring to the marketplace than those discovered by more conventional means. Parenthetically, several biotechnology-based companies are now in the process of establishing traditional pharmacology departments to support their overall goals and in many instances declare themselves to be en route to becoming pharmaceutical companies.

Apart from biotechnology *per se,* technology in the area of drug discovery has made substantial advances, from receptor binding assays to robotic techniques used in repetitive assay situations that facilitate the testing of new compounds [28], to molecular modeling [29], to various computer- and laser-based instrumentation that permits the study of intact cell function in vitro and non-invasive imaging.

2.5   Changing academic/industrial relationships

The interface between academic and industrial scientists has also changed, partly in regard to the quality of scientists joining the industry in the past decade and a half, and partly because of reductions in academic funding that have made industry a more attractive career alternative. This latter change has also engendered an increase in collaborative ventures between industry and academia, from major multi-million dollar grants to universities, to Upjohn's "mini-institute" concept, to traditional contract work of a limited time frame related to a specific project.

The interest of venture capitalists in pharmaceuticals, evidenced by companies such as Nova, Athena and Cambridge Neurosciences, and Pharmatec, together with governmental encouragement of academic/industrial liasons, has further broadened the scope of academic interest in industrial ventures. The trend has led to the establishment of many more consultantships and encouraged many govermental agencies to form the biological equivalent of the computer industry's "Silicon Valley". This trend has reached an apex with the consideration by the U.S. Federal Government to privatize the National Institutes of Health, a move that was the subject of heated debate and the ultimate in "Reaganomics".

3   The current competitive environment

To a major extent, the larger, research-based pharmaceutical companies, through their size, resources and traditional interest, have been the major source of new drugs. But this is changing as the type of company involved in drug-related research changes. In addition to the biotechnology based companies, others such as Marion have achieved considerable success by astutely licensing acquisitions rather than in-house research operations. Alternatively, Nova, the brainchild of venture capitalists and Solmon H. Snyder at Johns Hopkins School of Medicine, seeks to provide a commercial interface between academia and industry. Ventures such as the British Biotechnology Group, represent similar changes in Europe. Whatever the nature of the approach, these changes are having considerable influence on the industrial research environment in the U.S.A. and Europe.

Another factor in regard to small research organizations is their ability to compete in markets which the major companies would find unprofitable. Large companies like Merck, Glaxo, American Home Products (AHP) and so on, need to generate revenues in the area of $ 100 million to support the effort they would have to mount to launch a new product. A company the size of a Nova can however, theoretically market a compound with anticipated revenues of $ 10–50 million and, given their size and overheads, still maintain profitability.

However the greatest change in the competitive environment that has occurred and is occurring, is in relation to the Japanese industry. Many new drugs have had their origins in research facilities in Tokyo and Osaka and, in the not too distant future, Beijing and Shanghai are expected to contribute to the worldwide drug pool. Like many other industries, the pharmaceutical industry, is becoming a global presence. For instance, in recent years, over 50 % of the drug-related patents awarded by the U.S. Patent Office have been to non-U.S. sources while in the biotechnology area, of 1,232 patents issued in 1986, only 261 (21 %) were issued to U.S. companies [30].

While the fortunes of companies such as Smith Kline Beckman,

Squibb and Glaxo have been revitalized as the result of single compounds, other companies have, due to patent expiration or the loss of compounds due to the occurrence of adverse side effects, been forced to drastically refocus their research efforts. The expiration of diazepam and related patents of Hoffman-La Roche and the resultant generic competition led to a major cutback in their U.S. work force. This has set a major precedent in an industry where recession and other economic factors historically had minimal impact on a tradition of "life-time" employment. The consolidation in 1987 of AHP's research efforts into a single company, Wyeth-Ayerst, a restructuring of research efforts in the Searle Division of Monsanto and the strategic consolidation of worldwide research efforts by multinationals such as ICI and CIBA-GEIGY are all measures that have been taken to reduce spiraling research costs, increase efficiency and maintain a competitive advantage.

Research consolidation can also occur through merger or acquisition of separate research organizations. CIBA-GEIGY was formed from two separate Swiss companies in the early 1970s, Glaxo incorporated from Glaxo and Allen and Hanbury's in the same time period. Recently, A.H. Robins has been acquired by AHP and Sterling Drug by Eastman Kodak. Such corporate mergers/takeovers bring with them considerable change, not only in the melding of different corporate cultures [30] but also in regard to the tactical approach necessary to consolidate two entities into an efficient and productive whole. Management is typically faced with maintaining morale incentives while adopting an effective resource strategy that frequently involves staff outplacement to avoid duplication. While a difficult prospect, it is highly likely that such mergers will continue at an accelerated rate through the remainder of this century [20] bringing with them the need to effectively "manage change".

On a more mercenary note, the large amounts of realizable equity within the portfolios of some pharmaceutical companies have made them vulnerable targets for hostile takeovers from arbitrageurs and, more unexpectedly, other pharmaceutical companies [32]. The demise of Revlon as a pharmaceutical company attests to the potential for a rapid change in corporate interests when aggressive third parties become involved, causing research personnel to unexpectedly enter the marketplace.

While the older companies in the industry are refocussing, either by choice or by the market situation, their research approaches from a strategic viewpoint, newer companies are also entering the arena. Apart from Marion which, as noted, has been highly successful using licensing and limited research partnerships as a basis for obtaining new products, chemical and consumer based companies such as Monsanto, Proctor and Gamble, DuPont and Eastman Kodak have invested heavily in either acquiring or starting pharmaceutical divisions. In Europe, the establishment of Synthelabo as a joint venture of Nestle and L'Oreal and of Laboratories Pierre Fabre as a pharmaceutical offshoot of a perfumery house attest to the perceived potential for improved drug discovery. General Electric, Unilever, Fuji Photo and Philips N.V. have also been rumored to be interested in entering the pharmaceutical industry.

While current strategy appears to require that nearly all Japanese companies enter licensing or cooperative ventures to introduce their medications to Western markets, it is highly probable that before the end of the present century, some Japanese majors will enter the arena in their own right, to be followed perhaps by Chinese companies [19].

The probable impact of such events on management philosophy and directions may be underscored by the close parallels with the automotive industry. Toyota was unable to shape a niche for its automobiles in the United States in the mid-1950s. However, in the past two decades this Japanese manufacturer has become a major player in the global automotive industry and forced a dramatic restrategization of the Detroit automotive industry to remain competitive [33]. Whether the potential entry of Takeda, Yamonuchi, Kyowa Hakko and other Japanese pharmaceutical companies will have a similar effect on the way that the established pharmaceutical industry conducts its business, from research to marketing, remains to be seen. Most certainly it will not retard the formulation of joint marketing agreements between major companies, exemplified by the joint marketing of ranitidine, cifelin and cilazapril by Roche and Glaxo and, more recently, the agreement by Merck and the Stuart Division of ICI to cross license and to market (albeit under different trade names), aldose reductase and ACE-inhibitors.

In further illustration of these developing global relationships, McNeil is now part of the Janssen organization, and British Oxygen, under the name Anaquest, is increasing its pharmaceutical efforts in the U.S.

4   The changing technological basis of drug discovery?

The classical and mechanistic approaches to drug discovery have been briefly alluded to in the Introduction. The classical approach, depending to a major extent on whole animal models, is responsible for the discovery and introduction of many therapeutic entities. It is however perceived as being part of the "primitive period" of drug discovery, depending almost exclusively on gut instinct, serendipity and as such is antiintellectual. The mechanistic approach on the other hand represents the "new" high technology interface where the cutting edge of science meets with unmet medical need to give greater assurance of novel drugs.

The influx of the new scientific disciplines related primarily to molecular biology, and an increased emphasis on the contribution of "basic" research has changed the tenor of drug-related research. In certain instances, as in the case of radioreceptor assays [28], this has provided for a cheaper, more rapid means by which to assess large numbers of compounds for biological activity. Other benefits of the computer age such as molecular modeling, while being hailed as recipes for the easy and rapid design of new drugs, have as yet, due to limited data bases, failed to be realized. Similarly, the high expectations for biotechnology as a new way to find drugs has left many companies, some years after hiring a full roster of molecular biologists, wondering how best to integrate the talents of these scientists into ongoing research programs having expected a technology, existing within a vacuum, by some magical event, to create its own goals and products. It has been predicted [26] that the failure to bridge the apparent intellectual gulf between the "old" and "new" technologies may bring about the "end of the domestic (pharmaceutical) industry". On the other hand, Paul Janssen has commented [27] that "the importance of biotechnological methods should not be exaggerated at the expense of classical methods which will always be far more important". The realistic melding of new technologies with goal-oriented drug discovery projects is therefore a major challenge for research management. To ignore the potential that new technologies can bring to the applied research process would be foolhardy. However, at the present time, these technologies are more readily integrated as aids to drug discovery rather than serving as a means in themselves.

As molecular biologists replicate within the industry, one discipline crucial to a pharmaceutical company is becoming less plentiful. While biochemists, enzymologists, chemists, molecular biologists, immunologists, software experts, molecular modelers, physiologists and even business administrators are added to the staff of a research department, it is often the preclinical pharmacologist who can provide the necessary degree of perspective to turn an interesting and innovative research project into a product. This is not necessarily because of superior intellect, but rather because the goal orientation of this discipline is toward that of a drug rather than an interesting research compound *per se*. The difference in these two disciplines has been addressed in a somewhat different context by de Stevens [5]. In retrospect, many successful drugs have had as a crucial element in their development, the involvement of a pharmacologist who has been able to ask the appropriate questions and prevent a project from becoming side-tracked. Thus in incorporating new technologies into the drug discovery process, the Research Director has to find a process to allow their appropriate application, rather than their being the tail that wags the dog.

5   Selecting targets

The search for new drugs within a historical context is often deemed to be one of exponential difficulty. The "easy" drugs have been discovered and now it is left, somewhat unfairly, to the present day researcher to find the elusive needle in the haystack.

With a plethora of technology, a huge information base, more highly trained and expert scientists than in the past and a closer involvement in "cutting edge" research, such sentiments may appear inappropriate. But one may have any amount of talent and financial resources, yet through poor definition of goals fail in the search for new compounds. It is not sufficient to hire the best researchers and let them have free rein to do their "own thing" [34], in the hope that new drugs will emerge. A successful organizational structure dedicated to drug discovery is one where the proven researcher fits in and which has selected commercially viable targets.

The factors contributing to selection criteria have been reviewed in depth by Faust [35] and include some 24 items covering scientific, marketing and organizational factors. Major factors included:
  - synergy with ongoing programs
  - probability of achieving project goals

- impact on short- and long-term programs
- cost and time to completion
- competitive situation
- likelihood of exclusivity of discoveries
- prestige and image value to company
- projected sales and profits

In addition, moral responsibility to society, the flexibility of the existing research organization and the acquisition of experience can be other issues of importance influencing the choice of project.

Drugs have been divided into innovative and non-innovative, or novel and "me-too". This latter category, as already discussed, is one that creates a great deal of controversy. To the public and health care organizations, it has connotations of wasted funds [20], and to the scientist, a whiff of plagaristic and undemanding science. However, the approach has been far from unsuccessful. The "fast-follower" approach in the area of the histamine H_2-blockers has yielded two drugs, cimetidine and ranitidine, both having sales around the billion dollar a year mark with the third, famotidine, from Merck/Yamounchi, on its way to market. Clearly a large market that was thought not to exist at the time of Black's pioneering work in this area has had significant impact. In the 1950s, antacids and surgery were the prevalent therapies for gastric ulcers. The idea that a drug company would be able to sell a compound to treat ulcers when there was no apparent unmet medical need was considered ludicrous. It is estimated however [36], that many millions of dollars have been saved through the use of these drugs. Furthermore not only are Astra/Merck introducing omeprazole, a H^+/K^+ ATPase inhibitor that acts like the H_2-blockers by reducing gastric acid secretion but the therapy is now expanding to the treatment of reflux esophagitis. Similarly, in the face of successful antihypertensive therapy with β-blockers and diuretics as antihypertensives, the need for other blood pressure medications, namely the ACE inhibitors, was considered uncertain.

In both these instances however, the resultant drugs have revolutionized therapeutic treatments and have been highly successful commercially.

It is however relatively simple to identify unmet medical need when therapies are plainly inadequate or absent. For example in the fields of atherosclerosis, cancer, osteoarthritis, rheumatoid arthritis and associated inflammatory conditions, acquired immunodeficency diseases, cognitive malfunction, neurodegeneration and diseases of the gastrointestinal tract, there is a major need for new therapies. In addition improved psychotherapeutic drugs in the anticonvulsant, anxiolytic, antipsychotic and antidepressant areas as well as improved analgesic agents with the potency of morphine and its congeners but lacking the various side effects will aid in improving the quality of life.

Often, as is the case in many of the therapeutic areas mentioned above, lack of knowledge about the etiology of the disease state precludes the choice of an obvious mechanistic target. The researcher, as well as management, is then left with the choice of either waiting to see what develops in terms of basic knowledge or sanctioning a basic research effort in the hope that it will identify therapeutically viable targets. The latter approach is basically one of hypothesis testing where potent, selective, bioavailable compounds are needed before the suspicion of the involvement of a given target in a disease state can be tested. This approach can be further limited, when, as is often the case the available preclinical animal models are either limited or predictive only within the context of known compounds. This has been a problem in the area of antianxiety therapy where most animal models are predictive only for the benzodiazepines [37] and also in the areas of depression and schizophrenia where no reliable models exist. Within a historical context, it can be argued that animal models are overly relied upon and have limited predictive value. Therefore in using the hypothesis testing approach, one strategy is to take the most active compound identified and, following toxicological evaluation, introduce this into man as rapidly as possible.

The mechanistic approach has as a limitation that it is often predicated on already known therapeutic agents. Thus there is a danger of selecting an approach which is circumscribed and therefore limited to a target that will offer little in the way of improved therapy. However, before dismissing what may be termed this "me-too" type approach, it should be noted that many apparent second generation compounds have, on being introduced into the market, been found to have significant additional properties that were not apparent in preclinical evaluation. In analyzing the patient benefits of the first and second ACE inhibitors, captopril and enalapril, Davis [38] has noted that because of the efforts made by Squibb and Merck, respectively, to differentiate products acting by a similar mechanism, ACE-inhibitors were shown to improve the survival of patients with congestive heart failure.

In defining an unmet medical need, an issue of semantics can often arise. To the businessman and to the consumer, "novel" generally represents a new type of drug, either for a new indication or with demonstrably superior therapeutic efficacy. The mechanism by which such effects are produced is often immaterial. Thus, as an instance, a novel non-steroidal antiinflammatory drug (NSAID) acting by antagonizing the actions of the algesic mediator, bradykinin, that is apparently no better in preclinical models than a conventional cyclooxygenase inhibitor, would be judged on the basis of cost rather than novelty of mechanistic effect. When 100 aspirin cost in the region of $ 1.50, what benefit would be accrued from providing a similar type medication at ten, or even 100 times the cost? This presupposes however, that a bradykinin antagonist would not have any additional beneficial effects in the clinic that may make its cost worthwhile. This also underlines the need for a careful assessment of the therapeutic target for compounds acting by novel mechanisms such that drugs are truely "value added" rather than "me-too". To the research worker however, novel often means a new mechanism for modulating a given phenomenon thought to be disease-related. The distinction between what constitutes a genuine "me-too" drug and one that is unexpectedly found to have additional beneficial properties defies prediction without a broad knowledge of the pharmacological properties of the compound [39].

Knowledge of the mechanism of action of a drug can also permit the elucidation of potential mechanisms for side effects. Therefore, in modifying a compound and deriving its optimal structure activity profile, it is possible that activities known to be related to toxicity can be eliminated by astute chemical modifications. Cordes [40] has verbalized the "one step at a time" approach to compound optimization. Most, if not all medications required for chronic administration need to be orally active. In designing new compounds from existing entities, there is often an effort to design, concommitantly, compounds with greater activity which maintain oral activity. As in all scientific experimentation, focussing on two endeavors at the same time tends to both dilute and obfuscate the value of the resultant data. The philosophy described by Cordes is to achieve maximal activity/efficacy by establishing a substantial SAR before focussing subsequent efforts on oral activity of a compound, the rationale being that activity and bioavailability represent two distinct properties of a molecule.

To improve on the somewhat haphazard nature of the hypothesis testing, mechanistic approach there has been increased focus on disease diagnosis. The identification of disease markers or "analytes" permits not only the potential identification of the mechanistic cause of a disease and its attendant diagnosis, but also the means by which to monitor the effectiveness of drug therapy. As drugs of the future aim to treat the causes, rather than the symptoms of a disease, the role of diagnostics, either biological (immunoassays) or physical (PET and CAT scanning, bio-NMR) will assume increasing importance, a factor that will on the one hand increase the cost of diagnosis but also decrease the cost of medication. Several pharmaceutical companies including SKF, Abbott, Roche, DuPont, CIBA-GEIGY and Marion as well as several major Japanese companies have diagnostic divisions. Another factor that can affect the ability to compete in the diagnostic area is the availability of the instrumentation by which to measure analytes. For instance, the Beckman Division of Smith, Kline Beckman can offer SKF in house hardware technology as an additional tool in competing in the area of diagnostics. In a coordinated approach, the diagnostic tools, the drugs, the means by which to measure their therapeutic levels and correlate this with efficacy and disease remission, as well as the instrumentation provide a major competitive thrust.

The process of selecting a viable research project for a chosen therapeutic area usually involves the formulation of a proposal, similar in nature to that submitted to the NIH as a grant, that assesses unmet medical need, state of the art knowledge, scientific opportunity and commercial viability. It very often appears, however, as if the choice (or recommendation) is made by the elimination of other, more unattractive projects, rather than the championship of a good project to the objective exclusion of others. Exclusion criteria may even revolve around who else is adopting the same approach. As with basic research, the validity of a research project may be considerably enhanced if the topic is fashionable. If company X is not doing a certain project then this may often bear more weight in decision making than any scientific rationale or proprietary knowledge. For the bloody minded, the converse may also operate.

In evaluating a therapeutic approach or area, there is often a need, because of corporate product support requirements, to take as given that a proscribed therapeutic area has to be worked in, irrespective of whether an objective evaluation of the science, resources, market and so on may deem it as being potentially unproductive. For instance, in the area of antipsychotic research, progress has been slow [41], with chlopromazine-like dopamine receptor antagonists still being used in the clinic and representing a major portion of the research effort in this area in the industry over the past decade. Compounds acting through serotonergic mechanisms appear to be interesting clinical candidates although their clinical efficacy and potential for side effects remains to be determined. While regionally selective dopamine receptor antagonists appear efficacious in preclinical models and can be predicted to lack the ability to induce extrapyramidal side effects, a lack of knowledge of the properties responsible for these side effects and recurring hepatotoxicity in many promising drug candidates has precluded the introduction of better drugs. One dopamine receptor blocker, tiaspirone, appears however, to have overcome this hurdle and is close to being introduced.

While the dibenzodiazepine, clozapine, appears to be the most promising antipsychotic to replace the classical dopamine receptor blockers, the mechanism for its apparent lack of side effects has not been established in over a decade of feverish research effort in both academia and industry. Blood-related toxicity has prevented clozapine thus far from replacing the more traditional antipsychotics although the demonstrated efficacy of the compound has led to the reintroduction of this compound at the request of psychiatrists as an orphan drug, the incidence of agranulocyotosis being seen as an acceptable risk. Given situations where progress has been so slow and where, in 30 years, only one significant improvement in therapy has been found, the ethical nature of the research effort has been questioned [41]. When success appears so elusive one may question why effort driven by rote and reflex tends to take the place of either creative thought or the more difficult acknowledgement that there may be no pragmatic approach to a particular research area at a given moment in time. It is also of interest within this context that while dopamine autoreceptor agonists (3-PPP), adenosine receptor agonists and neurotensin and cholecystokinin receptor modulators are being examined as potentially novel antipsychotics, the animal models used for their evaluation are precisely those used to measure dopamine receptor blockade. Thus despite the novelty of the approach, the test procedures related to disease predictability have remained the same. While antipsychotics have been singled out, other CNS active agents such as anticonvulsants and antidepressants represent similar areas where research into mechanisms of action have yet to yield positive results let alone improved therapeutic agents. Again, there is a discrepancy between the behavioral and molecular approaches. In such circumstances resources should ideally be utilized elsewhere until conditions become more favorable to the initiation of a *bona fide* effort. Alternatively, the industrial organization can make the commitment to perform or fund the necessary basic research to increase knowledge in these areas.

The choice of the therapeutic area can occur either as a "top down" mandate from the Research Director based on corporate needs and a management response to these and can be followed by a similar approach to the choice of therapeutic target or by an analysis by Research Management of the proposals submitted by the scientists. The latter approach, however, ensures a greater degree of commitment from the scientists and also takes advantage of their individual knowledge.

6   Research organizations

As in most industries today, line organizations in the pharmaceutical industry are typified by being "flat", having a minimum of hierarchy. Most strive therefore, to put resources where they are most effective – in providing maximal "hands on" effort in the drug discovery progress. The degree to which this "hands on" philosophy exists in a research organization is a reflection of the history of that organization, i.e. what has been successful or unsuccessful in the past and what is perceived as being necessary for success in the future.

Some line organizations are controlled in an autocratic manner by a Research Director and are effective and, in some instances highly successful. However, the trend in recent years has been to separate line management administrative tasks from those related to research, to let the discoverers do the discovering while management performs monitoring, facilitating and mentoring roles. This has not abnegated the responsibilities or authority of the Research Director but rather allowed for a more synergistic team effort with ownership extending throughout the research organization. Therapeutic discovery units or disease area working groups typify this approach with varying degrees of autonomy, yet working through line management to address tactical, as opposed to scientific, issues. In some companies the emphasis is on all scientists, irrespective of seniority, maintaining contact with the bench by performing experiments. In others, senior staff are expected to direct research and draw conclusions from the work of their junior staff. In either utopian situation, the scientist is left dealing with the science for which he or she was trained.

The quality and expertise of the research personnel are significant factors in the structure of the research organization and as with governments, many differently designed organizations all appear to function equally well, with inevitable exceptions that may defy the imagination in terms of their innate potential for either success or failure.

The responsibilities of the Research Director differ from company to company. In some he or she has responsibility for the biological and chemical aspects of the drug discovery process, preclinical drug metabolism, formulation, toxicology and certain aspects of clinical testing, usually through early Phase II but occasionally to registration and beyond (i.e. product support). One individual is thus ultimately responsible for the allocation of resources and the decision making relevant to providing compounds with IND status. Irrespective of the degree and form of responsibility, once a target has been identified, the Research Director then has the mandate to implement the research effort. Given that the necessary skills and expertise are present within an existing organization, this challenge is to effectively direct these efforts in a manner again consistent with the corporate research philosophy and available resources.

In other organizations, research and development may be different entities with correspondingly different philosophies. Testa has commented [42], that "R and D ...overlap considerably. Schematically, research is a long-range task with the potential for considerable scientific benefits and a high risk of failure. Development is a short-range task with the potential for considerable material profits and a lower risk of failure". In such situations, the Research Director is then faced with additional problems in formulating tactical plans since key elements to success are outside his or her direct control while a synergistic interface crucial to a drug reaching the market may be totally lacking [43–45]. In newer research organizations, less bound by convention and history, a more entrepreneurial environment can exist, engendering a more flexible and less staid approach to the drug discovery process.

6.1   Line management

The line management approach is strictly hierarchical. The Research Director, by virtue of his or her authority, directs the research effort, often overseeing the day-to-day generation of data and its interpretation and then deciding what changes in direction may be necessary. Individuals working on a research project are assigned to that project as need be, their skills reflecting the needs of the project at a given point in time. Resource issues are normally dealt with by line management fiat, often with input from subordinates but often as the result of a single decision maker. In the line organization, a controlled research environment exists with the lines of responsibility and accountability usually well delineated. Thus management is responsible for the quality of the science ensuring, through frequent review processes and consistent hiring practices, that a culture is present congruous with the needs of the corporation. In many organizations, this becomes formalized into a "Center Of Excellence" concept, with the talent, expertise and experience for a given therapeutic area operating in a distinct geographical locale. Care must be taken at times however, to prevent such Centers from becoming incestuous. The working groups in such an organization revolve around distinct technical disciplines (Fig. 2), i.e. biochemistry, molecular biology, drug metabolism, etc. with each functional group responsible for several research projects. In these respects, project execution in a line organization is akin to that of the *ad hoc* structure described by de Stevens [5]. By adopting this approach, a maximum of flexibility can be achieved with highly skilled, technologically based groups being available as need for their support becomes evident [46].

One potential disadvantage of this approach is a general lack of ownership of the project with a corresponding reduction in motivation. Another is the vulnerability of the organization when a key individual leaves the company. However, on the positive side, resource management is usually easy to deal with and the organizational culture is visibly and effectively supported. In the *ad hoc* arrangement, however, the formation of interdisciplinary groups based on a common interest in a given project tend to be both informal and highly committed and as such more able to effect change and garner support in achieving the ongoing goals of a project. A management that can provide the environment for the effective formation of such groups within the line organization may have the best of all achievable worlds.

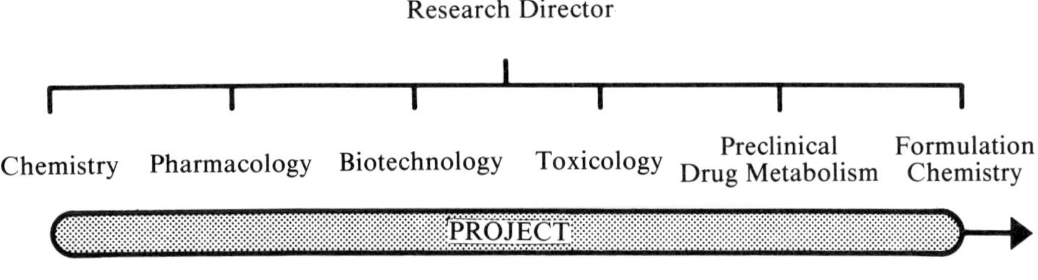

Fig. 2
The line/*ad hoc* organization

6.2   Project team

The project team has been described as a "task force created to solve a problem" [3]. It usually functions in relation to a specific therapeutic program approach as opposed to a complete therapeutic area [46]. The makeup of the project team and its limits of authority have as many permutations as there are colors. Some examples are: a single leader, having either authority or an advisory role, joint leadership by chemist and biologist, one subordinate to the other, line hierarchy taking precedent over project function, line management being purely admin-istrative, line management being the decision making body with input from the project team and so on.

In terms of makeup, the project team can include many different mixes of organizational skills. For instance at its most basic level, the project team would consist of the chemists and biologists involved in the pro-ject. These individuals are usually those involved in the synthesis and testing of compounds, and collectively, through a project leader or leader(s) set the day to day goals of the project and report on their progress to line management. As a project attains momentum, these in-dividuals could be augmented by representatives from preclinical drug metabolism (pharmacokinetics), pharmaceutical development, and clinical development (Fig. 3). As a compound becomes a candidate for safety evaluation, then representatives from toxicology and statistics are appropriate [45]. Once a compound is close to or has achieved the status of an Investigational Exemption for a New Drug (IND) and the European and Japanese equivalents, Regulatory Affairs would then become involved and at some point further in time, as Clinical Trials move through their various phases, Quality Control, Production and

Marketing would join the team. The precise overlaps between these various segments of the project team are of course a function of the organization in which they operate. In some companies, there are preclinical teams that hand over the "baton" to separate development and clinical teams as compounds progress through the development phases. In other companies, project teams may include clinicians who are responsible for the IND application and for clinical studies often through Phase II. The pros and cons of these various project team mixes are endlessly debatable. Keeping a project team operational solely at the preclinical level maintains the research expertise close to the bench, allowing those versed in the specialities of development to add their own unique efforts at an appropriate time.

This however, precludes research scientists from obtaining a broader perspective of the drug discovery process. When a clinician is part of

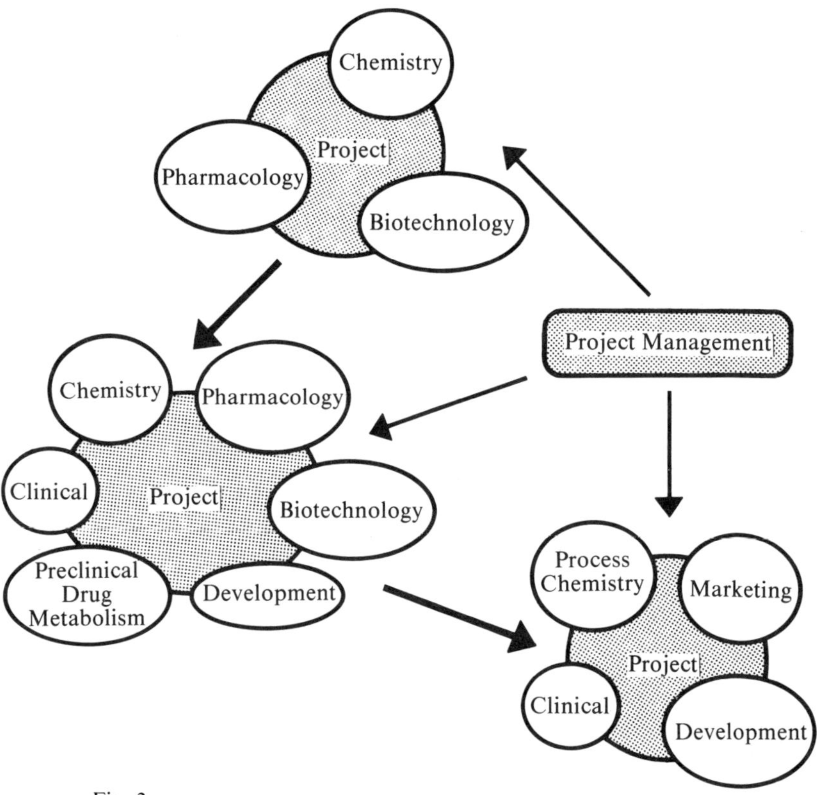

Fig. 3

The project team. Each stage represents an approximation of the relative contributions of key elements to the evolution of a typical project team.

the team, a project can move more smoothly through the preclinical/clinical transition with a maximum of synergy and without the need to run relay races and risk the "baton" being dropped.

The project team often results in a greater sense of ownership in a research project and an ability to create more interfaces across a variety of different functional groups resulting in a higher visibility for a project and the opportunity to delegate responsibility further down an organization. This becomes an ideal vehicle for exposing individuals to career challenges and ascertaining their organizational abilities without a permanent commitment away from their areas of scientific expertise. The project team system can also provide excellent rewards for the motivated individual by providing a constant intellectual challenge. In addition, to a large extent, the flexibility of a research organization can be maintained, especially in the case of "pinball" management, where those successful in achieving the goals of one project get as a reward, the opportunity to "play again" on another project team. Another positive facet of the project team system is that of multidisciplinary team building without "walls", a very necessary task when resources are limited. Research units, as well as individuals, can achieve a higher degree of flexibility through the project team, allowing line management to attend to the environment thus leaving the scientists to manage their science. In addition, entrepreneurship can be encouraged in a positive, synergistic manner with the goal attainment of the individual being of mutual benefit to team members, a sharing of success by being part of it. It is probably true that few compounds are discovered these days which are not the result of a team effort.

In an organization where resources are at a premium and where the team is formed across lines rather than as a self-contained multidisciplinary unit (Fig. 2), the project team concept can, through the enthusiasm and single-mindedness of project team leaders, generate an enormous amount of counterproductive friction. In such circumstances the line organization must implement a system of checks and balances to encourage such enthusiasm without ancillary resource sources being subjected to unrealistic work demands and subsequent "burnout". The allocation of sufficient resources to achieve the goals of the project (and the setting of realistic goals) by providing a critical mass during the strategic planning process can do much to ensure the success of a project and avoid such situations. Jack [3] has emphasized the need to limit the size of a project team to key personnel in order to maintain viability. This then avoids the inertia and "decision by committee" associated with large teams and the attendant need for meetings that are often "show and tell" in nature. In such circumstances, intellectual procrastination and socializing often take the place of measurable project progress. Form rather than substance.

In some companies professional project management groups exist, separate from the project teams. Acting in a support capacity, this provides the organization with a means to circumvent some of the more negative aspects of the project team concept while encouraging those factors that lead to success. In addition, especially where project teams are self-contained functional units, project management teams can provide a consistency of expertise and experience that ensures that documentation on a compound, and its progress through a flow chart process [6], are both uniform. A potential disadvantage however, may be to stifle scientific creativity and to require uniformity of documentation at the expense of progress.

6.3   Matrix

The matrix approach is the most literal extension of the project team where the therapeutic area and its attendant projects are multidisciplinary "minicompanies" [5] competing within a company for resources. The appeal of such an organization of which 3M and Johnson and Johnson are good examples [13] is that excellence and endeavor are rewarded on the basis of success and there is a constant motivation and ability to control resources in a highly effective manner. The maintenance of divisional integrity, the formation of new "minicompanies" when a therapeutic area becomes successful in its own right – as might occur in the case of an arteriosclerosis group forming as a biochemically-based offshoot of a more physiologically oriented cardiovascular group – and the ability to keep *bona fide* administrative personnel to a minimum are all attributes of the matrix system [46]. On the downside, as can occur with the project team concept, flexibility is often sacrificed, not only in terms of the expertise of the individuals within a matrix organization dedicated to one area but also in terms of the parent organization facing the choice of continually putting resources into one of these "minicompanies" or alternatively, disbanding them, usually with staff outplacements.

6.4   Critical mass and resource balancing

In allocating resources to a research organization, research management must balance the needs of competing projects that will ensure a well-balanced portfolio of compounds consistent with the identified needs of the company. To be successful however, it is not only the proper choice of target and the hiring and training of the best qualified individuals, but also the timing that is crucial. A perennial comment is that often, promising innovative research projects, because of their early status, get not the resources they need but those that are left over when other, more established projects are funded. When a product is obvious, even at the preclinical level, it is very difficult not to allocate resources to a well-established project to ensure that deadlines and goals are met, rather than investing resources in the uncertainty of initiatory projects. It is not unheard of that seemingly rationally planned, well thought out projects are, by a process of resource attrition, supported by half a person, clearly less than the optimal critical mass. This is of especial concern when it is often the initiatory projects that require the most support and offer the best opportunity for novel, rather than "me-too" compounds.

The optimal number of individuals required to work on a project in a therapeutic area and the ideal number of projects to support a therapeutic goal are issues that decide the allocation of resources and the ability of a research organization to work in a broad range of areas. For an established drug discovery project, 10–20 scientists may be optimal. A combination of two established projects and two initiatory projects could be considered an optimal mix for a success in discovery of a drug while at the same time having a significant presence in an area. The ratio of chemists to biologists is dependent on the stage of the project. One viewpoint is that a mature project would have two chemists to each biologist while an initiatory project may have no chemical support other than that required to produce reference standards for hypothesis testing. Another viewpoint is that Ph. D. level personnel should be matched on a 1:1 basis in both chemistry and biology with more technical support for the biologists than for the chemists.

In addition to actual project-related efforts, bench scientists may also be required to do non-project work to evaluate licensing candidates or support clinical and marketing efforts. As a compound nears a market launch, any new data generated using scientific methodologies that were not available at the time of the IND can serve to publicize the new drug. As previously noted [6], the involvement of a research scientist in drug discovery extends considerably beyond the actual work that led to the identification of a toxicological candidate. With research reports from *Nature, Science* and the *New England Journal Of Medicine* being reported regularly in the pages of the *New York Times,* publication is assuming an even more important role in supporting the introduction of new products into the market. Such efforts, however, divert resources from ongoing research projects and like mature projects can appear more important than novel research because of the product involvement or the time frames imposed on licensing evaluations. In some companies, where there is a history of product success in licensing, separate licensing departments exist that provide the necessary resources to evaluate outside compounds.

One final point in regard to the resource issue is that of flexibility. The assembly of all the necessary expertise to support a hypertension project, for instance, requires the commitment of personnel resources in this area for a period of anywhere between 3 and 8 years depending on the complexity and novelty of the tactical approach adopted. At the end of this time period, with the identity of the targeted clinical candidate and an appropriate backup, the Research Director is left with a choice of what to do with these resources. Should this group of researchers be directed to other projects? Or should they look to another therapeutic target in the same area? For instance, cardiotonics or treatments for myocardial ischemia? Or should they, because of their "cutting edge" knowledge, continue to work on new approaches to hypertension? These considerations assume greater proportions when corporate strategical plans require that whole areas be eliminated and research resources be diverted to start new projects or areas. Clearly a dilemma then arises that reflects on the flexibility of the organizational structure [35, 46]. One is left with either convincing the affected individuals to move to new areas or alternately, seek their fortunes elsewhere. In terms of the former, this is probably easiest in a line organization where the technical resources are not specifically project-related, less difficult in a project team based organization and nigh impossible in the matrix system. When the individuals identify their careers with a company rather than a research speciality, such problems can be overcome. While this is usually less of a problem with chemists than biologists, the loss of talented individuals, for whatever reason, negatively affects stability, even more so when the individuals affected have high visibility within the organization and with their peers at other companies.

Another facet of the flexibility issue is the balancing of resources when one area becomes "hot". This is of especial concern when a drug is close to being a therapeutic candidate and a highly competitive situation exists. The movement of necessary personnel between projects is not always viable, yet in achieving deadlines, it may be more applicable to increase a project-related group of 20 to a 100 or more, as circumstances require, rather than having 40 individuals totally dedicated for a longer time period with fluctuations in project-related workload. In some organizations, teams named in the vernacular of the police "SWAT" (Special Weapons And Tactics) teams can be established in the chemical area. Senior researchers with proven track records may then be free to input their skills to a project on a needed basis with management consent, thus providing critical extra resources without static commitment.

6.5   The Research/Development interface

Irrespective of whether Research and Development are part of one line organization or separate organizations, there can be considerable differences in philosophy and product goals. Ideally, marketing should identify unmet needs and work with Research and Development to identify projects that will yield products to treat the diseases targeted, this irrespective of mechanism. Given the 10–15-year lapse between the identification of a need and the introduction of the product to market, the value of predictability is a gamble at best that besets the whole industry. The example of the resistance to the H_2-blockers in light of the highly successful antacid therapies underscores the potential loss of opportunity that can occur when novelty is ignored. Yet, apart from possessing a crystal ball, it is difficult for any individual to know what is going to be a significant new drug for a medical need that appears to be met. Other than performing the impossible task of putting every compound that may treat a disease on the market to see whether it might have any significant advantages over existing medications, there appears to be no quantifiable way to assess what will be successful. Predicting unmet medical need based on a research portfolio that is 10–15 years out of date the day it is decided is a daunting task that is inevitably risky. Especially when those involved in the initial choices are often legatees who, due to the progression of time, are no longer with the organization when the consequences of their efforts bear fruit (or appear barren).

What can be done to reduce this risk is to have a synergistic interface between the various parties involved in the product process. This can only be done by a continuing dialog at the grass roots level where products emerge from research with the full knowledge and support of development and marketing. In at least one organization, a task force evaluating existing and potential problems in the R & D interface reporting directly to the Chairman of the company, outside the line organization, has proven to be an effective means by which to reduce unnecessary friction. In initiating a research project, full input from all three parties to what might be the best approach is essential. In this, research should realize that product, not interesting research tool, is the final point while development should give some credence to its own research staff in that they will strive to give them the best candidates that they can. In this whole process there is need for trust and mutual respect for the various skills and disciplines necessary to move a compound from test tube to the druggist's shelf. In such an environment where a proven track record of products exists, then development can be more comfortable in taking the "flyers" from research that may be the blockbuster drugs of the future.

Most companies have a large number of products in their developmental pipelines. Inevitably, the resources that are needed to bring these to the market are less than adequate. Accordingly, long time delays occur, especially when coordinating separate developmental organizations in different countries. As a result, any new candidates from research, irrespective of their potential become additional numbers in a pipeline which is already well loaded. This problem can be addressed by moving resources between the research and developmental departments. Theoretically, such a shift in resources will not affect the research aspects of the organization since the compounds which are moving through a finite pipeline represent the endpoint of a research effort which requires validation in the clinic. Thus, as the research effort reaches its endpoint, researchers can be transferred to development functions to oversee the compounds that they originally discovered. As the clinical trials provide evidence of the outcome of the research effort, especially in the case of compounds that are not successful, then the researchers working in development, finding themselves without compounds to develop, can move back to the laboratory to find successor candidates or start new projects.

While a pleasant scenario to balance research and development needs, this strategy can tend to emphasize development to the detriment of research. Inevitably, a focus on the short term at the expense of the future can lead to a mortgage on future viability. The loss of a clinical candidate, especially when the backup compound based on a mechanistic approach may suffer a similar fate, can incur time delays that double the time (and cost) of drug discovery and development. One way that has been suggested to cut down on the development overload is for research to have a "narrow window" to development [40]. Rather than introduce a number of compounds into the development pipeline, researchers should endeavor to identify the best compound possible, in terms of potential efficacy and bioavailability, and introduce this rapidly into development. While there is still an obvious need for a backup compound (a point that can sometimes be lost in the exigencies of resource priorities), in finding the best compound rather than settling for "second best", research can significantly aid the development process and the credibility of the research department.

Even when there is an agreement on the type of drug, failure to appreciate subtle nuances of the product's positioning in the marketplace, especially when dealing with "me-too's" can be extremely costly [45]. Resource management by effective interfacing and dialog at all levels can do much to reduce the inevitable risk as can the acknowledgement that friction is an inevitable occurrence in any concerted interaction between disciplines. Acceptance of this fact can do much to enhance the management process and avoid the establishment of a complacent organization.

6.6   Morale and motivation

In recent years as evidenced in part by the project team concept, there has been a trend to enhance the value of the individual research scientist by involving him or her more actively in the decision making process, thereby capitalizing on individual expertise and achieving a high degree of commitment and ownership. While this can tend to produce chaos where the insight and experience are lacking, it is able to provide a more effective research team, all the more so when the corporate philosophy is well established and can accordingly benefit from new blood by assimilation. The incentives for adopting this approach come, in part, from the folklore of Silicon Valley and from often related experiences from the major biotechnology-based companies where minimal management involvement has achieved an unusually high degree of commitment and success.

In reflecting on these success stories which often tend to treat with disdain the established research environment, one may wonder whether the dedication that results in stories of 18-hour work days and jammed parking lots at daybreak on a Saturday morning, may more cynically be viewed in terms of the incentives that an entrapreneurial, non-hierarchical approach to research can offer. These are often as traditional and a good deal more liberal than those existent within the established companies. The cooperative project team existent in such organizations in which self-interest is subjugated to project goals, rather than reflecting altruistic disinterest, may be more appropriately understood with the context of these individuals being stockholders in their own organizations. Unlike their counterparts in established companies whose opportunities for financial success (and failure) deviate little around a common norm, those in new companies have the ability to become millionaires (or bankrupts) as a result of success (or failure) in their efforts.

The role of the scientist within a business-oriented research organization offers a great deal of trial and tribulation to research management. Often the training that a scientist receives is conducive to creating an individual who perceives his or her personal goals as paramount over any related to a team effort and who also perceives as antithetical the constraints on the "freedom" of science imposed by business interests. As indicated, a strong organization can usually provide the culture to meld the needs and ambitions of such an individual without compromising his or her ideals or objectivity. Indeed it is precisely the goal orientation of the drug company that may harness such ideals in a more effective manner than academic research. However, maintaining creativity on a day-to-day basis within a business environment is a difficult challenge [47, 48]. The personal qualities critical to creativity in R & D scientists have been summarized [49] as: intrinsic motivation, ability and experience, risk orientation, social skills, persistence, curiosity, energy, intellectual honesty and the ability to impact team synergistics. A tall order especially when the physical environment has a considerable impact on the ability to encourage and harness these qualities. One approach that has had varying degrees of success is the establishment of a dual ladder system. Thus, in order to keep creative scientists working at the bench, rewards, both financial and peer-related, can be established for such individuals without their having to become administrators. The research ladder usually exists in some form of parity with that of the line research organization. Individuals high in this ladder are the role models for the commitment of an organization to scientific excellence and have to be identified with care. In organizations such as 3M, Corning Glass and Bell Laboratories, research scientists with three, two or no direct reports have been rumored to earn six figure salaries. To achieve similar status in a management or administrative ladder would require a considerable codification of budget and people responsibilities to the nth degree to accord similar status. The nature of the research emphasis, technological as opposed to biological in these companies may explain such largesse, the value of research being much more apparent because of the reduced time lines to product as compared to the drug discovery process. In general, however, the dual ladder is a perennial topic with which to find fault, with theoretical commitment often subject to resource priorities.

Aside from the availability of a true dual ladder system, the opportunity to publish research data, attend scientific meetings and have time to replenish creativity through nonaccountable research time can do much to enhance the scientists' status in the business environment. The failure to provide these opportunities will either destroy the innovative capability of the organization and/or enhance its staff turnover rate.

6.7   Managing research

The management of research where the self-motivation of individuals involves goal satisfaction and recognition in addition to pecuniary reward, is a difficult problem. Preventing moves into management by gifted scientists lacking administrative abilities, tolerating the nonconformist and other such aspects of chaos management [50] require that the manager at whatever level, while assuring that the company's interests are not ignored or abused, deal with each individual according to his or her obvious strengths and weaknesses. While it is entirely appropriate that those present within a *bona fide* business setting be aware of the subtle nuances of dress code and etiquette, to smother a gifted, innovative researcher to ensure conformity is a destructive process. Thus while the censure [4] of the "management by objectives" approach, typical of large organizations may appear anarchical, it is perhaps best interpreted in appreciating the fact that while no one is irreplaceable, some individuals can offer more to a research organization than the performance to the level of a somewhat abstract job description. Dealing with individuals on a case-by-case basis and accommodating differences and diversity is thus a necessary prequisite for the research manager who intends to successfully foster research projects [49]. Such management approaches must also engender support for the team approach. To favor "super stars" at the expense of other individuals whose talents are either not as visible but nonetheless indispensible, or held in check, can generate poor morale and a fragmentation of any cohesive focus. Within the context of the "new" organization defined by Drucker [51], synchrony and self-direction appear to be the trends for information-based organizations of the future. Crucial to such a concept is a high degree of personal autonomy that requires the goals of an organization to be consistent, few and highly visible. The flow of information in $R + D$ organizations will critically affect the type of organization, the individuals it employs and its overall effectiveness [46, 51].

7   Synergistic interfaces – Academic/industrial liasons and drug discovery

As noted, there has been a considerable change in the attitude of the academic scientist to his or her counterpart in industry in the past two decades. Where previously there was disdain for the industrial researcher as being an individual unable to compete for grants or hard money positions, today the industrial researcher is very much at the cutting edge of science. Many members of the Royal Society and the National Academy of Sciences work in industry, reflecting the quality and contribution that applied research can make to the broadening of scientific knowledge.

In examining the respective roles and contributions of academic and industrial scientists to the process of drug discovery, Maxwell [39] has noted that it is very much a symbiotic process, not only in terms of the funding process but also in the provision of the tools by which research advances are made. Industrial, mission-oriented research provides new drugs, often with new mechanisms of action. As these become public knowledge, either through their introduction as therapeutic agents or as research tools with no human potential, they provide the means by which researchers in academia can test hypotheses. In successfully doing so, new biological mechanisms can be identified which, in turn, become target sites for new drugs. Taber [52] has described this succinctly in that basic research deals with targets for drugs while drug discovery looks for drugs for targets. Maxwell has similarly analyzed the differences in research philosophies between academic and applied research endeavors. Academic pharmacologists are, in his view [39], accorded the "luxury" of "examining in relatively great detail few drugs" while "drug developers need to know the full spectrum of the pharmacology of their compounds . . . and relate it to already existing agents (and) are denied the additional luxury of intensively focussing their research efforts on a narrow front". Similar considerations by Black [34], have further delineated the respective research philosophies in a more dramatic manner into two classes of pharmacology termed exploratory and analytical.

Exploratory pharmacology, like the concept of academic research, is defined as the use of classified drugs to understand biological systems. The concept and need for drug classification is the object of analytical pharmacology, a theme proposed as central to drug-related research [53].

Analytical or applied, pharmacology involves the use of biological systems, preferably via a holistic approach, to define the molecular mechanism(s) of action of "chemical probes". The importance of this conceptual approach has resulted in the establishment of industrial units devoted to this theme, by Leff at Wellcome in the U.K. and Kenakin at Glaxo in North Carolina. Both these laboratories have made significant contributions to the science of pharmacology that go well beyond the topic of drug discovery.

At a self-described rhetorical level, Black has emphasized another major difference between the two defined branches of present day pharmacology, this being their relative degrees of reality. Analytical pharmacology deals "with real, existential, problems" while academic or basic research concerns itself with a continuous, self-perpetuating process that is "unencumbered by the trappings of reality". The implication of Sir James' perspective for research management is one of

scientific identity, namely in hiring the right type of person who is conceptually suited to do drug discovery, as opposed to basic research.
In many of the more successful organizations, there has been a trend to expand by hiring almost exclusively from academia, individuals whose perspective, based on their own experiences as researchers, may result in "wasted experience" [34]. This can result in a diminution in the complimentary basic research that is so necessary to fuel the drug discovery process. As the most qualified individuals leave academia for industry and reduce their involvement in the teaching and training of their successors, where do academicians of the future come from?
In reflecting on his two decades as one of the leading figures in drug research, Black also sees an increase in "quasi-academically oriented" studies in industry, a corresponding increase in the number of Ph. D. students, a weakening of research management, a less-disciplined and more single-minded attitude of the "journeyman scientist" and an increase in non-applied research in industry. While the term Churchillian may be too grandiloquent for such views, it will be interesting to see what comments historians of the pharmaceutical industry may have in regard to Black's comments when these are placed within the context of new product introduction in the years to come.
All three industry veterans therefore see a need for a delineation in the type of research done at universities and in industry and the need to maintain this distinction to allow each to perform its proscribed function.

7.1   The academic entrepreneur

In the past, the void between academia and industry prevented academia from benefitting financially from its own research efforts. In England, for instance, the technique of cloning for which Milstein received the Nobel prize, and which provided the genesis of the biotechnology industry, was not patented by the U.K. Medical Research Council. This discovery was therefore readily accessible at no cost to establish the worldwide biotechnology industry.
This attitude has changed to the extent that many scientists are now willing not only to interface with industry but also, in many instances, on the basis of even a single patent, to start their own companies.

7.2   Research funding from industry

Industry has traditionally funded academia, from relatively inexpensive studentships, to major research buildings, to universities. While such funding usually typified a concern for the social fabric of society, a willingness to reinvest profits for the good of future generations and, no doubt, some concern for tax incentives, more recent donations to universities and other seats of learning have been directly linked to business interests.

7.2.1   Focussed funding

The concept of "focussed" funding by the pharmaceutical industry has usually revolved around a specific problem for which the resources or expertise are not available in house. A sum of money is thus paid in expectation of a body of data that will assist in the characterization and/or development of a compound. The advantages of "fee" research enable the use of specialized resources in a cost effective manner and are highly beneficial to both payee and payor provided neither depends on the other to the point of exclusivity.

7.2.2   Open-ended funding

Open-ended research funding began in earnest in the mid 1970s with large chemical and pharmaceutical corporations donating many millions of dollars over 5–15-year periods to finance research of an undefined nature at leading biomedical universities. These companies have negotiated various types of contracts usually at the level of the "right-of-first refusal" of new products/technologies but also in relation to technology access. Sapienza [14] has cautioned that the over direction of such industrial grants by defining the specific nature of a project can inhibit the research it hopes to benefit from, thus turning the university into a contract laboratory. Conversely however, the granting of large amounts of unfettered money to an academic institution whose major functions are those of detached, basic research and teaching may represent an exceedingly poor return on investment somewhat akin to asking Le Corbusier to design a building without specifying its form or function. The questionable success of the Roche Institute for Molecular Biology [14] may highlight the danger of an industry

-funded basic research unit operating in isolation, both geographical and philosophical, from academic and industrial inputs, respectively.

7.2.3   Mini-institutes

Mini-institutes represent a mix between the two previously described funding approaches. These are not to be confused with the establishment of research units at universities as has occurred with the various Hoechst research units, the Parke Davis and Smith, Kline and Beckman research interactions at the University of Cambridge, U.K., the Squibb funding at the University of Oxford, the Astra and Sandoz Research Units in London, the Miles Institute at Yale and the Reckitt and Coleman Unit at Bristol, U.K., nor the various clinical research units at major medical centers that are funded by the industry, usually on an exclusive basis. Mini-institutes represent the funding of an independent research organization for a finite period with a defined therapeutic research focus. Members of these units are not company employees and remain as independent researchers and teachers. To date, two such entities have been publicly identified, both under the auspices of Upjohn as European Discovery Centers: a unit at Goteberg University researching into psychotropic drugs and the second, a joint Anglo-French project related to histamine receptors. Such ventures offer an ideal way to decentralize research while expanding research horizons via the utilization of small established research groups with expert leadership without the commitment to hire extra staff.

A major criticism of academic/industry interfaces, especially in the biomedical area, is that industry benefits at the expense of higher learning. This interface has been accused of impeding the objective pursuit of knowledge and the free dissemination of new research findings. Delays in publication, potential conflicts of interest where university staff hold major equity positions in venture capital companies and devote large portions of their time consulting with such entities are also legitimate points that need to be addressed in order that academia may continue to uphold its irreplaceable role in the fabric of society [54]. This increased focus on potential conflicts of interest in regard to the traditionally mutually beneficial interactions between industry and academia has probably arisen because of a marked increase in the profit motive. In the past, when interactions between universities and industry were usually restricted to major companies, few individuals became rich. Today, when the frequency but not the nature of such interactions has changed, the major difference is the fact that consultants can earn six figure retainers. This changes to overall perception of the benefits to be accrued from these interactions and engenders considerably more scrutiny with attendant ethical considerations.

7.3 Industry's role in the training of scientists

In the past decade there has been much concern about the training of those involved in drug-related applied research. The classical pharmacologist, as already noted, has become a rare breed to the extent that those countries where pharmacology is still taught as a discipline (England and Europe) loose their best graduates to industry, world-wide, as part of a brain drain. While working parties acting through various professional organizations in the U.S. have sought to redress this problem, pharmacologists are still few and far between in North America, a fact that will have major consequences for drug-related research in the near future.

There is little formal training within industry *per se*. In England, the government and industry sponsor CASE-type awards where students at the undergraduate or graduate level spend part of their period of academic training in industry. This permits the student to obtain exposure that will help in formalizing career objectives while allowing industry to participate in the training process. Various work/study programs of a similar nature exist throughout the U.S. although these are usually less formal in nature and usually occur between a company and a local university or when there is a research topic of mutual interest to both parties. Such interactions differ from the situation where doctoral training occurs within a company, a practice that can tend to weaken both the overall need of industry to conduct focussed research and of the individual to receive training in an environment that is primarily concerned with pure research and free from overt commercial considerations. Some compromises also need to be made to accommodate the extra time and resource allocations that are needed to provide the student with the tools and training necessary to obtain Ph.D. degrees comparable to those awarded in the very different university environment. Hopefully these should not negatively impact company-related efforts.

Many more postdoctoral fellowships now exist in industry. Rather than being exposed to different academic centers, Ph.Ds in increasing numbers forgo the opportunity of continuing their academic training to do "basic applied" research in industry. Indeed there appears to be a positive correlation between commercial success in the pharmaceutical industry and the number of postdoctoral fellows in residence although the latter has yet to be shown to be the cause of the former. A problem that emerges from this "quasi-academic" focus [34] is that many companies find themselves with two types of scientists; those doing drug-related research and those doing basic research. The integration of the product of these two endeavors is certainly easier than the management of the animosity that exists when one faction accuses the other of being either "ivory-tower elitists" and of no use to the company or "drug discovery drones" who are of no use to science. Problems also emerge when an industrial postdoctoral fellowship comes to an end since many more are trained than there are positions for. Presumably those individuals who are good continue their careers either in industry or academia, bringing to the latter a valuable perspective of goal-oriented research.

Two research areas unique to the industry are toxicology and drug delivery. Yet only very recently have these been the subject of intensive research efforts other than those related to the properties of a given compound on its way to market. The development of a body of theory to support these areas may be a legitimate cause for sponsoring basic research within the industry. Alternatively, specialized academic centers such as the Center for Bio-Pharmaceutical Sciences at Leiden University in The Netherlands may represent focal points for collaborative industry support to address global issues akin to that recently established for the U.S. electronics industry [55].

8   The quantifiable business of drug discovery?

A perennial problem in managing a pharmaceutical organization, as opposed to a drug discovery department, is the fine art of achieving a balance between business, in terms of the bottom line, and the support of research, an endeavor that often seems esoteric and a poor return on investment [16]. Not only can it appear unmanageable, but it lacks the predictability that is a given in other industries. In addition, many of the stages of bringing a product to market lie outside the control of the organization sponsoring the product, when that product is a drug. For instance, in returning to the comparisons with the automobile industry, there are highly publicized criteria that are mandatory to get a car onto the highway. It must have sufficient safety features, seat belts, brakes and the like, but in general, with few exceptions, the basic product has a proscribed passage onto the market with all aspects of the design, development and marketing processes lying within the control of the manufacturer. Even when problems are discovered when the product is marketed, the automobile can be recalled for modification rather than scrapped to start all over again. Thus there is a major, basic difference between products based on the outcome of either engineering or biological research. Many research projects, as great a proportion as 90 %, fail to yield product and even when they do, timing is a crucial factor. At one point about 5 years ago, 26 companies had research programs to identify leukotriene D_4 antagonists. Similarly there are currently 43 ACE inhibitors in the research pipelines of several companies in addition to the three already on the market. Given the economics of the market place, even when successful, many of these research endeavors will yield no profit to their organizations. Another feature that contrasts the pharmaceutical industry from the more traditional business enterprise is that much of the effort required to bring the product to market lies outside the control of the company. Thus clinical trials involve, by definition, disinterested third parties. The degree of governmental regulation is also far greater than that typically observed in business while prohibitive liability insurance can preclude a reasonable return on investment as has happened in regard to vaccine production.

In seeking to improve the uncertain nature of drug discovery, the pharmaceutical industry, like most other industries, has turned for advice to management consultants. The role of these individuals by analogy to other industries is usually to effect change [56]. As a benefit of their own experiences across a wide range of diverse industries, they are often able to improve inefficient line organizations and often bring about improvements in productivity and resource allocation. More often however, such entities are the instruments of change rather than the objective catalytic forces that their vocation may indicate [57], the corporate leaders of a given company having decided, based on their own instincts and experience as to what should occur within that company. The management consultants are then used as "objective" third parties to provide the subjective rationale to substantiate change. The value of both management consultants [5] and of the various types of management organization [34] has been questioned. In both instances little effect on the end product, new drugs, has been noted, while research management appears to have been weakened with a corresponding increase in individual self-interest. In considering that the fortunes of Smith Kline and French and those of Glaxo were reversed by the efforts of a single individual whose contributions, but not philosophy of drug discovery, are widely disseminated throughout the industry, it becomes apparent that the scientifically based research effort necessary to find and develop new drugs, and the equally valuable and quantifiable science of business administration that can work well in the reorganization of such venerable institutions (as in the case of the Bank of England), are two very distinct entities. A general lack of appreciation of the vagaries of nature can only serve to reinforce the lack of understanding by business of science in its literal sense. Until such time as management consultants seek to adopt within their ranks the experienced skill bases necessary to understand the uncertainties of intellectual endeavor and a less dogmatic, generic approach to organizational problems, one can only question the value of the theory that attempts to improve productivity in research-based organizations by applying maxims more appropriate for making widgets.

9   Future research organizations

In addressing a turbulent and unpredictable business environment [14, 17, 32, 50] drug companies are having to focus and consolidate their research efforts and resources. Accordingly, they will through success and attrition find their own niches in particular areas of the health care industry.

As with other industries, there is a growing need for more cooperative ventures between companies to more rapidly amass a body of knowledge that can be used as a stepping stone for future competitive research efforts. This has happened in the U.S. semiconductor industry [55] and to some extent in Japan under the auspices of MITI. To date, interactions between Hoechst and Bayer in Germany to work together on AIDS research [58], between Upjohn and Proctor and Gamble in relation to the hair restorant, minoxidil, and a "strategic alliance" between Squibb and Boehringer Ingelheim are the only published examples of such types of collaboration in the pharmaceutical industry. In the U.K., the government has set up, with industry, a jointly sponsored interface, "Link", involving Glaxo, Beecham and Wellcome, to facilitate the development of products and services from academia and government.

As further mergers in the industry are predicted [20, 32], it is probably that the industry of the future will comprise five distinct entities:

– multinational corporations providing innovative new drugs that fulfill unmet medical need – these organizations will be profitable and probably in time adopt speciality areas for their major research efforts
– fast follower companies – companies that are able to improve on first generation breakthrough drugs by their own expertise in effectively managing chemical and biological resources
– generic companies
– entrepreneurial companies – the research boutiques that are able, due to the smallness of their research operations to interface more appropriately with universities to identify new product candidates that can then be licensed to the larger companies with development resources.
– companies who have no research effort but depend on their development and marketing capabilities to bring licensed products to market.

Many organizations may find it appropriate to have a mix of such approaches in one organization with a good deal of the type of decentralization recommended by Sapienza [14] and documented by Peters and Waterman [13].

While it has been noted that Japanese companies predominately favor a joint venture approach in Western markets, the fact that this type of interaction is declining in Japan [59] may be a further indication of the increased competition that will emerge from this sphere of influence through the remainder of the 20th century.

In addition to the changing business environment the nature of the drugs that are now needed has changed. Diseases and conditions amenable to treatment by drugs are, by and large, treated. While pain killers have changed very little from the poppy extracts or oil of wintergreen of the 19th century, most individuals having pain where drugs are available, receive analgesics. On the other hand, patients with AIDS, Alzheimer's disease and cancer still await treatments comparable in effectiveness to the H_2-blockers in gastric ulceration. An increasing trend, worldwide, in holistic medicine, together with the prophylatic herbal remedies of Sino-Japanese Kampoh medicine will require the industry, like the medical portion of the health care industry, to move to preventing rather than finding palliatives for disease, returning to natural products as a source for new leads [60].

Such problems will face Research Directors as they and their colleagues incorporate decide on strategies for survival and success for the next 20 years or so. While the trend in the industry may be considered consolidation with concomitant diversification to ancillary areas of health care such as diagnostics for survival, many drug companies, rather than extending their commercial endeavors are instead focussing intently on the aspect where they feel they are qualified to do best, drug discovery [31]. Indeed against a backdrop of woe and uncertainty [61, 62] Merck was named the number one company in the U.S. in 1987 by the business magazine, *Fortune* [63].

In developing an organization poised for success in the drug discovery area, the real issue is that research, as noted previously by de Stevens [31], cannot "predict the rate or sequence of discovery, a consequence of the complexity of the process". On this very rational basis, it appears that the most successful organization for research will be one that is well managed, has high morale and scientific and ethical standards, a critical mass and a consistency in research philosophy, and an effort, that while neither abysmally poor or consistently stellar, is overall productive. The building of such a team will then provide the research organization with a cohesive, goal-oriented organization that is sufficiently flexible to take advantage of the unexpected findings that lead to novel therapeutic entities.

In this latter context, it is then necessary for the research manager not only to ensure that such an organization exists and continues to evolve, incorporating and contributing to advances in scientific knowledge, but also to engender the trust from corporate management both for him or herself and the research team. This will in turn allow the latter to trust management rather than focussing on maintaining visibility via publications, presentations and the like, in order to follow their own careers at the expense of the company. In the absence of a recipe for success, trust, engendered by consistent, honest leadership, flexibility engendered by a continuous objective evaluation of projects within the context of scientific advances, and good science predicated on the quality of recruiting and reputation of the organization represent the key elements for a research effort that will have as good a chance at success as any other. Organizations that are highly successful, such as Merck, have of course, the additional motivation that success can bring. As the pharmaceutical industry becomes increasingly more exposed to market forces and less traditional in its approaches to product identification, the final proof of success in effectively organizing for drug discovery, irrespective of philosophy, is not the number of patents or papers that an organization produces, nor the newness of their physical plant but rather the sales of products on the pharmacist's shelf. Planning to get it right the first time can be very beneficial [64]. However, the aim towards both cheap and innovative drugs, the latter based on expensive, time-consuming research efforts must be viewed as somewhat inconsistent [65] and ultimately, mutually exclusive.

10   Acknowledgments

While the views expressed are solely those of the authors we would like to thank our many colleagues in the pharmaceutical industry for providing food for thought, especially: Gene Cordes, George de Stevens, Heinz Gschwend, Bob Maxwell, Ron Robson and Bob Taber.

References

1   R. F. Hirschmann, Dolentium Hominum, *4*, 16 (1987).
2   C. J. Stetler, Pharmaceutical Executive, *5 (7)*, 54 (1985).
3   D. Jack, in *Decision Making in Drug Research* (ed. F. Gross), Raven, New York, p. 157 (1983).
4   A. B. Chivvis, D. J. Mackie and N. C. Selby, Pharmaceutical Executive, *7 (10)*, 50 (1987).
5   G. de Stevens, Prog. Drug. Res., *30*, 189 (1986).
6   M. Williams and J. B. Malick, in *Drug Discovery and Development* (ed. M. Williams and J. B. Malick), Humana, Clifton, NJ, p. 3 (1987).
7   C. R. Gannellin, J. Med. Chem., *24*, 913 (1981).
8   D. W. Cushman and M. A. Ondetti, Prog. Med. Chem., *17*, 43 (1980).
9   P. Deniker, in *Discoveries in Pharmacology, Vol. 1: Psycho- and Neuropharmacology* (eds. M. J. Parnham and J. Bruinvels,) Elsevier, Amsterdam, p. 163 (1983).
10   L. H. Sternbach, in *The Benzodiazepines: From Molecular Biology to Clinical Practice* (ed. E. Costa), Raven, New York, p. 1 (1983).
11   M. S. Eison, D. A. Taylor and L. A. Riblet, in *Drug Discovery and Development* (ed. M. Williams and J. B. Malick), Humana, Clifton, NJ, p. 387 (1987).
12   W. J. Sneader, *Drug Discovery: The Evolution of Modern Medicines*, Wiley, New York, (1985).

13 T. J. Peters and R. H. Waterman ،Jr, *In Search of Excellence,* Harper and Row, New York (1982).
14 A. Sapienza, quoted in the Economist, *305,* 67 (1987).
15 R. G. Denkewalter and M. Tishler, Prog. Drug. Res., *10,* 11 (1966).
16 H. Redwood, *The Pharmaceutical Industry · Trends, Problems and Achievements.* Oldwicks Press, Felixstowe, Suffolk, (1987).
17 W. Williams, New York Times, 24 February 1985, sec. 3, p. F1 (1985).
18 A. Wyck, The Economist, *302,* S1 (1987).
19 Scrip, *1270/1,* 11 (1988).
20 H. Schwartz, Scrip, *1268,* 16 (1987).
21 *The Bristol-Myers Report: Medicine in the Next Century,* Study No. 861018, Louis Harris and Associates, New York pp. i- xv (1987).
22 A. Artley, The Spectator, *259 (8318)* p. 20 (1987).
23 D. Callahan, *Setting Limits: Medical Goals in an Aging Society,* Simon and Schuster, New York (1987).
24 R. Shilts, *And The Band Played On,* St Martin's Press, New York (1987).
25 M. D. Dibner, Trends Pharmacol. Sci., *6,* 343 (1985).
26 R. N. Re, The Scientist, *1,* October 5th, 19 (1987).
27 P. Janssen, quoted in Scrip, *1258,* 9 (1987).
28 M. Williams and S. J. Enna, Ann. Rep. Med. Chem., *21,* 211 (1986).
29 Y. C. Martin and E. A. Danaher, in *Receptor Pharmacology and Function,* (ed. M. Williams, R. A. Glennon and P.B.M.W.M. Timmermans), Dekker, New York, p. 137 (1988).
30 M. D. Dibner, presentation at meeting *"New Systems in Drug Delivery"* Institute for International Research, San Francisco, CA, January, 1988.
31 G. de Stevens, Prog. Drug Res., *29,* 97 (1985).
32 W. L. Wall, Wall Street Journal, January 6, 4, (1988).
33 D. Halberstam, *The Reckoning,* Morrow, New York (1986).
34 J. W. Black, Br. J. Clin. Pharmacol. *22,* 5S (1986).
35 R. E. Faust, in *Drug Discovery* (ed. C. E. Hamner), CRC Press, Boca Raton, FL, p. 5, (1982).
36 R. W. Brimblecombe and C. R. Ganellin, in *Drug Discovery and Development* (ed. M. Williams and J. B. Malick), Humana, Clifton, NJ, p. 53 (1987).
37 P. Chopin and M. Briley, Trends Pharmacol. Sci., *8,* 383 (1987).
38 J. Davis, Scrip, *1262,* 22 (1987).
39 R. A. Maxwell, Drug Dev. Res., *4,* 375 (1984).
40 E. Cordes, presentation at Drug Discovery Management Group First Annual Meeting, PMA, Cherry Hill, New Jersey, September, 1987.
41 E. Costa, Dolentium Hominum, *4,* 40 (1987).
42 B. Testa, Adv. Drug Res., *13,* 1 (1984).
43 M. Williams and T. M. Glenn, Pharmaceutical Executive *7* (5), 34 (1987).
44 L. W. Chakrin and D. A. Byron, ibid, *7 (7)* 26, (1987).
45 J. A. W. Ansell, ibid, *7 (12),* 30 (1987).
46 R. D. Klimstra and J. Potts, Res. Tech. Manag. *31,* 23, (1988).
47 M. K. Badawy, Chem. Eng. News, *63,* 28 (1985).
48 J. B. Quinn, Harvard Business Review, May-June, 73 (1985).
49 T. M. Amabile and S. S. Gryskiewicz, *"Creativity in The R & D Laboratory",* Technical Report Number 30, Center for Creative Leadership, Greensboro, NC, May, 1987.
50 T. J. Peters, *Thriving On Chaos,* Knopf, New York, 1987.
51 P. F. Drucker, Harvard Business Review, January – February, 45 (1988).
52 R. I. Taber, presentation at Drug Discovery Management Group First Annual Meeting, PMA, Cherry Hill, New Jersey, September, 1987.
53 J. W. Black in *Perspectives on Receptor Classification,* Eds J. W. Black. D. H. Jenkinson and V. P. Gerskowitch, Liss, New York, p. 11 (1987).
54 M. A. Schwartz, Pharmaceutical Res. *4,* 83 (1987).

55 J. Walsh, Science, *239*, 248 (1988).
56 J. Peet, *"Outside Looking In"*, Management Consultancy Survey, The Economist, *306* (7537), p. 1 (1988).
57 N. Hayek, quoted in Forbes Magazine, January, 11, p. 258 (1988).
58 T. Jackson, Financial Times (Lond.), April 22, 1987, Section I, 14 (1987).
59 R. Maurer, quoted in Marketletter, February, 8th, 1988, p. 17 (1988).
60 S. L. Lee, *Medicinals And Other Chemicals From Plants: Status And Prospects*, Batelle Technical Inputs To Planning, Report No. 52, Columbus, OH (1987).
61 M. Wetherall, Nature (Lond.), *296*, 387 (1982).
62 P. O'Donnell, Scrip, *734*, 8 (1982).
63 E. Schultz, Fortune, *117*, 32 (1988).
64 C. E. Hamner, in *Drug Discovery* (ed. C. E. Hamner), CRC Press, Boca Raton, FL, p. 1, (1982).
65 D. A. Byron, presentation at meeting *"New Systems in Drug Delivery"*, Institute for International Research, San Francisco, CA, January, 1988.

Approaches to the rational design of bacterial vaccines

By Peter Hambleton, Stephen D. Prior and
Andrew Robinson

Public Health Laboratory Service, Centre for Applied Microbiology and
Research, Porton Down, Salisbury, Wilts. SP4 OJG, U. K.

1   Introduction

When Bacon wrote 'Man seeketh in society comfort, use and protection', he was not, as far as we know, referring to vaccines although his statement does encapsulate the essential features required of them. Vaccination has become an integral feature of life since the pioneering work of Jenner and Pasteur and has proved successful in diminishing the scourge of many infectious diseases. Many existing vaccines, developed on the basis of traditional, empirical principles, consist of killed or live attenuated (avirulent) microbes and act by inducing a broad immune response. By and large these vaccines have adequately fulfilled what was required of them although, in the case of enteric diseases, some have had little or no beneficial effect. In contrast a number of bacterial vaccines were designed to act by inducing specific immune protection, e. g. bacterial toxoid vaccines, although in general such vaccines were derived from relatively impure toxins.

In the last two decades or so there has been a decline in the rate of introduction of new vaccines, particularly those for bacterial diseases probably for the following reasons. Firstly, earlier vaccines dealt with the relatively 'easier' diseases leaving those more refractory to scientific investigation. Secondly, there has been the remarkable success of chemotherapeutic agents in controlling many bacterial diseases. Thirdly, increased concern about safety and product quality has given rise to demands for minimal risk vaccines that have a markedly reduced commercial viability. However, the increasing emergence of antibiotic resistant bacterial strains, the relative ineffectiveness of some traditionally made vaccines and the realisation of the potential of recent developments in biological sciences, are leading to a renaissance of interest in vaccines. The increasing need for novel, improved or refined vaccines has encouraged a more rational approach to vaccine formulation. Recent advances in the understanding of the mechanisms of pathogenesis of infectious diseases, in conjunction with developments in biochemistry, microbial physiology, protein chemistry and genetics, have made this rational approach to vaccine design both scientifically and commercially feasible.

The great majority of bacterial species that interact with man and animals are harmless, indeed many are beneficial, but a few possess biochemical characteristics that impart to them the distinguishing property of virulence or pathogenicity. These biochemical characteristics that enable microbes to cause disease are called *virulence determinants* [1]. Generally, in their primary interaction with the host, pathogenic bacteria must either adhere to the mucosa at the site of infection, perhaps after penetrating a mucous layer, or enter the tissue via wounds. Subsequently the organism must replicate within the host producing damage either locally at the site of infection or remotely by means of exotoxins. In addition the organism must interfere with, or avoid interaction with, host defence mechanisms. It is often the case that pathogenic organisms must possess several virulence determinants in order to establish disease in a susceptible host and loss of even one of these determinants may give rise to an attenuated or avirulent organism incapable of causing disease. A clear example of this is seen with *Bacillus anthracis,* the causative organism of the disease anthrax. Virulent strains of *B. anthracis* possess a capsular component, poly-D-glutamic acid, which inhibits host cellular immunity, and also excrete a lethal toxin [2]. The genetic determinants for both these virulence factors are carried on separate plasmids [3]. Strains that lack either the plasmid coding for capsule formation or that for toxin production are avirulent; moreover, strains lacking capsule can act as effective live vaccines. This demonstrates that successful prophylaxis or therapy of an infectious disease can be effected by neutralisation of one or more virulence determinants.

From the above example it is obvious that the recognition and understanding of bacterial virulence determinants could enable the selection of vaccine components required to neutralise specifically one or more key stages in pathogenesis. In attempting to study virulence factors it is, however, very important to recognise that since pathogenicity may result from possession of several virulence determinants and that virulence is only truly expressed *in vivo,* the identification of true determinants may be difficult. Attempts must be made to identify virulence factors from studies with diseased hosts; *in vitro* studies can then be assessed as to their relevance to the known characteristics of the infectious agent *in vivo* [1]. A good example of this approach is the expression of specific immunogenic outer membrane proteins (OMP) during *in vivo* growth of certain pathogens (e. g. *Pseudomonas aeruginosa* [4], see Section 4.4). Such OMPs are expressed *in vitro* by bacteria grown in iron-limiting conditions, but not when grown in iron-sufficient media. A recent patent [5] describes the use of a major iron-regulated OMP of *Neisseria gonorrhoea* as a vaccine.

2   Mechanisms of bacterial pathogenesis

To reiterate the point made previously, a prerequisite for the rational design of vaccines is a clear understanding of the variety of interactions of microbe and host that result in disease. The details and relative significance of the different stages of pathogenesis may well vary from one infectious disease to another but it is possible to identify broadly seven stages that are common to many infectious illnesses (Fig. 1).

It is clearly inappropriate to consider vaccination as a means of protecting against disease by interfering with the first and last of these stages viz. dissemination and deposition, since fundamentally these can only be prevented by physical and/or chemical means such as ventilation, filtration and disinfection. In contrast vaccination is generally an appropriate means of inhibiting any of the other stages, although the selection of vaccine components requires a clear understanding of the identity and significance of virulence factors involved.

The initial interactions of an infection will involve surface components of the bacterial envelope and host cell membranes, so bacterial envelope constituents that may be virulence determinants are obvious candidates as vaccine components. It is beyond the scope of this review to consider in detail the structure of the bacterial envelope (for

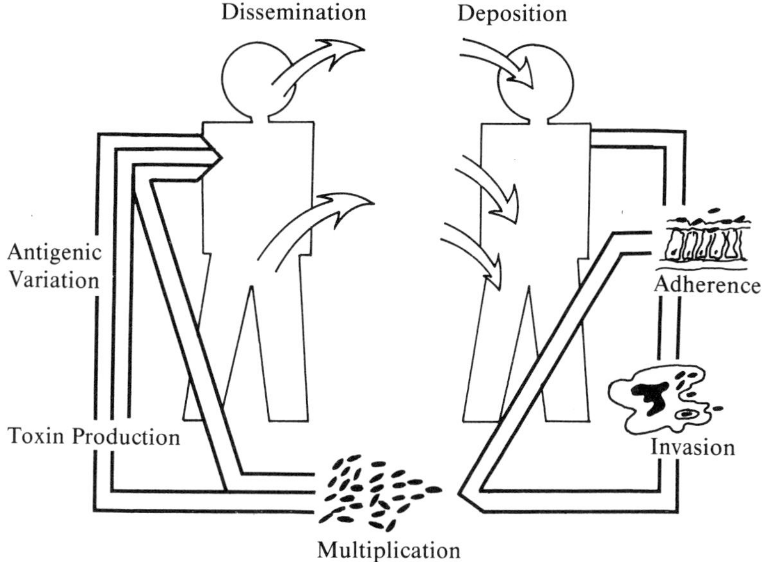

Figure 1
Some common stages in the development of infectious diseases.

which excellent reviews already exist [6]) and the importance of envelope components as adhesins is well established. For example, the ability of enteropathogenic bacteria to adhere to the host intestinal epithelium is the consequence of possessing specific envelope components such as the K 88 adhesion antigen of *Escherichia coli;* other antigens may even discriminate between different domains of the intestine. In some instances even though adhesion is a key primary feature of infection the nature of the adhesins may not be clearly understood. For example, virulent strains of *N. gonorrhoeae* cultured *in vitro* possess pili which promote adhesion of the bacteria to tissue culture cells, whilst avirulent strains lacking pili do not adhere as effectively to the tissue cells [7, 8]. Both strains will, however, adhere to human fallopian tube and endocervix epithelium. In instances such as these the identification of putative vaccine components may require considerable research effort.

The identification of bacterial components involved in other subsequent stages of pathogenesis such as the ability to penetrate into the host, to survive and replicate within the host, to counter host immune defence systems and to cause pathological damage to the host, might suggest other constituents to be included in vaccines of defined composition. Table 1 contains suggestions for vaccine components that might specifically interfere with particular stages of pathogenesis. The list is not intended to be complete but rather to highlight the concept of the selection of vaccine components based on the identification of those aspects of the pathogenic mechanism that are central to the establishment of disease in the host.

Table 1

Candidate components of vaccines targetted against particular pathogenic mechanisms

Pathogenic mechanism	Vaccine components
Adherence	Fimbriae, flagellae, capsule, OMP
Invasion of host tissues	Flagellae, capsule, outer envelope components, lipopolysaccharide
Evasion of host defence mechanisms	Capsule, lipopolysaccharide, toxins
Replication	Surface appendages, outer envelope components
Local/systemic toxin production	Endotoxin, exotoxin, enzymes
Antigenic variation	Common core antigens, multiple serotype antigens

Some vaccines comprising a single active component are effective in giving protection against disease. For example both tetanus and diphtheria toxoids successfully induce protective levels of circulating antibody which neutralise the potent exotoxins produced by *Clostridium tetani* and *Corynebacterium diphtheriae* respectively. However, in neither case does the vaccine appear to act by interfering with the processes of adhesion and replication. Indeed, in the case of *C. diphtheriae* the organisms are able to adhere to the faucial mucous membrane of vaccinees and replicate giving rise to the pseudomembranous lesion characteristic of diphtheria, although this no longer results in the recognised disease condition [9].

Selection and use of relevant experimental animal models to study pathogenicity and determine vaccine potency and/or toxicity, are essential to the process of vaccine development [9]. Ideally an animal model should closely mimic the essential features of naturally acquired infections and lend itself to assessment of the effects of vaccination. The importance of good animals models can clearly be seen in the development of vaccines against cholera and other enteropathogenic bacteria. Here, the ligated rabbit ileal loop test [10] has proved of fundamental benefit because it encompasses the primary pathogenic feature of many enteric diseases, i. e. the massive movement of fluid into the gut lumen, and so enables the effects of vaccine-induced immunity on the basic biochemical mechanisms to be directly assessed. Such a clear position does not hold for other diseases where the relevance of animal models which are used quite widely may be questionable. For example, the only routinely available test for protective potency of whole cell pertussis vaccines is the mouse intracerebral challenge test but this cannot be considered to mimic the natural respiratory-acquired infection. Also, the apparent potentiating effect of pertussis toxin that may result from changes in the mouse blood-brain barrier may give misleading data on the potencies of putative vaccines [11, 12]. Respiratory challenges [13, 14] might provide better alternative models even though the resulting colonisation may be asymptomatic.

3   Developments in vaccine manufacture

Recent advances in the enabling technologies of molecular biology will have considerable impact on the manufacture of rationally designed vaccines. The techniques of genetic engineering, protein chemistry and biochemical synthesis are being used in several areas of research related to the development of improved methods of and approaches to vaccine manufacture.

3.1   Chemically and genetically modified vaccine components

The use of inactivated toxins as vaccines represents a clear example of logical vaccine design. By administering chemically modified toxins (toxoids) having diminished toxicity but retaining their immunogenic properties, it is possible to induce protective levels of circulating antitoxins. Examples of successful toxoids include those derived from the exotoxins of *Clostridium botulinum, Cl. tetani* and *C. diphtheriae.* In each case inactivation is readily achieved by formaldehyde treatment. Toxoids administered to humans contain residual low levels of formaldehyde basically as a safeguard against the possibility of reversion to the toxin. Whilst the chance of significant reversion is slight it is nevertheless a real possibility [15] and the presence of formaldehyde may be associated with unpleasant transient side effects experienced by some vaccinees. Furthermore the use of formaldehyde is empirical and its mode of action is not fully understood. Better-defined and better-controlled means of detoxification are required to ensure the safety and full immunogenic potential of purified toxoids.
Recent studies by Shone *et al.* [16] with botulinum toxin type A have identified a fragment of the neurotoxin which appears to contain the active site region responsible for initial binding of the toxin to receptors on the presynaptic nerve surface. Enzymic cleavage of this fragment from the toxin, while rendering the molecule non-toxic, does not affect its immunogenicity. Hence, one has the potential to generate inactivated botulinum toxoids having no possibility of reversion to a toxic form. At present it is not possible to separate such fragmented toxoids from all traces of residual toxin, even though these are slight (C. C. Shone, unpublished data). However, once the sequences of intact toxin and the non-binding fragments are known it would be feasible to clone the gene sequence of the latter, so enabling production of a completely non-toxic, non-revertible toxoid vaccine.
The growth of molecular genetics and the advent of new techniques revolving around recombinant DNA technology is unparalleled in the life science field. The basis of this technology involves the use of defined nucleic acid vectors into which DNA/genes of interest can be in-

serted (DNA cloning). The result is a recombinant DNA product that can express and produce proteins and polypeptides in new and foreign hosts. Hence in theory the original barriers to working with "difficult" organisms can be overcome.

Alternatively, the ability to manipulate the DNA of pathogens could enable enhanced levels of expression of polypeptides or proteins in the natural host. Furthermore, site-directed mutagenesis could be used to yield proteins with enhanced immunogenicity and/or permanently abolished toxicity which should preclude the use of chemical inactivation of toxins and lead to safer stable vaccines.

The application of recombinant DNA is not, however, as simple as may appear from the above. In reality there are problems in isolating suitable strains to act both as hosts and as sources of DNA, in selecting vectors for transfer of DNA and, in particular, of obtaining expression of the proteins at the bacterial surface or secretion into the growth medium. In addition, there are other potential problems such as the tendency of some foreign proteins either to be hydrolysed rapidly or to be toxic for the engineered host and the possible failure of polypeptides of normally oligomeric proteins to be assembled into antigenic proteins. Finally, the failure of an engineered host correctly to glycosylate the 'foreign' protein may have important structural implications adversely affecting immunogenicity.

3.2   Live attentuated polyvalent vaccines

Perhaps one of the most exciting developments in bacterial vaccine design and manufacture is the use of attentuated strains which may be constructed to carry the genes for virulence determinants of other bacterial species. Of particular promise for their potential and safety are the attentuated strains of Salmonella species, namely the *gal*E strains which are incapable of converting glucose phosphate into the galactose component of LPS, and the *aro*A mutants which have a block in the synthesis of aromatic amino acids [17]. These bacteria can penetrate the mucous membrane of the host but fail to replicate sufficiently to establish an infection. However, the bacteria persist long enough to induce protection in the host against salmonella infection and, by the incorporation of extraneous genetic material (e.g. from shigella or the gene for the B subunit of *E. coli* heat labile toxin), could be used as oral vaccines to protect against several diseases. Because the induction of a protective immune response by these vaccines could preclude their subsequent use with other genetic inserts, it is perhaps important to construct vaccines to protect against as many diseases as possible concurrently. It is of course essential to ensure that any hybrid organisms are not endowed with properties which lead to the expression of novel pathogenic activities.

3.3   Synthetic peptides

In recent years there has been much discussion on the possible use of synthetic peptides as vaccines. In their recent article, Dorner and McDonel [18] noted several advantages of this approach for the development of novel vaccines. These included:

i)   elimination of the need to compose vaccines from inactivated preparations or avirulent strains of pathogenic microbes.

ii)   avoidance of contamination of the starting material by medium components from which antigenic proteins are prepared.

iii)   replacement, by organic synthesis of peptides, of biotechnological processes to produce antigenic proteins from genetically manipulated cells.

The basic concept of using peptides as vaccines involves the identification of the antigenic portions of the protein which have been shown to have protective capabilities. This is normally achieved by consideration of protein structure using amino acid (or DNA) sequences and identifying areas of the protein that are exposed on the surface and therefore having the potential of acting as antigenic sites. Synthesis of the peptide is then readily achieved using conventional methods and the efficacy examined. Alternatively, important peptides may be identified using fine immunochemical mapping of antigens with protective monoclonal antibodies. The principles of the use of synthetic peptides as vaccines and how these procedures may be improved have been considered in several recent publications [19–21].

The major problem encountered with synthetic peptide vaccines thus far is that short linear peptides corresponding to protective antigenic determinants are insufficient to provoke an effective antibody response. One means of overcoming this problem has been to link covalently the peptide to carrier proteins so as to create chimaeric antigens of sufficient size to elicit a good antibody response [19, 22].

Another problem is that these small peptides may not elicit an identical response in all individuals which could lead to some vaccinees being unprotected [23]; this problem would best be solved by using a cocktail of relevant peptides. Furthermore, as it is intended that many vaccines should protect whole populations by reducing the rate of transmission rather than protecting each individual, the use of a cocktail of peptides recognised by a large proportion of the population might be beneficial [24].

There is little doubt that in the near future synthetic peptide of vaccines will be valuable and increasingly used research tools for studying disease control but the long-term value of such work has yet to be proven and the first successful use of a human synthetic peptide vaccine is eagerly awaited.

3.4   Anti-idiotype vaccines

Some of the problems encountered in the development of defined vaccines e. g. the poor immunogenicity of polysaccharide antigens and short synthetic peptides, could perhaps be overcome by the use of anti-idiotype vaccines. The rationale here is that a particular epitope on an antigen is used to induce antibodies (AbI) which are subsequently used to induce the production of anti-AbI antibodies (AbII or anti-idiotypes). Some, but not all, AbII molecules will be structurally similar to the epitope on the original antigen (i. e. an internal image of the epitope) and can be used to induce the production of antibodies similar to AbI. In other words anti-idiotype antibodies behave as surrogate antigens [25, 26]. Such vaccines based on antigenic mimicry are an attractive proposition and there are several instances where they have been shown to induce protection [27]. However, the topic of anti-idiotype vaccines is in its infancy and considerable research will be required before their full potential can be assessed.

3.5   Adjuvant development

The role of immunopotentiating agents (adjuvants) in enhancing the prophylactic value of vaccines is well established. Freund's complete adjuvant (FCA), which contains killed mycobacteria in a water-in-oil emulsion as the active principal, is used to enhance the immune response of experimentally immunised animals since they give markedly elevated levels of immunity compared with those derived by administration of antigen alone. Freund's adjuvant, however, is unsuitable as a vehicle for human vaccines since it induces severe pathological lesions both at the site of injection and in the regional lymph nodes.

The most common use of adjuvants in humans involves the absorption of antigens onto insoluble aluminium salts which serve as an innocuous depot for the injected vaccine. Whilst the insoluble salts may act primarily by presenting a localised concentrate of antigens to the immune system and allowing only slow release of antigens, it is considered that they may also serve to initiate a non-specific immune reaction. Insoluble calcium salts are an alternative to aluminium which may prove more acceptable since they are natural constituents of the human body.

The effectiveness of the early examples of biological immunopotentiators, such as Freund's adjuvant and BCG, led to a search for alternative but innocuous immunopotentiating substances of biological origin and the recognition of bacterial components which may have considerable adjuvant potential. For example, lipopolysaccharides (LPS, endotoxin) derived from gram-negative bacterial cell envelopes have been shown to be capable of stimulating both humoral and cellular immunity. The immunopotentiating activity appears to be the consequence of several interactions within the immune system including direct action on B-cells, T-helper cells and macrophages, conversion of tolerogenic signals for both B- und T-cells to immunogenic ones and induction of cellular factors influencing T-cell function [28]. In addition to their high immunogenicity LPS induce a highly diversified pattern of pathophysiological effects including pyrexia, hypotension, changes in blood leucocyte counts, hypoferraemia and platelet aggregation [29] which may restrain the widespread application of LPS as vaccine adjuvants.

Studies to identify the mycobacterial component responsible for the marked immunopotentiating activity of FCA identified the diverse biological activities of the peptidoglycans [30]. It is now well established that most of these diverse biological activities of peptidoglycan are associated with the N-acetyl muramyl-L-alanyl-D-isoglutamine moiety (MDP; Fig. 2).

Synthetic muramyl dipeptides have been prepared which are capable of inducing the delayed hypersensitivity response characteristic of FCA but without the undesirable local reactions. The use of synthetic

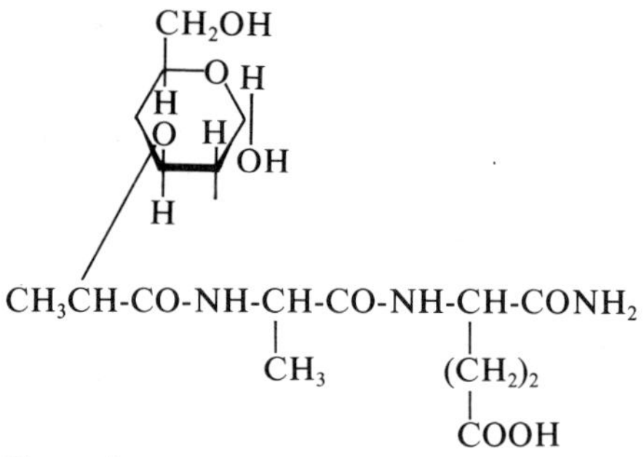

$$CH_2OH$$

Figure 2
Structural formula of MDP   N-acetyl muramyl-L-alanyl-D-isoglutamine.

muramyl-peptides to potentiate vaccines has been reviewed elsewhere [31, 32], and it is clear that they offer considerable possibilities for enhancing the activity of synthetic peptide vaccines. In particular, it is entirely possible that totally synthetic vaccines could induce cell-mediated immune responses. It is significant that a commercial company, Syntex, produces a synthetic MDP which is incorporated into several investigational new drugs (IND) currently under trial. Success in this area could dramatically extend the scope for vaccine design.

4   Improvement of existing vaccines

An examination of bacterial vaccines shows that these vary greatly in their efficacy and innocuity. In some instances considerable effort has been made to improve vaccines by application of the principles of rational design; in other cases considerable potential exists for improvements in vaccine safety and potency. The following examples have been selected to illustrate the use of rational design in the development of novel or improved vaccines. The list is not intended to be complete but rather to illustrate the principles by reference to different pathogenic processes.

4.1   Pertussis

Mechanism of pathogenesis and vaccine composition
Pertussis provides an excellent example of how increased knowledge of the mechanism of pathogenesis of the causative organism, *Bordetella pertussis,* has enabled the design of acellular vaccines. Such vaccines

are potentially more effective and less reactogenic than the currently used whole cell vaccines which have, nevertheless, proved markedly effective at reducing clinical pertussis [33]. The whole cell nature of the vaccines, however, means that they contain, in addition to those antigens which induce a protective immune response, components which are ineffective and/or which may even produce toxic effects. In a defined acellular vaccine which consists of one or more purified antigens, these toxic components would either be eliminated (in the case of lipopolysaccharide) or be detoxified.

The roles of a number of virulence components of *B. pertussis* in the pathogenic process are now fairly well established [33–35]. The primary stage in the pathogenesis of pertussis is the attachment of the bacteria to the ciliated respiratory epithelium of the child, allowing the organism to resist normal clearance mechanisms such as coughing and mucociliary flow. The attachment of the bacteria is a multifactorial process and a number of components of *B. pertussis* have been identified as potential adhesins. The first adhesin to be considered was filamentous haemagglutinin (FHA) which was originally thought to be derived from fimbriae [36]. Since the demonstration that FHA was not of fimbrial origin [37] its role as an adhesin has rested on studies of the adherence of *B. pertussis* to mammalian cells and the inhibition of this by antibody [36, 38–40]. FHA is an ideal vaccine candidate since it is easily isolated from the culture supernatants of *B. pertussis* in relatively high yields and is non-toxic, highly immunogenic and protective in a number of animal tests [33].

Having a less certain role in adhesion are the agglutinogens, which account for the serospecificity of *B. pertussis*. Agglutinogens 1, 2 and 3 are major antigens and agglutinogens 4, 5 and 6 relatively minor antigens. Whereas the nature of agglutinogen 1 is unknown, agglutinogens 2 and 3 have been shown to be fimbrial antigens [37, 41, 42]. Analysis by SDS–PAGE indicates that agglutinogens 2 and 3 have subunit molecular weights of 22,500 and 22,000, respectively. The complete nucleotide sequence of agglutinogen 2 has recently been determined [43]. The fimbrial nature of agglutinogen 2 has been confirmed by other laboratories [44, 45] but that of agglutinogen 3 has been confirmed by one group [45] and disputed by others [46–48]. At present it can only be stated that, depending on the serotype of the organisms, at least two types of fimbrial subunit can be isolated from *B. pertussis* and that these bear either agglutinogen 2 or agglutinogen 3/6. The fimbrial nature of the agglutinogens strengthens their case, on epidemiological considerations [33], for inclusion in acellular pertussis vaccines. Unfortunately an adhesive role for the agglutinogens has only been inferred from a study involving Vero cells [39] and has not been confirmed in one involving WiDr cells [40]. Isolated agglutinogens, however, protect mice against respiratory infection with *B. pertussis* [14, 44]. Pertussis toxin (PT) has also been implicated as an adhesin of *B. pertussis* [38].

Having established an infection on the respiratory epithelium of the child, the colonizing bacteria multiply to yield sufficient organisms to give rise to the toxic effects of the disease including local effects (ciliostasis and the sloughing of cilia and ciliated cells) and systemic effects (lymphocytosis and alterations in blood chemistry). The complex interplay of the many toxic components of *B. pertussis* in the production of the clinical symptoms of pertussis is not fully understood [34, 35]. However, a number of toxic components, e.g. lipopolysaccharide, tracheal cytotoxin and heat labile toxin, either are too poorly characterised, too difficult to isolate or fail to produce a protective immune response, and are therefore not considered to be of major interest in vaccine formulation. Adenylate cyclase has been considered as a potential vaccine component [49] but the principal component included to neutralise the toxic effects of *B. pertussis* is pertussis toxin (PT) in a suitably detoxified form. When active PT is injected into laboratory animals it produces a spectrum of biological effects many of which are similar to the systemic effects of pertussis in the child [33–35]. The role of PT as a major protective antigen has been confirmed by many experiments in which PT has been found to protect mice actively and passively against intracerebral and respiratory infections with *B. pertussis* [12, 14, 50–52].

Two approaches are thus available to the rational design of an acellular pertussis vaccine. Firstly, a vaccine could contain adhesins (FHA and agglutinogens) and be designed to inhibit the first stage of infection with *B. pertussis*. Secondly, if pertussis is a true exotoxinosis, PT should be sufficient to prevent the disease symptoms of clinical pertussis. Many researchers, however, believe that the most effective vaccines against pertussis will provide a dual mechanism of protection by inclusion of FHA and/or agglutinogens in combination with PT. Successful whole cell vaccines have been shown to contain all three antigens in immunogenic amounts.

Acellular vaccines

Much of the pioneering work on acellular pertussis vaccines has been undertaken in Japan in response to an urgent need to replace the whole cell vaccine. More than 20 million doses of acellular vaccines have been administered in Japan since 1981.

The Japanese vaccines consist primarily of PT and FHA co-purified by the procedure of Sato *et al.* [53]. Supernatants from cultures of *B. pertussis* are concentrated by ammonium sulphate precipitation and fractionated by sucrose density gradient centrifugation. Because the components are co-purified, the ratio of FHA to PT in the vaccines can vary from 1:1 to 9:1 depending on differences in the original cultures and in the particular fractions selected from the sucrose density gradients.

The incidence of pertussis in Japan is reported to have declined since the vaccines were introduced [54, 55]. The efficacy of acellular vaccines in protecting against clinical pertussis in household contacts appears to be about 80–90 % [53, 56, 57]. The vaccines induced high serum antibody titres to both PT and FHA, had only low pyrogenic activity and produced diminished local reactions when compared to whole cell vaccines [53, 55–57]. The USA Interagency Group to Monitor Vaccine Development, Production and Usage visited Japan and concluded that the Japanese vaccines could be considered to be safe and effective in 2-year-old children, but product-specific information on the safety and efficacy in infants was required [56, 57]. A Japanese FHA/PT vaccine has been compared with a plain whole cell vaccine and also with a PT-toxoid vaccine in Phase II clinical trials in Sweden in 6-month-old infants. The acellular vaccines were of lower local and systemic reactogenicity but produced higher antibody responses to PT even after two doses of vaccines. However, preliminary results of Phase III placebo-controlled, double-blind efficacy trials indicate that two doses of these acellular vaccines may not be as protective as three doses of whole cell pertussis vaccines (WHO meeting on phase III trials, Stockholm, December, 1987).

An acellular pertussis vaccine has been developed at PHLS, CAMR, which consists of equal amounts (10 μg/dose) of separately purified and formaldehyde-detoxified PT, FHA and agglutinogens. The PT and FHA are isolated from the supernatants of *B. pertussis* cultures by chromatography and the agglutinogens from the cells by homogenisation. All the components are pure as judged by SDS-PAGE (see

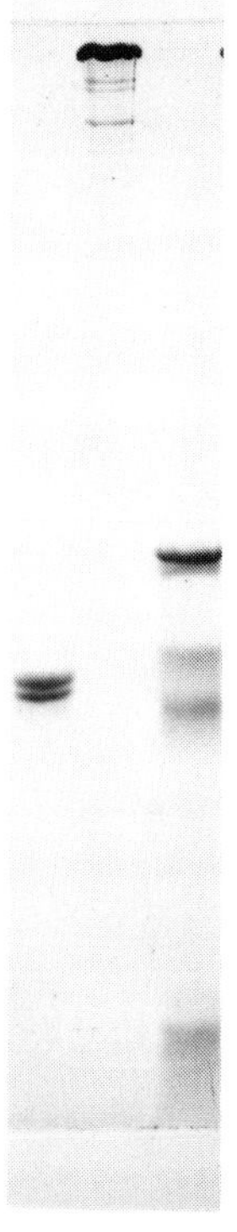

Figure 3
SDS-PAGE of the components of the CAMR acellular
pertussis vacine; [1] Agglutinogens 2 and 3;
[2] Filamentous haemagglutinin; [3] Pertussis toxin.

Fig. 3). Two fimbrial subunits are present corresponding to agglutinogens 2 (60 %) and 3 (40 %). The vaccine, when compared to a whole
cell vaccine, has a reduced LPS content as determined by the mouse
weight again and Limulus amoebocyte lysate tests. Detoxification of
the PT component is complete and, even after storage at 37° C for
several months, the vaccine has no activity in the mouse histamine sen-
sitisation test for active PT. In comparison with an adsorbed whole cell
vaccine, the acellular vaccine produces in mice equivalent or reduced
levels of antibody to FHA but much enhanced levels of antibody to PT
and agglutinogens as measured by ELISA. The antibodies to PT prevent leukocytosis in mice challenged with PT and neutralise the effects
of PT on Chinese hamster ovary cells. In phase I trials the vaccine had
low reactogenicity in adult volunteers and induced high antibody
responses to FHA, PT and agglutinogens [58].
In addition to the Japanese and CAMR vaccines and others of similar
composition, monocomponent PT vaccines have been advocated. Se-
kura has produced a vaccine from hydrogen peroxide detoxified PT
[57] and the Biken company has produced a vaccine consisting of

formaldehyde detoxified PT which has been assessed in the phase II and phase III trials in Sweden [57]. Although a PT-only vaccine is an attractive concept, there are several reasons why it may not be the optimal vaccine: (a) PT is not the only toxin of *B. pertussis;* (b) *Bordetella bronchiseptica* and *Bordetella parapertussis* elicit pertussis-like diseases but fail to produce PT; (c) other antigens have protective properties in animal tests; (d) PT may not be a primary adhesin of pertussis, and (e) vaccines produced from other purified bacterial toxins (e.g. cholera toxin – see below) have not been found to be highly effective.

The use of acellular vaccines in the prevention of pertussis has several potential advantages over the use of whole cell vaccines. Acellular vaccines are better defined and hence more amenable to effective quality control procedures. They may provide a more complete protection against both the infection and disease stages of pertussis and hence the number of vaccine doses required may be less. Finally, the acellular vaccines should be devoid of the reactogenicity of whole cell vaccines and hence, it is hoped, fully restore public acceptability of vaccination against this potentially serious disease.

4.2 Gonorrhoea and meningococcal meningitis

Although *Neisseria meningitidis* and *N. gonorrhoeae* are closely homologous with regard to virulence components, they produce markedly different diseases in the human host. It might thus be anticipated that the development of a successful vaccine against one disease could have little relevance to the design of a vaccine against the other. The initial stage of meningococcal infection is very similar, however, to the pathogenic process of gonorrhoea, except that the sites of infection differ [7, 8, 59–61]. Hence the development of a successful gonococcal vaccine might provide valuable information for the design of effective meningococcal vaccines.

Gonococci primarily attach to non-ciliated columnar epithelial cells of the genital tract and, after the establishment of close contact between mammalian and bacterial cells, the organisms become engulfed by the mammalian cell membrane. Following intracellular multiplication, the phagocytosed bacteria may penetrate the sub-epithelial tissues giving rise to the severe local inflammatory response that leads to the typical symptoms of gonorrhoea. In contrast, meningococci infect the mucous membrane of the nasopharynx which gives rise to asymptomatic carriage of the organism. The meningococcal carrier rate in normal populations can range from 5 to 30 % but be considerably higher during epidemics. Disease symptoms occur when the mucosal carriage leads to invasion of the blood stream and subsequent infection of distal organs, principally the meninges, giving rise to classical meningitis. The transition from the asymptomatic carrier state to one of the fastest known killers clearly distinguishes meningococcal from gonococcal infections. A vaccine assigned to protect against gonorrhoea must prevent the initial infection of the bacteria at the mucosal surface. For meningitis, however, this strategy of preventing colonisation may be effective but the essential stage to prevent is that leading from carriage to septicaemia.

The initial stage of infection of both gonococcal and meningococcal diseases is probably mediated by fimbriae (which are termed pili in the case of these pathogens). Gonococcal pili have been subjected to extensive immunochemical and genetical analysis in an attempt to unravel the complex mechanisms whereby the organism can produce a variety of structurally and antigenically distinct pili. This array of pilin (and other surface protein) variants enables the organism to evade the immune response developing in the host [8]. The gonococcal pili are composed of variable and conserved regions [7, 8, 59]. The variable regions are immunodominant and ensure that the host's immune-response is mainly directed against these "non-functional" regions. In contrast, the conserved regions, which are probably those responsible for the adherent function of pili, are less immunogenic. A synthetic peptide derived from a conserved region of various gonococcal pilin subunits, however, has been found to produce antibodies which inhibit the attachment of heterologous strains to human endometrial carcinoma cells [62]. Although other data have shown that some monoclonal antibodies directed against the conserved region do not block adhesion [63], this type of experiment raises the exciting possibility of a small common peptide providing an effective immune response against all strains and variants of gonococci. This is an attractive prospect since the alternative approach is to include, in a polyvalent vaccine, a variety of pilin subunits which will protect against all (or the major) strains of gonococci. The possibility that a distinct subunit at the tip of pili is responsible for adherence properties should also be considered. In addition to pili, outer membrane proteins of gonococci have been considered as important potential vaccine components.

Development of effective meningococcal vaccines is perhaps more complex than that of gonococcal vaccines. Prevention of meningococcal colonisation of the respiratory tract by vaccination would also protect against subsequent meningitis and it is interesting to note that there is some homology between meningococcal and gonococcal pili [64]. Indeed, a pilus peptide vaccine could in theory protect against meningococcal and gonococcal diseases. Most meningococcal vaccines are, however, designed to produce bactericidal antibodies which prevent the development of asymptomatic carriage into meningococcal disease. Since invasion of the blood stream depends on the lack of an effective immunity in the host and on the presence of a capsule on the bacteria, it is not surprising that much vaccine research has centred on the meningococcal serogroup capsular polysaccharides. Vaccines comprised of group A and C polysaccharides have been found to be effective, at least in older children and adults, and of low toxicity [7, 61]. Vaccines produced from the group B polysaccharide, which is a $(2 \rightarrow 3)$-α-linked homopolymer of N-acetyl-neuraminic acid, are, however, poorly immunogenic.

There have been a number of approaches made to try and resolve the problem of producing vaccines effective against the group B meningococcus. Lifely *et al.* [65] consider that the poor immunogenicity of the group B polysaccharide is due to suppression of the immune response when it is directed against short chains of $(2 \rightarrow 8)$-linked sialic acid determinants which are also found in a number of host sialogangliosides. Hence the immune response is directed against conformational determinants on the group B polysaccharide. It is thus essential to stabilise the three dimensional structure of the polysaccharide in solution to produce antibodies which are effective against cell-bound polysaccharide. This stabilisation and consequent improvement in immunogenicity has apparently been achieved by complexing the polysaccharide to outer membrane proteins and the addition of aluminium salts to the preparations [65].

An alternative approach to the development of a group B meningococcal vaccine, perhaps spurred on by the possibility of cross-reaction of anti group B polysaccharide antibodies with polysialoglycoproteins in foetal and neonatal brain tissue, has been the isolation and use of outer membrane proteins as vaccines. These proteins account for the various serotypes of meningococcal strains [66]. Although vaccines composed of outer membrane proteins, perhaps coupled to polysaccharides, have every prospect of being successful [67–70] it should be pointed out that these vaccines have been developed to some extent on an empirical basis and generally consist of several outer membrane proteins, the individual roles of which in meningococcal virulence are largely unknown. Before a meningococcal vaccine which is effective against all group B serotypes (and hopefully against all meningococcal strains) can be developed, a clearer understanding is required of the involvement of individual virulence components in the pathogenic process. The extent to which the components are conserved or varied during the course of infection should also be known [71]. Outer membrane proteins not normally expressed *in vitro* but produced *in vivo,* such as proteins induced by ion-restricted growth [72], may yet prove to be ideal vaccine components.

4.3   Enteric vaccines

Cholera

The disease caused by *Vibrio cholerae* is characterised by acute diarrhoea leading to rapid, and sometimes fatal, dehydration. Ingested organisms non-invasively colonise the gut lumen by means of attachment to the gut mucosa. A powerful enterotoxin (cholera toxin, CT, or choleragen) produced by the organisms is responsible for the major effects of the disease [73] although other extracellular proteins may contribute to the colonisation process [74].

The enterotoxin is a homogenous protein (Mr 84,000) composed of two subunits. The exterior subunit (choleragenoid or subunit B) which contains five identical moieties (of Mr 11,500) linked non-covalently, is the immunologically dominant portion of the toxin and is involved in binding of the toxin to cell host membranes. The interior subunit (A subunit) has a Mr of 28,000 and is composed of two fragments A_1 and A_2. A_1 is an enzyme that catalyses NAD-dependent ADP-ribosylisation of the GTP-binding protein responsible for regulating host cell adenylate cyclase activity [75, 76].

The process of pathogenesis of cholera has many similarities to that of pertussis in that both are non-invasive diseases and both produce potent exotoxins. Indeed the analogy can be taken further in that both exotoxins, CT and PT, act by causing enhanced cyclic AMP production in the target mammalian cells, although CT acts on the stimulatory arm and PT on the inhibitory arm of adenylate cyclase regulation

[77]. A difference between the two diseases, in addition to the different sites of infection, is in the role of these toxins in the disease process. CT alone has been shown to produce the typical cholera disease symptoms of pronounced fluid excretion with tissue damage [73, 78], whereas with pertussis, PT can only be assumed to be responsible for some disease manifestations, other toxins acting to produce the total clinical picture of pertussis [34, 35].

If CT is the only toxin involved in the disease process of cholera then detoxified purified CT or a non-toxic portion of the toxin, for example the immunodominant B subunit, should produce an ideal vaccine. In practise, however, vaccines consisting of formaldehyde- or glutaraldehyde-detoxified CT, or isolated B subunits, are poorly protective [78, 79]. Whole cell vaccines provide some degree of protection but the most effective vaccines would appear to consist of a combination of inactivated whole cells (or isolated components) supplemented with purified B subunits of CT [78, 80]. There is clearly a synergism between antibacterial and antitoxin antibodies and, being a non-invasive disease, the major protective response is that induced by the secretory IgA antibody of the gut mucosa [79–81]. Oral administration of a combined whole cell/B subunit vaccine has been found to be safe and protective in volunteers [78, 80].

Although the pathogenesis of cholera is primarily a consequence of toxin production, there are several other antigens with a role in pathogenesis and hence having potential for cholera vaccine development [82]. These include fimbriae, haemagglutinins, LPS, OMP and flagella.

The fimbriae (or pili) which occur on the surface of the bacterium are thought to have a role in overcoming the close-range electrostatic repulsion between cell surfaces by permitting adherence at a distance. The fimbriae purified from *V. cholerae* cells have a subunit Mr of approximately 16,000 [83]. It was also reported by these workers that this material possesses haemagglutinin activity suggesting that fimbriae may play a real role in colonisation. However, to date no evidence for protection by antibodies to fimbriae has been reported [84, 85].

In order to circumvent some of the drawbacks associated with killed whole cell vaccines considerable effort has been made into developing live attenuated strains for oral vaccination [78, 80]. Thus, the Texas Star-SR strain produces only the B subunit of CT and as a live vaccine has been found to confer significant protection in challenged volunteers [78]. However, the vaccine strain can induce mild diarrhoea in some recipients indicating either the existence of other toxic components in addition to CT or that the process of colonisation itself may induce diarrhoea. This latter point emphasizes that for any live attenuated vaccine it is essential that the genetic aberration which causes the diminished virulence be identified so that any prospect of reversion is eliminated.

Escherichia coli

Although perhaps not perceived as being such a problem as cholera, diarrhoea due to *E. coli* is common in most, if not all, developing nations, and diarrhoeal disease caused by this organism and other enteropathogenic bacteria constitutes a major worldwide health problem: thus, in 1975 some 17–18 million deaths were attributed to this cause [75]. The organisms that have been implicated in causing diarrhoea are pathogenic essentially because of their ability to produce enterotoxins. The genes for these proteins, which are thought to cause the disease symptoms, are carried on plasmid DNA and may occur singularly or in combination [86, 87, 88]. Two toxin types are implicated in disease and these are designated LT (heat-labile toxin) and ST (heat-stable toxin). Biochemical evidence has shown that the LT toxin is a homogenous protein similar to CT whereas the ST toxin is heterogenous in nature and comprises a group of small proteins in the range Mr = 2,000 – 10,000 [88]. The ST toxins act by stimulating guanylate cyclase activity which leads to fluid secretion in the gut.

As with cholera, the non-invasive nature of the organism makes the development of an effective vaccine difficult and has prompted several approaches to the development of vaccines that can induce appropriate immunity at the site of infection. The use of a toxoid vaccine has been attempted as has the use of killed whole cells, avirulent live cells and synthetic peptides [86, 89, 90]. More recently, genetic manipulation has been used to produce an inactivated *E. coli* LT having unaltered holotoxin configuration [91]. This toxoid when injected into pigs induced the formation of neutralising antibodies against native LT. The use of *E. coli* strains possessing plasmids carrying such modified toxin genes might provide a new approach to effective oral vaccines for this important class of infections.

4.4   Pseudomonas species

The Pseudomonads are frequently identified as pathogens in compromised hosts, such as those with thermal injury or undergoing therapy for malignant tumours, and it is a well-recognised agent of nosocomial disease [92, 93]. The virulence of Pseudomonas spp. is the product of the interaction of many variables involving both the pathogen and the host. Woods *et al.* [94] noted that damage to host tissue may occur directly through the action of bacterial products or indirectly through the induction of an overexuberant host response to these products. In some cases, in addition to cell associated factors, the organism may elaborate a number of exoenzymes including exotoxin A, proteolytic enzymes, phospholipase C and exoenzyme S, many of which have been implicated as pathogenic determinants [95, 96]. There have been reports of human immunity to Pseudomonas infection showing correlation to humoral antibody to these proteins, notably exotoxin A [97]. It has been shown that exotoxin A, which acts as an adenosine diphosphate ribosyl transferase in a similar nanner to diphtheria, cholera and *E. coli* LT toxins, acts to inhibit cellular protein synthesis by interfering with polypeptide translocation on messenger RNA [98]. The similarity of this toxin to toxins which have formed the basis of vaccines effective against other diseases has led to an increased interest in the possible use of exotoxin A as a component of Pseudomonas vaccines. Pavlovskis *et al.* [99] described the immunisation of mice with a chemically inactivated *P. aeruginosa* exotoxin A which, in the presence of the synthetic adjuvant MDP, induced high levels of anti-exotoxin antibody that afforded protection to burned and infected mice. Immunisation with toxoid alone did not induce antitoxin. A second approach to the use of exotoxin A has employed a mutant strain of *P. aeruginosa* that produces a non-toxic, immunologically cross-reactive protein (CRM) and has been designated CRM 66 [100]. The mutation is present in the DNA coding for the structural gene for exotoxin A and adversely affects enzymatic activity. The mutant toxin is, however, antigenically indistinguishable from its active counterpart and thus may be an ideal candidate as a component for development of a vaccine against Pseudomonas species.

A number of environmental factors have been shown to influence the production of pseudomonad antigenic determinants which may act as possible vaccine components. The yields of exotoxin A, elastase and alkaline phosphatase are independently regulated by iron [101]. Phosphate cations and the nature of the carbon source have been found to affect the production of phospholipase C [102]. It has recently been demonstrated that Pseudomonas species isolated directly from *in vivo* sites of infection exhibit immunogenic OMPs that are identical to those exhibited by the same species grown *in vitro* under iron-limited conditions [4, 103]. This has led to the suggestion that some of the OMP from iron-limited *in vitro* cultures of Pseudomonas may act as possible immunogens for vaccine development. It has recently been suggested that another protein (OMP F), which functions as a porin, may be used as a protective vaccine against pseudomonas infection [104]. This cell-surface protein is conserved and antigenically related in all serotype strains [105]. In addition, antibodies to OMP F have been found in the sera of both cystic fibrosis and burn patients [106, 107]. This would represent a different approach to vaccine formulation since porin proteins cannot be considered as being true virulence determinants as their function is concerned with the uptake of nutrients in both virulent and avirulent strains of bacteria.

Other researchers have shown convincing evidence for other antigens as vaccine candidates including lipopolysaccharide [108, 109]. Human immunity to *P. aeruginosa* has been correlated with humoral antibody to type-specific LPS [97] and antibody to LPS has been shown to be highly protective against *P. aeruginosa* infections in model systems [109, 110]. Attempts to use native LPS as a vaccine have been restricted by the high frequency of adverse reactions to immunisation, the need to include LPS from a number (up to 16) different serotype strains [111] and the need for multiple injections to evoke an optimal immune response [112, 113]. Alternatives to native LPS include high-molecular weight polysaccharide [114], detoxified LPS-protein conjugates [115] and polysaccharide-protein conjugates [116]. Conjugate vaccines appear particularly attractive because they present two protective antigens: the serotype specific component and the carrier protein which could be a Pseudomonas protein (eg exotoxin toxoid) or protein from another species. The use of a polysaccharide-tetanus toxoid conjugate vaccine was reported by Cryz *et al.* [117] and demonstrated the applicability of this approach.

4.5   Diphtheria

As already described in section 2 of this chapter the *C. diphtheriae* organisms are essentially non-invasive and disease results from the production of a potent extracellular toxin. Diphtheria toxin is produced only by strains of *C. diphtheriae* infected with lysogenic bacteriophages which carry within their genome the *tox* gene. Non-toxigenic strains can be made toxigenic by infection with any one of several morphologically and serologically distinct classes of corynephages carrying the *tox* gene. Thus, although the toxin is not coded for by the host bacterial DNA, the regulation and expression of the gene is under bacterial control. The toxin itself is synthesized as a precursor (Mr = 66,000–68,000) that is secreted into the culture medium [118] as a single polypeptide chain (Mr = 62,000) containing two disulphide bridges. This toxin is a pro-enzyme which can be activated by proteases to produce an adenosine diphosphate ribosyl transferase that is directly involved in the inhibition of cellular protein synthesis.

The diphtheria vaccine in current use consists of a purified preparation of formaldehyde-inactivated toxin absorbed onto aluminium salts for enhanced immunogenicity. This is highly effective in inducing antibodies that prevent disease even though immunity may not prevent acquisition or carriage of the organism. Improved vaccines might lead to preparations which are equally effective, less reactogenic and more amenable to rigorous quality control testing. One approach is to use material from mutant strains. For example several classes of cornynephage have been isolated that code for defective diphtheria toxin that is immunologically identical to the native toxin but which lacks enzymic activity [119, 120]. However, it has been noted that some of these proteins, which appear theoretically to be ideal vaccines, are considerably less effective than standard diphtheria toxoid in raising antitoxic antibodies [121]. More recently it has been shown that treatment of these non-active proteins with formalin (in a manner similar to the treatment used to produce standard toxoid from native toxin) enhances immunogenicity [122]. What advantages this may offer over formalin-inactivated native toxin as a vaccine is not readily apparent but clearly reversion would not be a major safety problem.

A second approach involves the use of recombinant DNA technology to produce synthetic peptides corresponding to fragments of the diphtheria toxin [123]. A small peptide analogous to the "loop" region of the toxin induced specific antibodies that were effective as vaccines in animal models [124]. This was the first report in which a totally synthetic vaccine, in the absence of an adjuvant, had induced effective protective immunisation against disease-causing bacteria. There is, however, still some way to go before this approach might yield an effective human vaccine since relatively high doses of synthetic antigen are needed to induce responses which are weak compared to those induced by much smaller doses of formolised toxoid. This is, however, a promising area of research and deserves continued support in view of its possible value in developing a new vaccine.

4.6   Tetanus

It is appropriate to briefly consider tetanus since it constitutes the third element of the ubiquitous triple D.T.P. vaccine, two elements of which have already been discussed. The pathogenic anaerobe *Cl. tetani,* produces an oxygen-labile membrane-damaging toxin (tetanolysin) but it is the potent neurotoxin (tetanus toxin, tetanospasmin) that gives rise to the widespread muscular spasms characteristic of the disease tetanus. Tetanospasmin comprises two polypeptide chains (Mr = 100,000 (A) and 50,000 (B)) which are joined by a single disulphide bridge. Both fragments are non-toxic in the disassociated form and fragment B is capable of binding to susceptible cell membranes and will compete with whole toxin for specific binding sites. Tetanus toxin acts primarily on the central reflex apparatus in the spinal cord, causing continual excitation of the motor neurones and hence spastic paralysis. The binding of tetanus toxin to neurones was recently reviewed by Montecucco [125].
The current vaccine is a formaldehyde-inactivated preparation of partially purified tetanospasmin; it is effective in inducing levels of circulating antitoxin that afford good protection. In consequence there would seem to be little need or scope for improvement except, perhaps, to eliminate the slight possibility of reversion. The development of improved pertussis and diphtheria vaccines could, however, generate pressure for an improved tetanus toxoid. In this context the elucidation of the sequence of tetanus toxins by the cloning of fragments of the toxin gene [126, 127] opens the door to studies on the structure and function of the molecule. As with *C. botulinum* toxin (see Section 3.1) these could facilitate the development of chemically or genetically in-

activated toxins having good immunogenicity or even, possibly, protective synthetic peptides.

It must, however, be pointed out that the currently used successful tetanus and diphtheria toxoid vaccines are produced from relatively impure toxin preparations and the extent to which contaminating proteins might contribute to protection is not known. Hence further purification of the toxins might yield vaccines which are less effective; although this has not been found to be the case with highly purified botulinum toxoid vaccines used in animals (P. Hambleton, unpublished data).

5 Conclusions

From the above discussion it is clear that there are several advantages (Table 2) to the design of vaccines based on an understanding of the mechanism of pathogenesis of the various diseases and the roles of virulence components in these pathogenic processes.

Table 2
Pros and cons of a rational approach to vaccine design

Advantages	Disadvantages
Stimulate optimal protective immunity	Low yields of antigen – high cost
Selection of best immunogens	High costs of purification
Exclusion of non-essential components	Over selectivity of antigen/epitope
Minimise side effects (reactogenicity)	Incorrect presentation of antigen
Uniformity and conformity of formulation and potency	Need for adjuvants
Cloned antigens – cost benefits	

Firstly, a rational approach offers the possibility of targetting the immune response against specific, critical stages of pathogenesis with the opportunity of stimulating an optimal immune response. This could be achieved by appropriate selection of antigenic determinants since, for example, remarkable selectivity has been found for different determinants on the same protein [128, 129].

Secondly, the strategy would allow the selection of those immunogenic components which stimulate a protective response with the exclusion of antigens which either are poorly immunogenic or are highly immunogenic but do not contribute at all to protection. One benefit of this, of course, would simply be to reduce the total amount of material present in a dose of vaccine. But a more positive advantage of the exclusion of non-essential substances could be the removal of components which may have deleterious side effects for the vaccinees.

Another particular benefit of the informed selection of vaccine components could be to bring conformity of vaccine formulation and potency. For example, it is now possible to prepare pertussis vaccines in which standardised amounts of highly purified antigens having defined immunogenicities are blended to a predetermined formulation, so minimising the possibility of excessive batch to batch variation.

Finally, the cloning of the genes for selected antigens offers the opportunity for enhanced expression and production of antigens or selected epitopes therefrom. This could offer considerable cost benefits where protective antigens are normally only poorly expressed in wild type organisms.

There are, however, potential disadvantages (Table 2) that should be borne in mind when considering a rational approach to vaccine design, particularly where purified component variations are concerned. Firstly there may be cost disadvantages; particularly where the purification of low yield or unstable antigens is concerned or where the purification procedure is costly and/or inefficient. There is also a danger of being too selective. For example, choosing a single peptide as a vaccine component would lead to restricted antibodies which might select for antigenic variants of the virulence determinant; the vaccine-induced antibodies would eventually no longer be protective. This would, however, appear to be a relatively unlikely occurrence in the case of antigens with multiple epitopes and/or multiple antigen vaccines.

Another potential disadvantage of the refined approach is that in many cases the selection of a single component may not be sufficient to afford complete protection against disease. There could even be a danger that whilst the vaccinated host might not succumb to disease, a carrier state might be induced resulting in perpetuation rather than control of the disease in communities. Clear examples of the value of multicomponent vaccines are seen in the attempts to develop simple acellular vaccines against cholera (and possibly pertussis) where there is a synergism between the protective effects of various antigens. This is not to argue that only vaccines possessing the totality of antigens found in the infecting parasite will ever be wholly effective, but simply that vaccines should comprise merely those protective antigens which are likely to tip the balance of the host-parasite relationship in favour of the host.

One further aspect to a rational approach to vaccines concerns the appositeness of the route of administration and hence the induced immune response. Generally, bacterial vaccines are given by injection either intramuscularly or into the dermal region which do not mimic many natural routes of infection. Whilst this may not be of concern when the aim is to induce circulating antibody or cell-mediated responses, it may result in low or even inadequate levels of topical immunity at the site of naturally acquired infection. In consequence, interest in the administration of vaccines by routes related to those of infection could lead to more effective protection by vaccines.

With these potential disadvantages borne in mind it is evident that the renaissance of interest in vaccine design will lead to the manufacture and widespread use of more effective and safe bacterial vaccines.

References

1   H. Smith: Symp. Soc. gen. Microbiol. *22*, 1 (1972).
2   P. Hambleton, J. A. Carman and J. Melling: Vaccine *2*, 125 (1984).
3   P. Turnbull: Abs. Hyg. Commun. Dis. *61*, R1 (1986).
4   G. H. Shand, H. Anwar, J. Kadurugamuwa, M. R. W. Brown, S. H. Silverman and J. Melling: Infect. Immun. *48*, 35 (1985).
5   Oregon State Board of Higher Education. U. S. Patent No. 790910.
6   J. B. Ward and R. C. W. Berkelely in: Microbial Adhesion to Surfaces, p. 47. Eds R. C. W. Berkeley, J. M. Lynch, J. Melling, P. R. Rutter and B. Vincent. Ellis Horwood, Chichester (1980).
7   E. C. Gotschlich: Ann. Inst. Pasteur/Microbiol. *136B*, 341 (1985).
8   J. E. Heckels: J. med. Microbiol. *18*, 293 (1984).
9   P. Hambleton and J. Melling in: Medical Microbiology, Vol. 3, p. 231. Eds C. S. F. Easmon and J. Jeljaszewicz. Academic Press Inc. London (1983).
10 S. N. De and D. N. Chatterje: J. pathol. Bacteriol. *66*, 559 (1953).
11 A. Robinson, L. A. E. Ashworth and L. I. Irons: Lancet *ii*, 108 (1982).
12 A. Robinson and L. I. Irons: Infect. Immun. *40*, 523 (1983).
13 L. A. E. Ashworth, R. B. Fitzgeorge, L. I. Irons, C. P. Morgan and A. Robinson: J. Hyg., Camb. *88*, 475 (1982).
14 A. Robinson, L. A. E. Ashworth, A. Baskerville and L. I. Irons in: Proceedings of Fourth International Symposium on Pertussis. Dev. Biol. Stand. *61*, 165–172 (1985).
15 D. W. Stainer and J. Cameron: Dev. Biol. Stand. *34*, 149 (1977).
16 C. C. Shone, P. Hambleton and J. Melling: Eur. J. Biochem. *151*, 75 (1985).
17 G. Dougan, C. E. Hormaeche and D. J. Maskell: Parasite Immunol. *9*, 151 (1987).
18 F. Dorner and J. L. McDonel: Vaccine *3*, 94 (1985).
19 M. F. Good, W. L. Maloy, M. N. Lunde, H. Margalit, J. L. Cornette, G. L. Smith, B. Moss, L. H. Miller and J. A. Berzofsky: Science *235*, 1059 (1987).
20 F. Borras-Cuesta, A. Petit-Camurdan and Y. Fedon: Eur. J. Immunol. *17*, 1213 (1987).
21 C. Leclerc, G. Przewlocki, M. Churze and L. Chedid: Eur. J. Immunol. *17*, 269 (1987).

22 M. J. Francis, G. Z. Hastings, A. D. Syred, B. McGinn, F. Brown and D. J. Rowlands: Nature *330,* 168 (1987).
23 D. Milich, A. McLachlan, G. Thornton and J. Hughes: Nature *329,* 547 (1987).
24 J. Rothbard: Nature *330,* 106 (1987).
25 K. C. McCullough and D. Langley: Vaccine *3,* 59 (1985).
26 I. M. Roitt, Y. M. Thanavala, D. K. Male and F. C. Hays: Immunol. Today *6,* 265 (1985).
27 M. Zanetti, E. Sercarz and J. Salk: Immunol. Today *8,* 18 (1987).
28 M. Davies in: Immunology of the Bacterial Cell Envolope, p. 271. Eds D. E. S. Stewart-Tull and M. Davies. J. Wiley & Sons Ltd., Chichester (1985).
29 A. C. McCartney and A. C. Wardlaw in: Immunology of the Bacterial Cell Envelope, p. 203. Eds D. E. S. Stewart-Tull and M. Davies. J. Wiley & Sons Ltd., Chichester (1985).
30 D. E. S. Stewart-Tull: A. rev. Microbiol. *34,* 311 (1980).
31 S. Kotani, I. Azuma, H. Takada, M. Tsujimoto and Y. Yamamura: Adv. exp. Med. Biol. *166,* 117 (1983).
32 H. Takada and S. Kotani in: Immunology of the Bacterial Cell Envelope, p. 119. Eds D. E. S. Stewart-Tull and M. Davies. J. Wiley & Sons Ltd., Chichester (1985).
33 A. Robinson, L. I. Irons and L. A. E. Ashworth: Vaccine *3,* 11 (1985).
34 A. Robinson, L. A. E. Ashworth and L. I. Irons in: Pathogenic Mechanisms in Respiratory Infections. Beecham Colloquia, Blenheim Palace (1987).
35 A. A. Weiss and E. L. Hewllett: A. Rev. Microbiol. *40,* 661 (1986).
36 Y. Sato, K. Izumiya, M.-A. Oda and H. Sato in: International Symposium on Pertussis, p. 51. Eds C. R. Manclark and J. C. Hill. U. S. Dept. Health Education and Welfare, Washington (1979).
37 L. A. E. Ashworth, L. I. Irons and A. B. Dowsett. Infect. Immun. *37,* 1278 (1982).
38 E. Tuomanen, A. A. Weiss, R. Rich, F. Zak and O. Zak in: Proceedings of Fourth International Symposium on Pertussis. Dev. Biol. Stand. *61,* 197 (1985).
39 A. R. Gorringe, L. A. E. Ashworth, L. I. Irons and A. Robinson: FEMS Microbiol. Lett. *26,* 5 (1985).
40 A. Urisu, J. L. Cowell and C. R. Manclark: Infect. Immun *52,* 695 (1986).
41 L. A. E. Ashworth, A. B. Dowsett, L. I. Irons and A. Robinson in: Proceedings of Fourth International Symposium on Pertussis. Dev. Biol. Stand. *61,* 143 (1985).
42 L. I. Irons, L. A. E. Ashworth and A. Robinson in: Proceedings of Fourth International Symposium on Pertussis. Dev. biol. Stand. *61,* 153 (1985).
43 I. Livey, C. J. Duggleby and A. Robinson: Molec. Microbiol. *1,* 203 (1987).
44 J. M. Zhang, J. L. Cowell, A. C. Steven and C. R. Manclark in: Proceedings of Fourth International Symposium on Pertussis. Dev. Biol. Stand. *61,* 173 (1985).
45 F. R. Mooi, H. G. J. van der Herde, A. R. ter Avest, K. G. Welinder, I. Livey, B. A. M. J. van der Zeijst and W. Gaastra: Microb. Pathog. *2,* 473 (1987).
46 E. J. Carter and N. W. Preston: J. med. Microbiol. *18,* 87 (1984).
47 A. C. Steven, M. E. Bisher, B. L. Trus, D. Thomas, J. M. Zhang and J. L. Cowell: J. Bacteriol. *167,* 968 (1986).
48 J. L. Cowell, J. M. Zhang, A. Urisu, A. Suzuki, A. C. Steven, T. Liu, T.-Y. Liu and C. R. Manclark: Infect. Immun. *55,* 916 (1987).
49 P. Novotny, A. P. Chubb, K. Cownley, U. A. Montarez and J. E. Beesley in: Proceedings of Fourth International Symposium on Pertussis. Dev. Biol. Stand. *61,* 27 (1985).

50 J. L. Cowell, Y. Sato, H. Sato, B. An der Lan and C. R. Manclark in: Seminars in Infectious Diseases, vol. IV, Bacterial Vaccines, p. 371. Eds J. B. Robbins, and J. C. Hill. Thiene-Stratton, New York (1981).
51 J. J. Munoz, H. Arai and R. L. Cole: Infect. Immun *32*, 243 (1981).
52 H. Sato and Y. Sato: Infect. Immun *46*, 415 (1984).
53 Y. Sato, M. Kimura and H. Fukumi: Lancet *i*, 122 (1984).
54 S. Isomura, S. Suzuki and Y. Sato in: Proceedings of Fourth International symposium on Pertussis. Dev. Biol. Stand. *61*, 531 (1985).
55 M. Kimura and N. Hikino in: Proceedings of Fourth International Symposium on Pertussis. Dev. Biol. Stand. *61*, 545 (1985).
56 G. R. Noble, R. H. Bernier, E. C. Esber, M. C. Hardegree, A. R. Hinman, D. Klein and A. J. Saah: J. Am. med. Assoc. *257*, 1351 (1987).
57 Workshop on Acellular Pertussis Vaccines. Transcript by U.S. Dept. Health and Human Services Sept. 22–24, 1986.
58 D. A. Rutter, L. A. E. Ashworth, A. Day, S. Funnell, F. Lovell and A. Robinson: Vaccine *6*, 29–32 (1988).
59 E. C. Gotschlich: Bull. W. H. O. *62*, 671 (1984).
60 I. W. DeVoe: Microbiol. Rev. *46*, 162 (1982).
61 J. E. Sippel: CRC crit. Rev. Microbiol. 267 (1981).
62 J. B. Rothbard, R. Fernandez, L. Wang, N. N. H. Teng and G. K. Schoolnik: Proc. natl Acad. Sci. USA *82*, 915 (1985).
63 M. Virji and J. E. Heckels: Infect. Immun. *49*, 621 (1985).
64 D. S. Stephens, A. M. Whitney, J. Rothbard and G. K. Schoolnik: J. exp. Med. *161*, 1539 (1985).
65 M. R. Lifely, C. Moreno and J. C. Lindon: Vaccine *5*, 11 (1987).
66 C. E. Frasch, W. D. Zollinger and J. T. Poolman: Rev. inf. Dis. *7*, 504 (1985).
67 C. E. Frasch: Eur. J. clin. Microbiol. *4*, 533 (1985).
68 C. E. Frasch, M. S. Peplar, T. R. Cole and J. M. Zahradnik: Seminars in Infectious Diseases, vol. IV, Bacterial Vaccines, p. 263 (1982).
69 W. D. Zollinger, R. E. Mandrell and J. McLeod Griffiss: Seminars in Infectious Diseases, vol. IV, Bacterial Vaccines, p. 254 (1982).
70 E. C. Beuvery, M. Witvlier, J. A. M. Timmermans, J. T. Poolman, C. T. P. Hopman, T. Teerlink, C. J. A. Speijers and L. H. J. C. Danse: Antonie van Leeuwenhoek *52*, 232 (1986).
71 C. R. Tinsley and J. E. Heckels: J. gen. Microbiol. *132*, 2483 (1986).
72 J. R. Black, D. W. Dyer, M. K. Thompson and P. F. Sparling: Infect. Immun. *54*, 710 (1986).
73 C. C. J. Carpenter, W. B. Greenough and R. Gordon in: Cholera, p. 129. Eds D. Barua and W. Burrows. W. B. Saunders, Philadelphia (1974).
74 D. R. Schneider and C. D. Parker: J. infect. Dis. *138*, 143 (1978).
75 R. A. Finkelstein: Prog. clin. Biol. Res. *47*, 133 (1980).
76 T. Yamamoto and T. Yokota: J. Bacteriol. *155*, 728 (1983).
77 A. J. Moss: Annls intern. Med. *101*, 653 (1984).
78 M. M. Levine, J. B. Kaper, R. E. Black and M. L. Clements: Microbiol. Rev. *47*, 510 (1983).
79 J. Holmgren and N. Lycke in: Development of Vaccines and Drugs Against Diarrhoea, p. 9. Eds J. Holmgren, A. Lindberg and R. Mollyby. Studentlitteratur, Lund, Sweden (1986).
80 A.-M. Svennerholm and J. Holmgren in: Development of Vaccines and Drugs against Diarrhoea, p. 33. Eds J. Holmgren, A. Lindberg and R. Mollby. Studentlitteratur, Lund, Sweden (1986).
81 N. F. Pierce, W. C. Cray and J. B. Sacci: Annls N. Y. Acad. Sci. *409*, 724 (1983).
82 P. A. Manning: Vaccine *5*, 83 (1987).

83   M. Ehara, M. Ishibashi, Y. Ichinose, M. Iwanaga, S. Shimotoris and T. Naito: Vaccine (in press, 1987).
84   W. Gaastra and F. K. De Graaf: Microbiol. Rev. *46*, 129 (1982).
85   E. C. Tramont and J. W. Boslego: Vaccine *3*, 3 (1985).
86   C. D. Ericsson, H. L. DuPont, P. Sullivan, E. Galindo, D. G. Evans and D. J. Evans: Annls Int. Med. *98*, 20 (1983).
87   L. V. Wood, L. E. Ferguson, P. Hogan, D. Thurman, D. R. Morgan, H. L. Dupont and C. D. Ericsson: Appl. environ. Microbiol. *46*, 328 (1983).
88   H. L. Dupont and L. K. Pickering in: Infections of the Gastrointestinal Tract, Microbiology, Pathophysiology and Clinical Features. Plenum, New York (1980).
89   H. W. Moon, A. L. Baetz and R. A. Giannella: Infect. Immun. *39*, 990 (1983).
90   A. M. Svennerholm and C. Ahren: Acta path. microbiol. immunol. scand. Sect. 6 *90*, 1 (1982).
91   A. N. Hobson, S. Harford, C. W. Dykes, J. J. Read and I. J. Halliday: Prot. Eng. *1*, 263 (1987).
92   J. V. Bennet: J. infect. Dis. 130, Suppl. 54–57 (1974).
93   M. I. Marks: J. Pediatr. *98*, 173 (1981).
94   D. E. Woods, M. S. Schaffer, H. R. Rabin, G. D. Campbell and P. A. Sokoh: J. clin. Microbiol. *24*, 260 (1986).
95   O. R. Pavlovskis and B. Wretlind in: Medical Microbiology, vol. 1, p. 97. Eds C. S. F. Easmon and J. Jeljaszewicz. Academic Press Inc., London (1982).
96   D. E. Woods and B. H. Iglewski: Rev. infect. Dis. *5*, 5715 (1983).
97   M. Pollack and L. S. Young: J. clin. Invest. *63*, 276 (1979).
98   B. H. Iglewski and D. Kabat: Proc. natl Acad. Sci. USA *72*, 2284 (1975).
99   O. R. Pavlovskis, D. C. Edman, S. H. Leppla, B. Wretlind, L. R. Lewis and K. E. Martin: Infect. Immun. *32*, 681 (1981).
100 S. J. Cryz, Jr, O. R. Pavlovskis and B. H. Iglewski in: Seminars in Infectious Diseases: Bacterial Vaccines, p. 128. Eds L. Weinstein and B. N. Fields. Thieme-Stratton, Inc., New York (1982).
101 M. J. Bjorn, O. R. Pavlovskis M. R. Thompson and B. H. Iglewski: Infect. Immun. *24*, 837 (1979).
102 R. M. Berka and M. L. Vasil: J. Bacteriol. *152*, 239 (1982).
103 M. R. Brown, H. Anwar and P. A. Lambert: FEMS Microbiol. Lett. *21*, 113 (1984).
104 J. M. Matthews-Greer and H. E. Gilleland, Jr: J. infect. Dis. *155*, 1282 (1987).
105 L. M. Multharia, T. I. Nicas and R. E. W. Hancock: J. infect. Dis. *146*, 770 (1982).
106 P. B. Fernandes, C. Kim, K. R. Cudy and N. N. Huang: Infect. Immun. *33*, 527 (1981).
107 H. Anwar, G. H. Shand, K. H. Ward, M. R. W. Brown, K. E. Alpar and J. Gowar: FEMS Microbiol. Lett. *29*, 225 (1985).
108 S. J. Cryz Jr, E. Furer and R. Germanier: Infect. Immun. *43*, 795 (1984).
109 G. W. Pier, H. F. Sidberry and J. C. Sadoff: Infect. Immun. *22*, 919 (1978).
110 J. A. Kazmierowski, H. Y. Reynolds, J. C. Kauffman, W. A. Durbin, R. G. Graw and H. B. Devlin: J. infect. Dis. *135*, 438 (1977).
111 J. M. Miler, J. F. Spilsbury, R. J. Jones, E. A. Rue and E. J. L. Lowbury: J. med. Microbiol. *10*, 19 (1977).
112 J. W. Alexander and M. Fisher: J. infect. Dis. *130*, 5152 (1974).
113 L. S. Young, N. Meyer and D. Armstrong: Annls intern. Med. *79*, 518 (1973).
114 G. B. Pier: J. clin. Invest. *69*, 303 (1982).
115 R. C. Sied Jr and J. C. Sadoff: J. biol. Chem *256*, 7305 (1981).

116   G. C. Tsay and M. S. Collins: Infect. Immun. *45*, 217 (1984).
117   S. J. Cryz Jr, J. C. Sadoff, E. Furer and R. Germanier: J. infect. Dis. *154*, 682 (1986).
118   W. P. Smith, P. C. Tai, J. R. Murphy and B. D. Davis: J. Bacteriol. *141*, 184 (1980).
119   T. Uchida, D. M. Gill and A. M. Pappenheimer Jr: Nature *233*, 8 (1971).
120   R. K. Holmes: J. Virol *19*, 195 (1976).
121   A. M. Pappenheimer, Jr, T. Uchida and A. A. Harper: Immunochemistry *9*, 891 (1972).
122   M. Porro, M. Saletti, L. Nencioni, L. Tagliaferri and I. Marsili: J. infect. Dis. *142*, 716 (1980).
123   F. Audibert, M. Jolivet, L. Chedid, R. Arnon and M. Sela: Proc. natl Acad. Sci. USA *79*, 5042 (1982).
124   F. Audibert, M. Jolivet, L. Chedid, J. E. Alouf, P. Boquet, R. Rivaille and O. Siffert: Nature *289*, 593 (1981).
125   C. Montecucco: Trends biochem. Sci. *11*, 314 (1986).
126   N. F. Fairweather, V. A. Lyness and D. J. Maskell: Infect. Immun. *55*, 2541 (1987).
127   U. Eisel, W. Jarausch, K. Goretzki, A. Henschen, J. Engels, U. Weller, M. Hudel, E. Habermann and H. Niemann: EMBO J. *5*, 2495 (1986).
128   R. M. Maizels, J. A. Clarke, M. A. Harvey, A. Miller and E. E. Sercarz: Eur. J. Immunol. *10*, 509 (1980).
129   E. E. Sercarz, R. L. Yowell, P. Turkin, A. Miller, B. A. Araneo and L. Adorini: Immunol. Rev. *39*, 108 (1978).

The chemistry of DNA modification by antitumor antibiotics

By Jed F. Fisher[a] and Paul A. Aristoff[b]

[a]Cardiovascular Diseases Research, and [b]Cancer and Infectious Diseases Research, The Upjohn Company, Kalamazoo, MI 49001, USA

1   Introduction

The known antitumor antibiotics are not ideal. Each has its shortcomings, and fails to reach a biochemically unique and metabolically essential locus of the tumor cell, and thus prove specifically lethal. This is a reflection on the fact that there are many different cancers, each of which with many biochemical similarities to the host. As a consequence, many of the currently effective antitumor agents are toxins in the broadest sense of the term, and do no better than to achieve an overall favorable balance in selective toxicity to the tumor cell. Against the most common cancers there is not a single agent where the balance tips decisively. The need for agents with greater selectivity remains. How is the objective to be achieved? Since the screening for antitumor agents is frequently done empirically, without prejudice as to a biochemical mechanism, one answer is suggested by the examination of their mechanistic similarities. And here it is seen that despite an enormous scope of structure, most of the clinically effective antitumor agents appear in some way to interact with cellular DNA. This interaction may be direct (for example, by covalent modification) or indirect (for example, by limiting deoxyribonucleotide biosynthesis). While it is easy to rationalize interference with DNA as resulting in an antitumor effect (particularly for rapidly growing cells), it is much more difficult to discern the biochemical basis for the selective toxicity. DNA itself is a heterogeneous target and its integrity ultimately is as important to the host as to the tumor. What then is the point of difference that provides for the antitumor effect? At this time no satisfactory answer exists, and it is arguably difficult even to speculate. Part of this difficulty arises from the broad expanse covered by the term "DNA": does the antitumor agent distinguish between mitochondrial and nuclear DNA; DNA synthesis and RNA synthesis; DNA sequence and DNA conformation; nucleosome and linker; DNA methylation and DNA repair; and so on. It is clear that the antitumor agents do not target the same mechanism, and, thus, there exist multiple opportunities of which to take advantage. To obtain a better antitumor agent is a difficult task. Only as a clearer understanding is provided for the *in vivo* effects of each agent will it become plain why DNA is so frequently the preferred target of the antitumor agents. For the present this topic is the next frontier to be reached by the constant updating and evaluation of our evolving knowledge.

What follows is a review of the chemistry of DNA modification by the important antitumor antibiotics. By choosing this emphasis the topic remains manageable. The focus is on an understanding of their mechanism of action and as a consequence detailed information on the discovery, development, new analogs, and clinical strategy are for the most part omitted. As a format for discussion the structures are divided into those requiring direct enzymatic participation to accomplish DNA modification (topoisomerase inhibition; reductive and oxidative activation) and those which do not. Several complementary reviews on this topic have appeared [1–4]. The literature cited emphasizes the events of the three-year-period up until late 1987.

2   Enzymatic activation of the antitumor antibiotics

Of the important antitumor antibiotics the majority take advantage of an ability to be recognized by enzymes in the environment of the DNA. This may have as its consequence the inhibition of an enzyme directly involved in manipulating transcriptionally active DNA (e. g., topoisomerase), or biooxidative activation (oxidation by a peroxidase to an electrophile), or bioreductive activation (reduction by a reductase to an electrophile). Among the topoisomerase inhibitors are the podophyllotoxins, the amsacrines, the ellipticines, the anthracenediones, and the anthracyclines. Although of these only the anthracyclines are antitumor antibiotics (that is, of microbial origin) topoisomerase inhibition has become of such importance to antitumor action that any review concerned with the antitumor antibiotic-DNA interactions would be compromised if the discussion omitted the podophyllotoxins and amsacrines. It is with these two classes that the best data for interpreting topoisomerase II inhibition have been obtained. In addition, the ellipticines, podophyllotoxins, and perhaps the amsacrines as well, may augment their antitumor activity by biooxidative activation, and this concept is introduced as well. The anthracyclines, as a class of tetrahydronaphthacene quinone intercalators, are enigmatic. It is now recognized that the principal mechanism exploited by the anthracyclines is also topoisomerase inhibition, perhaps augmented by bioreductive activation to an anthracycline-derived electrophile (a quinone methide) or by reduction of oxygen to initiate metal-dependent peroxidative fragmentation of the DNA backbone. The mitomycins act primarily by bioreductive activation in hypoxic tissue to quinone methide electrophiles. Since a single mitomycin molecule may be activated twice to provide two different quinone methides, it is possible for DNA strand cross-linking to result. The bleomycins are metal chelators containing an intercalating segment; by intercalation the metal chelate is held proximal to the DNA where reductase-mediated O_2 activation results in a complex series of radical rearrangements culminating in strand cleavage. In each instance the integrity of the DNA is lost and is responsible for cell death.

2.1   The podophyllotoxins: Topoisomerase II inhibition

Podophyllotoxin itself is a plant-derived antitumor agent and thus is not, in the strict sense of the term, an antitumor antibiotic. Nevertheless, the exception is worth making, for two reasons. The primary one is the relationship of this antitumor class to the topoisomerase II enzyme; the podophyllotoxins are among the best understood inhibitors of this enzyme. They are not unique in this inhibition, but share this enzyme as a target with other antitumor agents (amsacrines, ellipticines, anthracenediones, and anthracyclines). Secondly, the podophyllotoxins have more than one biological activity; this too is a general phenomenon and thus permits discussion of the interpretive difficulties that multiple mechanisms of action present.

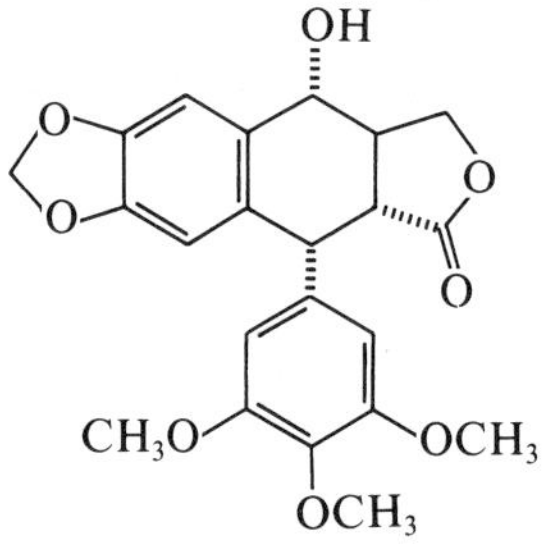

I

Podophyllotoxin

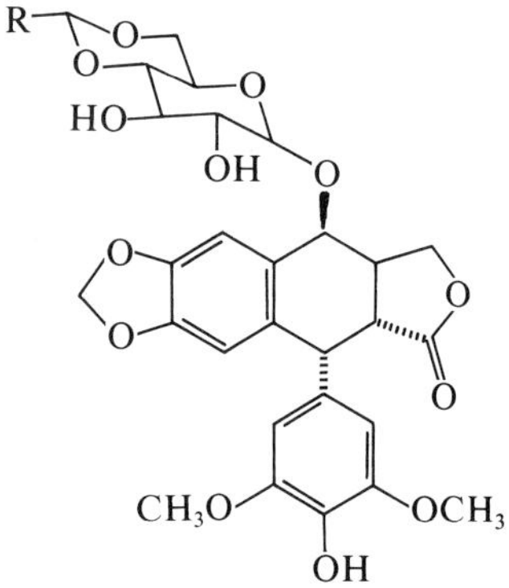

II, R = CH_3, Etoposide

III, R =  , Teniposide

Fig. 1

The early history of the podophyllotoxins is told well by Jardine [5] and is not duplicated here. Podophyllotoxin itself (I, Figure 1) is a moderately potent but highly toxic antitumor agent, having as a mechanism an inhibition of tubulin polymerization. In the course of synthetic modification this particular biological activity is lost entirely, and at the time etoposide (II) and teniposide (III) were synthesized, although both were potent antitumor agents, no knowledge of the biological mechanism was available. Regardless, the important synthetic alterations were clear and followed the suggestions of Loike and Horowitz [6–9]. The critical alteration is the C-4 glycoside, which in one stroke preserves the antitumor activity yet abolishes the effect on microtubules. The three additional alterations (C-4' demethylation; C-4 epimerization; and acetal formation) each enhance potency. Two such antitumor agents, etoposide and teniposide, have proven clinically useful in the treatment of cancer [10].

The search for a biological mechanism began with the observation that etoposide *in vivo,* but not *in vitro,* introduced DNA strand cleavage [6, 11]. With refinement in the alkaline elution technique for the measurement of DNA strand integrity, it was possible to prove that the concentration at which this occurred approached that at which the antitumor activity was manifest [12]. Simultaneously, Kohn and colleagues were observing similar phenomena for other antitumor agents [13]. The participation of an enzyme in this process was suggested by a number of observations, not the least of which is that the podophyllotoxins appear to have no ability to associate to or intercalate with DNA [14, 15]. In addition, the DNA strand breaks were not seen after heating, were reversed by high salt, and were irreversible only upon denaturing conditions. The involvement of topoisomerase II as the participant was suggested by Long [16] and independently confirmed by Ross and Liu [14, 15, 17, 18].

The topoisomerase enzymes [19–21] as targets for antitumor activity are now an entire field unto themselves [22–25]. Although of the two enzymes topoisomerase II occupies the major focus, it no longer stands alone, following the observation that a major consequence of the antitumor alkaloid camptothecin is inhibition of topoisomerase I [26–28]. The topoisomerase II will be discussed here from the particular perspective of the podophyllotoxins (with but brief mention of the complimentary studies with the antitumor amsacrines) and the anthracyclines (next section). Even though the topoisomerase enzymes are recent discoveries, the essential role of topoisomerase II, particularly in the cell growth cycle, is now well established. Briefly, this enzyme manipulates DNA structure during transcription in response to gene activation and inactivation. This is accomplished at two different levels, one being at the molecular and the other at the supra-molecular, as a protein component of the chromosomal "scaffold" [20, 29]. In this location, and by virtue of its catalytic activity, it participates in all phases of DNA change during cell growth, particularly the segregation of newly synthesized daughter DNA at the replication fork [30–33]. Its activity may be controlled by protein kinases [34]. At the molecular level the reaction catalyzed is sequential cleavage and resealing of both DNA strands. Prior to the resealing, strand passage through the break site is permitted. By this process, the topological state of the DNA is manipulated, as manifested by knotting/unknotting; supercoiling/relaxing; and catenation/decatenation of circular DNA. In the double-strand cleaved intermediate within the catalytic cycle, the enzyme preserves a covalent link to the strand by a phosphotyrosine bond to the 5' end. The cleavage reaction requires no cofactors (save Mg^{+2}), but the strand passage is ATP-dependent. Topoisomerase II is an enzyme required for cell reproduction and differentiation.

By what circumstance do the podophyllotoxins interfere with this catalyst, and how does this result in an antitumor effect? Neither question is simple, and in the case of the former an answer has been difficult to extract due to the general unavailability of purified enzyme. The overview is, however, in place. Using an alkaline filtration technique to measure DNA integrity, the presence of single- and double-strand breaks may be ascertained. In the presence of topoisomerase II inhibitors, protein-associated single- and double-strand breaks appear upon alkaline elution assay. The position of these cuts and the ratio of single to double-strand cleavage is dependent on the molecular structure of the inhibitor. Since this reaction is reversible (although not instantaneously so; resealing requires minutes after drug removal) the podophyllotoxins and the other inhibitors are not mechanism-based irreversible inhibitors, but entities which achieve a stabilization of the topoisomerase catalytic cycle at the point of strand cleavage(s). This stabilization occurs after single-strand cleavage or double-strand cleavage, depending on the antitumor agent, but prior to the strand passing event [14, 15, 17, 26]. The intercalator-induced DNA breaks are incapable of swiveling [35]. In support of the value of these observations to the actual *in vivo* occurrences are numerous parallels in overall behavior and kinetics. For example, upon examination of topoisomerase-mediated inhibitor cleavage of SV40 DNA *in vitro* and *in vivo*, identical cleavage sites were identified [18]. Likewise, teniposide is a more potent inhibitor *in vitro* than etoposide, as is also true for antitumor cytotoxicity [15, 36]. The evidence is compelling that a major consequence of cell exposure to these agents is an incapacitation of topoisomerase II as a catalyst.

Equally forceful arguments are provided by an examination of the amsacrine antitumor antibiotics (Figure 2) which also inhibit this enzyme. In contrast to the podophyllotoxins, the amsacrines are potent DNA intercalators. For this reason the two classes have frequently been used as complements in mechanistic study. Indeed it was the astonishing observation that the ortho isomer of meta-amsacrine (mAMSA, IV) was inactive both as a topoisomerase II inhibitor *and* as an antitumor agent [37] that stimulated the headlong development of topoisomerase as an antitumor target. The amsacrines are synthetic entities [38, 39] that intercalate by the reversible insertion of the acridine moiety between the base pairs of helical DNA [40–42]. As a relatively simple structure to synthesize, a breadth of analog structures is available and has been put to good use in investigating the structural basis for topoisomerase inhibition. There is no doubt that the strand breakage which accompanies *in vivo* amsacrine intercalation is a topoisomerase-mediated event [37, 43]. Antitumor potency correlates with the magnitude of the DNA association [44] and cytotoxicity parallels the extent of strand breakage [45, 46]. But intercalation alone is not sufficient. Strong intercalators within the amsacrine family (indeed intercalators

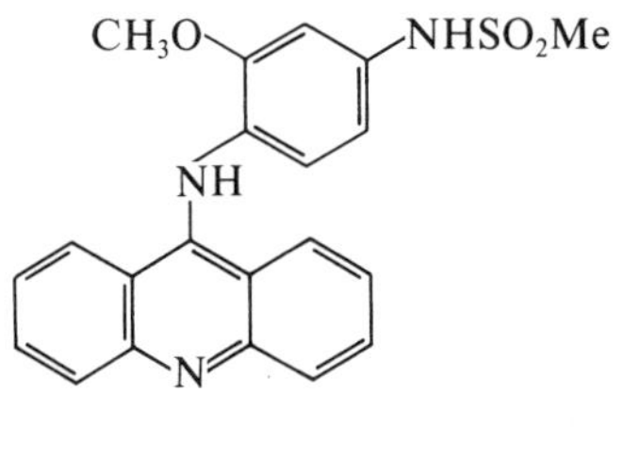

Fig. 2   IV. Amsacrine

in general) are neither topoisomerase inhibitors nor antitumor agents [47, 48]. In examining the antitumor amsacrines, Pommier *et al.*[47] observe that the point of topoisomerase cleavage in the DNA sequence is nearly independent of structural elements, and not at all dependent on a simple measure of intercalation (the unwinding angle) nor intrinsic GC or AT preference. While it is not yet possible to rigorously exclude a DNA conformational distortion by a proximal intercalator, the evidence is much more in the favor of a ternary DNA-inhibitor-enzyme complex. The identical conclusion is reached by an examination of podophyllin structure-activity behavior. All strand cleavages are topoisomerase-associated, and within several cell lines comparing etoposide and teniposide the cytotoxicity correlates with double-strand cleavage [49, 50]. There is not, however, a facile correlation between inhibition potency and cytotoxicity [50]. This is perhaps best understood in terms of the now general observation that different classes of topoisomerase inhibitors yield not simply different ratios of single to double-strand cleavage (amsacrines and podophylins give numerous single-strand breaks; ellipticines and anthracyclines very few) but different points of cleavage within the DNA sequence [51, 52]. It is reasonable then to presume that relative potency as an antitumor agent depends on both a relative balance in favor of double-strand cleavage, and in the location of the inhibition, beyond the usual factors (transport, metabolism, etc.).

At the level of the topoisomerase catalyst a working hypothesis has been postulated [35, 50]. The topoisomerase II is a dimer of identical subunits, and in the strand cleavage event covalent attachment to the two 5′ ends occurs. An inhibitor interrupts and stabilizes the catalytic cycle at a point between single-strand cleavage (one subunit inhibited by one inhibitor?) or double-strand cleavage (two subunits inhibited by two inhibitors?), and prevents strand passage. Since the podophyllins (non-intercalators) and amsacrines (intercalators) inhibit, and nonintercalating anthracyclines have an identical sequence preference for cleavage as intercalating anthracyclines [53], a model with an intimate association of both DNA and inhibitor with the enzyme is favored. This may occur either at the strand-cleaving active site or at the cleft that the enzyme is believed to require to accommodate the strand passage. The precise point at which cleavable complex stabilization occurs becomes dependent on both the structural details of the inhibitor and the precise sequence of the cloven strands.

How does topoisomerase inhibition result in tumor cell cytotoxicity? There is merit to the conclusion that when it happens, there is antitumor activity; and that if it happens in the right place on the DNA, potent antitumor activity may be attained. A more detailed answer is obviously beneficial in terms of rationally improving selectivity and potency, and considerable resources are now addressing this issue. The simple expectation that the topoisomerase is most active during cell cycle periods of high DNA synthesis is by and large fulfilled [33]. In CHO cells, for example, the topoisomerase content of the cells is elevated substantially during log phase growth, the quantity of protein-associated DNA cleavage increases with inhibitor concentration, and cytotoxicity parallels that increase [54, 55]. Generally speaking, the topoisomerase content of proliferating tumor cells is elevated over that found normally [55–58]. It is not, however, possible to make simple generalizations. There are cell lines which remain sensitive even after cell growth has plateaued, where little variation in topoisomerase content with cell cycle is seen, and where the extent of cleavage does not necessarily predict the relative cytotoxicity [59]. Once again it appears that apart from these factors it is also the location of the cleavable complex (which will be antibiotic structure-dependent) which contributes [60]. Perhaps the strongest evidence of a direct involvement of the topoisomerase in cytotoxicity comes from the study of antitumor resistant cells. Several cell lines resistant to the cytotoxic effects of one topoisomerase II inhibitor are also cross-resistant to other structurally unrelated topoisomerase II inhibitors [61–63]. In these resistant lines there is a corresponding decrease in the extent of cleavable complex formation, and in one instance this is attributable to the appearance of a second topoisomerase which no longer stabilizes a drug-inhibitor ternary cleavable complex [25, 64].

The evidence for topoisomerase II as a multi-drug target, and for an antitumor effect by enzyme-mediated covalent modification of DNA by strand cleavage, is strong. At the molecular level the outstanding question is the mechanism of the inhibition process, and the basis for cleavable complex stabilization. At the cellular level the relationships among structure, DNA sequence; and DNA consequence are imperfectly understood. The reward from such an undertaking is perhaps a rational basis to design therapeutic drug requirements which will result in the application of mechanistically complementary antitumor agents [65–66].

2.2   Biooxidative activation of the podophyllotoxins, ellipticines, and amsacrines

The three antitumor agents listed above share thematic similarities in that they are topoisomerase inhibitors and are capable of enzyme-dependent oxidation to an electrophile. Both may represent general mechanisms for DNA modification. As noted previously the majority of the topoisomerase inhibitors are generally redox active. The podophyllotoxins, ellipticines, and amsacrines are biooxidized; the anthracyclines and anthracenediones may be bioreduced. There is no basis as yet to believe that this is other than coincidental. Topoisomerase inhibition does not appear to involve a redox event; nor in the case of the podophyllotoxins are there strand cleavages that do not directly involve the topoisomerase [35]. But since the redox chemistry operates *in vivo,* and can result in direct covalent modification of DNA, its consequences are being pursued. As yet a biological justification has not developed. Against the expectation that one will be identified, a concise review of the redox activation by peroxidases of these three classes is provided.

Oxidative transformation of xenobiotics to a biologically active metabolite occurs frequently. A well-studied example is the sequential epoxidation, hydrolysis, and epoxidation of polycyclic aromatic hydrocarbons to the carcinogenic diol epoxides. The primary enzyme catalyst is cytochrome P-450. However, peroxidase involvement in this very same process, as well as others, is recognized [67–69]. By concerted action of these two enzyme classes a substantial alteration of the substrate molecule is possible, in particular to quinone-like electrophiles or to radicals. These might undergo a bimolecular reaction with DNA to achieve covalent modification, as occurs with the bioreductively activated mitomycin C (see below). The possibilities for etoposide biooxidative activation provide useful instruction. In the course of optimizing the antitumor activity of podophyllotoxin it was observed that removal of the methyl group on the 4′-oxygen resulted in a dramatic increase in cytotoxic potency. Subsequently a similar increase was seen for this change in the efficacy of topoisomerase inhibition [50]. Since the resulting functional group is a dimethoxyphenol, it is hardly surprising that etoposide can act as an electron or hydrogen atom donor in the presence of strong oxidants. This has been demonstrated directly [70]; in addition, etopside is an excellent inhibitor of lipid peroxidation [71], presumably by free radical chain termination by hydrogen atom donation. The important pathways from the perspective of DNA modification are etoposide oxidation by cytochrome P-450 and the peroxidases. Cytochrome P-450 accomplishes the oxidative demethylation of the dimethoxyphenol to give the methoxycatechol (diol) V as product [72, 73]. In a second oxidative cycle this is further oxidized to what is presumed to be the ortho-quinone (VI, Figure 3). There is circumstantial evidence that this has sufficient stability for both extraction and chromatographic detection [74], but in a heterogeneous milieu acts, as anticipated, as an electrophile. With excess reducing agent it is converted back to the catechol. There is also the possibility that the ortho-quinone reacts with nucleophiles by conjugate addition, to yield a substituted methoxycatechol. However, there are no direct experimental data establishing the chemical competence of this pathway. The ability of the ortho-quinone to inactivate both single- and double-stranded ΦX 174 DNA provides circumstantial evidence, as does the observation that etoposide cytotoxicity to a cell line increases with oxygen concentration [75]. The behavior of etoposide as

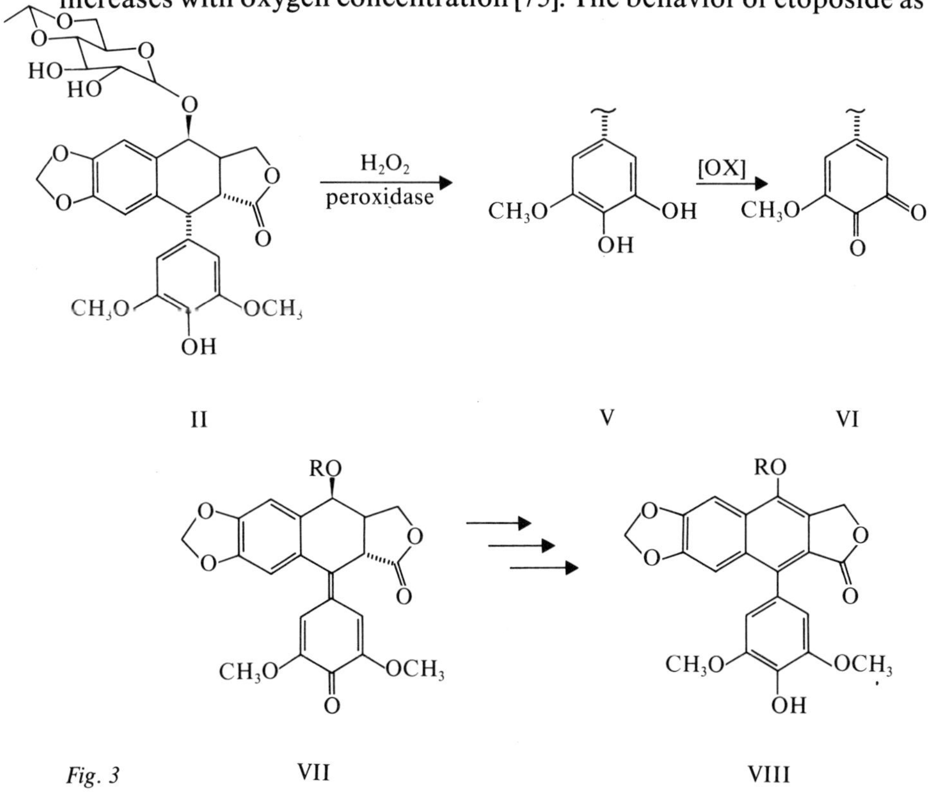

Fig. 3   VII   VIII

a reducing agent for peroxidases (both horseradish peroxidase and prostaglandin synthetase) is even more complex. Not only is oxidative demethylation and subsequent ortho-quinone formation seen (as occurs with cytochrome P-450), but a second pathway operates [76, 77] culminating in the production of the four-electron oxidized tetradehydroetoposide (VIII). On the basis of radiolabel studies, Naim *et al.* [72] find that while ortho-quinone alkylation of DNA does occur, the major alkylation pathway does not require loss of the C-4' methoxyl. Given the structure of the ultimate product the reasonable suggestion is made that nucleophile capture may occur at a quinone-methide-like intermediate (VII). Strong evidence for this proposal, as would be provided by the structure of an adduct, is awaited. As previously noted, there is no basis as yet for the belief that these processes contribute to the antitumor action. Since, however, prostaglandin synthetase has a very broad cellular distribution, it is to be presumed that such oxidative processes occur in virtually all cells exposed to the podophyllotoxins. Although careful study has failed to identify a contribution from redox chemistry to the topoisomerase inhibition by etoposide [35, 49] the possibility of a synergistic action to some aspect of the antitumor efficacy remains as speculation.

An identical situation exists for the amsacrines. These can be considered to contain two segments with separate functions: an intercalating acridine linked by the 1'-NH to the substituted anilino ring. The acridine is alone sufficient for antitumor activity, but for potent antitumor activity both are required. In their exhaustive manipulation of the amsacrine structure, Denny *et al.* [78, 79] have concluded that there is no useful antitumor amsacrine that lacks the oxidizable 1,4-diaminobenzene subgroup [80] and oxidation of this group *in vivo* most certainly occurs [81]. The primary metabolite isolated from rat bile is the glutathione conjugate (X, Figure 4), formed by cytochrome P-450 oxidation to the diiminoquinone (IX) and trapping of this electrophile by conjugate addition of glutathione [81]. This oxidation is also accomplished by Cu (II) *in vitro,* and metal-dependent oxidative cleavage of DNA results [82, 83]. The ortho AMSA isomer is less effective in this process, although the difference is modest (not the dramatic difference as there is for topoisomerase inhibition). Again, at this time there are no studies reporting a competence for DNA as a nucleophile in a putative biooxidative activation of the amsacrines as DNA alkylators. In summarizing the present perspective on this possibility, Jurlina *et al.* conclude that there is no basis to believe that the primary cytotoxic event of the amsacrine antitumor action is other than topoisomerase inhibition [80]. All that exists at present is the observation that all potent antitumor amsacrines retain an ability for quinone diimine formation, and that a trend exists among most of these for improved antitumor activity following the thermodynamic ease of quinone diimine formation [80]. And finally, an equally uninterpretable mix of topoisomerase II inhibition and biooxidative activation is to be found for the ellipticines. The history and present status of these semi-synthetic antitumor agents were recently reviewed [69, 84, 85]. Ellipticine itself is plant-derived and possesses modest antitumor ability; but is metabolized to give a 9-hydroxy derivative (among others) with a significant improvement in activity. Numerous additional synthetic alterations have been investigated; the first ellipticine to undergo clinical evaluation was the 9-hydroxy-N^2-methyl species celiptium C (XI). As is also true for the anthracyclines and amsacrines a strong positive correlation exists between potency and affinity for DNA intercalation [86, 87], and newer analogs are available with substituents (polyamines, analog dimers; [88, 89]) that improve upon this ability. The anticipation is that these

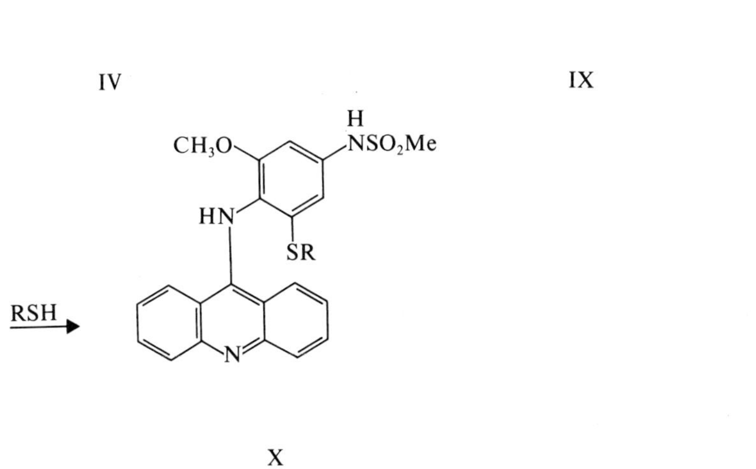

Fig. 4

will exhibit a superior therapeutic efficacy than the existing antitumor agents (as did not prove to be the case for celiptium). Most of the mechanistic studies are with celiptium, however, and from these it is evident that DNA modification may occur both by topoisomerase inhibition and oxidative alkylation. Celiptium is a topoisomerase inhibitor, as are the podophyllotoxins and amsacrines. The incorporation of the C-9 hydroxyl group dramatically enhances the extent of cleavable complex formation, as well as improving cytotoxicity [90, 91]. (The importance of the topoisomerase is, however, less clear in the newer ellipticine analogs [90, 92]). Topoisomerase inhibition as an antitumor property is discussed above and need not be repeated here. The additional consequence of the C-9 hydroxyl is that it allows oxidation to an electrophilic quinone monoimine (XII). Evidence that this is an *in vivo* process is unequivocal; both thiol conjugates and covalent DNA adducts are observed [93, 94]. While the structure of the DNA adduct is not yet identified, the properties of the modified DNA suggest alkylation of a primary amine within reach of the intercalation site [95]. The reactivity of the quinone imine under biomimetic conditions is now known. Once again the *in vivo* catalysts are the ubiquitous peroxidases [96, 97]. Numerous nucleophiles react with the quinone imine (pyridine, amino acids, alcohols, thiols), reflecting its potency as an electrophile. After an initial uncertainty the correct structure for these adducts has been established [84]. In each instance the preferred site of nucleophile attack is C-10 [96, 97]. With ribonucleotides as nucleophiles an unusual spiro ketal (XIII), obtained by sequential oxidation, trapping and oxidation is obtained (Figure 5). There is a bias for reaction at the vicinal ribose diols regardless of the ribonucleotide base (A, G, C, or U; [98]). A change in the adduct nature with polyribonucleotides is suggested [98]. Without a nucleophile present, or with deoxyribonucleotides where water acts as the nucleophile in the second cycle, the ortho-quinone (XIV) is detected [96]. Its potential as an oxidant and electrophile has not been reported. Presumably with DNA, the covalent attachment occurs via a reaction of the ortho-quinone or quinone imine with a nitrogen nucleophile of the bases, since the alcohols are not available. Regardless of structure, it is suggested by Dugue *et al.* [94] that ellipticine adduct formation may be cytotoxic as a result of a near complete inability of it to undergo repair (in cultured L 1210 cells). The biooxidative transformation of these three classes of topoisomerase II inhibitors presents interesting chemistry, but as yet chem-

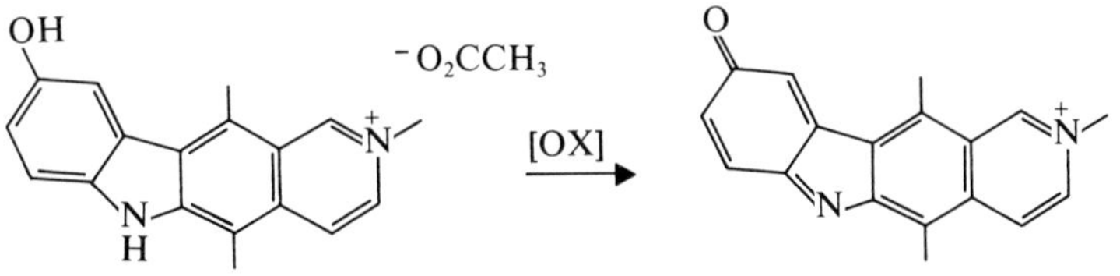

XI, Elliptinium Acetate

XII

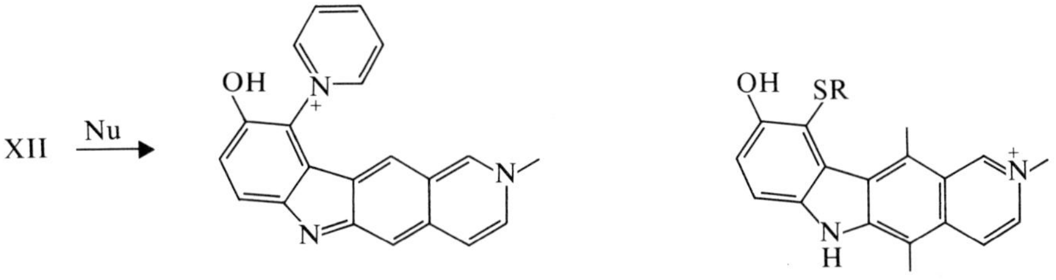

Nu = pyridine

Nu = RSH

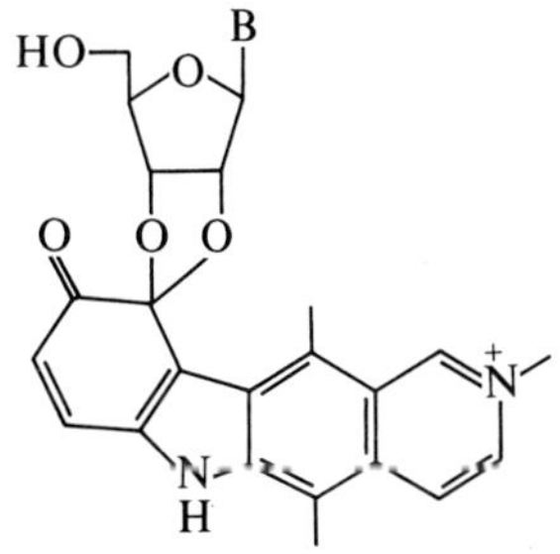

XIII, Nu = ribonucleoside

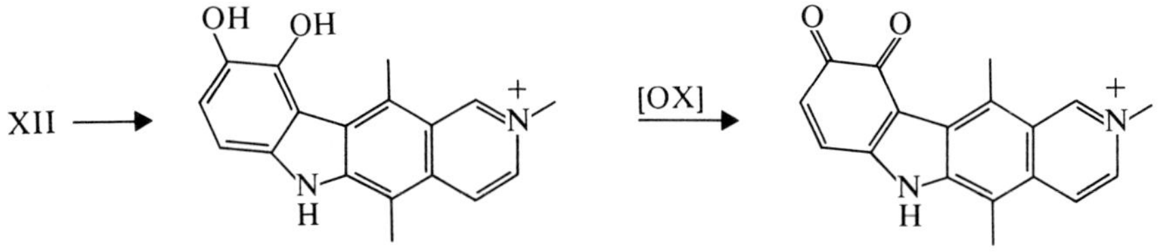

Fig. 5

Nu = H_2O

XIV

istry without a biological basis. It may only be speculated as to whether such a basis exists, or of what it might consist. This review is not the first to comment on the fortuitous coincidence of topoisomerase inhibition with the presence of oxidizable functional groups. Perhaps, since topoisomerase apparently requires a free sulfydryl for catalytic activity [99], cleavable complex stabilization may occur not by steric impedance of strand passage but by a redox event between the topoisomerase and the DNA-antitumor agent binary complex, resulting in modification of the sulfur as a labile adduct (in order to account for the reversibility of strand cleavage).

2.3   The anthracyclines: Topoisomerase inhibitors and bioreductive activation

The anthracyclines are arguably the most useful and most studied of the antitumor antibiotics. Yet paradoxically they are among the least well understood in terms of the molecular basis of their antitumor action. *In vivo* they demonstrate three fundamentally different mechanisms (DNA intercalation and topoisomerase inhibition; aerobic and anaerobic redox activation, and membrane association) each of which in the proper circumstance is capable of contributing to cytotoxicity. All efforts to completely disentangle these pathways to obtain simple generalizations have been unsuccessful [100]. At present the best hypothesis is that topoisomerase inhibition is an essential component of the antitumor cytotoxicity, while aerobic redox cycling is a major component of the host toxicity (particularly to the heart). Both processes involve DNA. In addition, there is anaerobic redox activation which provides a quinone methide intermediate, with the as yet undemonstrated potential to alkylate DNA. Such a mechanism does operate for mitomycin C (next section). Since the membrane interaction does not involve DNA [101, 102], a discussion of this phenomenon is omitted. The clinical efficacy of daunomycin (XV) against leukemia and adriamycin (XVI) against solid tumors has been known for two decades. While hardly optimal antitumor agents – both are highly cardiotoxic and poorly effective against many major cancers – they remain, by comparison to the others, among the most broadly useful. The efforts to improve upon these aspects have met with modest success [103–106], resulting in structures which retain comparable efficacy but with the prospect of diminished cardiotoxicity. These include both ad-

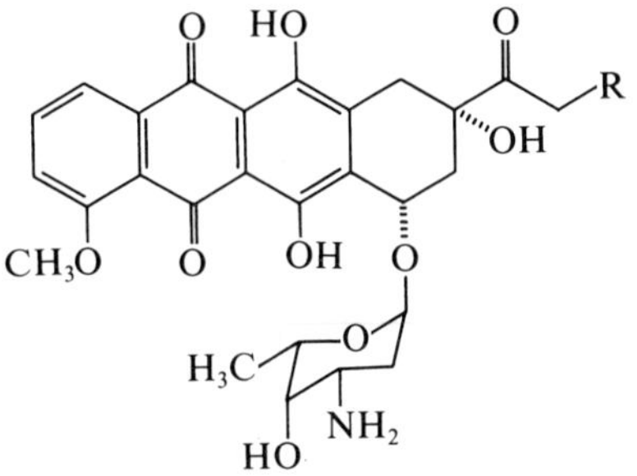

XV, R = H Daunomycin
XVI, R = OH Adriamycin

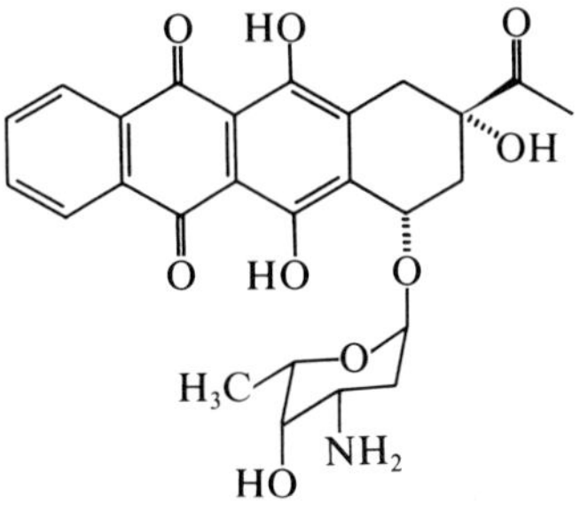

XVII, Idarubicin

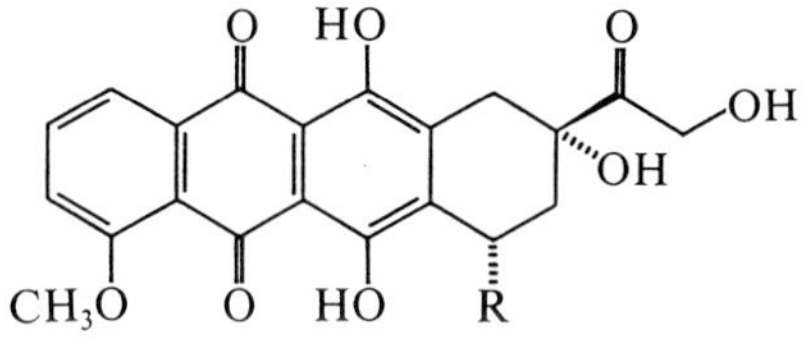

XVIII   R =

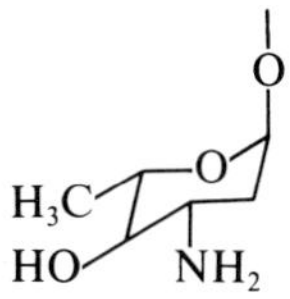

Epirubicin

XIX   R =

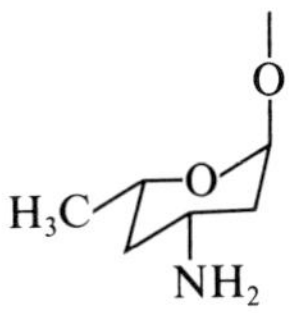

Esorubicin

Fig. 6

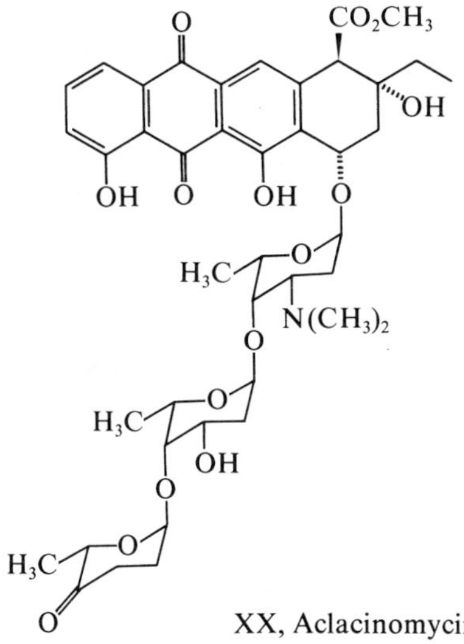

XX, Aclacinomycin

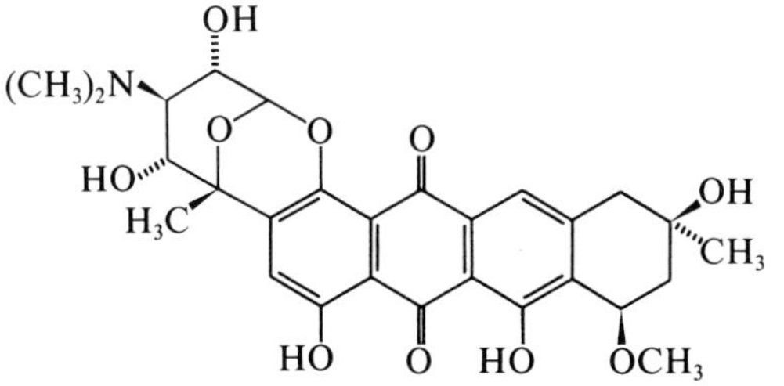

XXI, Menogaril

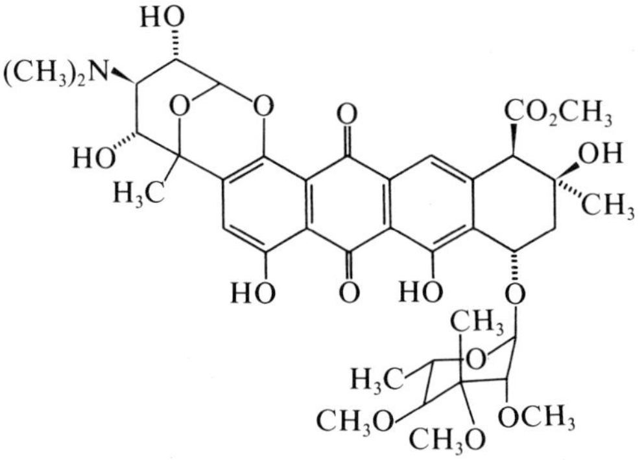

XXII, Nogalamycin

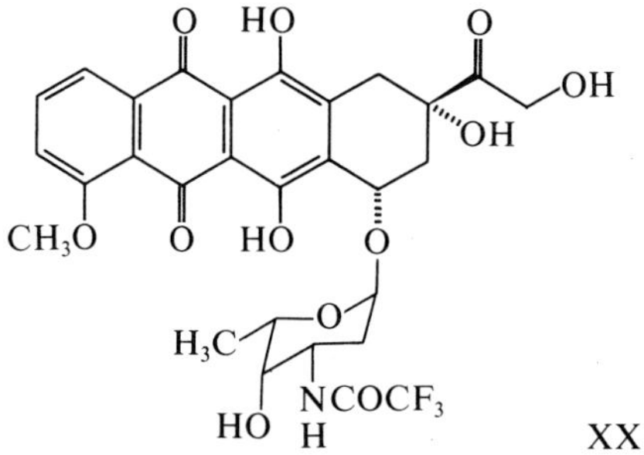

XXIII N-Trifluoroacetyladriamycin

Fig. 6

riamycin analogs (idarubicin (XVII), epirubicin (XVIII), and esorubicin (XIX) and congeners (such as aclacinomycin (XX) and menogaril (XXI)). Although these establish that the antitumor activity and cardiotoxicity are dissociable properties, and that the cardiotoxicity is the likely consequence of the redox ability of the anthraquinone sub-structure, efforts to abolish the redox capability have invariably diminished antitumor efficacy. While redox inactive anthracyclines preserve *in vitro* cytotoxicity to tumor cells [see 107–110], the most useful *in vivo* are those which retain the redox property, and hence synthetic modification of adriamycin has focused on alterations of the glycosidic ring in order to beneficially manipulate tissue distribution, kinetics and metabolism [104, 111, 112], or to incorporate latent functional groups capable of DNA alkylation ([113]; see below). This suggests that useful antitumor activity is attained by the synergistic action of different mechanisms (the obvious pair being topoisomerase inhibition and redox chemistry) although once again there are no definitive data. In the presentation of each of these processes the emphasis is placed upon the daunomycin/adriamycin pair. These have virtually identical *in vitro* characteristics (although daunomycin as the chemically more stable is often preferred) and, reflecting their present position in cancer chemotherapy, remain the anthracyclines of choice in mechanistic study.

The outstanding physical property of these two compounds is their affinity, as intercalators, for DNA. From the early efforts in adriamycin structure modification, the conclusion was reached that a predictive correlation existed between the antitumor activity and DNA affinity, measured by the association constant for complex formation [114–116]. While there are now exceptions (notably the *N*-acyladriamycins and menogaril), and even within the adriamycin series the correlation is blurred by the usual pharmacological variables, the evidence still justifies DNA association as a necessary event for tumor cytotoxicity. Among the anthracyclines adriamycin itself is one of the most tightly bound to DNA [116]. And apart from the *in vitro* physiochemical correlation, the most visible cellular consequence of adriamycin is the inhibition of DNA replication and RNA transcription [117–119] and this would appear to demand an effect on DNA. For these reasons anthracycline intercalation has been extensively examined, with the result that many of the kinetic and structural details are known.

With respect to intercalation, daunomycin and adriamycin are distinguished only by the slightly greater stability of the adriamycin-DNA complex. The molecular basis for this has been attributed to the ability of the C-14 hydroxy to position, by hydrogen bonding, a water molecule to within hydrogen-binding distance of the phosphate of the DNA backbone [120]. Under the usual *in vitro* conditions, little sequence preference for intercalation may be discerned [121, 122] with intercalation observed for both poly d(AT) and poly d(GC). The stability in heterogeneous DNA does, however, increase with increasing GC content [123–125] and a sequence preference appears under limiting drug conditions [126]. In accord with theoretical prediction [127] intercalation occurs to (normal) B-DNA [128, 129] at adjacent GC pairs (of variable order) flanked by an AT pair [126]. Since the preference coincides with sequence elements found frequently in promoter and enhancer regions, Chaires *et al.* [126] speculate that daunomycin may be guided by its intrinsic preference to a position where its intercalation will elicit the greatest effect. There is already evidence for binding to DNAse hypersensitive regions [130]. A self-complementary hexadeoxynucleotide with the preferred sequence was used by Wang *et al.* [120] in their crystallographic solution of the intercalation complex. Daunomycin occupies each of the GC pairs of the d(CpGpTpApCpG) duplex such that the tetrahydronaphthacenedione aglycone moiety rests at right angles to the long dimension for the base pairs. The methoxybenzene D ring protrudes from the major groove and projects into the solvent; the cyclohexene A ring falls into the minor groove at the other side. The amino sugar, necessary for effective intercalation, likewise falls into the minor groove but without any specific atomic contact. It is probable then that the requirement for the amino sugar is in stabilizing the initial contact between the DNA and anthracycline to allow insertion, and that following intercalation is only a secondary source of complex stability. Wang *et al.* note that the oligosaccharides found at C-7 of many other anthracyclines are easily accommodated into the daunomycin-oligonucleotide structure. The stability of the complex appears to derive from a hydrophobic effect between the ring structures, anchored by the exquisite positioning of the aceto and hydroxyl substitutents at C-9. The former rests within hydrogen bonding distance of two of the guanine nitrogens while the latter hydrogen bonds to a water molecule hydrogen bonded to the carbonyl of the cytosine, above the rings. The kinetics of complex formation are equally well studied [131–135], but do not provide particular insight. The binding phase is complex, consisting of a rapid association followed by two "isomerizations", one of which must correspond to insertion. Should the first of these isomerizations consist of insertion Chaires *et al.* [134] suggest that the second might represent a conformational adjustment or internal redistribution of the anthracycline. Of perhaps greater interest is the dissociation process. This is a relatively simple event for synthetic, homogeneous DNA, as evidenced by a single rate constant, but is at least two-exponential from heterogeneous DNA, possibly reflecting different sequence environments. The most intriguing aspect, however, is the magnitude of the dissociation rate constant(s). Fox *et al.* [133] and Chaires *et al.* [134] both have independently suggested that what may distinguish an antitumor intercalator from an intercalator without antitumor properties is a long residence time resulting from a slow dissociation process. This phenomenon was first noticed for actinomycin (see below), which is the slowest dissociating of the known intercalators. Even the anthracyclines represent a several order of magnitude difference relative to ethidium (as an example). Within the anthracycline family the entry with the slowest dissociation is nogalamycin (XXII). Due to its extensive functionalization in both the A and D rings, its intercalation demands the transient perturbation by hydrogen bond rupture of the DNA to generate a hole of sufficient size to allow insertion [133, 136]. As a result of this requirement nogalamycin shows sequence-dependent binding, internal migration and a multi-component exponential for dissociation. Fox *et al.* observe that it has the slowest dissociation and is the most strongly antibacterial of the anthracyclines tested, and suggest this as consistent evidence for a correlation between cytotoxicity and DNA residence.

Obviously for DNA to be entertained as a legitimate target of the anthracyclines, there must be evidence for a DNA-anthracycline interaction *in vivo*. Fortunately the intense fluorescence of this class has been put to use and has provided a consistent picture of the intracellular distribution [137]. At relatively low adriamycin levels (near what is attained during chemotherapy) the preferential location of the drug is the nucleus; at slightly higher levels the mitochondria are significantly targeted as well. Both organelles contain DNA and a persistent question is whether one DNA is more susceptible than the other. Even though the nucleus is preferentially occupied, anthracyclines are acutely toxic to the mitochondria [138] in a fashion permissive for DNA modification.

The mitochondrial DNA appears to represent an appealing target for a DNA-targeted drug. It is a closed circular structure of approximately 16000 base pairs (in man), without histone association and apparently without repair capability. Furthermore, its DNA is active throughout the cell cycle, and encodes for several components essential to the proper function of the mitochondrion, including various RNAs, cytochrome *b,* subunits of cytochrome *c* oxidase, ATPase and NADH dehydrogenase [139]. On the other hand, each cell contains numerous mitochondria and the DNA itself is (measured by molecular mass) a minor component of the organelle. The issue of mitochondrial DNA as the target for the anthracyclines (and other antitumor agents) remains unsettled. At present there is reason to believe that this DNA is capable of xenobiotic modification and that this modification can lead to gene disfunction and cellular mutation [139]. Among the xenobiotics known to modify this DNA are arene epoxides, aflatoxin epoxides and nitrosamines; the particular example of adriamycin has, however, provided conflicting data [140, 141]. Lim and Neims [141] could find no evidence of adriamycin-induced strand cleavage of the mitochondrial DNA, and little evidence of cleavage with bleomycin at physiologically relevant concentrations. It is suggested that neither drug is permeable to the mitochondrial membrane. Nevertheless the mitochondria of heart cells (the most studied) exhibit morphological and structural changes consistent with acute adriamycin-induced toxicity [101, 137, 138], providing evidence of organelle penetration. Ellis *et al.* [140] using different methods observe a loss of mitochondrial DNA integrity (formation of linear DNA) and suggest that adriamycin intercalation results in a termination of transcription at that point. The possibility also remains that cleavage is an aerobic redox event [142–145], or is related to topoisomerase inhibition. Regardless, there is no evidence directly implicating mitochondrial DNA damage as a lethal event. The acute toxicity to the mitochondrion is most probably a result of membrane association, given the abundance of cardiolipin and a particular affinity of the anthracyclines for this lipid [101, 137, 146] or a direct redox uncoupling of mitochondrial respiration [138, 142, 143]. Either hypothesis for the mitochondrial toxicity is reasonable, particularly in tissues (most notably the heart) where the mitochondria must satisfy an elevated energy demand. While an impaired function of the mitochondria in such a circumstance could clearly be disastrous, it is, however, not possible to assume this to be the cytotoxic event in a cancerous cell where mitochondrial integrity is not so obviously important.

If not the mitochondrial DNA than attention must focus on the nuclear DNA. The mechanistic possibilities for DNA impairment within the nucleus are no different than for the DNA of the mitochondria, and include intercalation, topoisomerase inhibition, and redox chemistry. It was noted in the introduction that DNA is a heterogeneous entity, even beyond the sequence variation used to encode the molecular working of the cell. This is particularly true of nuclear DNA. As knowledge of the higher order structuring of DNA is acquired, the effect of antitumor antibiotics may be studied, as has been done by Crothers and colleagues for the anthracyclines [147–149]. The behavior of a limited series of anthracycline structures with naked DNA, nucleosomal core particles, and a 175 bp nucleosome lacking histone H 1 has by and large confirmed common sense expectations. First, the intrinsic affinity of the anthracyclines decreases as nucleosome structure becomes increasingly more organized, without change in the relative affinity as a function of anthracycline structure. This may indicate a preference for intercalation into the linker DNA. Likewise the relative binding affinity is reflected in the extent of DNA synthesis inhibition, which is interpreted as the ability of the drug to compete against DNA polymerase for binding sites. This competition reflects either an inability on the part of the enzyme to deal with the presence of the intercalator or to an effect of the drug on the DNA conformation (such as by binding). In any case, the particular point is made that the efficacy of this inhibition may parallel the lifetime of the intercalated complex as directly measured by the complex dissociation rate constant [148]. Daunomycin-DNA interaction using the *lac* promoter and RNA polymerase as a probe led to such a small decrease in transcription initation and elongation as to suggest that the antitumor effect cannot be exerted upon RNA synthesis [147]. (This may not be true for the *N*-acyl adriamycins; see below.) Lastly, Sen and Crothers [149] have found that DNA binding drugs of completely different classes (actinomycin and daunomycin as intercalators and distamycin as a minor groove binder) inhibit the compaction of chromatin in a "remarkably similar manner". As all three inhibit DNA transcription *in vivo,* it is possible that this inhibition is a manifestation of drug binding to the more complex DNA structure as distinct from the simpler DNA fragments used in *in vitro* study.

The evaluation of adriamycin's effect on overall DNA structure and on the enzymes which operate upon the DNA is an ongoing endeavor. At present, attention has shifted from DNA polymerase inhibition to topoisomerase inhibition. This reflects not only curiosity concerning the new phenomenon but also excellent circumstantial evidence indicating a topoisomerase component to the antitumor action. The relative importance cannot be accurately assessed, nor can an objective distinction be made between mitochondrial and nuclear topoisomerases as targets. But since the anthracyclines are potent inhibitors of this enzyme and the topoisomerase must now be thought of as a target of antitumor agents in general, there is little reason to doubt that the effect is important, and perhaps even decisive. The capacity of adriamycin as a topoisomerase inhibitor was identified during a survey of the known antitumor agents following the recognition of topoisomerase II as an enzyme target [52].

Among the best evidence that topoisomerase inhibition is important to the antitumor ability is the behavior of adriamycin-resistant cells. In these there is lost the ability for topoisomerase-associated protein-DNA linkage, indicative of an alteration in topoisomerase structure or abundance [150, 151]. In addition several non-cross resistant adriamycin analogs, while reaching the same intracellular concentrations, have an increased tumor cytotoxicity which parallels a restoration of topoisomerase-dependent strand cleavage [151]. Further support of the topoisomerase as a component of the antitumor property is provided by the N-trifluoroacetyl adriamycins [152]. By trifluoroacylation of the glycosidic amine all ability to DNA intercalate is lost [152, 153]. It is to be presumed that without the positive charge on the glycoside the initial complex is neither sufficiently stable nor properly positioned for intercalation. The first N-acyl anthracyclines also contained an acyl substituent upon the primary hydroxyl of the C-9 sidechain, but it is now suggested that this ester is cleaved *in vivo* to provide N-trifluoroacetyl adriamycin (XXIII). This derivative does not bind to DNA, does not accumulate in the nucleus, but retains antitumor activity [152]. It differs from adriamycin (apart from DNA binding) by a change in membrane affinity and by a more potent inhibition of RNA polymerase. Since both anthracyclines have the net effect of diminished DNA and RNA synthesis the conclusion was either that the two attained this effect by different mechanisms, or that a common mechanism was unidentified. Eventually N-trifluoroacetyl adriamycin was examined as a topoisomerase inhibitor (perhaps prompted by the nonintercalating podophyllin inhibitors). The result of this somewhat implausible experiment was that N-trifluoroacetyl adriamycin was found to be not only a topoisomerase inhibitor but one with the same pattern for *in vitro* plasmid cutting as adriamycin [53]. This suggests that other anthracyclines (such as menogaril) with a poor DNA affinity and which do not concentrate in the nucleus may yet exert their antitumor effect through the topoisomerases. Is the topoisomerase the only mechanism? Some evidence suggests not. A direct correlation of anthracycline-induced strand cleavage and cytotoxicity is seen for some tumor cells [155] but not others [156]. Spadari *et al.* [157] conclude from the comparison of ten anthracyclines of variable cytotoxicity that the topoisomerase is an important but not exclusive component of the antitumor activity. This could indicate that a structure having only topoisomerase II inhibition as a biological mechanism may not necessarily prove to be an effective antitumor agent.

The other mechanisms available to the anthracyclines are redox-driven modification of lipids and DNA, and membrane association. Only the DNA aspect will be discussed here. It is necessary to divide the redox chemistry into its aerobic and anaerobic halves. In an oxygenated environment, the anthracyclines undergo adventitious reduction by several redox enzymes and proteins. One electron reduction of the anthracycline quinone functional group provides the semiquinone; two electron reduction provides the hydroquinone. The balance between the two is determined by the redox characteristics of the enzyme, and should two semiquinones encounter each other a favorable exchange of electrons occurs to provide one quinone and one hydroquinone (disproportionation). From the viewpoint of the aerobic chemistry a distinction between the semiquinone and hydroquinone as intermediates is unimportant, as both are kinetic and thermodynamically facile reducing agents toward molecular oxygen. (It is likely that it is the anions of semi and hydroquinone which are the actual reducing agents [158].) Thus the anthracycline functions catalytically, depleting the cellular reducing equivalents and consuming oxygen, by the process termed "redox cycling" [142, 143, 145, 159]. The immediate products of oxygen reduction are O_2- and H_2O_2, and in the presence of a suitable metal these are transformed into potent oxidants capable of DNA and lipid destruction (peroxidation). This is now a well-studied process, and indeed corresponds closely to the mechanism of yet another antitumor antibiotic, bleomycin (see below). Upon complete depletion of O_2, the hydroquinone (and perhaps semiquinone) persist sufficiently long for intramolecular rearrangement. The C-7 glycoside is ejected to provide an unstable intermediate, known generically as a quinone methide. As discussed below, this can act as both a nucleophile and as an electrophile, and presents the possibility of covalent DNA alkylation. There is no doubt that both the aerobic and anaerobic processes occur *in vivo*. The evidence suggests that the aerobic redox cycling is a major contributor to the (undesired) cytotoxicity, possibly explaining the particular toxicity of the anthracyclines to the heart [145]. In contrast there is no good evidence to a general importance of the anaerobic redox chemistry to any biochemical mechanism, other than the adventitious metabolism of the anthracyclines to the biologically inactive 7-deoxy aglycones. But in both processes there is circumstantial evidence intimating an involvement, in certain instances, in an antitumor effect. Accordingly the essential elements of each are described.

Redox cycling is neither a new process, nor a process limited to the anthracycline quinones. Rather, it is a broadly generalizable phenomenon which can occur *in vivo* with many redox-active functional groups (metals, nitro and quinone functional groups in particular). It has not, however, been until recently that the full implications of this process as a destructive consequence of adventitious xenobiotic metabolism have been appreciated. In brief, these redox processes can result in uncontrolled metal-dependent oxygen reduction, resulting in either the localized oxidative degradation of proximal molecules or the initiation of lipid peroxidation [160]. The essential element of both processes is the reductive, homolytic cleavage of hydrogen peroxide to yield the extraordinary oxidant $HO^{\bullet}$ ("Haber-Weiss" reaction). In order for an appreciable velocity *in vivo* a metal catalyst (Fe, Cu) is required, and the requirement for the homolytic reducing agent may be met by O_2 or, in the case of quinones, the reduced quinone itself [161]. The hydroxyl radical $HO^{\bullet}$ is reduced by hydrogen atom abstraction, and when this is done from the deoxyribose of DNA strand cleavage results. It is reasonable to assume that the magnitude of this effect will be in direct proportion to the facility of O_2 reduction *via* the anthracyclines, and attention has therefore focused on this process to account for the toxicity manifested by anthracyclines toward cells with elevated respiration rates. As discussed above, this hypothesis accounts well for the cardiomyopathy associated with the anthracyclines, likely reflecting a gross impairment of mitochondria function. Enzymes capable of this reductive activation are also found in the endoplasmic reticulum and the nucleus [162, 163]. These enzymes contain low potential flavin (exemplified by cytochome P-450 reductase and NADH dehydrogenase but apparently not [164, 165] xanthine oxidase) and iron (exemplified [166] by ferredoxin) centers. Generally speaking the redox active enzymes operate at the lower potential end of the redox spectrum (more strongly reducing) as the effective antibiotics tend to be relatively poor oxidants. At present attention has centered on NADH dehydrogenase (an enzyme with both flavin and iron-sulfur centers) as an anthracycline reducing catalyst due to its central role in mitochondrial respiration and general effectiveness [142, 143], and on NADPH cytochrome P-450 reductase due to its ubiquitous intracellular distribution and likewise general effectiveness [167, 168]. This latter enzyme has also been implicated as a catalyst for bleomycin reduction [169]; since this antitumor antibiotic is also reductively activated and exerts its activity by O_2-dependent metal-catalyzed DNA degradation, it may be safely assumed that all components are present at the DNA for the anthracyclines to operate in an identical fashion.

A complete set of circumstantial evidence is thus available implicating anthracycline redox cycling as a cytotoxic process. Following cell penetration the anthracycline encounters both the redox enzyme necessary for the reducing equivalents and the metal necessary to initiate peroxidation. The choice of metal is iron on the basis of the anthracyclines ability to directly chelate Fe(II) (reviewed in [170, 171]), the importance of iron as a catalyst of the Haber-Weiss type peroxidations, and its intracellular availability. Both the anthracyclines [172, 173] and bleomycin [174] are capable of directly mobilizing iron from ferritin (and indirectly via superoxide ion). An intrinsic affinity of such a chelate for DNA has been suggested on the basis of the structural similarity to the hydroxyquinone histochemical nuclear stains [144, 170, 175]. An exterior association to DNA appears to be required by the spatial inability of the intercalated daunomycin to accept a metal [120]. At the DNA surface enzyme-catalyzed reduction ensues, with the eventual passage of electrons to oxygen and metal-dependent homolytic peroxide cleavage. The resulting hydroxyl radical abstracts a hydrogen atom from the first molecule encountered; when this is the deoxyribose of

DNA chain fragmentation occurs by the process established for bleomycin (see below). There is substantial *in vitro* evidence for the effectiveness of this process [175–179] and even indications that this may result in covalent alkylation of DNA by the anthracycline itself, possibly as a result of radical-semiquinone recombination [180, 181]. The structures of these "adducts" are completely uncharacterized. Among the best evidence implicating aerobic redox chemistry as an undesirable cytotoxic effect (exemplified by damage to the heart sarcoplasmic reticulum and mitochondria) is the recognition that the heart is poorly prepared to suppress autoxidation, and that damage *in vitro* and *in vivo* may be reduced by redox inhibitors or scavengers (for discussion see [143, 170, 182–184]). An identical strategy has recently been used [185, 186] to establish that aerobic redox chemistry may beneficially contribute to the antitumor mechanism in some cell lines. Doroshow [185] demonstrated aerobic redox cycling within an Ehrlich tumor cell line and a significant reduction in the cytotoxicity toward this line of adriamycin and mitomycin C in the presence of antioxidant enzymes (catalase, superoxide dismutase); or in the presence of hydrogen atom donors (including dimethylsulfoxide or thiourea); or in the presence of iron chelators. Sinha *et al.*, [186] using human breast tumor cells, have made very similar observations. Here it is found that the sensitive and resistant cell lines contain substantial quantities of the low potential flavin enzymes known to participate in anthracycline redox cycling, and that the resistant cell line differed only by an order of magnitude increase in its ability to scavenge organic peroxides by glutathione peroxidase catalysis. It remains to be established whether the peroxidation in this antitumor process is randomly directed toward the intracellular constituents, or is specific for DNA degradation.

An equally ambiguous state of affairs exists for the anaerobic redox chemistry of the anthracyclines. Its nature is fundamentally different from the aerobic redox cycling in that it results in the direct formation of an intermediate with the potential for covalent bond formation, a quinone methide. There is unambiguous evidence that this occurs *in vivo*, as an appreciable quantity of the anthracycline metabolites consist of C-7 deoxy aglycones obtained by solvent protonation of the quinone methide. The steady state concentration of O_2 in hepatocytes (as an example) is not substantial [187] and the anthracyclines will, in any case, effect an immediate decrease in the oxygen concentration upon their arrival within the cell, by redox cycling. A connection between anthracycline structure, metabolism and quinone methide formation was first made by Moore, who has generalized this possibility to a host of quinone natural products [188]. A complementary approach, using the same physical chemical principle of reductive activation, has been an ongoing effort of Lin, Sartorelli and colleagues in the design of solid tumor active (anaerobic) antitumor agents (for recent references see [189, 190] and also the mitomycin discussion, below). The basic chemistry of this process as it pertains to the anthracyclines is shown in Figure 7. Without oxygen present as an oxidant, enzymatic reduction of the anthracycline produces (by one electron reduction) the semiquinone anion or (by two electron, proton reduction or by disproportionation) the hydroquinone anion. It has been established that the hydroquinone anion (**XXIV**) is unstable (and although unproven, the semiquinone anion may reasonably be thought to be similarly unstable) toward elimination of the C-7 glycoside, providing the quinone methide intermediate (**XXV**). The elimination reaction from the hydroquinone is rapid for the adriamycin family, and the quinone methide product is unstable. It reacts by conjugate addition of sulfydryl nucleophiles to C-7 [191], and by addition to strong electrophiles (such as aldehydes) from C-7 [192, 193]. It is by this latter behavior that the quinone methide reacts with H_3O^+ to give rise to the 7-deoxy aglycones as a major class of anthracycline metabolites. In principle this quinone methide may be thought of as a DNA-reactive entity, but its behavior with DNA on this score *in vitro* is ambiguous [181] and numerous efforts to

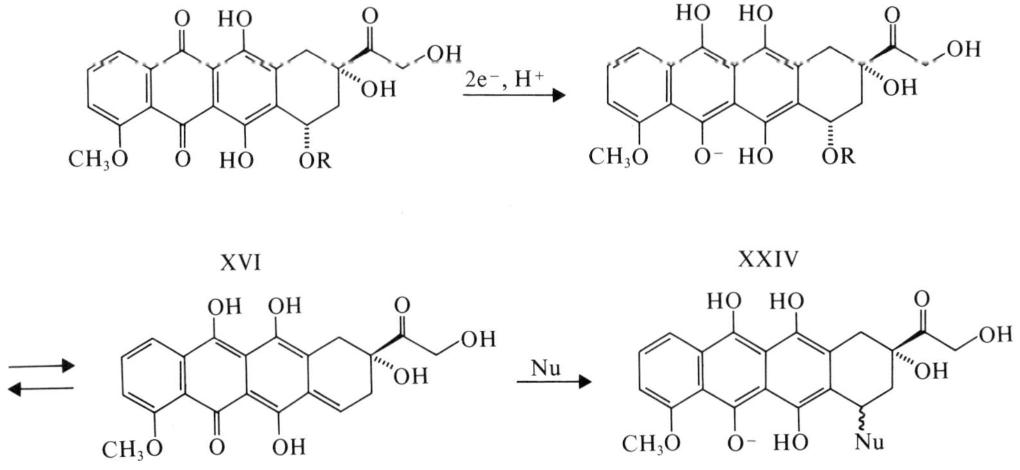

Fig. 7

obtain mononucleotide adducts have been unsuccessful. While there remains the possibility that reaction with DNA demands a particular set of conditions only to be satisfied *in vivo* [100], and there continue to be occasional reports of this reductive activation contributing to the antitumor property of the anthracyclines (see [194] for a recent discussion), there is no direct evidence implicating anaerobic redox chemistry as a general and important component of the anthracycline antitumor effect.

A coincidence that continues to attract comment is the prevalence of the quinone, as a functional group, in the antitumor agents [159, 184, 188, 195]. The obvious inference is that redox chemistry is a recurring component of the antitumor effect. But in the case of the anthracyclines there is no evidence suggesting redox chemistry as a necessary feature of the topoisomerase inhibition (unless the observation that most of the topoisomerase inhibitors are also redox active is to be taken as more than fortuitous); no compelling evidence for the anaerobic quinone methide chemistry in the formation of DNA lesions; and the understanding of the aerobic redox chemistry remains incomplete. Where there is to be found excellent circumstantial evidence to suggest the formation *in vivo* of covalent anthracycline-DNA lesions [196, 197], there is no *in vitro* experiment to provide even the suggestion of the chemistry leading to adduct formation. Yet, as a final topic, there remains for consideration the behavior of an adriamycin analog which suggests DNA adduct formation as having a beneficial antitumor consequence. The fusion of a 3-cyano-4-morphorinyl ring to the daunosa-

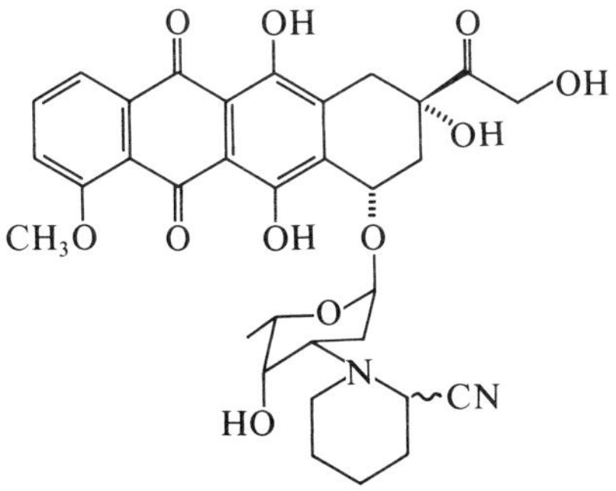

Fig. 8   XXVI

mine glycoside of adriamycin (XXVI, Figure 8) accomplished by Acton *et al.* [113, 198] increases substantially the *in vitro* potency relative to adriamycin; there is promise on the basis of initial *in vivo* studies that this potency carries over and as such may minimize cardiac damage and retain activity against anthracycline resistant cells [199]. The cyanoamine represents a latent functionality, capable of electrophilic imine formation by loss of cyanide as a leaving group. A virtually identical mechanism is used by the anthramycin antitumor antibiotics, where imine formation follows DNA association and leads to a covalent linkage between the drug and DNA (see below). The present evidence indicates that an identical process may occur for this adriamycin derivative [199–203] resulting in DNA cross-linking and correlating with the greatly enhanced inhibition of nucleic acid synthesis (relative to adriamycin). In what capacity the topoisomerase may be effected or quinone redox chemistry may contribute, and the actual structure of the "cross-link", await determination.

Despite the intensity and breadth of these continuing studies the anthracyclines retain their enigmatic character. While certainly capable of several unrelated chemical processes (intercalation; redox uncoupling; aerobic peroxidation and anaerobic quinone methide formation) which are capable (in the proper circumstance) of a cytotoxic effect, the perspective remains that the antitumor effect is attained by compromising DNA function. Whether this occurs primarily by topoisomerase inhibition, oxidative DNA degradation or covalent adduct formation remains uncertain. All these processes operate *in vivo* and may contribute. As the relative importance of each is assessed, a clearer understanding of the precise relationship between the anthracyclines and DNA will emerge, and it is to be expected that from this understanding a rational basis for manipulating anthracycline structure to optimize the desired antitumor behavior will become more evident.

2.4 Mitomycin C: Bioreductive activation

Of the enzyme-activated antibiotics only bleomycin rivals mitomycin C in the completeness with which its relationship to DNA is understood. While there remains much to accomplish, an *in vitro* simulation of the critical biological event has been achieved. It will now be possible to design more refined simulations, and to interpret these as well as the *in vivo* phenomena in terms of proven chemistry.

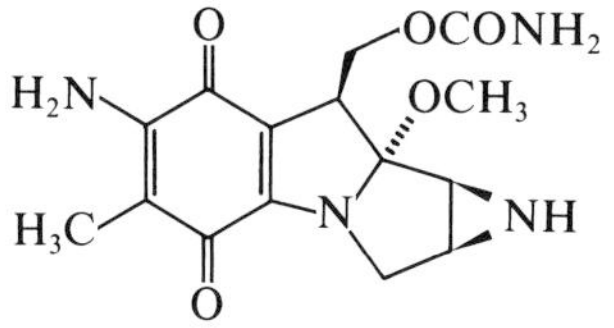

Fig. 9   XXVII, Mitomycin C

Enormous strides have been made from the point four years ago when
the absolute structural assignment of mitomycin C was revised. The
original assignment followed soon after the isolation of mitomycin C
some thirty years ago (for summaries of the early mitomycin studies
see [204–207]). Although the X-ray analysis correctly established the
connectivity of the atoms and relative stereochemistry, the proposed
absolute stereochemistry was to later present unprecedented mechanistic difficulty in order to harmonize with the biosynthetic pathway
[208]. This prompted a reexamination [209] and the reassignment of the
structure to that of the enantiomer of the first proposed structure. The
correct structure (XXVII, Figure 9) has been confirmed by the exciton
circular dichroism of a derivative [210], and with the correct structure
the stereochemical relationships among the members of the mitomycin
family have fallen into their correct place [211].
Mitomycin C was isolated on the basis of a broad antibacterial and antitumor spectrum. The biochemical basis for this activity was quickly
established as an inhibition of DNA synthesis. In the original efforts to
interpret this in terms of its structural elements, attention was drawn to
the aziridine ring as a potential alkylating moiety (analogous to the
mustards). However, the aziridine was found to be relatively unreactive; moreover a mechanism of simple DNA alkylation did not explain
the spontaneous renaturation of DNA isolated after mitomycin exposure. This phenomenon was known to imply a bis alkylation achieving
a cross-link between DNA strands. Clearly a more subtle mechanism,
requiring a cooperative relationship among its functional groups, was
required for mitomycin C. While the aziridine remained the most logical point for an initial alkylation, either the C-9a methoxyl or C-10
carbamate were possible leaving groups in the formation of a second
link. By the intensive effort of several laboratories the essentials of the
bis alkylation process during enzymatic reductive activation are now
established. But the full extent of mitomycin's biological activity is

every bit as complex as its closely interwoven functional groups would attest, and indeed two additional mechanisms exist as possible contributors to the antitumor action: aerobic reduction of O_2 and acid activation of the aziridine as a monoalkylating electrophile. Although reductive cross-linking remains the central focus as the lethal DNA modification, the understanding of each pathway's relative importance is not so complete as to permit neglect of the other two. In presenting these mechanisms the acid activation is examined first and is followed by the two redox processes.

The rudiments of the acid activation have been in place for at least a decade, mainly through the efforts of Lown, Remers, Tomasz, and Stevens (summarized in [212–217]). In the absence of other nucleophiles, an appreciable reaction velocity is sustained provided the pH is less than 5, with transformation of mitomycin C into a pair of indoloquinone (mitosene) diastereomers (XXVIII, XXIX; Figure 10). In these water has acted as a nucleophile in the aziridine opening, but the

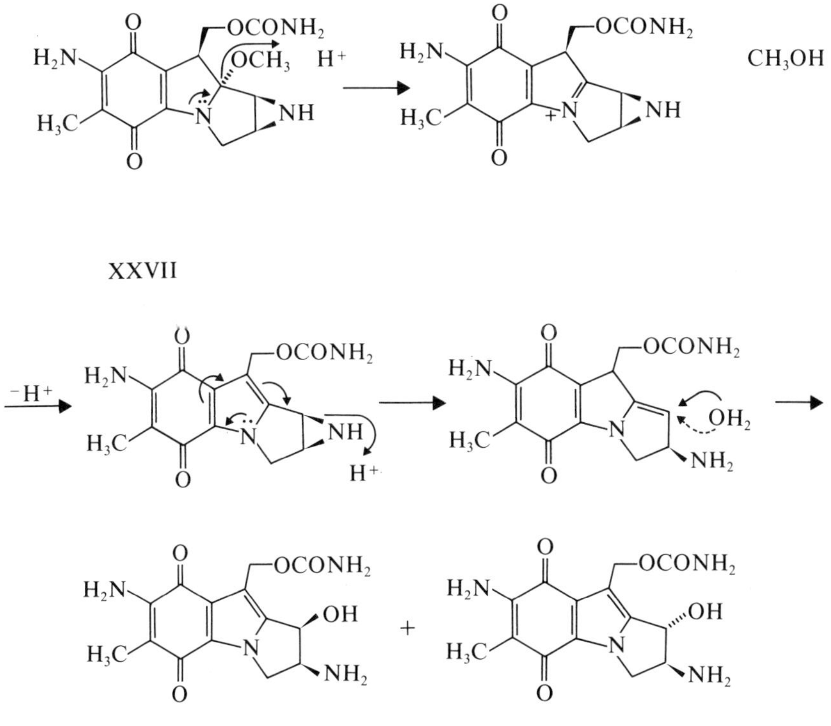

XXVII

Fig. 10   XXVIII   XXIX

appearance of two diastereomers excludes a simple, acid-catalyzed trans aziridine opening. The C-9a methoxyl is lost but without ensuing substitution, while acid hydrolysis of the carbamate requires much lower pH conditions. Thus, under acid-catalysis mitomycin C is capable of only one alkylation.

After some uncertainty, agreement has been reached as to the mechanism of this transformation. The aromatization to the mitosene and diastereomeric products indicate a multistep process. The rate-limiting step appears not to be acid-catalyzed aziridine cleavage, but acid-catalyzed expulsion of the C-9a methoxyl [214, 216] to the unstable imine. Once this is accomplished, all other events follow quickly. Tautomerization is fast enough to prevent nucleophile capture at the imine carbon (the microscopic reverse of the activating step). With assistance of the indole nitrogen (now in conjugation with the aziridine) and with the likely further assistance from a proton, the aziridine opens to a conjugated electrophilic imine. This third intermediate offers two stereogenic faces for nucleophile attack, both of which are used by water to provide the pair of final products. Generally speaking, there is a preference for *cis* (relative to the C-2 amino) capture of this imine by the nucleophile [217]. It is not known whether this is a steric or stereoelectronic bias. In neat trifluoroacetic acid Verdine *et al.* [218] have found that a single product results, corresponding to *cis* interception of the final imine by trifluoroacetate. It is suggested that in this instance the trifluoroacetate, after donating its proton to accomplish aziridine opening from the mitosene, simply has not the time to diffuse from the *cis* face before C-1 is available for capture. Although a dominance of *trans* product can result from other reaction circumstances, in no instance does the evidence suggest that simple aziridine opening is a major pathway. The most attractive explanation for this is the unusually weakly basic aziridine nitrogen (BH+ pKa ~ 2.5) of the mitomycins. The activation barrier for carbinolamine collapse remains the smaller of the two.

There is as yet no direct evidence that acid activation achieves covalent alkylation of DNA *in vivo*. Nevertheless this possibility has been thoroughly evaluated by the Nakanishi and Tomasz groups, and chemical precedent has been obtained. In phosphate buffer or in the presence of the nucleotides UMP or UTP, mitomycin acid activation provides the respective C-1 phosphate mitosene products [219]. With a substantial molar excess of the phosphate, the yields of the phosphate adducts are

reasonable (~ 40 %). Again, two diastereomers are produced with the *cis* dominating. With a dinucleotide nucleophile, d(GpC), the nature of the mitosene product changes to reflect primarily alkylation at N-7 and to a lesser extent N-2 of the guanine. Trapping by the phosphate oxygen is not consequential. Calf thymus DNA again provides guanosine adducts at nitrogen, but with a more equal distribution of N-2 and N-7 trapping [220]. The N-2 adduct (XXX) has a *trans* stereochemistry, consistent with the model for mitosene approach to the minor groove (see below). In contrast, both *cis* and *trans* N-7 adducts (XXXI) are produced. The *trans* adduct is the less abundant with the dinucleotide, but the more abundant with the polynucleotide. There again the *trans* preference is consistent with molecular modeling of the mitosene into the helical DNA. Tomasz *et al.* [220] suggest that the *cis* adduct arises from alkylation of locally denatured regions of DNA since unfavorable steric interactions result otherwise. The overall efficiency *in vitro* of mitomycin alkylation appears quite reasonable; with a 1:1 stoichiometry of mitomycin C to d(GpC) at pH 4, approximately 4 % of the guanine bases suffer alkylation over a 3-hour-period. With calf thymus DNA a somewhat lower yield results. Although most of the mitomcycin C is trapped by water and not by the DNA, simple exposure of DNA to an acidic solution of mitomycin C attains what may be significant alkylation of the deoxyguanosine residues.
Could this result in cell cytotoxicity? Tomasz *et al.* [220] conjecture an affirmative answer. In particular, they observe that the N-7 deoxyguanosine-mitosene adducts are hydrolytically labile and give rise to new products from two different solvolytic pathways. The first is depurination and yields the free guanine-mitosene *cis* and *trans* adducts. It is anticipated that this lesion would eventually undergo DNA repair should it result *in vivo*. The second hydrolysis occurs within the imidazolium ring of the deoxyguanosine to produce an N-7 alkyl formamidopyrimidine (XXXII, Figure 11). Partly due to the potential of such adducts to anomerize it is suggested that a "persistent lesion", no longer capable of DNA repair, may result. (This phenomenon is also believed to result from N-7 deoxyguanosine alkylation by mustards and aflatoxin epoxides, as examples.) Obviously without repair the entire transcriptional ability of the DNA could be compromised. As contributory circumstantial evidence it is noted that mitomycin C is effective against solid tumors [221], which are both lower in pH and hypoxic relative to normal cells. Kennedy *et al.* [222] have established a posi-

446 Jed F. Fisher and Paul A. Aristoff

tive correlation between lowered pH and decreased cell survival, and lowered pH and mitomycin-induced DNA cross-linking in an hypoxic tumor cell line. This suggests that pH is an important variable for mi-

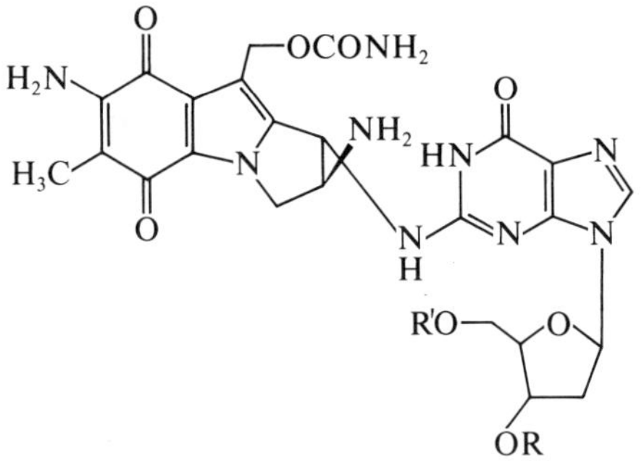

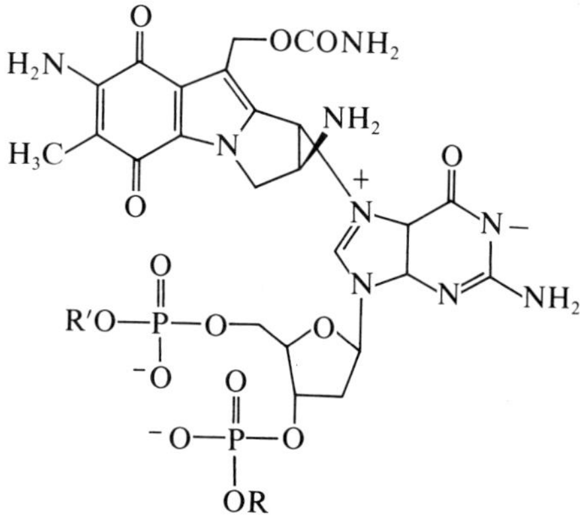

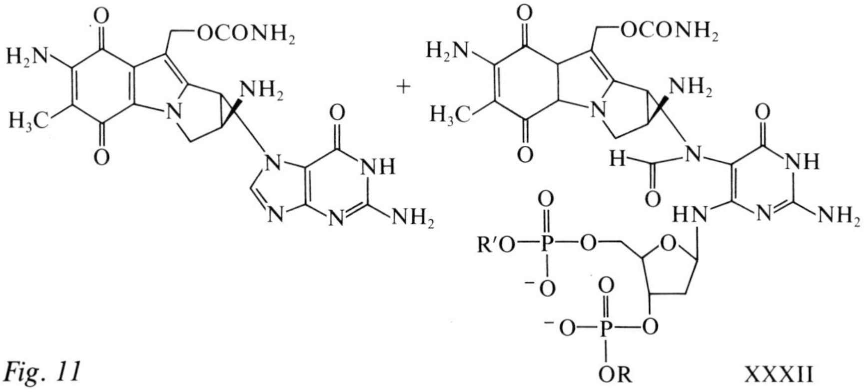

Fig. 11

tomycin toxicity. Even though acid activation is limited to monoalkylation between the mitomycin and the DNA, the hypoxia also permits reductive activation of the C-10 carbamate (see below) and there is no reason to suppose that a combination of acid activation and reductive activation cannot accomplish cross-linking. As noted previously for the antitumor agents there is no need to demand one single mechanism to explain the antitumor activity.

As with all quinones there exists the potential for a non-specific oxidation of redox enzymes, and passage of the electrons to O_2 in aerobic cells. Mitomycin C is not an exception. Even though it can be shown that mitomycin C is preferentially toxic to some hypoxic tumor cells (see below), it is toxic to some aerobic tumors [223] and there are therapeutically promising analogs which appear to have an enhanced aerobic toxicity [224, 225]. These observations are consistent with (but do not require) an O_2-dependent cytotoxic mechanism. This presumably would involve metal-dependent peroxidation resulting in oxidative strand cleavage of DNA. Since the rudiments of this process have been outlined above for the anthracyclines, and the details are presented for bleomycin (below), this topic will not be further elaborated here. The ability of mitomycin *in vitro* to catalyze O_2 reduction, and to initiate a metal-dependent peroxidation, is well established (reviewed in [224, 226, 227]). It is, however, difficult to imagine this as an efficient process *in vivo,* for the simple reason that mitomycin lacks an intercalating segment to obtain the close proximity to DNA during the redox chemistry. Such a segment is present in bleomycin. In principle the same argument applies to the reductive activation of mitomycin C discussed below, where the inital alkylation involves a bimolecular collision between an electrophilic mitosene and the minor groove of the DNA. In this instance there is a longer lifetime for the mitosene electrophile than the oxygen radical, and so the reductive alkylation is a more probable bimolecular event. Nevertheless metal-dependent peroxidation of DNA cannot be excluded as an important *in vivo* process in some instances. There is also the possibility that indirect effects of mitomycin-catalyzed O_2 reduction exist [228]. Last (as noted above for the anthracyclines) a sequence of mitomycin-catalyzed O_2 depletion followed by hypoxic bioreductive activation may be a particularly lethal sequence. Yet overall the data are more consistent with (either acid or hypoxic reductive) DNA alkylation as the primary cytotoxic process than O_2-dependent DNA strand cleavage [229].

The principal cytotoxic event with mitomycin is currently believed to be hypoxic reductive activation, on the basis of both proven chemistry and solid tumor susceptibility [239, 231]. Under the appropriate circumstances a two-fold activation of mitomycin C is achieved which results in intra and interstrand cross-linking of DNA. Of the two it is the interstrand cross-link which is thought to be more lethal. Several independent hypotheses as to the mechanism of this process have converged to provide a consensus of the events culminating in cross-link formation. Due to the complexity of this process, it is preferable to identify this consensus (Figure 12) with explanatory comments for each step, than to extract the mechanism from the disparate data. Mitomycin reductive activation proceeds only in the absence of oxygen. If even low concentrations of O_2 are present, the catalytic transfer of electrons from reducing agent to O_2 proceeds faster than the initial intramolecular transformation which compels the reductive activation process. Experimentally this is seen by the reisolation of intact mitomycin C after its use to catalytically deplete O_2 [224]. Upon attainment of anaerobicity the reductive activation ensues and is evidenced by a color change from the pale blue-grey of the oxidized mitomycin C to the deep purple of the oxidized mitosenes. The initial event triggered by mitomycin reduction is carbinolamine activation, as is also the case in the acid activation sequence. In the oxidized mitomycin the non-bonding electrons are sufficiently delocalized into the quinone as to require a much earlier involvement of a proton in the transition state to permit carbinolamine collapse. With quinone reduction at neutral pH, the non-bonding electrons are sufficiently available to the nitrogen as to permit the near instantaneous collapse of the carbinolamine, by CH_3O (H) expulsion, to the imine. How many electrons must be given to the quinone for this to happen? Surprisingly, this has turned out to be one of the more difficult mechanistic questions. There has been uncertainty as to whether one electron reduction (to the semiquinone) or two electron reduction (to the hydroquinone) is required to begin the activation process. In addition, the protonic state of the reduced species is also likely to be important. Although not proven it may be reasonably assumed that the radical or hydroquinone anions have greater reactivity than either the neutral radical or hydroquinone. A rigorous evaluation of the radical anion competence compared to the hydroquinone anion is experimentally difficult to obtain, since both exist concurrently as a result of the fluid interchange of electrons by reduced

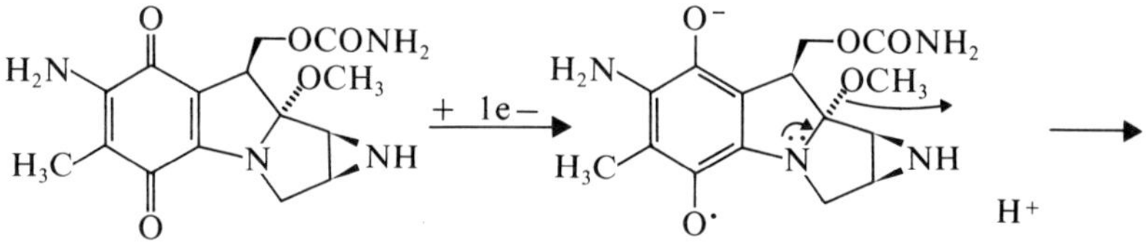

XXVII

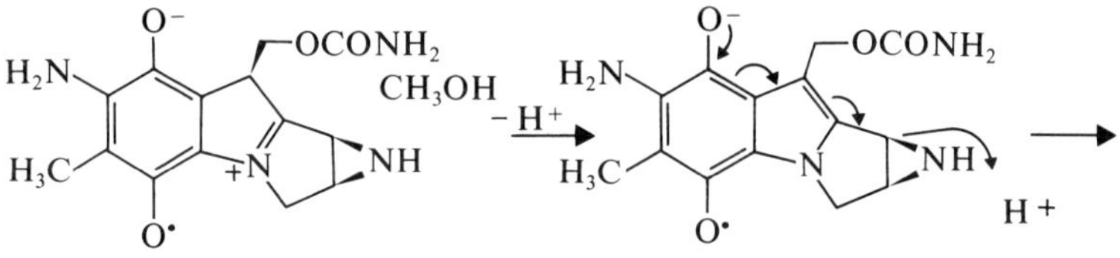

XXXIII

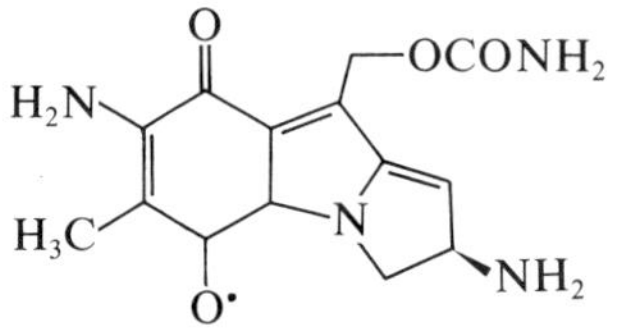

XXXIV

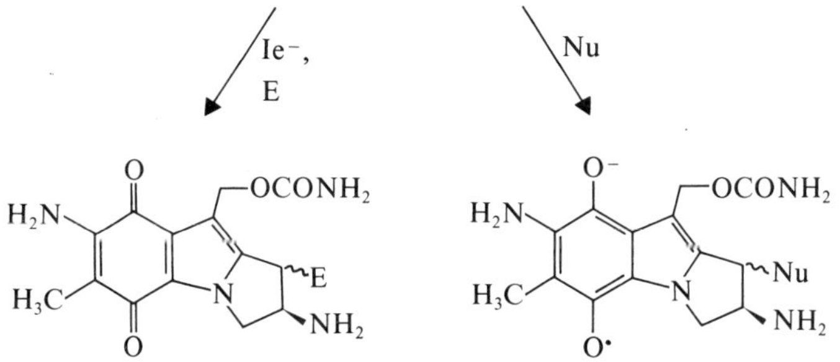

XXXV

E = H+,

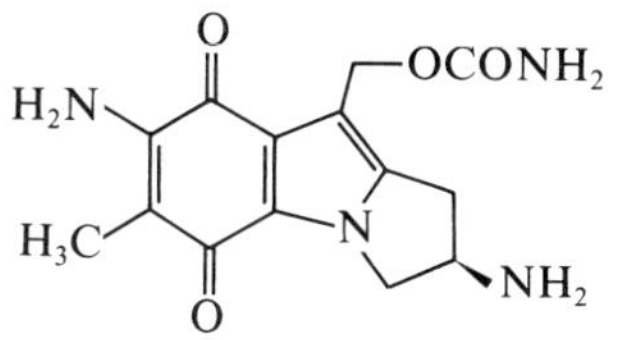

Fig. 12   Nu = OH,   XXVIII,   XXIX.

quinone disproportionation. In aqueous solution the equilibrium for quinones favors the conversion of a pair of semiquinones to one oxidized and one fully reduced (hydroquinone anion). Further, it is possible that both semiquinone and hydroquinone carbinolamine activation take place and that any observed reaction velocity is the sum of the respective rate constant and concentration velocities. Without specific knowledge as to the precise rate constants and equilibria (as is the case for mitomycin C) all combinations remain possible. Fortunately from the perspective of the ultimate inquiry, the requirements for bis alkylation of DNA, a distinction between the two is not an important issue.

Nevertheless there are experimental observations on this issue that require summary. In much of the literature a preference for semiquinone reductive activation of mitomycin is expressed, based on the appearance of O_2 during aerobic mitomycin recycling, and the assumption is that this requires a one electron reduced quinone. This assumption is unwarranted. A hydroquinone may be just as willing as a semiquinone to yield up its electrons one at a time. Since O_2 interception of the reduced mitomycin species is more rapid than the intramolecular carbinolamine collapse, it could be argued that the slower event is either direct carbinolamine collapse from the semiquinone, or disproportionation to the more reactive hydroquinone. No basis exists to differentiate between these possibilities for mitomycin C. The extraordinary difficulty in finding a basis to make this distinction is well illustrated by the exhaustive electrochemical studies of Andrews *et al.* [232]. As with most quinones, it is possible by careful electrochemical titration in aprotic solvent to attain selective formation of either the semiquinone anion or hydroquinone dianion. Although mitomycin is by no means an ordinary quinone, Andrews *et al.* overcame many of the difficulties and constructed an anaerobic electrochemical flow cell, having an anaerobic aqueous quench, capable of these reductions. Both mitomycin C semiquinone anion and hydroquinone dianion were generated in buffered dimethylformamide and upon anaerobic aqueous quench were converted to the mitosenes (indicative of reductive activation; see below). Only in the case of the semiquinone anion was a product distribution resembling that of the enzymatic activation attained. On this basis the conclusion was drawn that one electron activation is the dominant mode of bioreductive activation. Unfortunately, this conclusion is premature. Since the aqueous quench was not a "stopped quench" (nor can it be made to be) the possibility of disproportionation prior to activation cannot be excluded. Secondly, the hydroquinine dianion is not a biologically relevant entity, likewise presents the uncertainty of disproportionation, and may suffer more than the semiquinone anion from a transient increase in pH during the quench process, well beyond that seen in the *in vivo* activation. Even the potential required for dianion formation (-1.45 V. vs. Ag/AgCl) may introduce interpretive uncertainty as a consequence of competitive electrochemical reduction elsewhere in the molecule. An alternative electrochemical approach likewise does not provide more than suggestive evidence in favor of a semiquinone initiation. Kohn *et al.* [233] observe that electrochemical reduction of mitomycin C in CH_3OH provides cleanly the semiquinone anion (reduction potential is independent of $[H^+]$), and that reductive activation ensues. (In H_2O, however, reduction is a $2e^-$, H^+ process to yield the hydroquinone anion [234].) While the pleasant simplicity of this system tempts the choice of a simple answer, the temptation is to be resisted. The electrochemical data do not provide a basis for the distinction.

The least equivocal examination of this issue is found in the continuing examination of the stability of reduced mitomycins in organic media by Danishefsky *et al.* [235–238]. In the first of these studies the observation was made that both mitomycin B (like mitomycin A, a C-7 methoxy mitomycin) and mitomycin C are reversibly reduced in pyridine to the neutral hydroquinone. Of the two the mitomycin C hydroquinone is the more O_2-labile. But the complete anaerobic stability of these hydroquinones was unanticipated, and established that reductive activation need not be the inevitable consequence of mitomycin reduction. Subsequently it was established in the mitomycin A series that if disproportionation were permitted to the semiquinone, reductive activation ensued [236, 238]; exclusion of hydroquinone activation on both kinetic and thermodynamic terms is possible [237]. Although mitomycin A is an excellent (but not perfect) model for mitomycin C, and the aqueous pyridine of these model studies an excellent (but not perfect) model for neutral water, the conclusion that the mitomycin A semiquinone is competent to initiate the reductive activation process is inescapable. The initial step of this process (Figure 12) reflects this conclusion. In passing, however, it is noted that a preliminary report by Butler and Hoey [184] on the pulse radiolysis reduction of mitomycin C claims only the hydroquinone is competent for reductive activation.

It is the act of reducing the quinone which permits loss of the methoxyl moiety (by analogy to the acid activation the beneficial participation of a proton is presumed) to generate the imine (XXXIII). It is not the irreversibility of the methoxyl loss, but the rapid and irreversible tautomerization of this imine to the indolo semiquinone which drives the reaction forward. The imine is an electrophile in its own right and has on a few occasions been intercepted by nucleophiles [218, 238]. Since, however, tautomerization occurs preferentially, C-9a is not perceived as a viable locus for covalent attachment to DNA. Rather it is the fourth step of the sequence that provides the opportunity. Whereas reduction of the quinone permits methoxyl departure, the formation of the enamine brings the aziridine into conjugation with the reduced quinone and so permits the latter's assistance in its cleavage to the quinone methide (XXXIV). This methide differs from that obtained by acid activation only by the oxidation state of the quinone. And, like with that intermediate, a direct observation has not been made. Its nature is inferred from the ultimate product composition. At neutral pH without other nucleophiles the reduced quinone methide is trapped by water to provide, after oxidation, the same two *cis* and *trans* 1-hydroxy mitosenes (XXVIII and XXIX) as are obtained directly from mitomycin acid hydrolysis. Again it is the appearance of these two diastereomers which indicates a quinone methide intermediate with a discreet lifetime. The second observation in support of this last conclusion is a change in product distribution at acidic pH (≤ 5). The two hydroxymitosenes are no longer dominant, and are replaced by the product (XXXV) corresponding to proton (electrophile) capture of the quinone methide at C-1 [239, 240]. This dual character (electrophile, nucleophile) for this quinone methide has been seen previously with the anthracycline quinone methide. That the mitomycin quinone methide undergoes preferential reaction as an electrophile at neutral pH (the opposite of the anthracyclines) indicates an intrinsically greater reactivity for DNA, which has a preponderance of nucleophilic functional groups. Not surprisingly it is the electrophilic character at C-1 in this quinone methide by which the first alkylation of the DNA is accomplished.

Since *in vitro* mitomycin activation is accomplished by many reducing agents, the reactivity of the resulting quinone methide is one of the best-studied aspects of the reductive process [218, 241, 242]. In general, C-1 expresses electrophilic reactivity to a variety of good nucleophiles.

While generating the quinone methide and achieving its reaction with nucleophiles has been generally easily accomplished, in the case of nucleotide nucleophiles the most difficult aspect has been the assignment of adduct structure and achieving two-fold rather than mono mitomycin activation. This difficulty is reflected in a misassignment of structure to what is now known as an N-2 guanosine adduct (discussed in [216] and the still outstanding discrepancies between the Tomasz-Nakanishi and Shudo groups as to the ability of the quinone methide to alkylate elsewhere on guanosine and at the other bases [244–246]. At present the issue of enzymatic or chemical reductive alkylation of mitomycin C in the presence of heterogeneous DNA appears settled [244, 247]. As noted above, both *in vitro* and *in vivo* activation provide a preponderance of monoalkylated (rather than cross-linked, dialkylated) DNA-mitosene adducts. Tomasz *et al.* [243, 244] have isolated the major adduct by extensive degradation and purification and have identified it as the (*trans*)-*N*2-(2″β, 7″-diaminomitosen-1″α,yl) 2′-deoxyguanosine (XXXVI). This assignment limits the mechanistic possibilities considerably and permits the sequence for its formation to be described with some confidence. Mitomycin C (which does not intercalate DNA [248]) is activated by an enzyme in an hypoxic region near DNA, and undergoes sequential rearrangement to the quinone methide. Under the electrostatic attraction of the minor groove, this occupies the minor groove so as to place the mitosene C-1 within bonding distance to the N-2 of the guanine base. Covalent bond formation follows. The minor groove of β-DNA as the target is consistent with a large number of chemical studies and graphical simulations [244, 249]. The steric limitations of the complex dictate the formation of a *trans* linkage between N-2 and the mitosene. Note that the initial bond formation results in the formation of a reduced quinone within the mitosene (Figure 12); alkylation becomes irreversible only when a subsequent transformation converts the mitosene adduct to the quinone oxidation level. All in all this reaction is not difficult to accomplish. Although the requirements for the reaction are particular (or at least they appear so from the perspective of our meager knowledge of the *in vivo* DNA environment), once these are in place (minor groove, mitomycin nearby, a competent enzyme catalyst, hypoxia) the covalent modification is the expected outcome.

But at this point a knowledge of what follows, leading to cross-link formation, is once again uncertain. Not just by the process of elimination but also by graphical simulation [249, 250] the locus for the second attachment must be C-10. It follows that a second process of reductive activation must exist in order for this to be accomplished. Only recently has the situation clarified sufficiently to allow reasonable conjecture of how this second cycle operates. Although there remain arguments to the contrary [247], there is now reason to believe that the second cycle begins precisely where the first ends. The straightforward conclusion was obscured for some time by the apparent inability to duplicate C-10 activation *in vitro*. While this could occur inadvertently [218, 237, 238, 240, 241], the circumstances which permitted this to occur were not apparent; under many *in vitro* conditions the reductive activation stops at the mitosene C-1 mono adduct stage. The principal explanation is now known. Attention was called above to the observation that addition of a nucleophile to the C-1 quinone methide results in the restoration of a reduced quinone. If these reducing equivalents are preserved for activation by transfer to an unactivated mitomycin (this is thermodynamically favorable) *via* disproportionation, then it is possible to propogate an autocatalytic reaction. A sub-stoichiometric requirement of reducing agent for mitomycin activation has now been seen in several circumstances [233, 237, 238] and has unequivocally been established as the dominant kinetic process under "ordinary" mitomycin activation [251]. Since mitosene reduction potentials are much lower than mitomycin, unless a reducing agent is chosen having the capacity to effect both mitomycin *and* mitosene activation, the rapid phase of the reaction will terminate at the monoactivated level. In other words, to initiate C-10 functionalization it is necessary to achieve mitosene reduction. If the aziridine and carbamate moieties are both available, a partitioning to both products may be observed [238]; otherwise carbamate loss as a leaving group in the process of forming a new C-10 based quinone methide (XXXVI, Figure 13) will occur [251, 252]. As with the previous methide it is possible for a proton to competitively intercept and destroy its alkylating potential [232, 252]. With this understanding, Tomasz *et al.* were able to design an experimental protocol achieving extensive bis adduct formation; structural analysis indicated the major product to be the bis C-1, C-10, N-2 guanosine species derived from both interstrand and intrastrand cross-linking (XXXVII). Gratifyingly the adduct structure is easily accommodated by graphical simulation [249]. Finally, the identical adduct is isolated from rat DNA following mitomycin injection. The chemical mechanisms delineated above indeed have a biological relevance.

There remain several secondary issues to resolve concerning mitosene activation to effect carbamate loss. Among these are the relative electrophilicity of the quinone methide; its relative partitioning between nucleophile and electrophile (H$^+$) capture; and its mechanistic linkage to the first activation process. This latter issue directly asks for the source of the reducing equivalents required for the second quinone methide's formation. Are these provided by nucleophile attack to the C-1 methide, and the relative distribution of bis and mono adducts determined by the competition between intramolecular ejection of the carbamate and intermolecular leakage of the electrons (to O$_2$, other quinones, proteins, etc.)? Or is it possible for an enzyme to directly activate the oxidized mono adduct? This would, of course, require the enzyme to reduce the mitosene while covalently attached to the DNA. The difficulty of this task is unproven but since the probable identity of the enzyme is cytochrome P-450 reductase, it is an experimentally testable point.

The evidence connecting mitomycin reductive alkylation to its antitumor ability is circumstantial, but convincing. As summarized above, the three requirements for a cross-linking effect are mitomycin availability to the DNA, the enzyme catalyst, and hypoxia. In addition, it is necessary for the cross-link to be lethal. A brief discussion of each is provided. There is little doubt that mitomycin is ultimately made available as an alkylating agent to DNA, if for no other reason than mitomycin adducts are a consequence of its *in vivo* administration. Obviously, however, the particular issue of how it is transported into the cell and then distributed to the organelles is of more than casual concern, particularly in order to understand the biochemical consequences of its action. (Likewise an understanding of antibiotic efflux is important to understanding resistance.) Unfortunately mitomycin is no better understood than the other antitumor antibiotics. There is a report that the enhanced toxicity of porfiromycin (aziridine *N*-methyl mitomycin C) to hypoxic tumor cells is due to its preferential sequestration [253]. We can only conclude with the truism that the cytotoxicity is entirely dependent on the antibiotic reaching its target. The requirement for an enzyme catalyst also is satisfied. Mitomycin reductive activation occurs in both the nucleus and microsome *in vitro* [254, 255], and certainly in the mitochondrion as well. A principal catalyst is identified as cytochrome P-450 reductase [223] although there are other enzymes, but not including xanthine oxidase [231] and DT dia-

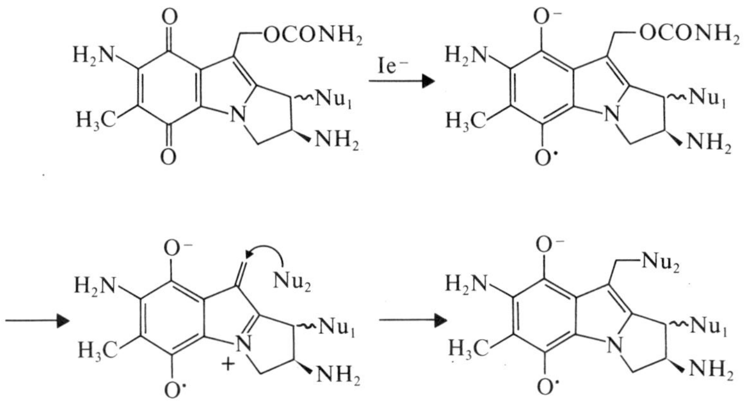

XXXVI

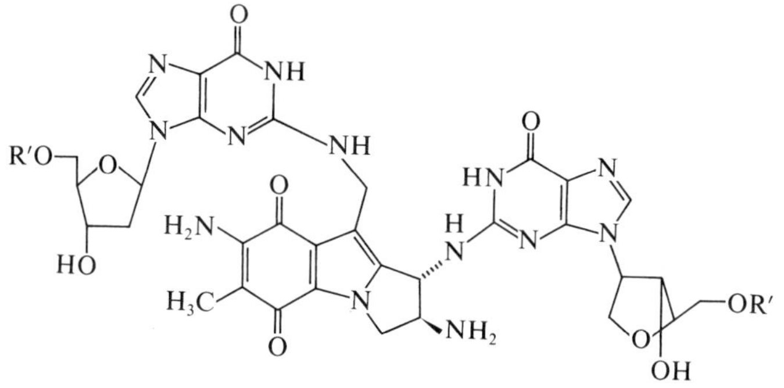

Fig. 13   XXXVII

phorase, that contribute. The importance of the bioreductive alkylation to the antitumor effect has been specifically established in a
mitomycin C-resistant colon carcinoma, where administration of a mitomycin analog capable of the reductive activation confers cytotoxicity [256]. Given this circumstance, it is not surprising that the hypoxic
tumor is more susceptible to mitomycin. Its relative anaerobicity may
be the major point of difference exploited by this antibiotic. It may be
less toxic to the better oxygenated host cell, and more selectively toxic
to the poorly vascularized hypoxic cell [230, 231]. In any case, the
correlation among hypoxia, DNA cross-linking, and cytotoxicity is
convincing. In Chinese hamster ovary cells, mitomycin cytotoxicity
decreased from anaerobic conditions to 1 % O_2, in an approximately
parallel manner with the quantity of DNA cross-links [257]. Dorr *et al.*

[258] observe that interstrand cross-links correlate with cytotoxicity in both L1210 and human myeloma cells, and not with strand breakage. Strand breakage was minimal at cytotoxic mitomycin concentrations and suggests the unimportance of aerobic redox cycling. Virtually identical observations are reported by Sartorelli *et al.* [225, 259]. That cytotoxicity is a consequence of interstrand cross-links may reflect the inability of the lesion to undergo repair [228, 231]. Regardless, the reductive activation of mitomycin C is lethal to the cell.

All of the events as they are now described bear a strong resemblance to the original expectations of Iyer and Szybalski some twenty-five years ago [260]. Mitomycin C is reductively activated in two separate stages, each generating an alkylating intermediate that by its spatial and electrostatic complimentarity to the minor groove [261] is poised to tether together the DNA strands. As an antitumor antibiotic it is complementary to the O_2-dependent agents and topoisomerase inhibitors, neither of which are effective against hypoxic tumors [231]. Its limitation may be the very factor that confers hypoxia; a tumor which has poor vascularity, and, hence, low O_2 will also be poorly accessible to the drug. But as the understanding of anaerobic bioreductive alkylation improves, the capacity to rationally deal with these problems improves [189, 190]. The entire concept of mitosene quinone methide tautomerization has been incorporated into a synthetic molecule [262]. In addition, it has been possible to expand the antitumor breadth of poorly active mitomycin derivatives by manipulation of quinone redox potential and leaving group stability [263, 264]. With a working hypothesis for the chemical mechanism in place, the opportunity for the rational exploration of structure and consequence has been attained.

2.5   Bleomycin: Oxidative DNA degradation

The bleomycins are a family of metal-chelating glycopeptides which exert an antitumor action by the O_2-dependent and metal-dependent degradation of DNA. Numerous bleomycins have been obtained in the quest for optimal antitumor activity, both by isolation from *Streptomyces* and by semi-synthesis. Shown in Figure 14 as structure (XXXVIII) is bleomycin A_2, the major compound utilized clinically. In achieving DNA oxidative degradation three separate domains of this molecule act in a cooperative fashion. The carbohydrates may be important to the selective accumulation of the bleomycin into the cancerous cell. The right-hand bisthiazole and its positively charged amide substituent promote stable complex formation with DNA, and

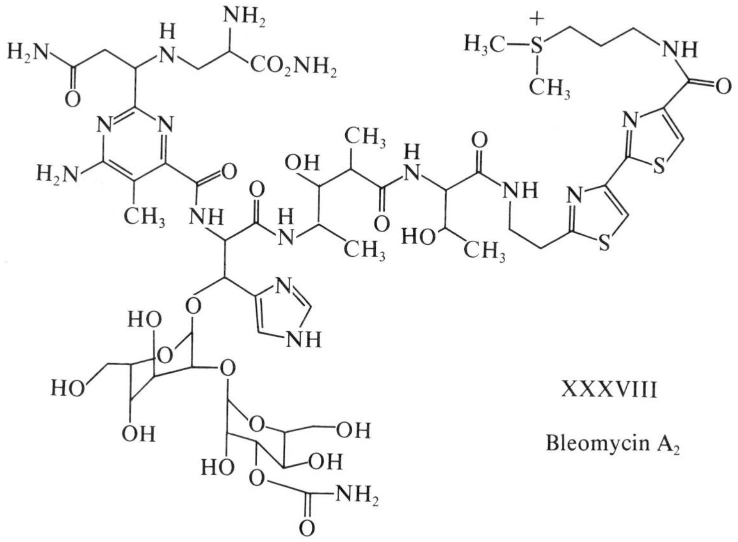

XXXVIII

Bleomycin A_2

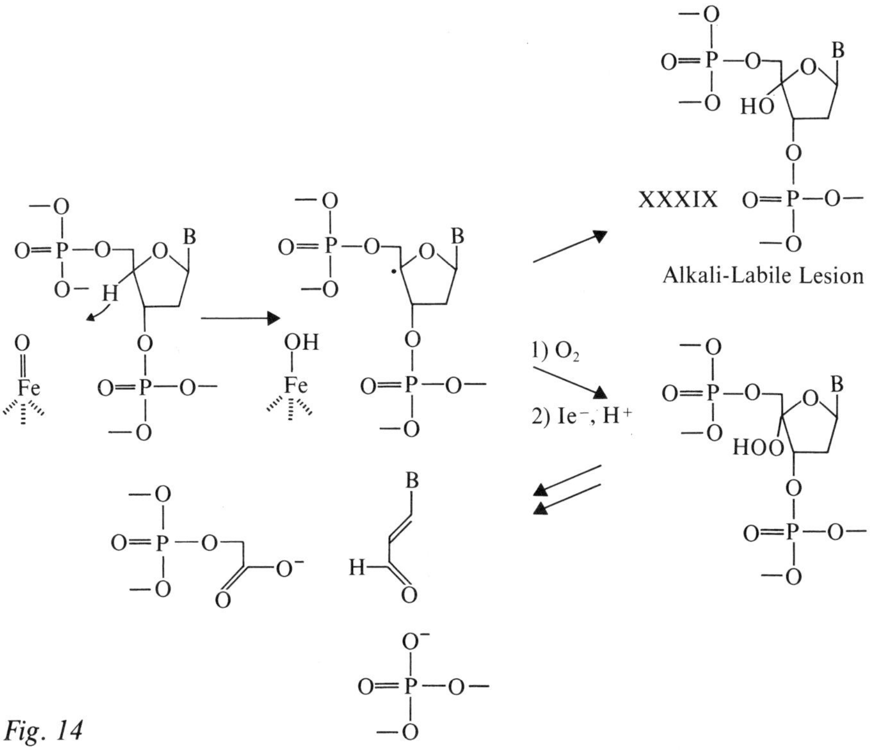
XXXIX
Alkali-Labile Lesion
1) O_2
2) Ie^-, H^+

Fig. 14

the central array of nitrogenous ligands provide a metal binding site. Although a number of metals may be shown capable of occupying this site, the one of principal focus is Fe(II). Molecular oxygen is accepted as a ligand of the Fe(II)-bleomycin complex whereupon one electron reduction accomplishes oxygen bond cleavage and oxidant formation. Reduction of the oxidant occurs at the expense of the DNA deoxyribose, by hydrogen atom transfer, and culminates in DNA strand cleavage. At present more is known about this strand cleavage mechanism than the structure of the critical activated bleomycin intermediate. It is nevertheless clear that the chemistry bears more than just a passing resemblance to that of cytochrome P-450 monooxygenation, and it is for this reason and its clinical importance that bleomycin continues as a major focus of antitumor antibiotic chemistry. It has been reviewed on a regular basis (most recently in [265–268]) and the entire topic of metal-dependent DNA degradation has been comprehensively covered by Stubbe and Kozarich [269]. An effort to duplicate this coverage will not be made here, but rather to impart the salient aspects of the relationship between the mechanism of DNA degradation by bleomycin and antitumor activity.

In a fundamental respect the molecular action of mitomycin C and bleomycin are identical. While the former achieves anaerobic reductive alkylation and the latter aerobic oxidative strand cleavage, both require the adventitious participation of a reducing enzyme. While the *in vitro* reducing agent for the Fe(II)-O_2 bleomycin complex need not be an enzyme (Fe(II) and thiols are often used), and likewise there is no evidence requiring an enzyme catalyst *in vivo* (as opposed to, for instance, cellular Fe(II) and glutathione), the ability of bleomycin to catalytically activate molecular oxygen [270, 271] suggests enzymatic participation. There is no doubt that a number of microsomal enzymes [273] in addition to cytochrome P-450 reductase, as first identified by Kilkuskie *et al.* [169], are competent catalysts, consistent with the relatively positive potential estimated for metal-bleomycin chelates [275]. These enzymes are present in sufficient quantity within the nucleus to account for the requisite DNA damage [276]. All together this reinforces the perception of the nucleus as a more biochemically heterogeneous environment than was formerly thought, and underscores the availability of redox enzymes to participate in the enzymatic activation processes important to the action of so many of these antitumor antibiotics.

An appropriate beginning is the issue of bleomycin tissue selectivity in terms of its mechanism of action, DNA degradation. In comparison with the chemical mechanism, the biological mechanism is not as well understood, and the efforts to empirically improve the tissue selectivity by analog synthesis have not been wholly successful. This is not to say that continued effort is futile. It is now, for example, recognized that an enzyme (termed bleomycin hydrolase) is widely distributed in all tissues and catalyzes the hydrolytic cleavage of the carboxamide of the β-aminoalaninamide of the "northern" (pyrimidoblamic) segment. The resulting carboxylic acid lacks the DNA degradative ability and hence antitumor efficacy. The potential role this enzyme occupies in determining cellular toxicity and relative resistance was recognized immediately after its discovery by the Umezawa group [256, 266]. Two tissues unusually sensitive to bleomycin are the skin and lung, and indeed it is a potentially lethal pulmonary fibrosis which limits the clinical effectiveness of this drug. This enzyme can be perceived as a ubiquitous cytoprotective agent, and successful analog development should exploit the differences (if any) between the substrate specificity of the tumor and lung enzymes [266]. A major limitation has been the difficulty in obtaining substantial quantities of enzyme. Pure enzyme has been isolated from two tissues [277, 278] and from this accomplishment the eventual cloning of this enzyme may be expected. By providing mild access to the carboxylic acid, the re-functionalization with amines to provide analogs with improved activity may be accomplished with a synthetic ease and on a scale not presently possible.

Bleomycin analog development has focused on alternatives to the N^1[3-(dimethylsulfonio)propyl] moiety of the bis thiazole [279, 280]. It is evident that such changes may be beneficial, as evidenced by the improved squamous cell carcinoma activity of bleomycin B_2 (having an N^1-[4-[aminoiminomethyl)amino]butyl]) and the lowered pulmonary toxicity of peplomycin (having an N^1-[3-[(1-phenylethyl)amino)propyl]). The retention of cationic character is consistent with the participation of this segment in the DNA association. Alteration within the central region is generally not possible within the function group constraints of the natural product and the undesirability of changing the metal ligands. Similar synthetic difficulties are presented in manipulating the carbohydrate segment. While it is not required for the DNA oxidizing event, there is evidence suggesting an importance in tumor cell uptake [281] and in the structure of the DNA complex, as reflected in a change in position for DNA cleavage [282].

The metal specificity and structure of the metal-bleomycin complex are the most controversial topics of the DNA oxidation process. The bleomycins from the fermentation contain copper and are purified as the blue copper chelates. In order to prevent necrosis at the injection site the copper is removed in the drug formulation, leaving open the question of the relative importance of copper and iron to the mechanism. An unequivocal answer as to copper's ability toward oxygen activation has been difficult to obtain as an apparent consequence of the difficulty in excluding adventitious iron contamination. While it is certain that stable Cu(II) chelation occurs *in vivo* and that enzymatic reduction of the complex to Cu(I) can occur [169], the capacity of Cu(I) to activate molecular oxygen has been addressed by experiments yielding negative [283–285] and positive [286, 287] conclusions. The Cu(II)-bleomycin complex is more difficult to reduce than the Fe(III) complex [274, 275], and both reduced metal-O_2 complexes undergo relatively rapid autoxidation with release of O_2 [285, 288, 290]. Further, there is agreement that when both metals are present they interact together, although the nature of this interaction is described in different terms, depending on whether it is believed that the Cu(I)-O_2 complex may be further transformed into a DNA oxidant. An answer will not be attempted here. Discussion may be found in the recent references and in the appraisal by Stubbe and Kozarich [269].

The strand cleaving ability of the Fe(II)-O_2 complex is not in doubt and will comprise the remaining discussion. Whether intracellular iron is available to the bleomycin does not appear to be an issue for contention, although the nature of an intracellular pool of free iron and its availability have come under increased focus in terms of the broader issue of lipid peroxidation. It was noted previously that bleomycin has been seen to abstract iron from ferritin [174], and it is highly probable that these same peroxidative events which liberate iron [160, 290] may be initiated by bleomycin itself. Ekimoto *et al.* [291] have made the logical suggestion that the pulmonary toxicity of the bleomycins is primarily a lipid peroxidative effect. The precise chelate structure of the iron complex continues also to be an issue of discussion. Considerable information on the structure of numerous metal-bleomycin and bleomycin analogs has accumulated, supporting the importance of the imidazole, pyrimidine, and β-aminoalanine nitrogens. However, the precise geometry of the complex and relative importance of other potential ligands (pyrimidine amides, sugar carbamate) is not as clear.

The complexity of this issue exceeds the present opportunity for discussion. Oxygen binding, activation, and DNA oxidation have proven more rewarding and comprise the remainder of the section.

Efficient DNA cleavage is accomplished by the cooperative abilities of bleomycin to form a stable DNA complex and then to activate oxygen at the metal coordination site. A reasonable presumption is that O_2 activation follows DNA complexation. Activation without DNA is easily done but presents an opportunity for bleomycin self-destruction by oxidative degradation of itself. The exquisite balance among metal chelation, non-productive O_2 activation, and DNA complexation which ultimately allow a catalytically active bleomycin to complex to an essential DNA locus can be at best poorly described, and is considered after the chemistry of DNA oxidation has been discussed. For DNA oxidation the reagents required for bleomycin are the metal (such as Fe(II)), O_2, and a reducing agent [292]. Subsequent to the metal chelation, reversible binding of O_2 occurs to provide an intermediate which may be visualized in terms of an Fe(II)-O_2, Fe(III)-O_2 equilibrium [289, 293, 294]. From this "activated" bleomycin is obtained by one electron reduction, at the expense of reduced metals, thiols or enzymes. Alternatively, activated bleomycin may be obtained by oxidation of Fe(III)-bleomycin by peroxides; this approach has proven of value in understanding the chemistry of this intermediate and in establishing mechanistic similarity to the active oxidant generated from O_2 during cytochrome P-450 catalysis. Further discussion on this latter aspect is available elsewhere [268, 269, 295–298]. A molecular formulation of the structure of this intermediate is uncertain, even on such a fundamental issue as to whether separation of the two oxygen atoms has occurred. The best interpretation of the data is that cleavage has occurred and the intermediate is a high valent iron oxene as may be represented as $Fe^V = O$ [269, 299]. In this respect the bleomycins are similar to the heme mono-oxygenases, but whereas enzymatic oxygen transfer is exceedingly rapid (as the substrate is already bound and in position), under *in vitro* conditions the activated bleomycin has a reasonable lifetime (minutes at 0 °C). All kinetic [293] and product [268, 269] studies agree that the activated oxygen is held until an acceptor is encountered, as opposed to being lost as HO$^\bullet$. Although multiple products form when DNA is the acceptor (reducing agent), the evidence is also in agreement that the initial event is common to all products, and that this is hydrogen atom abstraction at C-4' of the ribose at the point of eventual strand cleavage (Figure 14). The resulting carbon radical may be intercepted by a second O_2 molecule [303] to provide the C-4' hydroperoxyradical, to initiate one of two possible pathways to the final products. On this pathway this radical is further reduced, possibly by the very same enzyme as had participated in the activation. The set of final products from this intermediate are known to be a phosphoglycolate 3' terminus; a base propenal; and a 5'-phosphate terminus [300, 304–306]. A reasonable pathway connecting the hydroperoxide to these intermediates may be conjectured (see [269] for a complete discussion of difficulties and alternatives). The ribose ring is opened by a fragmentation involving loss of the 2'-hydrogen as a proton, alkene formation accompanied by C-3'-C-4' bond cleavage, and heterolytic cleavage of the dioxygen bond with departure of HO-. Hydrolysis of the vinyl phosphate would provide the 5'-phosphate terminus product and accomplish strand breakage. The resulting aldehyde could eliminate the more stable β-substituent, the carboxylate leaving group, and so produce the remaining two products. Alternatively [269, 309], the peroxide may accomplish oxygen insertion (Criegee-type rearrangement) to the hemiketal, which would unravel to produce the identical products. The feasibility of hydrogen labeling studies to distinguish among all possibilities is currently being evaluated [269]. The second set of bleomycin products is believed to result from a different fate for the 4'-radical than O_2 interception, and to correspond to a released base and oxidatively transformed sugar [268, 307, 308]. Of the mechanistic possibilities the one which reconciles best with the structure of the oxidized sugar product is the one proceeding through a C-4' alcohol. This may be reasonably thought to arise via oxygenation by the bleomycin-oxygen intermediate. Straightforward collapse of the C-4' alcohol (hemiacetal) to the C-4' ketone, and anomeric oxygen as leaving group, allows the eventual loss of the nucleotide base. Equilibrium reclosure provides an alkaline-labile, C-4' hydroxyribose abasic site XXXIX [307–309]. Upon alkaline treatment elimination of the 5'-end occurs to accomplish strand cleavage [307, 308]. While there is now excellent evidence indicating that these two pathways comprise the major routes taken by the activated bleomycin, there is also equally excellent evidence indicating that these are not necessarily the only routes. Such a conclusion would be reasonable even without the experimental evidence. First, since bleomycin functions catalytically there is no limit to a single hydrogen atom abstraction during the degradation process. Additional activation cycles may divert the above mechanisms at any stage. It is this catalytic ability which allows bleomycin to achieve double-strand cleavage, particularly as it exhibits an enhanced affinity for the initial cleavage site and thus an increased possibility for oxidation of the second strand at this site [310]. Since the position of hydrogen atom abstraction is dictated only by the relative positioning of the oxygen of the activated bleomycin [269, 308], and this is determined by the particular conformation of the bleomycin-DNA complex, there can be no assurances that the mechanism of the second cleavage in any way resembles the first. The well-established ability of DNA intercalators to modulate bleomycin efficiency [311] and the heterogeneity of product formation upon manipulation of DNA conformation [300, 312, 313] forcefully indicate a relationship between mechanism and DNA structure. The unanswered question is the importance of this relationship to the antitumor effect.

The association of bleomycin to DNA remains a topic of continuing study. While the bis thiazole rings are clearly major contributors to the overall stability of the complex, whether these participate as classic intercalators, partial intercalators, or in minor groove binding remains uncertain [268, 269, 314]. One complication is that metal complexation results in a conformational adjustment that alters the nature of the complex [268]; hence it is necessary to use an inert metal or to use iron and exclude oxygen. Since the action of bleomycin as an oxidant results in strand breakage an easier experiment has been the sequence preference for cleavage. With linear DNA oligomers there is a preference for strand scission at GC and GT and for double-strand scission at pyr·G·C·pur (and its complement) [269, 315, 316]. With more heterogeneous DNA these preferences remain, with all GC and GT sequences retaining a finite possibility for cleavage [317, 318]. Fox *et al.* [319] establish a clear preference for PyGPy and make the important observation that the polynucleotide sequences protected by the inert cobalt-bleomycin are very similar to those cleaved by iron-bleomycin.

The relationship of these observations to bleomycin as the antitumor agent has been hardly illuminated at all. Indeed, the essential features of metal-dependent DNA degradation have been incorporated into an array of structures [269] only bleomycin of which is (as yet) an antitumor agent. Like all of the others, the answer is to be found by understanding why selective toxicity is expressed toward the cancer cell. The answer may be as simple as selective uptake or found at the much more

complicated level of the dynamic relationship among the antitumor
agent, the DNA, and the DNA environment. Unfortunately these do
not lend themselves to discreet in vitro dissection. For bleomycin an
examination of its specificity for transcriptionally active (DNase hypersensitive) DNA indicates that a preference exists but that it is much
less pronounced than with other antitumor agents [320]. Bleomycin
cleavage is diminished by DNA methylation [321] and it is suggested
that this may provide selective therapeutic action through a preference
for transcriptionally active, undermethylated DNA. There remains also
as a final possibility that bleomycin has no intrinsic preference for DNA
degradation of normal versus transformed cells but that it is selectively
toxic due to the relative availability of oxygen, iron, and reducing agents
combined with diminished bleomycin hydrolase activity.

3   Direct DNA modification by the antitumor antibiotics

The previous examples require the direct participation of an enzyme in
the DNA-modifying event. In contrast there are antitumor agents
which act directly upon DNA, without the intimate cooperation of an
enzyme. These agents contain latent functionality which is used from
within a drug-DNA complex to alter DNA function. While the antitumor effect which results may involve enzymes, the absence of an enzyme in the lesion-forming step sets these apart from the previous
classes of antitumor agents discussed in terms of mechanism. Many of
these agents (for example, the mustards and cisplatinum) are not, however, antibiotics so will not be discussed further. The two classes to
which the major attention is directed in this review are the actinomycins and CC-1065. For the former, the relevant DNA complex is
reached by intercalation and for the latter by minor groove association
prior to covalent alkylation. This difference in DNA binding and
modification underscores once again the breadth of mechanism exploited for an antitumor effect.

3.1   The actinomycins

Although of quite limited clinical value, actinomycin D (XL, Figure 15) provides the benchmark in the study of DNA intercalation.
The dynamics of its interaction with DNA have been studied by all

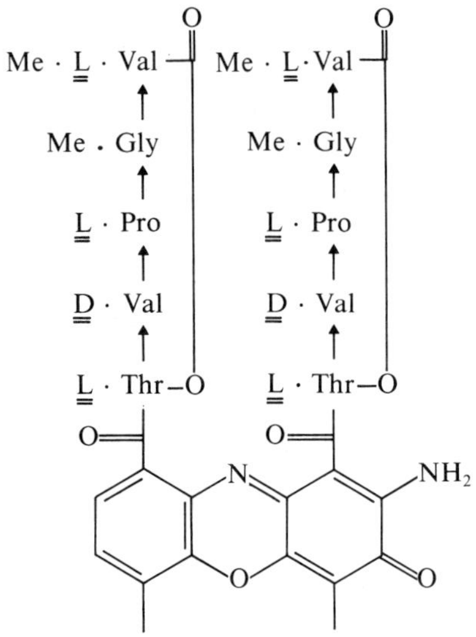

Fig. 15   XL, Actinomycin D.

known techniques and constitute a literature far greater than can be summarized here. Actinomycin is a rare example of a neutral intercalator and possibly as a consequense of the absence of a positive charge, its DNA association and dissociation are kinetically complex. Indeed, it shares with the anthracyclines the phenomenon of slow dissociation, which (as discussed previously) has been suggested by Crothers and colleagues as the important determinant in establishing antitumor activity for an intercalator. It is probable that this stability of the DNA complex also limits the clinical usefulness of actinomycin D, which shows poor selectivity for the tumor relative to the host cell. Not surprisingly, DNA intercalation is also consistent with the basic observation relating to the mechanism of action. At its cytotoxic concentration actinomycin is a potent inhibitor of DNA-dependent RNA elongation catalyzed by RNA polymerase. While the molecular details of this inhibition are elusive – why is this a specific consequence of actinomycin intercalation but not of others – the molecular details of the intercalation are well established. A terse summary of these aspects follows.
The ability of actinomycin to intercalate is not in doubt. Actinomycin inserts into duplexed DNA of virtually any sequence, and indeed this proclivity has created a certain amount of uncertainty within the literature as the characteristics of the intercalated complex are very sequence-dependent. As a result, independent studies using DNA oli-

gomers or polymers of different sequence would yield contradictory conclusions, preventing the development of a consensus. At the present time the physiologically relevant sequence preference of actinomycins are not in any doubt whatsoever, and it is to be expected that the consensus will emerge from ongoing study. In brief, at low ratios of drug to DNA actinomycin shows an absolute preference for insertion between the d (GpC) of a d (AGCT) or d (TGCA) duplex (summarized in [322–326]). Evidence for intercalation is long-standing and includes classic kinetic and thermodynamic studies [324, 327] as well as more recent crystallographic [328–330] and NMR [325, 331–333] investigation. In all instances an overall model for intercalation having insertion of the 2-amino-phenoxazin-3-one ring and placement of the peptide macrocycles into the minor groove (up and down) has been confirmed. Since the intercalating segment is unsymmetrical, in the NMR studies with self-complementary duplexes the anticipated loss of symmetry is in fact observed. Overall the model for this complex [334] is quite close to that proposed by Sobell from his pioneering crystallographic studies [338] and consistent with the experimental observation that the intercalation is primarily an entropically driven process. The dramatic d(GC) specificity is proposed to arise from good hydrogen bond interactions between the N-3 and 2-amino of guanine with the threonine amide NH and carbonyl, respectively [330, 334]. The overall conformation of the peptide macrocycles is little changed upon intercalation from that found in solution [333], and it is suggested that the remaining amino acids present a hydrophobic inner face which protects these hydrogen bonds from displacement by water [330]. An initial evaluation by computer simulation is fully consistent with this suggestion [334] and provides for the two lactone macrocycles a specific role in intercalation stabilization (as is required by the poor intercalating ability of actinomycins with alterations within these rings) which is not otherwise obvious from the intercalation model (no obvious hydrogen bonding interactions). By the cooperative action of these hydrogen bonds and the entropic benefit of driving off water from the 2-aminophenoxazinone chromophore a stable, d(GC)-specific intercalated complex is formed.

The kinetics of the intercalation are complex, particularly for heterogeneous DNA. In order to obtain a consistent picture of the kinetic processes, it has been necessary to use simple oligomers with the preferred binding sequence, or high base pair to drug ratios with heterogeneous DNA to allow sequence selection [324, 326, 335]. The association process is even under these circumstances multi-step. The discreet chemical changes that occur at each step remain open to speculation. Possibly, these correspond to movement along the DNA or to conformational adjustments at the site of intercalation. The observation that the macrocycles do not suffer a change in conformation upon intercalation does not exclude the possibility of a change in attaining the final, stable complex. There are several close atomic contacts (notably between the N-methyl of the N-methylvaline and the neighboring base) which may require conformational mobility in order to both minimize contact and displace solvent from the minor groove. Dissociation, in contrast, from a homogeneous intercalated complex is a single exponential, and the multi-exponential kinetics seen for heterogeneous DNA is reasonably attributed to binding site sequence heterogeneity. For neither association nor dissociation is there evidence for DNA unwinding or melting as an essential step. Actinomycin dissociation from the d(TGCA) or d(AGCT) site is exceedingly slow ($t_{1/2} \sim 30$ min) and, as noted previously, has been suggested by Crothers as the key physical property of the antitumor intercalators.

If the inference that DNA is the target of the actinomycins leading to the antitumor property is accepted, the question to be addressed is a mechanism for the cytotoxicity. A reasonable but hardly proven pair of assumptions is that actinomycin cytotoxicity is due to a single mechanism, regardless of whether the cell is cancerous or not, and that it is intercalation which results in a lethal inhibition of an enzymatic activity. An obvious candidate is RNA polymerase [324, 336]. It is not unreasonable to further presume that the lengthy residence time sufficiently delays RNA polymerase recognition or transcription as to cause cell death. Whether this is accomplished simply by intercalation preventing strand separation or by some other effect (actinomycin is among the most potent of all intercalators in altering DNA conformation; [337, 338]) remains unknown. As mechanistic explanations these suggestions are hardly profound, but defensible as the best explanations currently available.

A most recent development suggests a second enzyme as a target. It has been known for several years that actinomycin elicits a protein associated-DNA strand break exposed upon alkaline elution [339, 340]. This phenomenon is precisely that seen with topoisomerase II inhibitors; however, actinomycin does not inhibit topoisomerase II. The obvious possibility then to be excluded is topoisomerase I, the more so since the discovery of this enzyme as the target of the antitumor agent camptothecin. The relevant experiments have now been done by Trask and Muller [341] with a positive outcome. At concentrations of actinomycin at which RNA transcription is terminated, actinomycin is found to stabilize the covalent topoisomerase-enzyme complex. This is fully consistent with the proven sensitivity of rRNA transcription to actinomycin *in vivo,* and the known requirement for topoisomerase I in this process. The preliminary data suggest that actinomycin impairs the ability of topoisomerase I to reseal the single-strand breaks necessary to accomplish strand passage [341]. As is also true for the topoisomerase II inhibitors, there is independent evidence indicating that the actinomycins are redox-active and initiate redox cycling [342, 345]. A relationship between the two phenomena is speculative, and there is no evidence establishing the redox activity as important to the antitumor activity. For the present the topoisomerase I-inhibitory activity appears as the most promising lead for an antitumor mechanism for the actinomycins, and conceivably the related antitumor agent echinomycin [346, 347]. Should this be proven it will be necessary in a future review to move the actinomycins from those antibiotics acting directly of themselves against DNA, to those requiring enzymatic activation.

3.2   CC-1065 and its analogs

In 1978 the isolation of CC-1065 (XLI, Figure 16), an extremely potent cytotoxic agent, from *Streptomyces zelensis* was reported by investigators from The Upjohn Company [348]. Although CC-1065 proved difficult to purify and characterize [349], its structure [350] eventually succumbed to X-ray analysis [351]. This same compound was subsequently reported from two additional laboratories [352, 353]. The chemistry [354, 355] and mechanism of action [355, 356] of the natural product were recently reviewed. A summary of the biological data, particularly as they pertain to interaction with DNA, of CC-1065 and several of its analogs follows.

CC-1065 is a very potent antitumor agent both *in vitro* and *in vivo. In vitro* against various tumor cell lines CC-1065 is several orders of magnitude more cytotoxic than adriamycin, even more potent than either actinomycin D or maytansine [357]. CC-1065 retains significant activity in the *in vitro* human tumor cloning assay at 0.1 ng/ml in a variety of tumor explants [358]. While CC-1065 is only a very weak mutagen in the Ames *Salmonella* mutation assay, the compound is a potent mutagen in mammalian (V79) cells with chromosomal aberrations being noted at very low doses, although no DNA single-strand breaks or DNA protein cross-links are observed [359]. However CC-1065, like several other antitumor agents (e. g., adriamycin) is only weakly mutagenic at biologically relevant doses *in vitro* and is only significantly genotoxic at 100 µg/kg in the *in vivo* micronucleus test [359]. *In vivo* in murine antitumor models CC-1065 is acutely toxic to the mice at doses greater than 200 µg/kg. At lower doses (down to 50 µg/kg) the compound is active in a number of the standard murine tumor models, particularly when both drug and tumor are administered intraperitoneally; however, CC-1065 is not curative in any of these tumor models [349, 360]. Treatment of non-tumor-bearing mice with CC-1065 uncovered an unusual delayed form of lethal toxicity at all doses which prolong life in tumor-bearing mice [361]. Even at subtherapeutic doses in the mice (e. g. a single i. v. dose of 12.5 µg/kg) delayed deaths, tentatively attributed to hepatotoxicity, are noted 45–50 days post dosing [361].

Evidence has accumulated to suggest that DNA is the cellular target for CC-1065. In both L1210 and B16 cells treated with CC-1065, DNA synthesis is greatly inhibited relative to RNA synthesis or protein synthesis [357, 362]. In analogy to a number of alkylating agents, CC-1065 is most cytotoxic to mitotic cells; however, surprisingly the S phase cells are most resistant to the drug [363]. Premixing of CC-1065 with calf thymus DNA abolishes the potent cytotoxicity of the agent against L1210 cells [357]. In contrast little effect is noted upon pretreatment of the CC-1065 with RNA or albumin. Differential circular dichroism studies clearly demonstrate that CC-1065 requires double-stranded B-form DNA for binding and that AT regions are preferred over GC regions [357, 362, 364]. More recent spectroscopic studies indicate that CC-1065 may bind to DNA by at least three mechanisms involving two reversibly bound and one irreversibly bound species [365]. Rapid and irreversible binding to DNA is also supported by size exclusion chromatographic studies of the CC-1065-DNA complex [362]. Studies with supercoiled DNA indicate that CC-1065 does not bind intercalatively to DNA nor does it cause single- or double-strand breaks at 37 °C [362]. In fact, binding of CC-1065 to DNA polymers causes marked stabilization of the double helix with a large increase in the

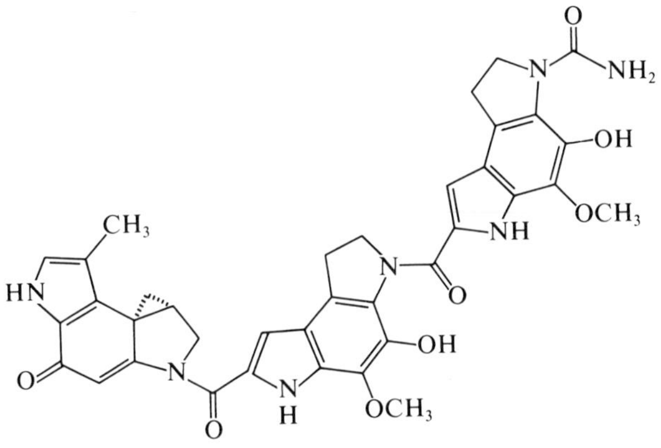

Fig. 16   CC-1065   (XLI)

DNA-helix melting temperature, reduction of the ability of ethidium bromide to intercalate and unwind supercoiled DNA, and the inhibition of S1 nuclease digestion of the DNA [357, 362].

Evidence suggests initial reversible binding followed by irreversible binding (covalent bonding) of CC-1065 to DNA selectively within the minor groove of DNA. This binding, which appears saturated at 7-11 base pairs per molecule of drug, inhibits alkylation by nitrosoureas known to alkylate sites in the minor groove of DNA [362]. Furthermore, CC-1065 is able to displace the non-covalent minor groove binder netropsin. Competitive binding experiments with netropsin show initial inhibition of CC-1065 binding to the DNA [351]. However, netropsin is not able to displace CC-1065 already bound to DNA. Minor groove binding is also suggested by the finding that CC-1065 binds equally well to T4 DNA, which has the major groove glycosylated (and thus hindered) [362].

Ultimately N-3 of adenine was identified as the target of alkylation by the drug. Thermal treatment of the CC-1065 modified DNA results in the release of a chemically modified CC-1065 chromophore covalently attached to an adenine-derived moiety [366, 367]. Subsequently, NMR studies suggested compound XLII (Figure 17) as the CC-1065-DNA adduct [366]. The structure of the adduct suggests the mechanism shown in Figure 17 wherein the methylene group of the cyclopropane moiety on the left-hand segment is attacked by the N-3 nitrogen of adenine, resulting in ring opening of the cyclopropane moiety and formation of the phenol. Analog studies suggest that the cyclopropane ring cleavage is an acid-catalyzed process [368].

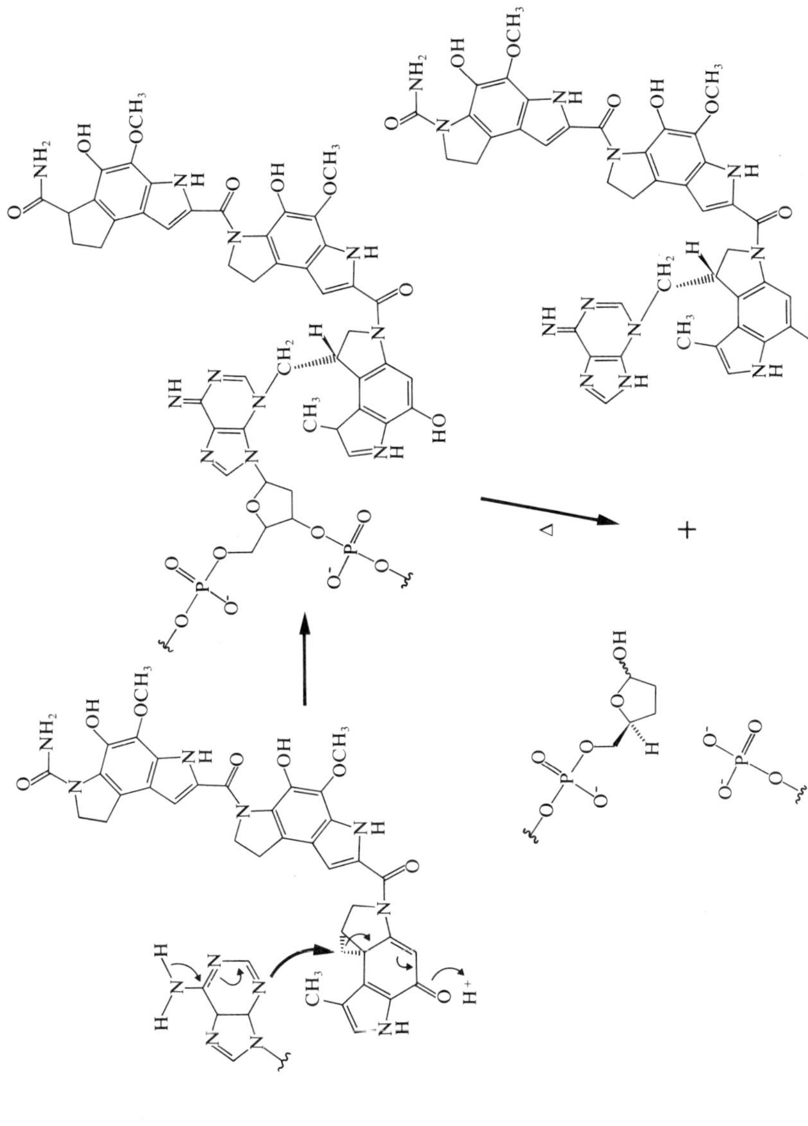

XLII

Fig. 17

The thermal depurination of N-3 alkylated adenines allowed the use of CC-1065 as a Maxam-Gilbert type sequencing agent to reveal the site of CC-1065 alkylation of a known sequence of DNA. This was used to identify, from the alkylation sites on a number of DNA restriction fragments, two consensus sequences for CC-1065. These are 5'-PuNTTA-3' and 5'-AAAAA-3', with the 3' adenine site being the point of CC-1065 covalent attachment (Pu = purine, N = any base) [366, 367]. Additional studies with synthetic duplex oligonucleotides support this remarkable sequence selectivity [369, 370]. NMR studies support the hypothesis that the drug aligns itself in the minor groove extending in the 3' to 5' direction with the polar convex edge of CC-1065 pointing away from the DNA duplex [371].

Computer modeling confirms that in the CC-1065-(N-3-adenine)-DNA complex the CC-1065 molecule should overlap about 5 base pairs [366]. Modeling of the covalently attached CC-1065 molecule in the consensus sequences was originally used to propose the absolute stereochemistry of the compound [366]. A recent X-ray study on a chlorinated derivative of CC-1065 confirmed the assignment of stereochemistry [372]. In agreement with the oligomer studies [370], more sophisticated modeling studies [373] also suggest that CC-1065 should prefer binding to DNA at sites which are substituted with either AA or TT on the 5'-side of the covalently modified adenine. To explain the sequence selectivity it is proposed that the neighboring AT pairs are stabilized by a favorable electrostatic effect upon reaction of CC-1065 with the N-3 of adenine [373]. Furthermore, the modeling suggests that with neighboring GC pairs the binding of the drug to adenine is disfavored because of the greater distortion of the DNA helix that arises from the steric hindrance of the 2-amino group on guanine [373]. Computer modeling [365] of the CC-1065-DNA complex also indicates that the covalent adduct of CC-1065 (and unreacted CC-1065) can attain the same close van der Waals contacts between the adenine C2 hydrogens and CC-1065-CH groups that have been observed with the netropsin-DNA complex [374].

A consequence of covalent attachment of CC-1065 to the DNA may be local conformational distortion of the DNA. Recent results with a site-directed adduct of CC-1065 in a 117 base pair segment of DNA [375] employing DNase I footprinting and restriction enzyme analysis indicated that the drug produces an asymmetric effect of DNA structure which extends more than one helix turn to the 5' side of the covalent

binding site [376]. Studies with the site-directed CC-1065 adduct indicate a slight enhancement of methidiumpropyl-EDTA-iron (II) cutting to the 5′ side of the CC-1065 adduct on the strand to which it is attached and to the 3′ side of the adduct on the opposite strand [375].
It is reasonable to assume that the CC-1065-DNA covalent adduct is
recognized by appropriate repair mechanisms. Addition of CC-1065
to human fibroblast cells produces a prolonged depletion of the nico-
tinamide adenine dinucleotide (NAD) pool which is blocked by 3-ami-
nobenzamide, thus indicative of a process involving poly (ADP-ri-
bose) synthesis in response to a repair-induced DNA strand breakage
[377]. However, it should be noted in these experiments that a relatively high dose of CC-1065 is employed [377].
Since effort to prevent the delayed toxicity of CC-1065 in mice by
several biochemical procedures (e. g. co-treatment with N-acetyl-
cysteine) failed [361], synthetic analogs were sought which might
retain much of the potency and at least the same level of activity of the
natural product but be devoid of its unusual toxicity [378]. These
efforts proved highly successful. Since these analogs also provide valuable information relative to the mechanism of action of this class of
compounds they will be discussed in some detail.

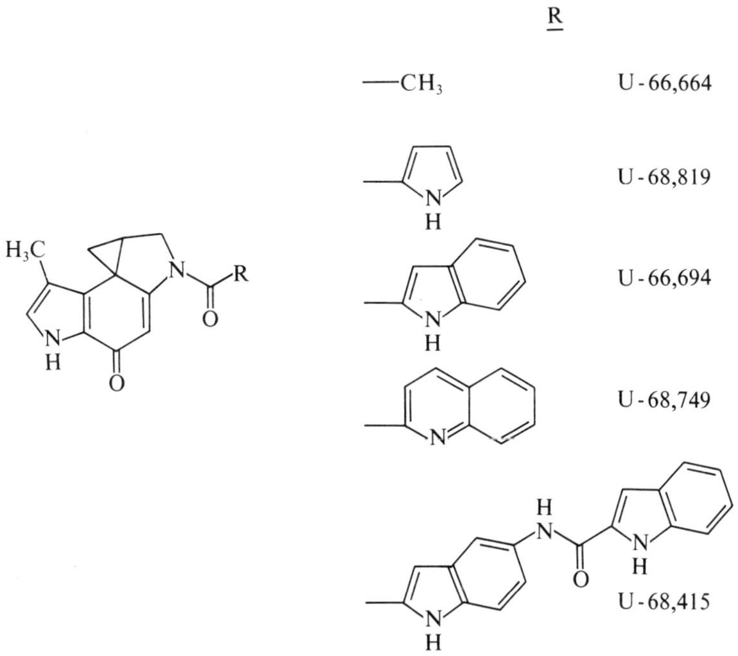

Fig. 18

A number of analogs of CC-1065 containing an unmodified lefthand segment conjugated to alkyl, aromatic, or heteroaromatic groups through an N-acyl linkage have been reported by the Upjohn laboratories, most prepared by total synthesis. In an early series of racemic analogs (structures shown in Figure 18), a correlation was noted between *in vitro* potency and DNA binding (as measured by differential circular dichroism), with the more cytotoxic compounds exhibiting higher DNA binding [378–381]. However, several analogs, with limited DNA binding (and low potency), such as U-68,749, exhibit excellent *in vivo* activity [378, 379]. In terms of DNA synthesis inhibition the following order applies: U-68,415 > U-66,694 > U-68,819 > U-66,664, which exactly parallels their level of DNA binding, L1210 cell growth inhibition, and *in vivo* antitumor potency, as well as their structural complexity [380, 381]. Interestingly, these analogs displayed similar DNA base sequence specificity to CC-1065 [378, 382]. In addition, with these four compounds, as with CC-1065, DNA synthesis is inhibited ten times more than RNA synthesis with protein synthesis being the least inhibited [381, 382]. These studies suggest that cellular DNA is probably a major site of action with the analogs, and that the more complex the structure (e. g. U-68,415), approaching that of CC-1065 at least in length, the more favorable is the interaction with DNA.

A recently reported study [368] employing a larger number of CC-1065 analogs indicates:

1) a mechanism involving acid catalysis in the cyclopropane ring opening step is probably operative since biological potency parallels chemical reactivity towards acid-catalyzed nucleophilic attack, with acyl substitution on the left-hand fragment being optimal for both properties

2) that possibly because of conformational constraints five-membered ring heteroaroyl substituents attached to the left-hand segment of CC-1065 are more cytotoxic and show more favorable DNA binding relative to six-membered ring substituents (e. g. U-66, 694 as compared to U-68,749)

3) that DNA interaction and potency is in general enhanced by extension of the indole chain (e. g. as in U-68,415) suggestive of an important hydrophobic contribution being involved in the drug-DNA binding [368].

In fact, U-68,415 is comparable to CC-1065 in terms of *in vitro* and *in vivo* potency, and it exhibits about one-third of binding to DNA as

CC-1065 as measured by differential circular dichroism [378]. *In vivo* U-68,415 is significantly more active than CC-1065. It is highly curative in a number of murine leukemia models following a single intravenous dose of the drug [378, 381]. Like CC-1065, U-68,415 binds irreversibly (i. e. bonds) to double-stranded DNA with AT specificity.

U-68,415 is a racemate. Preparation of the optically active species indicates that all of the biological activity can be attributed to U-71,184 (Figure 19), i. e. the isomer containing the same configuration as CC-1065 [380, 383]. The enantiomer of U-71,184, i. e. U-71,185, is at least two orders of magnitude less cytotoxic *in vitro,* is inactive *in vivo* at doses where U-71,184 is highly curative, and shows much less interaction with DNA *in vitro* [382, 383]. Like CC-1065, U-71,184 binds to double stranded B-DNA, is extremely potent *in vitro* and *in vivo* though much more efficacious than CC-1065, but, unlike CC-1065, does not cause delayed death in mice [380, 383]. Studies with U-71,184 (and U-66,694) indicate a different phase specific toxicity than CC-1065. Unlike CC-1065, which is most cytotoxic to cells in M and early G_1, U-71,184 is most toxic to cells in late G_1 [384]. In Chinese hamster ovary cells a significantly higher dose of U-71,184 is necessary to inhibit DNA synthesis by 90 % than to kill cells or inhibit growth by 90 % [384]. At relevant cytotoxic doses of U-71,184, cell death may be due to unbalanced cell growth caused by low level inhibition of DNA synthesis combined with stimulation of RNA and protein synthesis [384].

Replacement of the indole portion(s) of U-71,184 with benzothiophene and benzofuran moieties has also been reported [385]. These analogs also show significant DNA interaction, and one closely related analog, U-73,975 (Figure 19), looks particularly promising. Like U-71,184, U-73,975 does not cause delayed lethality in mice, and *in vitro* and *in vivo* both analogs are equipotent (and bind roughly equally to DNA); however, the benzofuran derivative U-73,975 appears to be consistently superior to U-71,184 against a number of solid tumors in mice, where excellent activity is seen even with a single i. v. injection [386].

Like U-71,185 (the enantiomer of U-71,184), the enantiomer of U-73,975, (U-76,579) is biologically inactive and does not readily bind to DNA. In contrast, it was recently found that the enantiomer of CC-1065 [387] is roughly equicytotoxic *in vitro* and approximately equipotent and active *in vivo* as the natural product [386, 387]; however, it does not show the delayed death phenomenon in mice characteristic

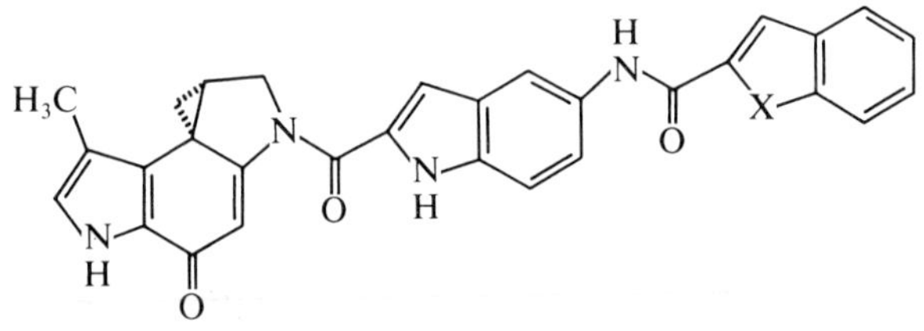

U-71,184:   X = NH
U-73,975:   X = O

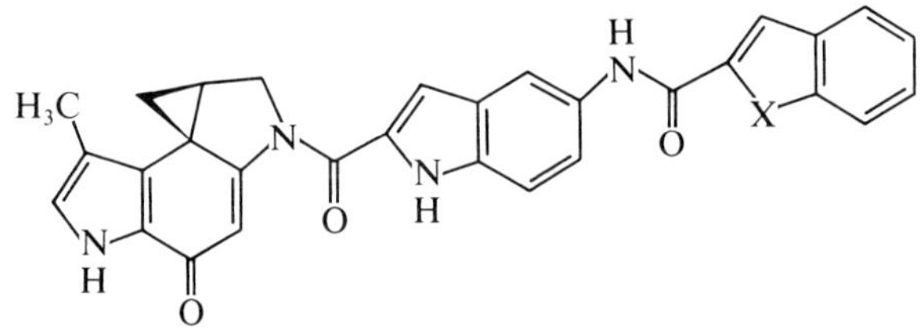

Fig. 19

of CC-1065 [388]. The enantiomer of CC-1065 avidly binds to DNA (presumably at the N-3 of adenine) but in the 3'-orientation relative to the covalently modified adenine, (opposite to CC-1065) [388]. In addition, the enantiomer of CC-1065 appears to have different preferred binding sequences than the natural product [388].

Studies with the optically pure forms of U-66,664, U-66,694, and U-68,415 revealed an unexpected and mechanistically important result. The key finding is that all three compounds alkylate exactly the same subset of adenines on DNA, thus demonstrating that the left-hand segment of CC-1065 (i. e. U 66,664) is responsible for the DNA-sequence specificity for this class of molecules [389]. Moreover, the left-hand segment causes the same asymmetric effect on local DNA structure (as measured by altered DNase I cleavage patterns) as larger molecules, including U-71,184 and CC-1065 [389]. For the most part, the sequences alkylated by CC-1065 are the same as those alkylated by the optically pure forms of U-66,664, U-66,694, and U-68,415; however, several adenines alkylated by CC-1065 are not alkylated by the three analogs and vice versa [389]. Thus, these results overall demonstrate that the specificity of covalent bonding of CC-1065 analogs at the molecular level is controlled, not by non-covalent interactions of the B- and C-subunits with the minor groove, but by sequence-dependent reactivity of adenines with the alkylating left-hand segment. The middle and right-hand segments of these analogs (and in CC-1065), probably by enhancing non-covalent interactions with the DNA prior to covalent bond formation, apparently serve mainly to increase the rate of the alkylation reaction with the appropriately reactive adenines in the DNA [389]. However, the left-hand segment is sufficient for the covalent bonding, and it is proposed that only certain adenine containing DNA sequences can undergo the conformational distortion necessary to allow irreversible binding (i. e. covalent bonding) to the cyclopropyl ring of left-hand segment in CC-1065 and its analogs [389]. A circular dichroism study of the binding of CC-1065 to various synthetic DNA polymers also suggests that reversible as opposed to irreversible binding is influenced by small differences in DNA secondary structure, and that covalent alkylation at certain adenines is dependent on favorable DNA conformation [390].

Recently, U-78,057 (Figure 20) the tetradesoxy analog of CC-1065 was reported [391, 392]. This compound is found to more closely resemble CC-1065 than U-71,184 in its biological properties, although its differential circular dichroism with DNA is similar to that of U-71,184 [391]. CC-1065 contains middle and right-hand benzodipyrrole segments [393] giving the molecule a right-hand twist with the hydrophobic concave edge capable of reversible non-covalent interaction with DNA in the minor groove. A similar situation exists with U-78,057 and U-71,184. It is instructive to note that all three compounds exhibit similar potency, though U-71,184 and U-78,057 bind less avidly than CC-1065 to DNA [391]. Of significant interest is the finding that U-78,057 causes the delayed death phenomenon characteristic of CC-1065, indicating that the fused pyrrolidine rings, and not the *o*-catechol units of the natural product, are sufficient for this unusual toxicity [391]. Furthermore, U-78,057 and CC-1065 appear to have identical DNA sequence alkylation specificity [389]. This analog thus provides an important clue for understanding the molecular basis for the delayed toxicity of CC-1065.

A series of bifunctional, potentially cross-linking, CC-1065 analogs composed of two of the left-hand segments of CC-1065 linked together

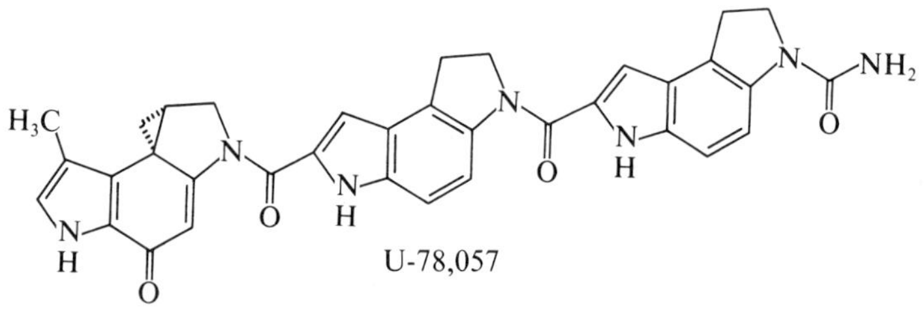

Fig. 20

by methylene chains of varying length has also been reported [394]. Although members of this group of analogs have significantly greater *in vitro* and *in vivo* potency (and efficacy) than the natural product, evidence that DNA is their target has been equivocal [394]. For example, in contrast to CC-1065, preincubation of these agents with DNA does not readily abolish their cytotoxic potency. Several of these analogs do in fact cause a significant amount of thermally induced double-strand breaks relative to CC-1065 (which caused only single-strand breaks); however, overall, these same compounds are only about 1 % as effective at inducing any strand breakage relative to CC-1065 despite their enhanced potency relative to the parent antibiotic [394]. Clearly more work needs to be done to understand this series of analogs.

4   Other DNA-interactive antitumor antibiotics

In this section several additional selected classes of antitumor antibiotics which appear to significantly interact with DNA will be briefly discussed. No attempt is made to cover all the many different antitumor natural products whose mechanism of action may involve DNA. Similarly, the discussion of each of the classes will be limited to a few pertinent points.

The pyrrolo(1,4)benzodiazepines represent another class of potent antitumor antibiotics which, like CC-1065, appear to covalently bind in the minor groove of DNA in a non-intercalative fashion [395]. The mechanism of action of this class of compounds, which includes tomaymycin, sibiromycin, anthramycin, neothramycin A, and neothramycin B, has recently been reviewed [356]. Like CC-1065, the pyrrolo(1,4)benzodiazepines form non-distortive covalent DNA adducts which are helix-stabilizing [396, 397]. Unlike CC-1065, the pyrrolo(1,4)benzodiazepines show a preference for binding to poly(dG)-poly(dC) [397] and form a reversibly bound covalent adduct [398]. As evidenced by DNA footprinting studies the pyrrolo(1,4)benzodiazepines bind most strongly to 5'PuGPu (where Pu is a purine base) [399]. The mechanism proposed for the interaction of these agents with DNA involves nucleophilic attack by the exocyclic 2-amino group of guanine in the minor groove of B form DNA on the imine generated from the carbinolamine of the drug [400]. This is illustrated in Figure 21 with tomaymycin (XLIII) reacting with DNA. Note that potentially two adducts can form differing in the stereochemistry at the linkage site.

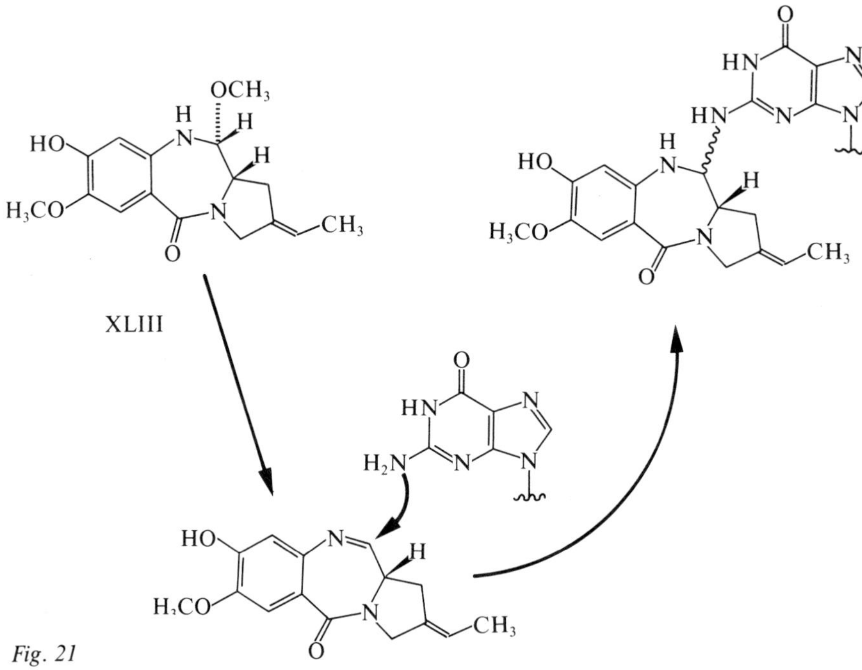

XLIII

Fig. 21

Spectroscopic evidence exists for both forms [401]. This model is further supported by NMR studies on an anthramycin oligomer adduct [402, 403] and by modeling studies [404–406] suggesting stabilization of the linkage product by the DNA.

Somewhat mechanistically related to the pyrrolo(1,4)benzodiazepines are the saframycins (e. g. XLIV and XLV) safracins (i. g. XLVI), and naphthyridinomycin (XLVII) [407–409]. These antitumor antibiotics also contain a chemically reactive carbinolamine or its equivalent [410] and appear (at least for the saframycins) to react with the 2-amino group of guanine [411]. However, like the mitomycins, the saframycins [411, 412] and naphthyridinomycin [413–415] appear to react with DNA via a bioreductive alkylation process involving reduction of quinone to the hydroquinone. A related compound containing a masked carbinolamine, quinocarcin (XLVIII), like the saframycins [410] generates free radicals while inhibiting DNA synthesis [415]. A citric acid salt of quinocarcin, quinocarmycin citrate, which is itself an antibiotic, shows promising in vivo activity in mouse tumor models [417].

Neocarzinostatin, a member of a family of protein antitumor antibiotics, appears to interact with DNA by an entirely different mechanism. Neocarzinostatin, which has been evaluated clinically in Japan, specifically damages the deoxyribose moiety, causing single-strand

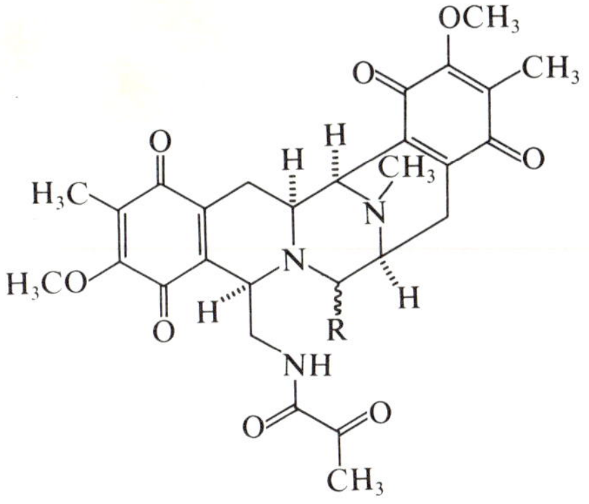

XLIV   Saframycin A   (R = CN)
XLV Saframycin S   (R = OH)

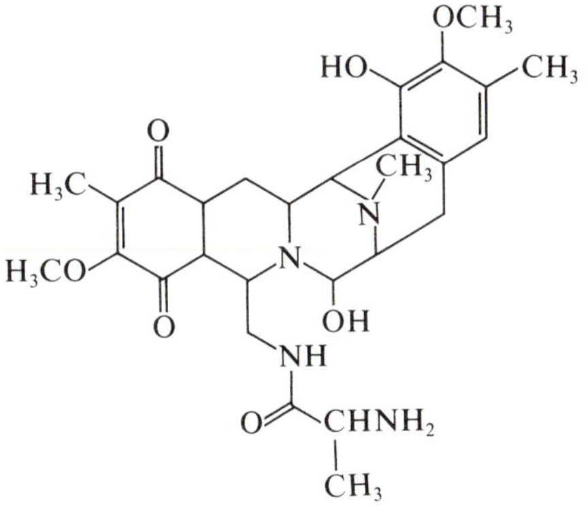

XLVI   Safracin B

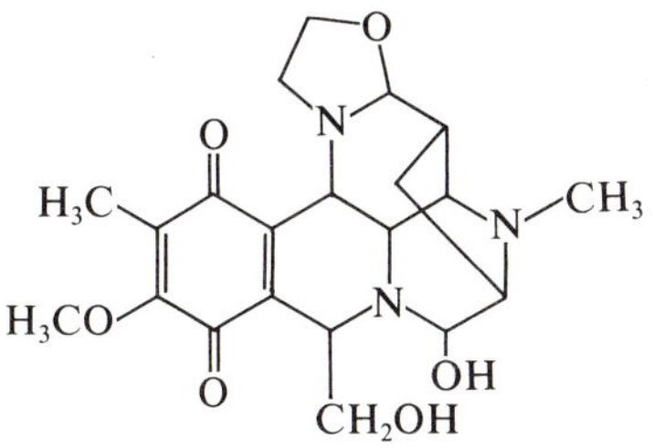

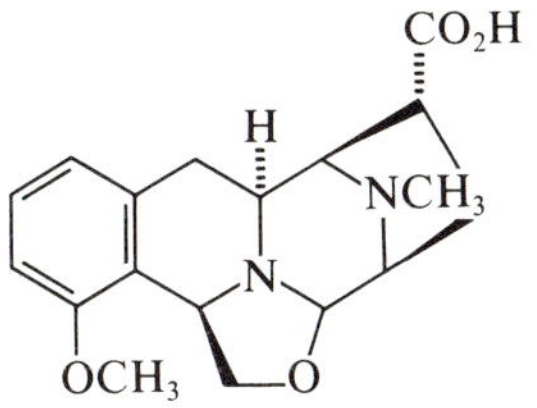

Fig. 22   XLVII   XLVIII

breaks via superoxide free radical generation [418, 419]. The mechanism of action of neocarzinostatin has been extensively reviewed [420–423]. Neocarzinostatin consists of two separable portions, a protein and a non-peptide chromophore [424]. The chromophore has recently been assigned the structure XLIX (Figure 23) [425] and appears to contain all the cytotoxicity and DNA-cleaving activity of the natural product [426, 427]. The primary structure of the protein portions has also been determined [428–430] and it seems to serve to stabilize the chromophore which is otherwise quite prone to decomposition [431–433]. Reversible binding to DNA in the minor groove may occur via intercalation of the naphthyl portion of the chromophore which then places the nonaromatic portion of XLIX in the minor groove of DNA [434] with electrostatic binding of the amino sugar to a phosphate group on the DNA backbone [421]. In an oxygen-independent

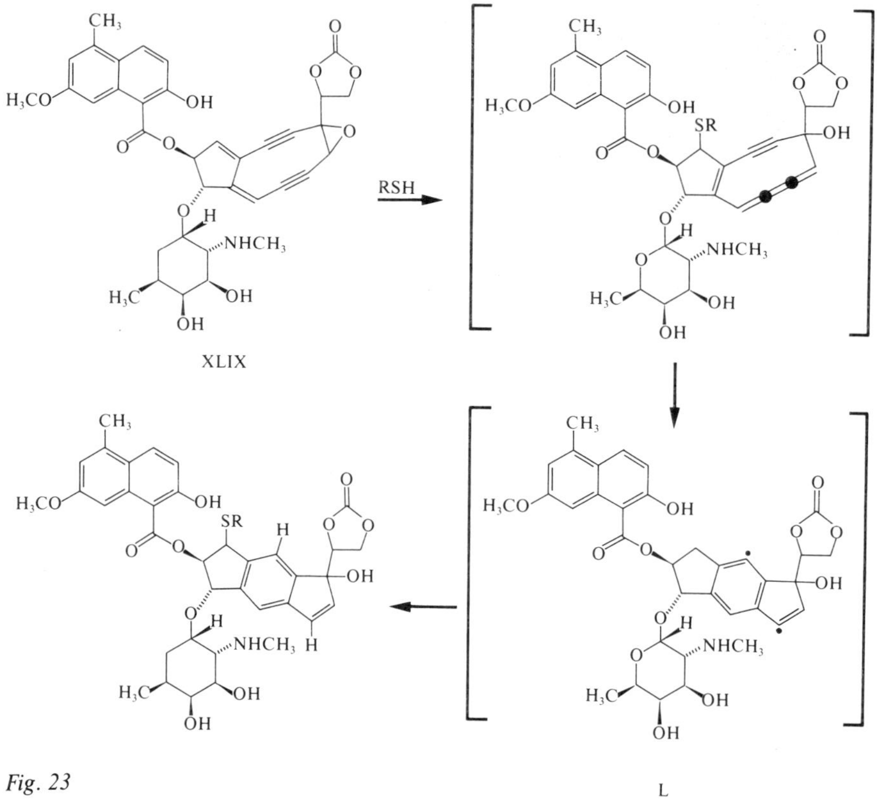

XLIX

RSH

L

Fig. 23

but thiol-promoted reaction [435–437] the neocarzinostatin chromophore is activated to selectively cleave DNA at dA and dT residues
[438] to give a 5′-aldehyde DNA fragment, the carbonyl oxygen of the
aldehyde coming from molecular oxygen [439]. It has been proposed
[440] that activation of XLIX gives a free radical intermediate which
then abstracts a nearby 5′ hydrogen atom from the DNA sugar backbone. Reaction of the 5′-sugar-radical with oxygen then leads ultimately to backbone cleavage and aldehyde formation. The activation
of the chromophore to give the diradical intermediate L was recently
postulated to proceed as shown in Figure 23 [441].
A somewhat analogous scheme has been proposed for the mechanism
of action of the esperamicins (e. g. compound LI, Figure 24) [442] and
the closely related compounds, the calichemicins [443], two recently reported series of interesting and extremely potent antitumor antibiotics
[444–446]. These antibiotics also seem to be closely related to the verac-

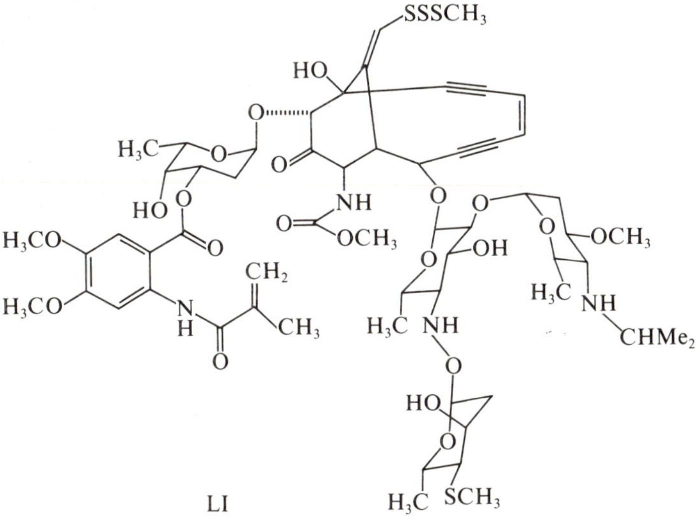

Fig. 24   LI

tamycins [447–452] and FR-900405 and FR-900406 [453–455]. As illustrated in Figure 25, cleavage of the esperamicin trisulfide in LII could lead to intermediate LIII which can undergo an intramolecular Michael reaction to give LIV [442]. In compound LIV, the olefin geometry is now suitable for cyclization to the phenylene diradical LV which is then capable of hydrogen atom extraction from the sugar phosphate

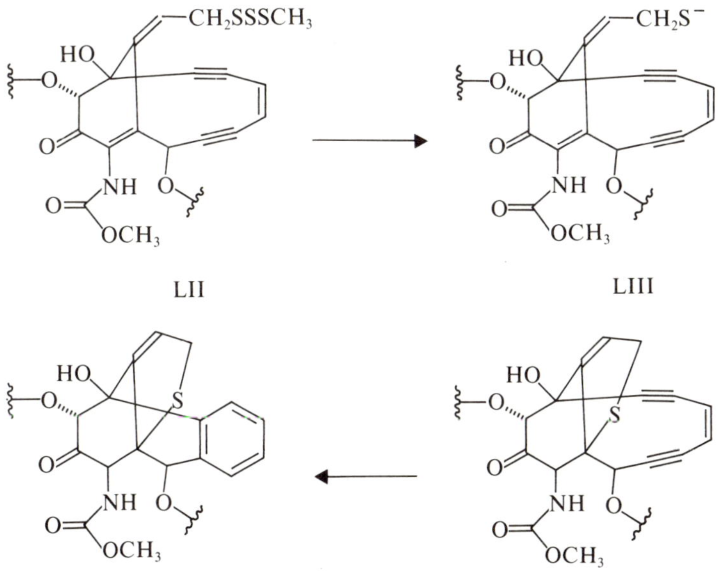

Fig. 25   LIV   LV

DNA backbone [442]. Intermediate LV could cause the DNA double-strand breaks that are observed with esperamicin by simultaneous cleavage on each strand of the DNA [456]. Esperamicin A_1 (LI) is an extremely potent antitumor antibiotic being approximately 5–10 times more potent than CC-1065 [457]. In murine antitumor models esperamicin also showed good antitumor activity [457]; however, the future for this class of agents is presently clouded due to reports of delayed toxicity at optimal antitumor doses [458].

Acknowledgements

We are grateful to M. A. Mitchell and M. A. Warpehoski for their corrections and suggestions concerning the review. We thank P. D. Johnson, C. A. Meister, M. E. Meissner, and especially C. A. Kiewiet for the preparation of the manuscript.

References

1   K. T. Douglas: Chem. Ind. (London), 738 (1984); 766 (1984).
2   J. W. Lown: Molecular Aspects of Anti-Cancer Drug Action, p. 283. Eds. S. Neidle and M. Waring, Macmillan, London 1983.
3   W. A. Remers: Antineoplastic Agents, pp. 83–261. Ed. W. A. Remers, Wiley, New York 1984.
4   K. Hemminki and D. Ludlum: J. Nat. Cancer Inst. 73, 1021–1028 (1984).
5   I. Jardine: Anticancer Agents Based on Natural Products, pp. 314-351. Eds. J. Cassady and J. Douros, Academic, New York 1980.
6   J. D. Loike and S. B. Horowitz: Biochemistry 15, 5543–5448 (1976).
7   B. H. Long, S. Musial and M. Brattain: Biochemistry 23, 1183-1188 (1984).
8   L. S. Thurston, H. Irie, S. Tani, F. Han, Z. Liu, Y. Cheng and K. Lee: J. Med. Chem. 29, 1547–1550 (1986).
9   R. S. Gupta, P. Chenchaiah and R. Gupta: Anti-Cancer Drug Design 2, 1–12 (1987).
10 P. J. O'Dwyer, B. Leyland-Jones, M. Alonso, S. Marsoni and R. Wittes: N. Eng. J. Med. 312, 692–700 (1985).
11 D. Roberts, S. Hillard and C. Peck: Cancer Res. 40, 4225–4231 (1980).
12 A. J. Wozniak and W. Ross: Cancer Res. 43, 120–124 (1983).
13 L. A. Zwelling, S. Michaels, L. Erickson, R. Ungerleider, M. Nichols and K. Kohn: Biochemistry 20, 6553–6563 (1981).
14 G. L. Chen, L. Yang, T. Rove, B. Halligan, K. Tewey and L. Liu: J. Biol. Chem. 259, 13560–13566 (1984).
15 W. E. Ross, T. Rove, B. Glisson, J. Yalowich and L. Liu: Cancer Res. 44, 5857–5860 (1984).
16 B. H. Long and A. Minoccha: Biochem. Biophys. Res. Commun. 122, 165–170 (1984).
17 T. Rowe, G. Kupfer and W. Ross: Biochem. Pharmacol. 34, 2483–2487 (1985).
18 L. Yang, T. Rowe and L. Liu: Cancer Res. 45, 5872–5876 (1985).
19 L. F. Liu: Critical Rev. Biochem. 15, 1–24 (1985).

20 G. North: Nature 316, 394–395 (1985).
21 J. Wang: NCI Monograph 4, 3–6 (1987).
22 W. E. Ross: Biochem. Pharmacol. 34, 4191–4195 (1985).
23 L. A. Zwelling: Cancer Metastasis Rev. 4, 263–276 (1985).
24 G. L. Chen and L. F. Liu: Annu. Rep. Med. Chem. 21, 257–262 (1986).
25 R. B. Lock and W. E. Ross: Anti Cancer Drug Design 2, 151–164 (1987).
26 Y. H. Hsiang, R. Hertzberg, S. Hecht and L. F. Liu: J. Biol. Chem. 260, 14873–14878 (1985).
27 M. R. Mattern, S. Mong, H. Bartus, C. Mirabelli and S. Crooke: Cancer Res. 47, 1793–1798 (1987).
28 T. C. Rowe, E. Couto, D. Kroll: NCI Monograph. 4, 49–53 (1987).
29 W. C. Earnshaw and M. Heck: J. Cell Biol. 100, 1716–1725 (1985).
30 S. DiNardo, K. Voelkel and R. Sternglanz: Proc. Nat. Acad. Sci. U.S.A. 81, 2616–2620 (1984).
31 W. G. Nelson, L. F. Liu and D. Coffey: Nature 322, 187–189 (1986).
32 W. G. Nelson and D. Coffey: NCI Monograph. 4, 23–29 (1987).
33 L. Yang, M. Wold, J. Li, T. Kelley and L. F. Liu: Proc. Nat. Acad. Sci. U.S.A. 84, 950–954 (1987).
34 N. Sahyoun, M. Wolf, J. Besterman, T. Hsieh, M. Sander, H. LeVine, K. Chang and P. Cuatrecasas: Proc. Nat. Acad. Sci. U.S.A. 83, 1603–1607 (1986).
35 K. W. Kohn, Y. Pommier, D. Kemgan, J. Markovits and J. Covey: NCI Monograph. 4, 61–71 (1987).
36 Y. Matsushima, F. Kanzawa, N. Miyazawa, Y. Sasaki and N. Saijo: Anti-cancer Res. 6, 921–924 (1986).
37 E. M. Nelson, K. M. Tewey and L. F. Liu: Proc. Nat. Acad. Sci. U.S.A. 81, 1361–1365 (1984).
38 B. Marshall and R. Ralph: Adv. Cancer Res. 44, 267–293 (1985).
39 W. A. Denny and B. C. Baguley: Anti-Cancer Drug Design 2, 61–70 (1987).
40 S. Neidle, G. Webster, B. Baguley and W. Denny: Biochem. Pharmacol. 35, 3915–3921 (1986).
41 W. A. Denny and L. Wakelin: Cancer Res. 46, 1717–1721 (1986).
42 L. P. G. Wakelin, G. Atwell, G. Rewcastle and W. Denny: J. Med. Chem. 30, 855–861 (1987).
43 J. Minford, J. Filipski, K.W. Kohn, D. Kerrigan, M. Mattern, S. Michaels, R. Schwartz and L. Zwelling: Biochemistry 25, 9–16 (1986).
44 B. C. Baguley, W. Denny, G. Atwell and B. Cain: J. Med. Chem. 24, 520–525 (1981).
45 W. A. Denny, I. Roos and L. Wakelin: Anti-Cancer Drug. Design 1, 141–147 (1986).
46 M. Bakic, M. Beran, B. Andersson, L. Silberman, E. Estey and L. Zwelling: Biochem. Biophys. Res. Commun. 134, 638–645 (1986).
47 Y. Pommier, J. Covey, D. Kerrigan, W. Mattes, J. Markovits and K. Kohn: Biochem. Pharm. 36, 3477–3486 (1987).
48 T. C. Rowe, G. Chen, Y. Hsiang and L. F. Liu: Cancer Res. 46, 2021–2026 (1986).
49 D. Kerrigan, Y. Pommier and K. Kohn: NCI Monograph. 4, 117–121 (1987).
50 B. H. Long: NCI Monograph. 4, 123–127 (1987).
51 K. M. Tewey, G. Chen, E. Nelson and L. Liu: J. Biol. Chem. 259, 9182–9187 (1984).
52 K. M. Tewey, T. Rowe, L. Yang, B. Halligan and L. Liu: Science 226, 466–468 (1984).
53 R. Silber, L. Liu, M. Israel, A. Bodley, Y. Hsiang, S. Kirschenbaum, T. Sweatman, R. Seshadri and M. Potmesil: NCI Monograph. 4, 111–115 (1987).

53a   N. Maniar, A. Krishnan, M. Israel and T. A. Samy: Biochem. Pharmacol. 37, 1763–1772 (1988).

54   D. M. Sullivan, B. Glisson, P. Hdoges, S. Smallwood-Kentro and W. Ross: Biochemistry 25, 2248–2256 (1986).

55   D. M. Sullivan, M. Latham and W. Ross: Cancer Res. 47, 3973–3979 (1987).

56   L. A. Zwelling, E. Estey, L. Silberman, S. Doyle and W. Hittelman: Cancer Res. 47, 251–257 (1987).

57   J. Markovits, Y. Pommier, D. Kerrigan, J. Covey, E. Tilchen and K. Kohn: Cancer Res. 47, 2050–2055 (1987).

58   W. G. Nelson, K. Cho, T. Hsiang, L. Liu and D. Coffey: Cancer Res. 47, 3246–3250 (1987).

59   E. Estey, R. Adlakha, W. Hittelman and L. Zwelling: Biochemistry 26, 4338–4344 (1987).

60   J.-F. Riou, M. Vilarem, C. Larsen, E. Multon and G. Riou: NCI Monograph. 4, 41–47 (1987).

61   B. Glisson, R. Gupta, P. Hodges and W. Ross: Cancer Res. 46, 1939–1942 (1986).

62   Y. Pommier, D. Kerrigan, R. Schwartz, J. Swack and A. McCurdy: Cancer Res. 46, 3075–3081 (1986).

63   M. K. Danks, J. Yalowich and W. Beck: Cancer Res. 47, 1297–1301 (1987).

64   F. H. Drake, J. Zimmerman, F. McCabe, H. Bartus, S. Per, D. Sullivan, W. Ross, M. Mattern, R. Johnson, S. Crooke and C. Mirabelli: J. Biol. Chem. 262, 16739–16747 (1987).

65   M. Bakic, D. Chan, B. Andersson, M. Beran, L. Silberman, E. Estey, L. Ricketts and L. Zwelling: Biochem. Pharm. 47, 1560–1565 (1987).

66   F. M. Muggia and G. McVie: NCI Monograph. 4, 129–133 (1987).

67   L. J. Marnett and T. Eling: Reviews in Biochemical Toxicology, pp. 132–172. Eds. E. Hodgson et al. Elsevier, Vol. 5 1985.

68   P. R. Ortiz de Montellano and L. Grab: Biochemistry 26, 5310–5314 (1987).

69   B. Meunier: Biochemie 69, 3–9 (1987).

70   J. M. S. vanMaanen, C. DeRuiter, P. Kootstra, J. DeVries and H. Pinedo: Free Radical Res. Commun. 1, 263–272 (1986).

71   B. K. Sinha, M. Trush and B. Kalyanaraman: Biochem. Pharmacol. 34, 2036–2040 (1985).

72   N. Haim, J. Nemec, J. Roman and B. Sinha: Biochem. Pharmacol. 36, 537–536 (1987).

73   J. M. S. van Maanen, J. deVries, D. Pappie, E. van den Akker, V. Lafleur, J. Retel, J. van der Greef and H. Pinedo: Cancer Res. 47, 4658–4662 (1987).

74   B. K. Sinha and C. Myers: Biochem. Pharmacol. 33, 3725–3728 (1984).

75   B. A. Teicher, S. A. Holden and C. M. Rose: J. Nat. Cancer Inst. 75, 1129–1133 (1985).

76   M. Broggini, C. Rossi, E. Benfenati, M. D'Incalci, R. Fanelli and P. Gariboldi: Chem.-Biol. Interact. 55, 215–224 (1985).

77   N. Haim, J. Nemec, J. Roman and B. Sinha: Cancer Res. 47, 5835–5840 (1987).

78   G. J. Atwell, G. Rewcastle, B. Baguley and W. Denny: J. Med. Chem. 39, 664–669 (1987).

79   G. Rewcastle, B. Baguley, G. Atwell and W. Denny: J. Med. Chem. 30, 1576–1581 (1987).

80   J. L. Jurlina, A. Lindsay, J. Packer, B. Baguley and W. Denny: J. Med. Chem. 30, 473–480 (1987).

81   D. D. Shoemaker, R. Cysyk, P. Gormley, J. DeSouza and L. Malspeis: Cancer Res. 44, 1939–1945 (1984).

82   A. Wong, C. Huang and S. Crooke: Biochemistry 23, 2939–2945 (1984); 2946–2952 (1984).

83   A. Wong, H. Cheng and S. Crooke: Biochem. Pharmacol. 35, 1071–1078 (1986).

84 V. K. Kansal and P. Potier: Tetrahedron 42, 2389–2408 (1986).

85 C. Auclair: Arch. Biochem. Biophys. 259, 1–14 (1987).

86 C. Malvy and S. Cros: Biochem. Pharmacol. 35, 2264–2267 (1986).

87 L. Larue, M. Quesne and J. Paoletti: Biochem. Pharmacol. 36, 3563–3569 (1987).

88 C. Auclair, A. Pierre, G. Voison, O. Pepin, S. Cros, C. Colas, J. Saucier, B. Verschvere, P. Gros and C. Paoletti: Cancer Res. 47, 6254–6261 (1987).

89 P. Leon, C. Garbay-Jaureguiberry, M. Barsi, J. LePecq and B. Roques: J. Med. Chem. 30, 2074–2080 (1987).

90 Y. Pommier, J. Minford, R. Schwartz, L. Zwelling and K. Kohn: Biochemistry 24, 6410–6416 (1985).

91 M. J. Vilarem, J. Riou, E. Multon, M. Gras and C. Larsen: Biochem. Pharmacol. 35, 2087–2095 (1986).

92 J. Markovits, Y. Pommier, M. Mattern, C. Esnault, B. Roques, J. LePecq and K. Kohn: Cancer Res. 46, 5821–5826 (1986).

93 M. Maftouh, Y. Amiar and C. Picard-Fraire: Biochem. Pharmacol. 34, 427–428 (1985).

94 B. Dugue, C. Auclair and B. Meunier: Cancer Res. 46, 3828–3833 (1986).

95 C. Auclair, B. Dugue, B. Meunier and C. Paoletti: Biochemistry 25, 1240–1245 (1986).

96 V. K. Kansal, S. Funakoshi, P. Mangeney, B. Gillet, E. Guittet, J. Lallemand and P. Potier: Tetrahedron 41, 5107–5120 (1985).

97 G. Meunier, J. Bernadou and B. Meunier: Biochem. Pharmacol. 36, 2599–2604 (1987).

98 G. Pratviel, J. Bernadou, T. Ha, G. Meunier, S. Cros, B. Meunier, B. Gillet and E. Guittet: J. Med. Chem. 29, 1350–1355 (1986).

99 B. S. Glisson, S. Smallwood and W. Ross: Biochem. Biophys. Acta 783, 74–79 (1984).

100 B. R. J. Abdella and J. Fisher: EHP, Environ. Health Perspect. 64, 3–18 (1985).

101 E. Goormaghtigh and J. Ruysschaert: Biochim. Biophys. Acta 779, 271–288 (1984).

102 T. G. Burke and R. R. Tritton: Biochemistry 24, 5972–5980 (1985).

103 J. R. Brown and S. H. Imam: Prog. Med. Chem. 21, 169–236 (1984).

104 F. Arcamone: Cancer Res. 45, 5995–5999 (1985).

105 R. B. Weiss, G. Sarosy, K. Clagett-Carr, M. Russo and B. Leyland-Jones: Cancer Chemother. Pharmacol. 18, 185–187 (1986).

106 A. M. Casazza: Cancer Treatment Rep. 70, 43–49 (1986).

107 J. W. Lown, S. Sondhi, S. Yen, J. Plambeck, J. Peters, E. M. Acton and G. Gordon: Drugs Exp. Clin. Res. 10, 735–744 (1984).

108 E. M. Acton: Drugs Exp. Clin. Res. 11, 1–8 (1985).

109 J. H. Peters, G. Gordon, D. Kashiwase and E. Acton: Cancer Res. 44, 1453–1459 (1987).

110 J. W. Lown and S. Sondhi: J. Org. Chem. 50, 1413–1418 (1985).

111 B. Barbieri, F. Giuliani, T. Bordoni, A. M. Casazza, C. Geroni, O. Bellini, A. Suarato, B. Gioia and F. Arcamone: Cancer Res. 47, 4001–4006 (1987).

112 F. Formelli, R. Carsana and C. Pollini: Cancer Res. 47, 5401–5406 (1987).

113 B. I. Sikic, M. Ehsan, W. Harker, N. Friend, B. Brown, R. Newman, M. Hacker and E. Acton: Science 228, 1544–1546 (1985).

114 F. Zunino, A. Casazza, G. Pratesi, F. Formelli and A. DiMarco: Biochem. Pharmacol. 30, 1856–1858 (1981).

115 F. Arcamone: Med. Chemistry 16, 1–41 (1980).

116 G. Capranico, C. Soranzo and F. Zunino: Cancer Res. 46, 5499–5503 (1986).

117 F. Arcamone: Med. Res. Rev. 4, 153–188 (1984).

118 H. Fritzsche, U. Wahnert, J. Chaires, N. Dattagupta, F. Schlessinger and

D. Crothers: Biochemistry 26, 1996–2000 (1987).

119   M. Israel, J. Idriss, Y. Koseki and V. Khetarpal: Cancer Chemother. Pharmacol. 20, 277–284 (1987).

120   A. H.-J. Wang, G. Ughetto, G. Quigley and A. Rich: Biochemistry 26, 1152–1163 (1987).

121   M. W. Van Dyke, R. Hertzberg and P. Dervan: Proc. Nat. Acad. Sci. U.S.A. 79, 5470–5474 (1982).

122   K. R. Fox and M. Waring: Nucleic Acids Res. 14, 2001–2014 (1986).

123   J. B. Chaires, N. Dattagupta and D. Crothers: Biochemistry 21, 3933–3940 (1982).

124   M. Robbie and R. Wilkins: Chem.-Biol. Interact. 49, 189–207 (1984).

125   M. B. Jones, U. Hollstein and F. Allen: Biopolymers 26, 121–135 (1987).

126   J. B. Chaires, K. Fox, J. Herrera, M. Britt and M. Waring: Biochemistry 26, 8227–8236 (1987).

127   K.-X. Chen, N. Gresh and B. Pullman: J. Biomol. Struct. Dyn. 3, 445–466 (1985).

128   J. B. Chaires: Biochemistry 25, 8436–8439 (1986).

129   J. B. Chaires: J. Biol. Chem. 261, 8899–8907 (1986).

130   C. K. Panda, K. Choudhury and R. Neogy: Chem.-Biol. Interact. 57, 65–72 (1987).

131   W. Forster and E. Stutter: Int. J. Biol. Macromol. 6, 114–124 (1984).

132   B. M. Gandecha, J. R. Brown and M. Crampton: Biochem. Pharmacol. 34, 733–736 (1985).

133   K. R. Fox, C. Brassett and M. Waring: Biochim. Biophys. Acta 840, 383–392 (1985).

134   J. B. Chaires, N. Dattagupta and D. Crothers: Biochemistry 24, 260–267 (1985).

135   C. R. Krishnamoorthy, S. Yen, J. Smith, J.W. Lown and W. D. Wilson: Biochemistry 25, 5933–5940 (1986).

136   D. A. Collier, S. Neidle and J. Brown: Biochem. Pharmacol. 33, 2877–2880 (1984).

137   K. Nicolay, J. Fok, W. Voorhout, J. Post and B. de Kruijff: Biochim. Biophys. Acta 887, 35–41 (1986).

138   H. Proumb and I. Petrescu: Prog. Biophys. Mol. Biol. 48, 103–125 (1986).

139   J. W. Shay and H. Webin: Mutation Res. 186, 149–160 (1987).

140   C. N. Ellis, M. Ellis and W. Blakemore: Biochem. J. 245, 309–312 (1987).

141   L. O. Lim and A. Neims: Biochem. Pharmacol. 36, 2769–2774 (1987).

142   K. J. A. Davies and J. Doroshow: J. Biol. Chem. 261, 3060–3067 (1986).

143   J. H. Doroshow and K. Davies: J. Biol. Chem. 261, 3068–3074 (1986).

144   C. Myers, L. Gianni, J. Zweier, J. Muindi, B. Sinha and H. Eliot: Fed. Proc. 45, 2792–2797 (1986).

145   T. G. Burke, C. Pritsos, A. Sartorelli and T. Tritton: Cancer Biochem. Biophys. 9, 245–255 (1987).

146   K. Nicolay, R. Van der Neut, J. Fok and B. deKruijff: Biochim. Biophys. Acta 819, 55–65 (1985).

147   D. C. Straney and D. Crothers: Biochemistry 26, 1987–1995 (1987).

148   H. Fritzsche, U. Wahnert, J. Chaires, N. Dattagupta, F. Schlessinger and D. Crothers: Biochemistry 26, 1996–2000 (1987).

149   D. Sen and D. Crothers: Biochemistry 25, 1503–1509 (1986).

150   G. J. Goldenberg, H. Wang and G. Blair: Cancer Res. 46, 2978–2983 (1986).

151   G. Capranico, A. Riva, S. Tinelli, T. Dasdia and F. Zunino: Cancer Res. 47, 3752–3756 (1987).

152   M. Israel, J. Idriss, Y. Koseki and V. Khetarpal: Cancer Chemother. Pharmacol. 20, 277–284 (1987).

153   L. F. Pearlman, R. Chuang, M. Israel and H. Simpkins: Cancer Res. 46, 341–346 (1986).

154 R. Y. Chuang, L. F. Chuang and M. Israel: Biochem. Pharmacol. 35, 1293–1297 (1986).

155 D. P. Evans, R. Meyn and S. Tomasovic: Cancer Chemother. Pharmacol. 18, 137–139 (1987).

156 G. Capranico, C. Sorazo and F. Zunino: Cancer Res. 46, 5499–5503 (1986).

157 S. Spadari, G. Pedrali-Noy, F. Focher, A. Montecucco, T. Bordoni, C. Geroni, F. Giuliani, G. Ventrella, F. Arcamone and G. Ciarrocchi: Anticancer Res. 6, 935–940 (1986).

158 R. L. Blankespoor, D. Schutt, M. Tubergen and R. De Jong: J. Org. Chem. 52, 2059–2064 (1987).

159 G. M. Cohen and M. d'Arcy Doherty: Brit. J. Cancer 55, (Suppl. VIII), 40–52 (1987).

160 B. Halliwell and J. Gutteridge: Biochem. J. 219, 1–14 (1984).

161 H. Nohl and W. Jordan: Bioorg. Chem. 15, 374–382 (1987).

162 N. R. Bachur, M. Gee and R. Friedman: Cancer Res. 42, 1078–1081 (1982).

163 E. G. Mimnaugh, K. Kennedy, M. Trush and B. Sinha: Cancer Res. 45, 3296–3304 (1985).

164 M. P. Goren, N. Ahmed and A. Tereba: Cancer Res. 47, 1924–1929 (1987).

165 R. Wallin: Cancer Lett. 30, 97–101 (1986).

166 J. Fisher, B. Abdella and K. McLane: Biochemistry 24, 3562–3571 (1985).

167 S. S. Pan, L. Pederson and N. Bachur: Mol. Pharmacol. 19, 184–186 (1981).

168 T. Komiyama, T. Kikuchi and Y. Sugiura: J. Pharmacobio. Dyn. 9, 651–664 (1986).

169 R. E. Kilkuskie, T. Macdonald and S. Hecht: Biochemistry 23, 6165–6171 (1984).

170 C. Meyers, J. Muindi, G. Batist, N. Haim and B. Sinha: Cancer Chemother. 7, 57–75 (1985).

171 H. Beraldo, A. Garnier-Suillerot, L. Tosi and F. Lavelle: Biochemistry 24, 284–289 (1985).

172 E. J. F. Demant: FEBS Lett. 176, 97–100 (1984).

173 C. E. Thomas and S. Aust: Arch. Biochem. Biophys. 248, 684–689 (1986).

174 N. Norskow-Lauritsen, E. Demant and P. Ebberson: Biochem. Pharmacol. 36, 2685–2686 (1987).

175 H. Eliot, L. Gianni and C. Myers: Biochemistry 23, 928–936 (1984).

176 J. Muindi, B. Sinha, L. Gianni and C. Myers: Mol. Pharmacol. 27, 356–365 (1985).

177 L. Gianni, J. Zweier, A. Levy and C. Myers: J. Biol. Chem. 260, 6820–6826 (1985).

178 E. Mimnaugh, M. Trush and T. Gram: Biochem. Pharmacol. 35, 4327–4335 (1986).

179 J. M. C. Gutteridge and G. Quinlan: Biochem. Pharmacol. 34, 4099–4103 (1985).

180 B. K. Sinha, M. Trush, K. Kennedy and E. Mimnaugh: Cancer Res. 44, 2892–2896 (1984).

181 K. B. Wallace and J. Johnson: Mol. Pharmacol. 31, 307–311 (1987).

182 J. H. Peters, G. Gordon, D. Kashiwase, J. W. Lown, S. Yen and J. Plambeck: Biochem. Pharmacol. 35, 1309–1323 (1986).

183 W. J. M. Hrushesky, R. Olshefski, P. Wood, S. Meshmick and J. Eaton: Lancet i, 585 (1985).

184 J. Butler and B. Hoey: Br. J. Cancer 55, 53–59 (1987).

184a B. M. Hoey, J. Butler and A. Swallow: Biochemistry 27, 2608–2614 (1988).

185 J. H. Doroshow: Proc. Nat. Acad. Sci. U.S.A. 83, 4514–4518 (1986).

186 B. K. Sinha, A. Katki, G. Batist, K. Cowan and C. Myers: Biochemistry 26, 3776–3781 (1987).

187 D. A. Gewirtz and S. Yanovich: Biochem. Pharmacol. 36, 1798–1798 (1987).

188 H. W. Moore, R. Czerniak and A. Hamdan: Drugs Exper. Clin. Res. 12, 475–494 (1986).
189 T.-S. Lin, L. Wang, I. Antonini, L. Cosby, D. Shiba, D. Kirkpatrick and Sartorelli: J. Med. Chem. 29, 84–89 (1986).
190 I. Wilson, P. Wardman, T.-S. Lin and A. Sartorelli: Chem.-Biol. Interact. 61, 229–240 (1987).
191 K. Ramakrishnan and J. Fisher: J. Med. Chem. 29, 1215–1221 (1986).
192 D. L. Kleyer and T. Koch: J. Am. Chem. Soc. 106, 2380–2387 (1984).
193 M. Boldt, G. Gaudiano and T. Koch: J. Org. Chem. 52, 2146–2153 (1987).
194 N. Willmott and J. Cummings: Biochem. Pharmacol. 36, 521–526 (1987).
195 B. Pullman: Int. J. Quantum. Chem. 13, 95–105 (1986).
196 B. M. Kacinski and W. Rupp: Cancer Res. 44, 3489–3492 (1984).
197 B. Lambert, P. Laugaa, B. Roques and J. LePecq: Mut. Res. 166, 243–254 (1986).
198 E. M. Acton, G. Tong, C. Mosher and R. Wolgemuth: J. Med. Chem. 27, 638–645 (1984).
199 K. Wassermann, L. Zwelling, T. Mullins, L. Silberman, B. Anderson, M. Bakic, E. Acton and R. Newman: Cancer Res. 46, 4041–4046 (1986).
200 J. B. Johnston, L. Pugh and A. Begleiter: Cancer Res. 47, 4076–4080 (1987).
201 M. I. Jesson, J. Johnston, C. Anhalt and A. Begleiter: Cancer Res. 47, 5935–5938 (1987).
202 J. Westendorf, G. Groth, G. Steinheider and H. Marquardt: Cell Biol. Toxicol. 1, 87–101 (1985).
203 J. Westendorf, M. Aydin, G. Groth and H. Marquardt: Arch. Pharmacol. 335, 27 (1987).
203a J. H. Peters, G. R. Gordon, H. Nolen, M. Tracy and D. Thomas: Biochem. Pharmacol. 37, 357–360 (1988).
204 R. W. Franck: Prog. Chem. Org. Nat. Prod. 38, 1–45 (1978).
205 W. A. Remers: The Chemistry of Antitumor Antibiotics, pp. 221–276, Wiley, New York 1979.
206 W. A. Remers: Anticancer Agents Based on Natural Product Models, pp. 131–145. Ed. J. Douros, Academic, New York 1980.
207 J. H. Beijnen, H. Lingeman, H. VanMunster and W. Underberg: J. Pharm. Biomed. Anal. 4, 275–295 (1986).
208 U. Hornemann: Antibiotics (N. Y.) 4, 295–312 (1981).
209 K. Shirahata and N. Hirayama: J. Am. Chem. Soc. 105, 7199–7200 (1983).
210 G. Verdine and K. Nakanishi: J. Chem. Soc., Chem. Commun., 1093–1095 (1985).
211 U. Hornemann and M. J. Heins: J. Org. Chem. 50, 1301–1302 (1985).
212 M. Tomasz and R. Lipman: J. Am. Chem. Soc. 101, 6063–6067 (1979).
213 J. Rebek, S. Shaber, Y. K. Shue, J. C. Gehret and S. Zimmerman: J. Org. Chem. 49, 5164–5174 (1984).
214 B. S. Iyengar and W. A. Remers: J. Med. Chem. 28, 963–967 (1985).
215 U. Hornemann, P. Keller and K. Takeda: J. Med. Chem. 28, 31–36 (1985).
216 R. A. McClelland and K. Lam: J. Am. Chem. Soc. 107, 5182–5186 (1985).
217 U. Hornemann, K. Iguchi, P. Keller, H. Vu, J. Kowalski and H. Kohn: J. Org. Chem. 48, 5026–5033 (1982).
218 G. L. Verdine, B. F. McGuinness, K. Nakanishi and M. Tomasz: Heterocycles 25, 577–587 (1987).
219 M. Tomasz and R. Lipman: J. Am. Chem. Soc. 101, 6063–6067 (1979).
220 M. Tomsasz, R. Lipman, M. Lee, G. Verdine and K. Nakanishi: Biochemistry 26, 2010–2027 (1987).
221 H. O. Douglas, P. Lavin, A. Goudsmit, D. Klassen and A. Paul: J. Clin. Oncol. 2, 1372–1377 (1984).
222 K. A. Kennedy, J. McGurl, L. Leondaridis and O. Alabaster: Cancer Res. 45, 3541–3547 (1985).

223 S. R. Keyes, P. Fracasso, D. Heimbrook, S. Rockwell, S. Sligar and A. Sartorelli: Cancer Res. 44, 5638–5643 (1984).

224 C. A. Pritsos and A. Sartorelli: Cancer Res. 46, 3528–3532 (1986).

225 P. M. Fracasso and A. Sartorelli: Cancer Res. 46, 3939–3944 (1986).

226 H. Nakano, K. Sugioka, M. Nakano, M. Mizokami, H. Kimura, S. Tero-Kubota and Y. Ikegami: Biochem. Biophys. Acta 796, 285–293 (1984).

227 K. Hamana, K. Kawada, K. Sugioka, M. Nakano, S. Tero-Kubota and Y. Ikegami: Biochem. Int. 10, 301–309 (1985).

228 D. L. Kimpel and A. Sagone: Biochem. Pharmacol. 33, 3479–3484 (1984).

229 R. T. Dorr, T. Bowden, D. Alberts and J. Liddil: Cancer Res. 45, 3510–3516 (1985).

230 A. C. Sartorelli: Biochem. Pharmacol. 35, 67–69 (1986).

230a A. C. Sartorelli: Cancer Res. 48, 775–778 (1988).

231 K. A. Kennedy: Anti-Cancer Drug Design 2, 181–194 (1987).

232 P. A. Andrews, S.-S. Pan and N. Bachur: J. Am. Chem. Soc. 108, 4158–4166 (1986).

233 H. Kohn, N. Zein, X. Lin, J.-Q. Ding and K. M. Kadish: J. Am. Chem. Soc. 109, 1833–1840 (1987).

234 G. M. Rao, A. Begleiter, J. W. Lown and J. A. Plambeck: J. Electrochem. Soc. 124, 199–202 (1977).

235 S. Danishefsky and M. Ciufolini: J. Am. Chem. Soc. 106, 6424–6425 (1984).

236 S. Danishefsky and M. Egbertson: J. Am. Chem. Soc. 108, 4648–4650 (1986).

237 M. Egbertson and S. Danishefsky: J. Am. Chem. Soc. 109, 2204–2205 (1987).

238 M. Egbertson, S. Danishefsky and G. Schulte: J. Org. Chem. 52, 4424–4426 (1987).

239 M. Tomasz and R. Lipman: Biochemistry 20, 5056–5061 (1981).

240 N. Zein and H. Kohn: J. Am. Chem. Soc. 108, 296–297 (1986).

241 M. Bean and H. Kohn: J. Org. Chem. 48, 5033–5041 (1983).

242 M. Bean and H. Kohn: J. Org. Chem. 50, 293–298 (1985).

243 M. Tomasz, R. Lipman, G. Verdine and K. Nakanishi: Biochemistry 25, 4337–4344 (1986).

244 M. Tomasz, D. Chowdary, R. Lipman, S. Shimotakahara, D. Verio, V. Walker and G. Verdine: Proc. Nat. Acad. Sci. U.S.A. 83, 6702–6706 (1986).

245 Y. Hashimoto and K. Shudo: EHP, Environ. Health Perspect. 62, 219–222 (1985).

246 N. Zein and H. Kohn: J. Am. Chem. Cos. 109, 1576–1577 (1987).

247 S.-S. Pan, T. Tracki and N. Bachur: Mol. Pharmacol. 29, 622–628 (1986).

248 D. K. Kaplan and M. Tomasz: Biochemistry 21, 3006–3013 (1982).

249 S. N. Rao, U. C. Singh and P. Kollman: J. Am. Chem. Soc. 108, 2058–2068 (1986).

250 M. Tomasz, R. Lipman, D. Chowdary, J. Pawlak, G. Verdine and K. Nakanishi: Science 235, 1204–1208 (1987).

250a M. Tomasz, A. Chawla and R. Lipman: Biochemistry 27, 3182–3187 (1988).

251 D. M. Peterson and J. Fisher: Biochemistry 25, 4077–4084 (1986).

252 N. Zein and H. Kohn: J. Am. Chem. Soc. 108, 296–297 (1986).

253 S. R. Keyes, S. Rockwell and A. Sartorelli: Cancer Res. 47, 5654–5657 (1987).

254 K. A. Kennedy, E. Mimnaugh, M. Trush and B. Sinha: Cancer Res. 45, 4071–4076 (1985).

255 K. A. Kennedy, S. Sligar, L. Polonski and A. Sartorelli: Biochem. Pharmacol. 31, 2011–2022 (1982).

256 S. Chakrabarty, Y. Daniels, B. Long, J. Willson and M. Brattain: Cancer Res. 46, 3456–3458 (1986).

257 R. S. Marshall and A. Rauth: Cancer Res. 46, 2709–2713 (1986).

258 R. T. Dorr, T. Bowden, D. Alberts and J. Liddil: Cancer Res. 45, 3510–3516 (1985).
259 S. R. Keyes, S. Rockwell and A. Sartorelli: Cancer Res. 45, 3642–3645 (1985).
260 V. N. Iyer and W. Szybalski: Proc. Nat. Acad. Sci. U.S.A. 50, 355–362 (1963).
261 W. A. Remers, S. Rao, U. Singh and P. Kollman: J. Med. Chem. 29, 1256–1263 (1986).
262 E. A. Oostveen and W. N. Speckamp: Tetrahedron 43, 255–262 (1987).
263 J. C. Hodges, W. A. Remers and W. Bradner: J. Med. Chem. 24, 1184–1191 (1981).
264 B. S. Iyengar, W. A. Remers and W. Bradner: J. Med. Chem. 29, 1864–1868 (1986).
265 H. Umezawa: In: Anticancer Agents Based on Natural Product Models (J. Cassady and J. Douros, Eds.), Academic. pp. 148–166 (1980).
266 H. Umezawa: Rev. Infect. Diseases 9, 147–164 (1987).
267 S. M. Hecht: Fed. Proc. 45, 2784–2791 (1986).
268 S. M. Hecht: Acc. Chem. Res. 19, 383–392 (1986).
269 J. Stubbe and J. Kozarich: Chem. Rev. 87, 1107–1136 (1987).
270 L. F. Povirk: Biochemistry 18, 3989–3995 (1979).
271 H. Sugiura, R. Kilkuskie, L. Chang, L. Ma, S. Hecht, G. van der Marel and J. VanBoom: J. Am. Chem. Soc. 108, 3852–3854 (1986).
272 M. R. Cirolo, R. Magliozzo and J. Peisach: J. Biol. Chem. 262, 6290–6295 (1987).
273 I. Mahmutoglu, M. Scheulen and H. Kappus: Arch. Toxicol. 60, 150–153 (1987).
274 D. Melnyk, S. Horwitz and J. Peisach: Biochemistry 20, 5327–5331 (1981).
275 D. Melnyk, S. Horwitz and J. Peisach: Inorg. Chim. Acta 138, 75–78 (1987).
276 I. Mahmutoglu and H. Kappus: Biochem. Pharmacol. 36, 3677–3681 (1987).
277 C. Nishimura, N. Tanaka, H. Suzuki and N. Tanaka: Biochemistry 26, 1574–1578 (1987).
278 S. M. Sebti, J. DeLeon and J. Lazo: Biochemistry 26, 4213–4219 (1987).
279 H. Umezawa: In: Anticancer Agents Based on Natural Products (J. Cassady and J. Douros, Eds.), Academic Press, pp. 147–166 (1980).
280 W. J. Vloon, C. Kruk, U. Pandit, H. Hofs and J. McVie: J. Med. Chem. 30, 20–24 (1987).
281 L. H. DeReimer, C. Meares, D. Goodwin and C. Diamanti: J. Med. Chem. 22, 1019–1023 (1979).
282 H. Sugiyama, R. Kilkuskie, L. Chang, L. Ma, S. Hecht, G. van der Marel and J. VanBoom: J. Am. Chem. Soc. 103, 3852–3854 (1986).
283 T. Suzuki, J. Kuwahar and Y. Sugiura: Biochemistry 24, 4719–4721 (1985).
284 K. Takahashi, T. Takita and H. Umezawa: J. Antibiot. 40, 348–353 (1987).
285 K. Takahaski, T. Takita and H. Umezawa: J. Antibiot. 40, 542–546 (1987).
286 G. M. Ehrenfeld, L. Rodriquez, S. Hecht, C. Chang, V. Basus and N. Oppenheimer: Biochemistry 24, 81–92 (1985).
287 G. M. Ehrenfeld, J. Shipley, D. Heimbrook, H. Sugiyama, E. Long, J. van Boom, G. van der Marel, N. Oppenheimer and S. Hecht: Biochemistry 26, 931–942 (1987).
288 N. J. Oppenheimer, C. Chang, L. Chang, G. Ehrenfeld, L. Rodriquez and S. Hecht: J. Biol. Chem. 257, 1606–1609 (1982).
289 S. Goldstein and G. Czapski: Free Radical Res. Commun. 2, 259–270 (1987).
290 H. Asakuka, H. Okubu and M. Hori: J. Antibiot. 38, 244–252 (1985).
291 H. Ekimoto, K. Takahashi, A. Matsuda, T. Takita and H. Umezawa: J. Antibiotics 38, 1077–1082 (1985).

292 E. A. Sausville, J. Peisach and S. Horwitz: Biochemistry 17, 2740–2746 (1978).
293 R. M. Burger, J. Peisach and S. Horwitz: J. Biol. Chem. 256, 11636–11644 (1981).
294 R. M. Burger, T. A. Kent, S. Horwitz, E. Munch and J. Peisach: J. Biol. Chem. 258, 1559–1564 (1983).
295 G. Padbury and S. Sligar: J. Biol. Chem. 260, 7820–7823 (1985).
296 N. Murugesan and S. Hecht: J. Am. Chem. Soc. 107, 493–500 (1985).
297 D. C. Heimbrook, R. Mulholland and S. Hecht: J. Am. Chem. Soc. 108, 7839–7840 (1986).
298 L. E. Rabow, J. Stubbe, J. Kozarich and J. Gerlt: J. Am. Chem. Soc. 108, 7130–7131 (1986).
299 R. M. Burger, J. Blanchard, S. Horwitz and J. Peisach: J. Biol. Chem. 260, 15406–15409 (1985).
300 L. Giloni, M. Takeshita, F. Johnson, C. Iden and A. Grollman: J. Biol. Chem. 256, 8608–8615 (1981).
301 J. C. Wu, J. Kozarich and J. Stubbe: J. Biol. Chem. 258, 4694–4697 (1983).
302 J. C. Wu, J. Kozarich and J. Stubbe: Biochemistry 24, 7562–7568 (1985).
303 G. H. McGall, L. Rabow, J. Stubbe and J. Kozarich: J. Am. Chem. Soc. 109, 2836–2837 (1987).
304 R. Burger, J. Peisach and S. Horwitz: J. Biol. Chem. 257, 8612–8614 (1982).
305 S. Vesugi, T. Shida, M. Ikehara, Y. Kobayashi and Y. Kyogoku: Nucleic Acids Res. 12, 1581–1592 (1984).
306 N. Murugeson, C. Xu, G. Ehrenfeld, H. Sugiyama, R. Kilkuskie, L. Rodriquez, L. Chang and S. Hecht: Biochemistry 24, 5735–5744 (1985).
307 L. Rabow, J. Stubbe, J. Kozarich and J. Gerlt: J. Am. Chem. Soc. 108, 7130–7131 (1986).
308 H. Sugiyama, C. Xu, N. Murugesan, S. Hecht, G. van der Marel and J. vanBoom: Biochemistry 27, 58–67 (1988).
309 I. Saito, T. Morii and T. Matsuura: J. Org. Chem. 52, 1008–1012 (1986).
310 T. J. Keller and N. Oppenheimer: J. Biol. Chem. 262, 15144–15150 (1987).
311 L. Strekowski, A. Strekowska, R. Watson, F. Tanious, L. Nguyen and W. Wilson: J. Med. Chem. 30, 1415–1420 (1987).
312 K. Ueda, S. Kobayashi and T. Komano: Nucleic Acids Res. Symp. Ser. (13th) 16, 197–200 (1985).
313 C. R. Krishnamoorthy, D. Vanderwall, J. Kozarich and J. Stubbe: J. Am. Chem. Soc. 110, 1008–1009 (1988).
314 J. P. Henichart, J. Bernier, N. Helbecque and R. Houssain: Nucleic Acids Res. 13, 6703–6717 (1985).
315 H. Sugiyama, R. Kilkuskie, S. Hecht, G. van der Marel and J. vanBoom: J. Am. Chem. Soc. 107, 7765–7767 (1985).
316 C. K. Mirabelli, A. Ting, C. Huang, S. Mong and S. Crooke: Cancer Res. 42, 2779–2785 (1982).
317 V. Murray and R. Martin: Nucleic Acids Res. 13, 1467–1481 (1985).
318 V. Murray and R. Martin: J. Biol. Chem. 260, 10389–10391 (1985).
319 K. R. Fox, G. Grigg and M. Waring: Biochem. J. 243, 847–851 (1987).
320 R. P. Beckmann, M. Agostino, M. McHugh, R. Sigmund and T. Beerman: Biochemistry 26, 5409–5415 (1987).
321 R. Hertzberg, M. Caranfa and S. Hecht: Biochemistry 24, 5285–5289 (1985).
321a R. Hertzberg, M. Caranfa and S. Hecht: Biochemistry 27, 3164–3174 (1988).
322 V. A. Aivasashvilli and R. Beabealasvilli: FEBS Lett. 160, 124–128 (1983).
323 J. J. Duffy and T. Lindell: Biochem. Pharmacol. 34, 1854–1856 (1985).
324 D. R. Phillips and D. Crothers: Biochemistry 25, 7355–7362 (1986).
325 E. V. Scott, R. Jones, D. Banville, G. Zon, L. Marzilli and W. Wilson: Biochemistry 27, 915–923 (1988).
326 F.-M. Chen: Biochemistry 27, 1843–1848 (1988).

327 M. J. Waring: *In:* The Molecular Basis of Antibiotic Action (F. Gales *et al.,* Eds.), pp. 314–333, Wiley (1981).
328 H. M. Sobell: Prog. Nucleic Acid Res. Mol. Biol. 13, 153–190 (1973).
329 F. Takusagawa, B. Goldstein, S. Youngster, R. Jones and H. Berman: J. Biol. Chem. 259, 4714–4715 (1984).
330 F. Takusagawa: J. Antibiot. 38, 1596–1604 (1985).
331 T. R. Krugh and M. Nuss: *In:* Biological Applications of Magnetic Resonance (R. Schulman, Ed.) Chapter 3, Academic (1979).
332 D. G. Reid, S. Salisburg and D. Williams: Biochemistry 22, 1377–1385 (1983).
333 S. C. Brown, K. Mullis, C. Levenson and R. Shafer: Biochemistry 23, 403–408 (1984).
334 T. P. Lybrand, S. Brown, S. Creighton, R. Shafer and P. Kollman: J. Mol. Biol. 191, 495–507 (1986).
335 S. C. Brown and R. Shafer: Biochemistry 26, 277–282 (1987).
336 H. M. Sobell: Proc. Nat. Acad. Sci. U.S.A. 82, 5328–5331 (1985).
337 H. Simpkins and L. Pearlman: Biochim. Biophys. Acta 783, 293–300 (1984).
338 S. Cacchione, R. Canera and M. Savino: Biochim. Biophys. Acta 867, 229–233 (1986).
339 W. E. Ross, D. Glaubinger and K. Kohn: Biochim. Biophys. Acta 562, 41–50 (1979).
340 J. M. K. Fox, T. Byrne and W. Woods: Biochem. Pharmacol. 34, 2741–2747 (1985).
341 D. K. Trask and M. Muller: Proc. Nat. Acad. Sci. U.S.A. 85, 1417–1421 (1988).
342 R. K. Sehgal, S. Sengupta, D. Waxman and A. Tauber: Anti-Cancer Drug Des. 1, 13–25 (1985).
343 R. K. Sehgal, D. Greenwood and S. Sengupta: Bioorg. Chem. 13, 110–120 (1985).
344 S. Sengupta, C. Kelly and R. K. Sehgal: J. Med. Chem. 28, 620–628 (1985).
345 H. Nakasawa, N. Bachur, F. Chou, M. Mossoba and P. Guitierrez: Biophys. Chem. 21, 137–143 (1985).
346 B. J. Forster, K. Clagett-Carr, D. Shoemaker, M. Suffness, J. Plowman, L. Trissel, C. Greishaber and B. Leyland-Jones: Invest. New Drugs 3, 403–410 (1986).
347 X. Gao and D. Patel: Biochemistry 27, 1744–1751 (1988).
348 L. J. Hanka, A. Dietz, S. A. Gerpheide, S. L. Kuentzel and D. G. Martin: J. Antibiot. 31, 1211 (1978).
349 D. G. Martin, C. Biles, S. A. Gerpheide, L. J. Hanka, W. C. Krueger, J. P. McGovren, S. A. Mizsak, G. L. Neil, J. C. Stewart and J. Visser: J. Antibiot. 34, 1119 (1981).
350 D. G. Martin, C. G. Chidester, D. J. DuChamp and S. A. Mizsak: J. Antibiot. 33, 902 (1980).
351 C. G. Chidester, W. C. Krueger, S. A. Mizsak, D. J. DuChamp and D. G. Martin: J. Am. Chem. Soc. 103, 7629 (1981).
352 D. E. Nettleton, Jr., J. A. Bush and W. T. Bradner: U.S. Patent 4,301,248 (1981).
353 V. I. Frolova, A. D. Kuzovkov, A. I. Chernyshev, M. M. Taig, L. P. Ivanitskaya, T. G. Terentyeva and I. P. Fomina: Antibiotiki 27, 483 (1982).
354 V. H. Rawal, R. J. Jones and M. P. Cava: Heterocycles 25, 701 (1987).
355 V. L. Reynolds, J. P. McGovren and L. H. Hurley: J. Antibiot. 39, 319 (1986).
356 L. H. Hurley and D. R. Needham-VanDevanter: Acc. Chem. Res. 19, 230 (1986).
357 L. H. Li, D. H. Swenson, S. L. F. Schpok, S. L. Kuentzel, B. D. Dayton and W. C. Krueger: Cancer Res. 42, 999 (1982).

358 B. K. Bhuyan, K. A. Newell, S. L. Crampton and D. D. Von Hoff: Cancer Res. 42, 3532 (1982).

359 P. R. Harbach, R. J. Trzos, J. H. Mazurek, D. M. Zimmer, G. L. Petzold and B. K. Bhuyan: Mutagenesis 1, 407 (1986).

360 G. L. Neil, G. L. Clarke and J. P. McGovren: Proc. Am. Assoc Cancer Res. 22, 244 (1981).

361 J. P. McGovren, G. L. Clarke, E. A. Pratt and T. F. DeKoning: J. Antibiot. 37, 63 (1984).

362 D. H. Swenson, L. H. Li, L. H. Hurley, J. S. Rokem, G. L. Petzold, B. D. Dayton, T. L. Wallace, A. H. Lin and W. C. Krueger: Cancer Res. 42, 2821 (1982).

363 B. K. Bhuyan, S. L. Crampton and E. G. Adams: Cancer Res. 43, 4227 (1983).

364 W. C. Krueger, L. H. Li, A. Moscowitz, M. D. Prairie, G. Petzold and D. H. Swenson: Biopolymers 24, 1549 (1985).

365 W. C. Krueger, D. J. DuChamp, L. H. Li, A. Moscowitz, G. L. Petzold, M. D. Prairie and D. H. Swenson: Chem.-Biol. Interactions 59, 55 (1986).

366 L. H. Hurley, V. L. Reynolds, D. H. Swenson, G. L. Petzold and T. A. Scahill: Science 226, 843 (1984).

367 V. L. Reynolds, I. J. Molineux, D. J. Kaplan, D. H. Swenson and L. H. Hurley: Biochemistry 24, 6228 (1985).

368 M. A. Warpehoski, I. Gebhard, R. C. Kelly, W. C. Krueger, L. H. Li, J. P. McGovren, M. D. Prairie, N. Wicnienski and W. Wierenga: J. Med. Chem. 31, 590 (1988).

369 D. R. Needham-VanDevanter, L. H. Hurley, V. L. Reynolds, N. Y Theriault, W. C. Krueger and W. Wierenga: Nucleic Acids Res. 12, 6159 (1984).

370 N. Y. Theriault, W. C. Krueger and M. D. Prairie: Nucleosides and Nucleotides 6, 327 (1987).

371 T. A. Scahill, N. D. Brahme, W. C. Krueger, N. Y Theriault and W. Wierenga: Proc. Am. Assoc. Cancer Res. 27, 251 (1986).

372 D. G. Martin, R. C. Kelly, W. Watt, N. Wicnienski, S. A. Mizsak, J. W. Nielsen and M. D. Prairie: J. Org. Chem. 53, xxxx (1988).

373 K. Zakrzewska, M. Randrianarivelo and B. Pullman: Nucleics Acids Research 15. 5775 (1987).

374 M. L. Kopka, C. Yoon, D. Goodsell, P. Pjura and R. E. Dickerson: Proc. Nat. Acad. Sci. USA 82, 1376 (1985).

375 D. R. Needham-VanDevanter and L. H. Hurley: Biochemistry 25, 8430 (1986).

376 L. H. Hurley, D. R. Needham-VanDevanter and C.-S. Lee: Proc. Natl. Acad. Sci. USA 84, 6412 (1987).

377 M. K. Jacobson, D. Twehous and L. H. Hurley: Biochemistry 25, 5929 (1986).

378 W. Wierenga, B. K. Bhuyan, R. C. Kelly, W. C. Krueger, L. H. Li, J. P. McGovren, D. H. Swenson and M. A. Warpehoski: Advances in Enzyme Regulation, Vol. 25, pp. 141–151. E. G. Weber. Pergamon Press, Oxford, 1986.

379 M. A. Warpehoski, R. C. Kelly, J. P. McGovren and W. Wierenga: Recent Advances in Chemotherapy, pp. 570–571. Ed. J. Ishigami. University of Tokyo Press, Tokyo, 1985.

380 M. A. Warpehoski, R. C. Kelly, J. P. McGovren and W. Wierenga: Proc. Am. Assoc. Cancer Res. 26, 221 (1985).

381 L. H. Li, T. L. Wallace, T. F. DeKoning, M. A. Warpehoski, R. C. Kelly, M. D. Prairie and W. C. Krueger: Investigational New Drugs 5, 329 (1987).

382 G. L. Petzold, W. C. Krueger, D. H. Swenson, T. L. Wallace, M. D. Prairie and L. H. Li: Proc. Am. Assoc. Cancer Res. 26, 225 (1985).

383 M. A. Warpehoski: Tetrahedron Lett. 27, 4103 (1986).

384   E. G. Adams, G. J. Badiner and B. K. Bhuyan: Cancer Res., 48, 109 (1988).
385   M. A. Warpehoski, R. C. Kelly, W. C. Krueger, P. A. Aristoff and L. H. Li: Proc. Am. Assoc. Cancer Res. 27, 252 (1986).
385   T. F. DeKoning, W. C. Krueger, T. L. Wallace, M. D. Prairie, M. A. Warpehoski, R. C. Kelly and L. H. Li: Proc. Am. Assoc. Cancer Res. 28, 262 (1987).
387   R. C. Kelly, I. Gebhard, N. Wicnienski, P. A. Aristoff, P. D. Johnson and D. G. Martin: J. Amer. Chem. Soc. 109, 6837 (1987).
388   J. P. McGovren, C.-S. Lee, L. H. Hurley, L. H. Li and R. C. Kelly: Proc. Am. Assoc Cancer Res. 28, 262 (1987).
389   L. H. Hurley, C.-S. Lee, J. P. McGovren, M. A. Warpehoski, M. A. Mitchell, R. C. Kelly and P. A. Aristoff: Biochemistry 27, 3886 (1988).
390   W. C. Krueger and M. D. Prairie: Chem.-Biol. Interactions 62, 281 (1987).
391   M. A. Warpehoski and V. S. Bradford: Tetrahedron Lett. 29, 131 (1988).
392   D. L. Boger and R. S. Coleman: J. Org. Chem. 53, 695 (1988).
393   Y. Enomoto, Y. Furutani, H. Naganawa, M. Hamada, T. Takeuchi and H. Umezawa: Agric. Biol. Chem. 42, 1331 (1978).
394   M. A. Mitchell, D. H. Swenson, T. L. Wallace, M. G. Williams, G. L. Petzold, P. A. Aristoff, P. D. Johnson and L. H. Li: Proc. Am. Assoc. Cancer Res. 28, 270 (1987).
395   L. H. Hurley and R. Petrusek: Nature 282, 529 (1979).
396   L. H. Hurley: J. Antibiot. 30, 349 (1977).
397   K. W. Kohn, D. Glaubiger and C. L. Spears: Biochim. Biophys. Acta 361, 288 (1974).
398   L. H. Hurley, G. Gairola and M. Zmijewski: Biochim. Biophys. Acta 475, 521 (1977).
399   R. P. Hertzberg, S. M. Hecht, V. L. Reynolds, I. J. Molineux and L. H. Hurley: Biochemistry 25, 1249 (1986).
400   R. L. Petrusek, G. L. Anderson, T. F. Garner, Q. L. Fannin, D. J. Kaplan, S. G. Zimmer and L. H. Hurley: Biochemistry 20, 1111 (1981).
401   M. D. Barkley, S. Cheatham, D. E. Thurston and L. H. Hurley: Biochemistry 25, 3021 (1986).
402   D. E. Graves, C. Pattaroni, B. S. Krishnan, J. M. Ostrander, L. H. Hurley and T. R. Krugh: J. Biol. Chem. 259, 8202 (1984).
403   D. E. Graves, M. P. Stone and T. R. Krugh: Biochemistry 24, 7573 (1985).
404   W. A. Remers, M. Mabilia and A. J. Hopfinger: J. Med. Chem. 29, 2492 (1986).
405   K. Zakrzewska and B. Pullman: J. Biomolec. Struct. Dyn. 4, 127 (1986).
406   S. N. Rao, U. C. Singh and P. A. Kollman: J. Med. Chem. 29, 2484 (1986).
407   T. Arai and A. Kubo: The Alkaloids, Vol. 21, pp. 55–100. Academic Press, New York, 1983.
408   Y. Ikeda, H. Matsuki, T. Ogawa and T. Munakata: J. Antibiot. 36, 1284 (1983).
409   T. Hayashi, T. Okutomi, S. Suzuki and H. Okazaki: J. Antibiot. 36, 1228 (1983).
410   K. Kishi, K. Yazawa, K. Takahashi, Y. Mikami and T. Arai: J. Antibiot. 37, 847 (1984).
411   J. W. Lown, A. V. Joshua and J. S. Lee: Biochemistry 21, 419 (1982).
412   K. Ishiguro, K. Takahashi, K. Yazawa, S. Sakiyama and T. Arai: J. Biol. Chem. 256, 2162 (1981).
413   M. J. Zmijewski, K. Miller-Hatch and M. Goebel: Antimicrob. Agents Chemother. 21, 787 (1982).
414   M. J. Zmijewski and M. Mikolajczak: Pharm. Res. 77 (1985).
415   M. J. Zmijewski, K. Miller-Hatch and M. Mikolajczak: Chem.-Biol. Interactions 52, 361 (1985).
416   F. Tomita, K. Takahoshi and T. Tamaoki: J. Antibiot. 37, 1268 (1984).

417 K. Fujimoto, T. Oka and M. Morimoto: Cancer Res. 47, 1516 (1987).
418 L. S. Kappen, T. E. Ellenberger and I. H. Goldberg: Biochemistry 26, 384 (1987).
419 D.-H. Chin and I. H. Goldberg: Biochemistry 25, 1009 (1986).
420 I. H. Goldberg: Basic Life Sci. 38, 231 (1986).
421 I. H. Goldberg, L. S. Kappen, L. F. Povirk and D. H. Chin: Drugs Exptl. Clin. Res. 12, 495 (1986).
422 I. H. Goldberg, T. Hatayama, L. S. Kappen, M. A. Napier and L. F. Povirk: Molecular Actions and Targets for Cancer Chemotherapeutic Agents, pp. 163–191, Academic Press, New York, 1981.
423 H. Maeda: Anticancer Res. 1, 175 (1981).
424 Y. Koide, F. Ishii, D. Hasuda, Y. Koyama, K. Edo, S. Katamine, F. Kitame and N. Ishida: J. Antibiot. 33, 342 (1980).
425 K. Edo, M. Mizugaki, Y. Koide, H. Seto, K. Furihata, N. Otake and N. Ishida: Tetrahedron Lett. 26, 331 (1985).
426 L. S. Kappen, M. A. Napier and I. H. Goldberg: Proc. Natl. Acad. Sci. USA 77, 1970 (1980).
427 Y. Koide, A. Ito, K. Edo and N. Ishida: Chem. Pharm. Bull. 34, 4425 (1986).
428 B. W. Gibson, W. C. Herlihy, T. S. A. Samy, K.-S. Hahm, H. Maeda, J. Meienhofer and K. Biemann: J. Biol. Chem. 259, 10801 (1984).
429 K. Kuromizu, S. Tsunasawa, H. Maeda, O. Abe and F. Sakiyama: Arch. Biochem. Biophys. 246, 199 (1986).
430 K. Hirayama, T. Ando, R. Takahashi and A. Murai: Bull. Chem. Soc. Jpn. 59, 1371 (1986).
431 L. F. Povirk and I. H. Goldberg: Biochemistry 19, 4773 (1980).
432 L. S. Kappen and I. H. Goldberg: Biochemistry 19, 4786 (1980).
433 K. Edo, H. Sato, K. Saito, Y. Akiyama, M. Kato, M. Mizugaki, Y. Koide and N. Ishida: J. Antibiot. 39, 535 (1986).
434 D. Dasgupta and I. H. Goldberg: Biochemistry 24, 6913 (1985).
435 L. F. Povirk and I. H. Goldberg: J. Biol. Chem. 258, 11763 (1983).
436 T. A. Beerman, R. Poon and I. H. Goldberg: Biochem. Biophys. Acta 475, 294 (1977).
437 L. S. Kappen and I. H. Goldberg: Nucleic Acids Res. 5, 2959 (1978).
438 M. Takeshita, L. S. Kappen, A. P. Grollman, M. Eisenberg and I. H. Goldberg: Biochemistry 20, 7599 (1981).
439 D.-H. Chin, S. A. Carr and I. H. Goldberg: J. Biol. Chem. 259, 9975 (1984).
440 L. S. Kappen and I. H. Goldberg: Nucleis Acid Res. 13, 1637 (1985).
441 A. G. Myers: Tetrahedron Lett. 28, 4493 (1987).
442 J. Golik, G. Dubay, G. Groenewold, H. Kawaguchi, M. Konishi, B. Krishnan, H. Ohkuma, K. Saitoh and T. W. Doyle: J. Amer. Chem. Soc. 109, 3462 (1987).
443 M. D. Lee, T. S. Dunne, C. C. Chang, G. A. Ellestad, M. M. Siegel, G. O. Morton, W. J. McGahren and D. B. Borders: J. Amer. Chem. Soc. 109, 3466 (1987).
444 M. Konishi, H. Ohkuma, K. Saitoh, H. Kawaguchi, J. Golik, G. Dubay, G. Groenewold, B. Krishnan and T. W. Doyle: J. Antibiot. 38, 1605 (1985).
445 J. Golik, J. Clardy, G. Dubay, G. Groenewold, H. Kawaguchi, M. Konishi, B. Krishnan, H. Ohkuma, K. Saitoh and T. W. Doyle: J. Amer. Chem. Soc. 109, 3461 (1987).
446 M. D. Lee, T. S. Dunne, M. M. Siegel, C. C. Chang, G. O. Morton and D. B. Borders: J. Amer. Chem. Soc. 109, 3464 (1987).
447 S. W. Mamber, W. G. Okasinski, C. D. Pinter and J. B. Tunac: J. Antibiot. 40, 1044 (1987).
448 D. W. Fry, J. L. Shillis and W. R. Leopold: Invest. New Drugs 4, 3 (1986).
449 J. H. Wilton, C. D. Rithner, G. C. Hokanson and J. C. French: J. Antibiot. 39, 1349 (1986).

450   J. B. Tunac, B. D. Graham, S. W. Mamber, W. E. Dobson and M. D. Lenzini: J. Antibiot. 38, 1337 (1985).
451   J. H. Wilton, G. C. Hokanson and J. C. French: J. C. S. Chem. Commun. 919 (1985).
452   R. H. Bunge, T. R. Hurley, T. A. Smitka, N. E. Willmer, A. J. Brankiewicz, C. E. Steinman and J. C. French: J. Antibiot. 37, 1566 (1984).
453   M. Iwami, S. Kiyoto, M. Nishikawa, H. Terano, M. Kohsaka, H. Aoki and H. Imanaka: J. Antibiot. 38, 835 (1985).
454   S. Kiyoto, M. Nisikawa, H. Terano, M. Kohsata, H. Aoki, H. Imanaka, Y. Kawai, I. Uchida and M. Hashimoto: J. Antibiot. 38, 840 (1985).
455   S. Kiyoto, T. Shibata, Y. Kawai, Y. Horio, O. Nakayama, H. Terano, M. Kohsaka, H. Aoki and H. Imanaka: J. Antibiot. 38, 955 (1985).
456   B. H. Long, J. J. Catino, S. T. Musial, D. M. Francher and A. M. Casazza: Proc. Am. Assoc. Cancer Res. 28, 312 (1987).
457   J. E. Schurig, W. C. Rose, H. Kamei, A. P. Florczyk, Y. Nishiyama, A. R. Farwell and W. T. Bradner: Proc. Am. Assoc. Cancer Res. 28, 310 (1987).
458   J. P. Thomas, S. G. Carvajal, H. L. Lindsay, R. V. Citarella, R. E. Wallace, M. D. Lee and F. E. Durr: Intersci. Conf. Antimicrob. Agents Chemother. 26, 138 (1986).

Index Vol. 32

The references of the Subject Index are given in the language of the respective contribution.
Die Stichworte des Sachregisters sind in der jeweiligen Sprache der einzelnen Beiträge aufgeführt.
Les termes repris dans la Table des matières sont donnés selon la langue dans laquelle l'ouvrage est écrit.

Index of Titles
Verzeichnis der Titel
Index des titres
Vol. 1–32 (1959–1988)

Author and Paper Index
Autoren- und Artikelindex
Index des auteurs et des articles
Vol. 1–32 (1959–1988)

Teaching tropical medicine *18*, 35 (1974)	Prof. Dr. K. M. Cahill Tropical Disease Center, 100 East 77th Street, New York City 10021, N.Y., USA
Anabolic steroids *2*, 71 (1960)	Prof. Dr. B. Camerino Director of the Chemical Research Laboratory of Farmitalia, Milan, Italy Prof. Dr. G. Sala Department of Clinical Chemistry and Director of the Department of Pharmaceutical Therapy, Farmitalia, Milan, Italy
Immunosuppression agents, procedures, speculations and prognosis *16*, 67 (1972)	Dr. G. W. Camiener Research Laboratories, The Upjohn Company, Kalamazoo, Michigan, USA Dr. W. J. Wechter Research Head, Hypersensitivity Diseases Research, The Upjohn Company, Kalamazoo, Michigan, USA
Dopamine agonists: Structure-activity relationships *29*, 303 (1985)	Joseph G. Cannon The University of Iowa, Iowa City, Iowa 52242, USA
Analgesics and their antagonists: recent developments *22*, 149 (1978)	Dr. A. F. Casy Norfolk and Norwich Hospital and University of East Anglia, Norwich, Norfolk, England
Chemical nature and pharmacological actions of quaternary ammonium salts *2*, 135 (1960)	Prof. Dr. C. J. Cavallito Professor, Medicinal Chemistry, School of Pharmacy, University of North Carolina, Chapel Hill, North Carolina, USA Dr. A. P. Gray Director of the Chemical Research Section, Neisler Laboratories Inc., Decatur, Illinois, USA
Contributions of medicinal chemistry to medicine – from 1935 *12*, 11 (1968) Quaternary ammonium salts – advances in chemistry and pharmacology since 1960 *24*, 267 (1980)	Prof. Dr. C. J. Cavallito Professor, Medicinal Chemistry, School of Pharmacy, University of North Carolina, Chapel Hill, North Carolina, USA
Changing influences on goals and incentives in drug research and development *20*, 159 (1976)	Prof. Dr. C. J. Cavallito Ayerst Laboratories, Inc., New York, N. Y., USA
Über Vorkommen und Bedeutung der Indolstruktur in der Medizin und Biologie *2*, 227 (1960)	Dr. A. Cerletti Direktor der medizinisch-biologischen Forschungsabteilung der Sandoz AG, Basel, Schweiz

Cholesterol and its relation to atherosclerosis *1*, 127 (1959)	Prof. Dr. K. K. Chen Department of Pharmacology, University School of Medicine, Indianapolis, Indiana, USA Dr. Tsung-Min Lin Senior Pharmacologist, Division of Pharmacologic Research, Lilly Research Laboratories, Indianapolis, Indiana, USA
Effect of hookworm disease on the structure and function of small bowel *19*, 44 (1975)	Prof. Dr. H. K. Chuttani Prof. Dr. R. C. Misra Maulana Azad Medical College & Associated Irwin and G. B. Pant Hospitals, New Delhi, India
The psychotomimetic agents *15*, 68 (1971)	Dr. S. Cohen Director, Division of Narcotic Addiction and Drug Abuse, National Institute of Mental Health, Chevy Chase, Maryland, USA
Implementation of disease control in Asia and Africa *18*, 43 (1974)	Prof. Dr. M. J. Colbourne Department of Preventive & Social Medicine, University of Hong Kong, Sassoon Road, Hong Kong
Structure-activity relationships in certain anthelmintics *3*, 75 (1961)	Prof. Dr. J. C. Craig Department of Pharmaceutical Chemistry, University of California, San Francisco, California, USA Dr. M. E. Tate Post Doctoral Fellow, University of New South Wales, Department of Organic Chemistry, Kensington, N. S. W., Australia
Contribution of Haffkine to the concept and practice of controlled field trials of vaccines *19*, 481 (1975)	Dr. B. Cvjetanovic Chief Medical Officer, Bacterial Diseases, Division of Communicable Diseases, WHO, Geneva, Switzerland
Antifungal agents *22*, 93 (1978)	Prof. Dr. P. F. D'Arcy Dr. E. M. Scott Department of Pharmacy, The Queen's University of Belfast, Northern Ireland
Some neuropathologic and cellular aspects of leprosy *18*, 53 (1974)	Prof. Dr. D. K. Dastur Dr. Y. Ramamohan Dr. A. S. Dabholkar Neuropathology Unit, Grant Medical College and J. J. Group of Hospitals, Bombay 8, India
Autonomic dysfunction as a problem in the treatment of tetanus *19*, 245 (1975)	Prof. Dr. F. D. Dastur Dr. G. J. Bhat Dr. K. G. Nair Department of Medicine, Seth G. S. Medical College and K. E. M. Hospital, Bombay 12, India

A review of advances in prescribing for teratogenic hazards *29*, 121 (1985)	E. Marshall Johnson, Ph. D. Daniel Baugh Institute, Jefferson College, Thomas Jefferson University, 1020 Locust Street, Philadelphia, PA 19107
A comparative study of bitoscanate, bephenium hydroxynaphthoate and tetrachlorethylene in hookworm infection *19*, 70 (1975)	Dr. S. Johnson Department of Medicine III, Christian Medical College Hospital, Vellore, Tamilnadu, India
Tetanus in Punjab with particular reference to the role of muscle relaxants in its management *19*, 288 (1975)	Prof. Dr. S. S. Jolly Dr. J. Singh Dr. S. M. Singh Department of Medicine, Medical College, Patiala, India
Virulence-enhancing effect of ferric ammonium citrate on *Vibrio cholerae* *19*, 546 (1975)	Dr. I. Joó Institute for Serobacteriological Production and Research 'HUMAN', WHO International Reference Centre for Bacterial Vaccines, Budapest, Hungary
Toxoplasmosis *18*, 205 (1974)	Prof. Dr. B. H. Kean The New York Hospital – Cornell Medical Center, 525 East 68th Street, New York, N. Y., USA
Tabellarische Zusammenstellung über die Substruktur der Proteine *16*, 364 (1972)	Dr. R. Kleine Physiologisch-Chemisches Institut der Martin-Luther-Universität, 402 Halle (Saale), DDR
Experimental evaluation of anti-tuberculous compounds, with special reference to the effect of combined treatment *18*, 211 (1974)	Dr. F. Kradolfer Head of Infectious Diseases Research, Biological Research Laboratories, Pharmaceutical Division, Ciba-Geigy Ltd., Basle, Switzerland
The oxidative metabolism of drugs and other foreign compounds *17*, 488 (1973)	Dr. F. Kratz Medizinische Kliniken und Polikliniken, Justus-Liebig-Universität, Giessen, BR Deutschland
Die Amidinstruktur in der Arzneistofforschung *11*, 356 (1968)	Prof. Dr. A. Kreutzberger Wissenschaftlicher Abteilungsvorsteher am Institut für pharmazeutische Chemie der Westfälischen Wilhelms-Universität Münster, Münster (Westfalen), Deutschland
Present data on the pathogenesis of tetanus *19*, 301 (1975) Tetanus: general and pathophysiological aspects; achievements, failures, perspectives of elaboration of the problem *19*, 314 (1975)	Prof. Dr. G. N. Kryzhanovsky Institute of General Pathology and Pathological Physiology, AMS USSR, Moscow, USSR

Mechanism of action of anxiolytic drugs *31*, 315 (1987)	T. Mennini S. Caccia S. Garattini Istituto di Ricerche Farmacologiche "Mario Negri", Via Eritrea 62, 20157 Milan, Italy
Pathogenesis of amebic disease *18*, 225 (1974) Protozoan and helminth parasites – a review of current treatment *20*, 433 (1976)	Prof. Dr. M. J. Miller Tulane University, Department of Tropical Medicine, New Orleans, Louisiana, USA
Synopsis der Rheumatherapie *12*, 165 (1968)	Dr. W. Moll Spezialarzt FMH Innere Medizin – Rheumatologie, Basel, Schweiz
On the chemotherapy of cancer *8*, 431 (1965) The relationship of the metabolism of anticancer agents to their activity *17*, 320 (1973) The current status of cancer chemotherapy *20*, 465 (1976)	Dr. J. A. Montgomery Kettering-Meyer Laboratory, Southern Research Institute, Birmingham, Alabama, USA
Der Einfluss der Formgebung auf die Wirkung eines Arzneimittels *10*, 204 (1966) Galenische Formgebung und Arzneimittelwirkung. Neue Erkenntnisse und Feststellungen *14*, 269 (1970)	Prof. Dr. K. Münzel Leiter der galenischen Forschungsabteilung der F. Hoffmann-La Roche & Co. AG, Basel, Schweiz
A field trial with bitoscanate in India *19*, 81 (1975)	Dr. G. S. Mutalik Dr. R. B. Gulati Dr. A. K. Iqbal Department of Medicine, B. J. Medical College and Sassoon General Hospital, Poona, India
Comparative study of bitoscanate, bephenium hydroxynaphthoate and tetrachlorethylene in hookworm disease *19*, 86 (1975)	Dr. G. S. Mutalik Dr. R. B. Gulati Department of Medicine, B. J. Medical College and Sassoon General Hospital, Poona, India
Ganglienblocker *2*, 297 (1960)	Dr. K. Nádor o. Professor und Institutsdirektor, Chemisches Institut der Tierärztlichen Universität, Budapest, Ungarn
Nitroimidazoles as chemotherapeutic agents *27*, 163 (1983)	Dr. M. D. Nair Dr. K. Nagarajan Ciba-Geigy Research Centre, Goreagon East, Bombay 400063

Immuno-diagnosis in filarial infection *19*, 128 (1975)	Prof. Dr. T. Sawada Dr. K. Sato Dr. K. Takei Department of Parasitology, School of Medicine, Gunma University, Maebashi, Japan Dr. M. M. Goil Department of Zoology, Bareilly College, Bareilly (U. P.), India
Quantitative structure- activity relationships *23*, 199 (1979)	Dr. A. K. Saxena Dr. S. Ram Medicinal Chemistry Division, Central Drug Research Institute, Lucknow, India
Advances in chemotherapy of malaria *30*, 221 (1986)	Anil K. Saxena Mridula Saxena Division of Medicinal Chemistry, Central Drug Research Institute, Lucknow 226001, India
Phenothiazine und Azaphenothiazine als Arzneimittel *5*, 269 (1963)	Dr. E. Schenker Forschungschemiker in der Sandoz AG, Basel, Schweiz Dr. H. Herbst Forschungstechniker in den Farbwerken Hoechst, Frankfurt a. M., Deutschland
Antihypertensive agents *4*, 295 (1962)	Dr. E. Schlittler Director of Research of CIBA Pharmaceutical Company, Summit, New Jersey, USA Dr. J. Druey Director of the Department of Synthetic Drug Research of CIBA Ltd., Basle, Switzerland Dr. A. Marxer Research Chemist of CIBA Ltd., Basle, and Lecturer at the University of Berne, Switzerland
Die Anwendung radioaktiver Isotope in der pharmazeutischen Forschung *7*, 59 (1964)	Prof. Dr. K. E. Schulte Direktor des Instituts für Pharmazie und Le- bensmittelchemie der Westfälischen Wilhelms-Universität Münster, Münster (Westfalen), Deutschland Dr. Ingeborg Mleinek Leiterin des Isotopen-Laboratoriums, Institut für Pharmazie und Lebensmittelchemie der Westfälischen Wilhelms-Universität Münster, Münster (Westfalen), Deutschland

Nonsteroid antiinflammatory agents *10,* 139 (1966)	Dr. C. A. Winter Senior Investigator Pharmacology, Merck Institute for Therapeutic Research, West Point, Pennsylvania, USA
A review of the continuum of drug-induced states of excitation and depression *26,* 225 (1982)	Prof. Dr. W. D. Winters Departments of Pharmacology and Internal Medicine, School of Medicine, University of California, Davis, California 95616, USA
Basic research in the US pharmaceutical industry *15,* 204 (1971)	Dr. O. Wintersteiner The Squibb Institute for Medical Research, New Brunswick, New Jersey, USA
Light and dark as a "drug" *31,* 383 (1987)	Anna Wirz-Justice Psychiatric University Clinic, Wilhelm Klein Strasse 27, CH-4025 Basel, Switzerland
The chemotherapy of amoebiasis *8,* 11 (1965)	Dr. G. Woolfe Head of the Chemotherapy Group of the Research Department at Boots Pure Drug Company Ltd., Nottingham, England
Antimetabolites and their revolution in pharmacology *2,* 613 (1960)	Dr. D. W. Woolley The Rockefeller Institute, New York, USA
Noise analysis and channels at the postsynaptic membrane of skeletal muscle *24,* 9 (1980)	Dr. D. Wray Lecturer, Pharmacology Department, Royal Free Hospital School of Medicine, Pond Street, London NW3 2QG, England
Krebswirksame Antibiotika aus Actinomyceten *3,* 451 (1961)	Dr. Kh. Zepf Forschungschemiker im biochemischen und mikrobiologischen Laboratorium der Farbwerke Hoechst, Frankfurt a.M., Deutschland Dr. Christa Zepf Referentin für das Chemische Zentralblatt, Kelkheim (Taunus), Deutschland
Fifteen years of structural modifications in the field of antifungal monocyclic 1-substituted 1H-azoles *27,* 253 (1983)	Dr. L. Zirngibl Siegfried AG, Zofingen, Switzerland
Lysostaphin: model for a specific anzymatic approach to infectious disease *16,* 309 (1972)	Dr. W. A. Zygmunt Department of Biochemistry, Mead Johnson Research Center, Evansville, Indiana, USA Dr. P. A. Tavormina Director of Biochemistry, Mead Johnson Research Center, Evansville, Indiana, USA